Rainfall-Induced Soil Slope Failure

Stability Analysis and Probabilistic Assessment

Rainfall-Induced Soil Slope Failure

Stability Analysis and Probabilistic Assessment

Lulu Zhang ▲ Jinhui Li ▲ Xu Li
Jie Zhang ▲ Hong Zhu

CRC Press is an imprint of the
Taylor & Francis Group, an **informa** business
A SPON PRESS BOOK

CRC Press
Taylor & Francis Group
6000 Broken Sound Parkway NW, Suite 300
Boca Raton, FL 33487-2742

First issued in paperback 2018

ISBN-13: 978-1-4987-5279-4 (hbk)
ISBN-13: 978-0-367-13901-8 (pbk)

Library of Congress Cataloging-in-Publication Data

Names: Zhang, Lulu, author.
Title: Rainfall-induced soil slope failure : stability analysis and probabilistic assessment / authors, Lulu Zhang, Jinhui Li, Xu Li, Jie Zhang, and Hong Zhu.
Description: Boca Raton : Taylor & Francis, 2016. | English language translation. | Includes bibliographical references.
Identifiers: LCCN 2015042673 | ISBN 9781498752794 (hard cover)
Subjects: LCSH: Landslide hazard analysis--Statistical methods. | Landslides--Risk assessment--Statistical methods. | Rainfall probabilities.
Classification: LCC QE599.2 .Z4613 2016 | DDC 551.3/07--dc23
LC record available at https://lccn.loc.gov/2015042673

Visit the Taylor & Francis Web site at
http://www.taylorandfrancis.com

and the CRC Press Web site at
http://www.crcpress.com

Contents

PART II
Probabilistic assessment 227

Foreword

This book, *Rainfall-Induced Soil Slope Failure: Stability Analysis and Probabilistic Assessment* by Lulu Zhang, Jinhui Li, Xu Li, Jie Zhang, and Hong Zhu, is a welcome addition that provides an engineering methodology for a challenging problem of interest the world over. Virtually every country encounters serious hazards associated with slope instability triggered by rainfall conditions. The problem is complex, and there is need to bring together a number of physically based disciplines in order to provide an engineered approach to addressing risks associated with property and life loss.

This book brings together two primary application areas of engineering that have been individually successful within geotechnical engineering: the analysis of the stability of natural slopes and the modeling of moisture movement through saturated–unsaturated soil systems. The ground surface forms the interface between imposed climatic conditions and the unsaturated soils near the ground surface. The authors are well aware of the extensive research that has been undertaken in various parts of the world related to modeling moisture infiltration. The infiltration component becomes particularly difficult to assess when the soils at the ground surface contain secondary structural features such as fissures and cracks. The engineering characterization of the ground surface characteristics is particularly well presented in this book. The authors are particularly successful in addressing the many components related to the assessment of rainfall-induced landslides.

The book combines the complex topics of finite element numerical modeling of moisture infiltration and seepage into unsaturated soils along with the analysis of the stability of a slope that is subjected to stochastic phenomena. When addressing issues related to rainfall-induced landslides, there is need to not only understand unsaturated soil behavior (i.e., near-ground-surface soil conditions), but also be able to explain how the required input data can be obtained through practical means. The authors have also made a significant contribution in synthesizing research information related to the estimation of unsaturated soil properties for analysis purposes.

The authors have brought together the many components related to rainfall-induced landslides. I foresee this book as being well received by practicing engineers and researchers. I believe that this book will also form an important reference book that will be in demand in university and other engineering libraries. The authors have made a significant contribution to engineering practice through the synthesis of the information presented in this book.

Delwyn G. Fredlund
University of Saskatchewan
and
Golder Associates Ltd., Saskatoon, Canada

Foreword

Several excellent books, such as *Unsaturated Soil Mechanics in Engineering Practice* by Delwyn Fredlund, Harianto Rahardjo, and Murray Fredlund (published by Wiley) and *Hillslope Hydrology and Stability* by Ning Lu and Jonathan Godt (published by Cambridge University Press), cover a wide spectrum of unsaturated soil mechanics and its applications to slope engineering. This new book, *Rainfall-Induced Soil Slope Failure: Stability Analysis and Probabilistic Assessment* by Lulu Zhang, Jinhui Li, Xu Li, Jie Zhang, and Hong Zhu, is a new addition that focuses on rain-induced slope failures from a soil mechanics approach. The authors of this book are all very active young researchers in the field of study.

This book concisely presents the fundamental mechanics behind rain-induced soil slope failures: infiltration and evapotranspiration in soil slopes, slope stability under rainfall infiltration, and hydro-mechanical coupled deformation. The concepts and analyses are presented clearly from the first principle with a strong physical sense. Some advanced topics, such as plastic analysis of hydro-mechanical coupled seepage and deformation in unsaturated soil slopes, are approached in a simple manner with hands-on tutorials, which is ideal for graduate students. The analysis of the stability of slopes with cracks is an excellent addition—practitioners are aware of the vital importance of cracks, but few have the know-how to quantify cracks on slopes and their effects on slope stability. This book provides the overall fundamental knowledge to deal with climate–slope interactions and resulting slope failures.

This book also systematically presents the uncertainties involved in the characterization and analysis of rain-induced slope failures. Methods for reliability analysis of slopes under rainfall, including probability concepts, identification of sources of uncertainty, stochastic finite element method, and probabilistic model calibration, are introduced. This is a unique feature of this book—it is perhaps the only book that addresses both the mechanics and uncertainties behind rain-induced slope failures.

I believe that *Rainfall-Induced Soil Slope Failure: Stability Analysis and Probabilistic Assessment* will serve as an excellent reference for educators, university graduate students, researchers, and practitioners.

Limin Zhang
The Hong Kong University of Science and Technology

Preface

Rainfall-induced landslides pose a significant threat to the public safety in many parts of the world especially in tropical or subtropical mountainous areas. The research field of rainfall-induced landslides has been developed tremendously during the past 20 years attributed to the advancement of unsaturated soil mechanics. Yet a great number of uncertainties are involved in the analysis of slope stability under rainfall conditions. Measured soil properties are subject to inherent spatial variability, measurement error, and statistical error because of a limited number of samples. The predicted performance of a slope from a geotechnical model may deviate from reality because of uncertainties of input parameters and systematic error associated with the prediction model. Using probabilistic approaches, different types of uncertainties can be accounted for in a systematic and quantitative way. The information from various sources can be utilized systematically.

The aim of this book is thus to bring together the perspectives of both geomechanics and reliability to reflect the complicated mechanism and the tremendous uncertainties underlying the problem of rainfall-induced slope failure. The book clearly presents infiltration analysis and stability analysis methods based on geomechanics approaches. The methods of geotechnical reliability and probability which can be used to address the related uncertainties are also presented. This book is an essential reading for researchers and graduate students who are interested in slope stability, landslides, geohazards, and risk assessment in fields of civil engineering, engineering geology, and earth science. Professional engineers and practitioners in slope engineering and geohazard management can use it as a supplementary reading material. Knowledge of undergraduate soil mechanics, statistics, and probability is a prerequisite for the readers.

This book consists of nine chapters and can be divided into two parts. In the first part (Chapters 2 through 6), we focus on the failure mechanisms of rainfall-induced slope failure. Chapter 2 presents commonly used conceptual models and analytical solutions of modeling infiltration into the ground and discusses important issues related to the numerical modeling. Chapters 3 and 4 present the stability analysis methods based on limit equilibrium methods and coupled hydro-mechanical models, respectively. Chapters 5 and 6 focus on the infiltration and stability of natural soil slopes with cracks and colluvium materials, respectively. In the second part of the book (Chapters 7 through 9), the reliability methods and probabilistic approaches are presented. In Chapter 7, the effect of uncertainties of essential soil properties related to the stability of unsaturated soil slopes is characterized. In Chapter 8, the effect of spatial variability of soil properties on an unsaturated soil slope is discussed. Finally, probabilistic calibration for infiltration and slope stability is presented in Chapter 9.

This book also presents major outcomes of the research conducted by the authors in the past 10 years. We express our highest gratitude to the late Professor Wilson H. Tang (Department of Civil and Environmental Engineering, The Hong Kong University of

Science & Technology). The authors are grateful to Professor Limin Zhang (Department of Civil and Environmental Engineering, The Hong Kong University of Science & Technology) for his support and help during all these years. We are also grateful to Prof. D. G. Fredlund (Professor Emeritus at the University of Saskatchewan and geotechnical engineering specialist at Golder Associates) for his great help in our research study.

Lulu Zhang
Shanghai Jiao Tong University

Jinhui Li
Harbin Institute of Technology

Xu Li
Beijing Jiaotong University

Jie Zhang
Tongji University

Hong Zhu
Hong Kong University of Science and Technology

MATLAB® is a registered trademark of The MathWorks, Inc. For product information, please contact:

The MathWorks, Inc.
3 Apple Hill Drive
Natick, MA 01760-2098 USA
Tel: +1 508 647 7000
Fax: +1 508 647 7001
E-mail: info@mathworks.com
Web: www.mathworks.com

List of symbols

B	Variance between the mean values of Markov chains
b_x, b_y, b_z	Body forces in the x-, y-, and z-directions
c'	Effective cohesion of a saturated soil
C	Specific storage capacity of unsaturated soil
C( )	Covariance/autocovariance function
COV	Coefficient of variation
C_v	Coefficient of consolidation
$\mathbf{C_X}$	Covariance matrix of $\mathbf{X}$
d	Distance from the origin to the limit state surface line in standardized random field space
d	Number of observed data points
D	Depth of slip surface
$\mathbf{D}^{ep}$	Elastoplastic stress–strain matrix
df	Degrees of freedom of Student's t distribution
$D_{\max}$	Kolmogorov–Smirnov (K–S) D statistic
D_w	Depth of the ground water table
e	Void ratio
E	Interslice normal force
E	Young's modulus
E[.]	Expectation operator
F_s	Factor of safety
$F(X)$	Cumulative density function (CDF) of X
$f(\boldsymbol{\theta})$	Prior density function of $\boldsymbol{\theta}$
$f(\boldsymbol{\theta}\|\cdot)$	Posterior density function of $\boldsymbol{\theta}$
$f_{\mathbf{X}}()$	Joint probability density function of the random variable vector $\mathbf{X}$
g	Gravitational acceleration
G	Shear modulus
$g()$	Prediction model
G_s	Specific gravity of soil particle
$g(\mathbf{X})$	Performance function
$g'(\mathbf{X})$	Response surface function to approximate the performance function $g(\mathbf{X})$
$\mathbf{h}$	Vector of nodal pressure heads
h	Pore pressure head
H	Total head or hydraulic head
H	Slope height
H_B	Elastic modulus for soil structure with respect to pore-water pressure in Biot (1941) coupled formulation

H_B'	Coefficient of pore-water content change due to normal stresses in Biot (1941) coupled formulation
H_F	Elastic modulus for soil structure with respect to a change in matric suction in Fredlund and Rahardjo (1993) coupled formulation
H_m	Elevation difference between the starting point and the lowest point of deposition of the mass movement
h_s	Suction head
i	Infiltration rate or capacity
I	Cumulative infiltration
I	Rainfall intensity in I-D threshold curve equation
$\mathbf{I}$	Unit matrix
I_p	Cumulative infiltration at the moment of ponding
I_p	Plasticity index
I_s	Cumulative infiltration on a sloping ground
$I(\mathbf{x})$	Indicator function of the Monte Carlo simulation
$J(\cdot\|\cdot)$	Jumping distribution or transition kernel of the Markov chain
$\mathbf{k}$	Tensor of coefficient of permeability
k	Coefficient of permeability
k	Normalization constant for the posterior density function in Bayesian theory
k_s	Saturated coefficient of permeability
k_w	Coefficient of permeability in the wetted zone of the Green–Ampt model
L	Total depth of a soil
$\mathbf{L}$	Lower triangular decomposed matrix of the covariance matrix ($\mathbf{LL}^T=\mathbf{C}$)
$L(\cdot\|\cdot)$	Likelihood function
L_c	Length of the aperture
L_m	Horizontal distance between the starting point and lowest point of deposition of the mass movement
M	Slope of the critical state line
M	Number of Markov chains
m_1^s	Coefficient of volume change of soil skeleton with respect to a change in $(\sigma_{\text{mean}} - u_a)$ in Fredlund and Rahardjo (1993) coupled formulation
m_2^s	Coefficient of volume change of soil skeleton with respect to a change in $(u_a - u_w)$ in Fredlund and Rahardjo (1993) coupled formulation
m_1^w	Coefficient of volume change of pore water with respect to a change in $(\sigma_{\text{mean}} - u_a)$ in Fredlund and Rahardjo (1993) coupled formulation
m_2^w	Coefficient of volume change of pore water with respect to a change in $(u_a - u_w)$ in Fredlund and Rahardjo (1993) coupled formulation
M_{col}	Slope of collapse surface
n	Porosity of the soil
n	Number of segments along the slip surface
n	Dimension of the input model parameter vector $\boldsymbol{\theta}$
N	Normal force on the base of the slice
N	Intercept of normal consolidation line in the v-$\bar{p}$ plane
n_c	Crack porosity

N_{chain}	Length or number of samples of a Markov chain
p	Mean total stress
$\bar{p}$	Mean net stress
p'	Mean effective stress
$\bar{p}_0$	Preconsolidation stress
p_f	Probability of failure
$\hat{p}_f$	Estimation of the probability of failure p_f
p_{at}	Atmospheric pressure
p^c	Reference stress
p'_{cs}	Mean effective stress at the critical state
P_{clay}	Percentage of clay
p_s	Intercept of the ellipse of yield surface at the $\bar{p}$ axis at suction s
P_{sand}	Percentage of sand
q	Rainfall intensity
q	Deviator stress
Q	Load
R	Radius or the moment arm of a circular slip surface
R	Resistance
R_B	Coefficient of pore-water content change due to pore-water pressure in Biot (1941) coupled formulation
R_{stat}	Gelman–Rubin convergence diagnostic
s	Scale of the inverse chi-square distribution
S	Degree of saturation
$S(.)$	Spectral density function of a random field
S_e	Effective degree of saturation
s_f	Suction head at the wetting front of the Green–Ampt model
S_m	Mobilized shear force at the base of a slice
S_r	Residual degree of saturation
$s(\mathbf{x})$	Sampling function of the importance sampling method
t	Time
$\mathbf{T}$	Diagonal and upper matrix of triangular Cholesky decomposition of the covariance matrix
t_{debris}	Thickness of debris
t_p	Time of ponding
T_s	Surface tension of water
u	Displacement components along the x-direction
U	Random number from a uniform distribution
u_a	Pore-air pressure
$(u_a - u_w)$	Matric suction
u_w	Pore-water pressure
V	Vulnerability
$\mathbf{v}$	Vector of flow rate
v	Displacement components along the y-direction
v	Specific volume of a soil
var[.]	Variance operator
V_m	Volume of the sliding mass
w	Displacement components along the z-direction
W	Total weight of the slice
W	Average of the within-chain variances of Markov chains
$\mathbf{W}^{ep}$	Elastoplastic suction-strain matrix

w_w	Gravimetric water content
$w(\mathbf{x})$	Weighting function of the importance sampling method
X	Interslice shear forces
X	Random variable
$\mathbf{X}$	Vector of random variables
$\mathbf{x}$	Location coordinates vector or the vector of a spatial variable
x^*	Design point in the original random variable space
$\mathbf{Y}$	Vector of model output
$\hat{\mathbf{Y}}$	Vector of observed response
y^*	Design point in the standardized random variable space
Y_i	Standardized (dimensionless/reduced) random variable of the random variable X_i
$\mathbf{Z}$	Correlated standard normal random field
z	Elevation head; coordinate of elevation; vertical coordinate
z_w	Depth of wetting front
α_B	Coefficient measures the ratio of the water volume squeezed out to the volume change of the soil in Biot (1941) coupled formulation
α_c	Biot's hydro-mechanical coupling coefficient
α_c	Angle of contact between water and soil particle
α_s	Slope angle
α_{slice}	Angle between the tangent to the base of slice and the horizontal direction
β	Reliability index
β_w	Compressibility of water
Γ	Intercept of the critical state line in the v–ln p' plane
$\Gamma^2(.)$	Variance reduction factor function
γ_t	Total unit weight of the soil
γ_w	Unit weight of water
$\gamma(.)$	Semivariogram
$\gamma_{xy}, \gamma_{yx}, \gamma_{yz}, \gamma_{zy}, \gamma_{xz}, \gamma_{zx}$	Components of the shear strain tensor
δ	Scale of fluctuation
δ	Coefficient of variation
$\boldsymbol{\varepsilon}$	Strain tensor
ε	Vector of residual errors
ε_v	Volumetric strain
$\varepsilon_x, \varepsilon_y, \varepsilon_z$	Normal strain components along the x-, y-, and z-directions
κ	Unloading stiffness parameter due to net stress loading
κ_s	Swelling stiffness parameter due to suction unloading
λ	Slope of the critical state line in the v–ln p' plane
λ_i	ith eigenvalue of the covariance matrix
λ_n	nth positive root of equation $\sin(\lambda L) + (2\lambda/\alpha)\cos(\lambda L) = 0$
μ	Mean value
$\boldsymbol{\mu}_{\mathbf{X}}$	Mean value vector of $\mathbf{X}$
Θ_d	Dimensionless water content
$\boldsymbol{\theta}$	Vector of random input model parameters
θ_r	Residual volumetric water content
θ_s	Saturated volumetric water content
θ_w	Volumetric water content
$\boldsymbol{\sigma}$	Total stress tensor
σ	Total stress

σ	Standard deviation
$(\sigma - u_a)$	Net normal stress
σ_{mean}	Mean total normal stress
$\sigma_{\text{mean}} - u_a$	Mean net normal stress
σ_n	Total normal stress on the slip surface
$\sigma_n - u_a$	Net normal stress on the slip surface
σ_x	Standard variation of the variable x
$\sigma_x, \sigma_y, \sigma_z$	Normal stresses in the x-, y-, and z-directions
σ'	Effective stress
σ^s	Suction stress
τ	Lag or separation distance between two spatial locations
τ_f	Shear strength
τ_n	Mobilized shear stress parallel to the segment
$\tau_{xy}, \tau_{xz}, \tau_{yx}, \tau_{yz}, \tau_{zx}, \tau_{zy}$	Components of the shear stress tensor
ν	Poisson's ratio
π	Osmotic suction
ρ	Correlation coefficient
ρ_w	Density of water
$\boldsymbol{\rho}_{\mathbf{X}}$	Correlation matrix of $\mathbf{X}$
Φ	Matric flux potential
$\Phi(\cdot)$	Cumulative distribution function (CDF) of the standard normal distribution
Φ_s	Steady-state matric flux potential
ϕ'	Effective friction angle of a soil
$\phi()$	Probability density function (PDF) of the standard normal distribution $N(0,1)$
ϕ^b	An angle indicating the rate of increase in shear strength related to matric suction
ϕ'_{col}	Friction angle of the collapse surface
ϕ'_{cs}	Friction angle of the critical state line
χ	The parameter with the value between zero and unity in Bishop's effective stress for unsaturated soils
ψ	Soil suction
ψ_{aev}	Air-entry value (AEV) of the soil
ψ_r	Residual suction
ω	Angular frequency
∇^2	Laplace operator or Laplacian

Part I

Stability analysis methods

Chapter 1

Introduction

1.1 SLOPE FAILURES UNDER RAINFALL AND FAILURE MECHANISMS

Rainfall-induced landslides are common in many regions under tropical or subtropical climates (Lumb, 1962; Brand, 1984; Fourie, 1996; Glade et al., 2006; Sidle and Ochiai, 2006; Schuster and Highland, 2007). Figure 1.1 shows the global landslide susceptibility map of rainfall-induced landslides produced by National Aeronautics and Space Administration (NASA) with combination of surface landslide susceptibility and a real-time space-based rainfall analysis system (Hong and Adler, 2008). The red and orange indicate regions with high-potential landslide risk include the Pacific Rim, the Alps, the Himalayas and South Asia, Rocky Mountains, Appalachian Mountains, and parts of the Middle East and Africa. Based on the historical records, most catastrophic landslides and debris flows have occurred in China, India, Japan, Singapore, the United States, Italy, Brazil, and Venezuela. Here, some of the most notable landslides in different regions are briefly introduced.

Based on Huang (2007), 46% of large-scale landslides occurred in China during the twentieth century are triggered by rainstorms (Figure 1.2). The rainfall-induced landslides are common, especially in South China. Among the most notable landslides occurred in South China were the landslide events that occurred in 1972 and 1976 at Sau Mau Ping and Po Shan Road in Hong Kong. The 1972 Sau Mau Ping landslide, which occurred on June 18, 1972, during a severe rainstorm, destroyed many huts in a temporary housing area, killing 71 people. This landslide was followed a few hours later by the Po Shan Road landslide, which demolished a 12-story apartment building, killing 67 people (Figure 1.3). In August 1976, a loose fill slope failure 40 m away from the previous landslide occurred in Sau Mau Ping and 18 people were killed. The 1972 and 1976 landslide events resulted in the establishment of the Geotechnical Control Office of Hong Kong (renamed as Geotechnical Engineering Office in 1991).

In the United States, the west coastal states of California, Oregon, and Washington are regions with high risk of rainfall-induced landslides. One of the most notable landslides triggered by the rainstorms was the March 4, 1995, La Conchita landslide (Figure 1.4a) in the small residential community of La Conchita at the northwest of Los Angeles. This landslide was a reactivation of an ancient complex earth flow in marine sediments and badly damaged nine houses (Jibson, 2005). Because the slide moved at only a moderate rate (tens of meters in a few minutes), there were no casualties. On January 10, 2005, the left side of the 1995 La Conchita landslide remobilized due to a heavy rainfall. A high-velocity debris flow at an estimated velocity of 10 m/s destroyed 13 houses (Figure 1.4b). The flow overwhelmed a steel and timber retaining wall that had been constructed at the toe of the 1995 landslide in an attempt to keep landslide debris off the road (Jibson, 2005). Another devastating landslide was the one occurred in Oso, Washington, on March 22, 2014. The collapse of the

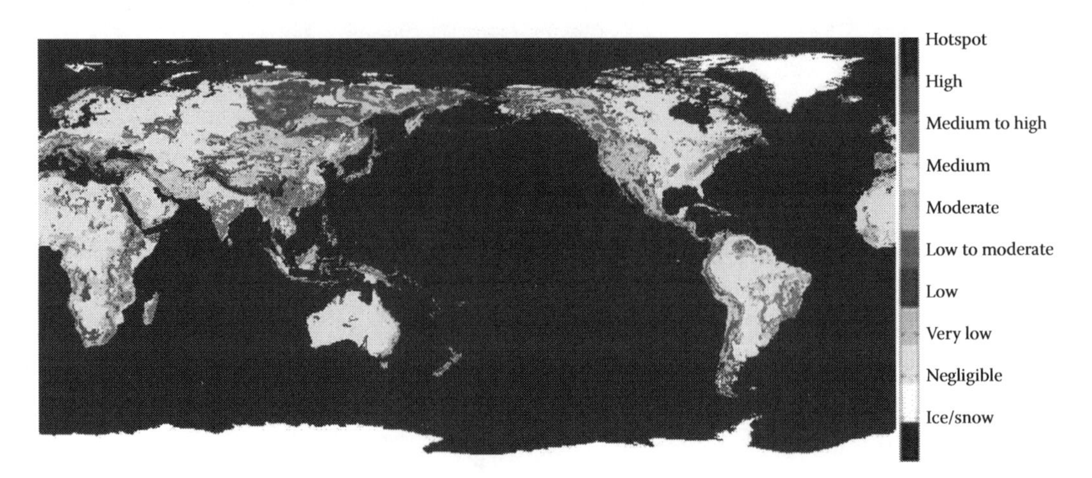

Figure 1.1 Global susceptibility map of rainfall-induced landslides. (Adopted from *International Journal of Sediment Research*, 23, Hong, Y., and Adler, R. F., Predicting global landslide spatiotemporal distribution: Integrating landslide susceptibility zoning techniques and real-time satellite rainfall estimates, 249–257. Copyright 2008, with permission from Elsevier.)

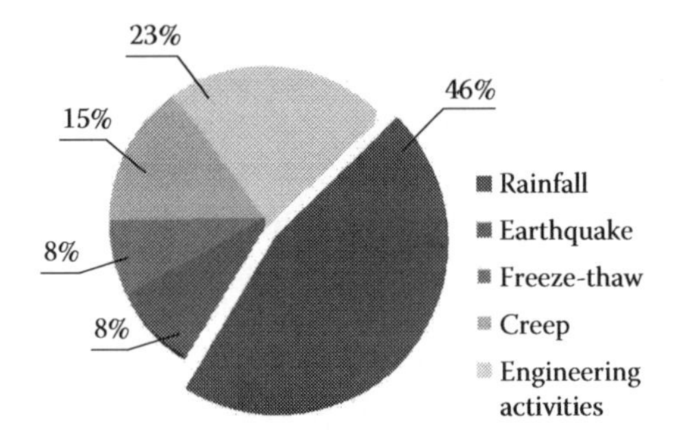

Figure 1.2 Mechanisms of large-scale landslides in China during the twentieth century.

Figure 1.3 The June 18, 1972, Po Shan Road landslide. (Adopted from the website of Hong Kong Slope Safety, http://hkss.cedd.gov.hk/hkss/eng/photovideo_gallery.aspx. With permission from Information Services Department, the Government of Hong Kong Special Administration Region.)

Figure 1.4 The landslides occurred in La Conchita, California. (a) The March 4, 1995, rainfall-triggered La Conchita, California, landslide. (Courtesy of Robert L. Schuster, US Geological Survey.) (b) The January 10, 2005, remobilization of part of the 1995 landslide. (Courtesy of Mark Reid, US Geological Survey.)

Figure 1.5 The landslide occurred in Oso, Washington, DC, on March 22, 2014. This photo was taken on March 27, 2014. (Courtesy of Jonathan Godt, US Geological Survey.)

hillside generated a massive mudflow that rushed across the valley of the North Fork of the Stillaguamish River (Figure 1.5). The slide and mudflow engulfed about 50 residences, buried an important highway, and created a landslide dam that temporarily blocked the river. Precipitation in February and March in the area was 150%–200% of the long-term average, and contributed to landslide initiation (Iverson et al., 2015).

The rainfall-induced slope failures may occur on natural slopes in a variety of materials including residual and colluvial soils (Fourie, 1996; Crosta and Frattini, 2003; Dai et al., 2003). Fill slopes, cut slopes, and embankments may also be prone to this type of slope failure (Day and Axten, 1989; Chen et al., 2004; Briggs et al., 2013). The slope failures are normally shallow with a depth of failure less than 3 m, above the groundwater table, and generally of small volume on steep soil slopes of 30°–50° (Johnson and Sitar, 1990; Dai et al., 2003). Deep-seated landslides after rainfall are also occasionally reported (Gerscovich et al., 2006; Prokešová et al., 2013). Shallow landslides are typically induced by intense rainfall, while deep-seated landslides require rainfall with a longer duration. Multiple landslides may occur in one region under a single rainstorm event. Most of the regionally distributed landslide events were accompanied by debris flows. Some of the areas that are sensitive to regional landslides induced by rainfall include Fukushima Prefecture and Hiroshima Prefecture, Japan (Wang et al., 2002, 2003); Alps region in Italy (Crosta and Frattini, 2003; Cascini et al., 2008); Río de Janeiro and Petropolis areas, Brazil (Da Costa Nunes et al., 1979); and Southwestern China (Cui et al., 2005).

It is generally recognized that rainfall-induced landslides are caused by changes in pore-water pressures and seepage forces (Zhu and Anderson, 1998; Gerscovich et al., 2006; Kitamura and Sako, 2010; Zhang et al., 2011; Fredlund et al., 2012; Lu and Godt, 2013). Two distinct failure mechanisms have been observed and analyzed for rainfall-induced landslides (Collins and Znidarcic, 2004). In the first mechanism, significant build-up of positive pressures was observed in a low area on the slope or along the soil–bedrock interface. Movements along the sliding surface may lead to liquefaction along this surface, resulting in rapid movements, long run-out distances, and finally a complete liquefaction of the failed mass (Wang and Sassa, 2001). The stress path of soils inside the slopes subject to rainfall can be described by a constant shear stress path (Anderson and Sitar, 1995). In the second mechanism, the soil is in the unsaturated state, and slope failure is mainly due to rainfall infiltration and a loss in shear strength when soil suctions are decreased or dissipated (Fredlund and Rahardjo, 1993; Fourie et al., 1999). Cascini et al. (2010) classified the rainfall-induced shallow landslides as slide, slide to flow (slides turning into flows), and flowslide based on the acceleration of the failed mass and suggested that the failure and postfailure stages should be separately analyzed. In this book, we will mainly focus on the second mechanism and slope instability in the failure stage.

1.2 RECENT ADVANCES AND HOT RESEARCH TOPICS

The research field of rainfall-induced landslides has been developed tremendously during the past 30 years along with the development of unsaturated soil mechanics. Infiltration plays a significant role in the instability of slopes under rainfall conditions. To provide simplified solutions of the complex problem of infiltration, theoretical and empirical models based on a wetting front concept have been developed (e.g., Green and Ampt, 1911; Horton, 1938; Holtan, 1961; Mein and Larson, 1973; Sun et al., 1998; Beven et al., 2004), among which the model by Green and Ampt (1911) is widely used. Solutions of the second-order partial differential equation, which combines the Darcy's law and the equation of continuity, are considered robust for flow and infiltration in saturated and unsaturated soils. Analytical methods solve the differential equation by applying various transformations or green functions and result in explicit solutions of pressure head or water content (Srivastava and Yeh, 1991; Basha, 2000; Iverson, 2000; Chen et al., 2001; Serrano, 2004; Yuan and Lu, 2005; Huang and Wu, 2012). The disadvantage of analytical methods is that analytical solutions often include infinite series or integrals that need to be evaluated numerically and can only be derived for simplified problems involving linearization of governing equations, homogeneous or layered soils, simplified geometries, initial and boundary conditions (Zhan and Ng, 2001; Šimůnek, 2006). Numerical studies can incorporate more sophisticated and advanced models of soil hydraulic properties and are widely used to simulate the seepage and infiltration in soil slopes under rainfall conditions (Ng and Shi, 1998; Gasmo et al., 2000; Tsaparas et al., 2002; Blatz et al., 2004; Zhang et al., 2004; Rahardjo et al., 2007; Lee et al. 2009; Rahimi et al., 2010).

The infinite slope stability analysis and the limit equilibrium method of slope stability based on methods of slices are often adopted for slope stability analysis under rainfall condition (Ng and Shi, 1998; Rahardjo et al., 2001; Tsaparas et al., 2002; Wilkinson et al., 2002; Blatz et al., 2004; Rahardjo et al., 2007; Cascini et al., 2010). The extended Mohr–Coulomb failure criterion (Fredlund et al., 1978) is usually used together with incorporated pore-water pressure distribution from infiltration analyses to consider the effect of matric suction.

The seepage and stress-deformation in unsaturated soil slopes are actually interacted and require a coupled hydro-mechanical modeling. A number of formulations and constitutive models based on either the Bishop's effective stress concept (Bishop, 1959) or independent stress state variables (Fredlund and Morgenstern, 1977) have been proposed in the literature (Lloret et al., 1987; Alonso et al., 1990; Fredlund and Rahardjo, 1993; Wheeler and Sivakumar, 1995; Cui and Delage, 1996; Thomas and He, 1998; Loret and Khalili, 2002; Chiu and Ng, 2003; Georgiadis, 2003; Wheeler et al., 2003; Sun et al. 2007; Gens et al., 2008; Zhang and Ikariya, 2011; Sheng et al., 2008). Because of the complexity of the governing equations, numerical methods are usually adopted (Cho and Lee, 2001; Smith, 2003; Cheuk et al., 2005; Ye et al., 2005; Zhang et al., 2005; Borja and White 2010; Cascini et al., 2010; Kim et al., 2012; Lu et al., 2012; Xiong et al., 2014; Jamei et al., 2015). The analytical solutions for the coupled hydro-mechanical modeling are very limited (Wu and Zhang, 2009).

The assumption of uniform porous media is often not valid for natural soils as a variety of heterogeneities such as fractures, cracks, macropores of biotic origin, and interaggregate pores (Novak et al., 2000; Li and Zhang, 2010, 2011) may exist. The presence of cracks in a slope decreases the shear strength of the slope soils and increases the hydraulic conductivity of the soils. Water-filled cracks may lead to additional driving forces. In addition, the cracks may form a part of the slip surface when landslide occurs. The tubular voids or passageways (known as soil pipes) and bedrock may contribute to the rapid development of positive pore-water pressure and trigger landslides (Pierson, 1983; Gerscovich et al., 2006). Landslides which occurred in natural slopes with colluviums should also be investigated distinctly as colluviums are a special natural material that is composed of both soil and gravels and has a wide and sometimes multimodal gradation.

Substantial progress has been made in the investigation of heterogeneous unsaturated soils (Morris et al., 1992; Novak et al., 2000; Li et al., 2009; Peron et al., 2009; Li and Zhang, 2010, 2011; Zhou et al., 2014). However, the limited comparison studies of measured and calculated responses (Alonso et al., 2003; Lan et al., 2003; Huat et al., 2006; Sako et al., 2006; Trandafir et al., 2008; Hamdhan and Schweiger, 2013) imply that further efforts are necessary to better understand the mechanisms of landslides in natural soil slopes.

One reason for the discrepancy between field responses and results of theoretical model is that a large amount of uncertainties are involved in the analysis of rainfall infiltration and slope stability, for example, geological formation, spatial variability, uncertainty in boundary conditions and initial conditions, measurement errors, and the error associated with the prediction model. A probabilistic method provides a systematic and quantitative way to account for the uncertainties involved in a complicated geotechnical system. The stability of a slope can be measured by not only a safety factor but also a reliability index, which is related to the probability of satisfactory performance of the slope (Sivakumar Babu and Murthy, 2005; Zhang et al., 2005, 2014a; Penalba et al., 2009; Srivastava et al., 2010; Santoso et al., 2011; Park et al., 2013; Tan et al., 2013; Zhu et al., 2013;

Ali et al. 2014, Cho, 2014; Dou et al., 2015). The information from various sources can be utilized systematically to evaluate errors of prediction models through model calibration (Zhang et al., 2010, 2014b).

1.3 OUTLINE OF THE BOOK

This book presents a comprehensive view of rainfall-induced slope failures from both geomechanical and probabilistic perspectives to reflect the complicated mechanisms and the tremendous uncertainties underlying the problem. First, we concentrate on the failure mechanisms of rainfall-induced slope failure (Chapters 2 through 6). Chapter 2 presents commonly used conceptual models and analytical solutions of infiltration and discusses important issues related to numerical analyses. Chapters 3 and 4 present the stability analysis methods based on limit equilibrium methods and coupled hydro-mechanical models. Chapters 5 and 6 focus on the infiltration and stability of natural soil slopes with cracks and colluvium materials. In the second part of the book (Chapters 7 through 9), the stability evaluation methods based on probabilistic approaches are presented. In Chapter 7, the effect of uncertainties of essential soil properties that can influence stability of unsaturated soil slopes are characterized. In Chapter 8, the effect of spatial variability on unsaturated soil slope is discussed. Finally, probabilistic model calibration for infiltration and slope stability models is presented in Chapter 9.

REFERENCES

Ali, A., Huang, J., Lyamin, A. V., Sloan, S. W., Griffiths, D. V., Cassidy, M. J., and Li, J. H. (2014). Simplified quantitative risk assessment of rainfall-induced landslides modelled by infinite slopes. *Engineering Geology*, 179, 102–116.

Alonso, E. E., Gens, A., and Delahaye, C. H. (2003). Influence of rainfall on the deformation and stability of a slope in over consolidated clays: A case study. *Hydrogeology Journal*, 11(1), 174–192.

Alonso, E. E., Gens, A., and Josa, A. (1990). A constitutive model for partially saturated soils. *Géotechnique*, 40(3), 405–430.

Anderson, S. A., and Sitar, N. (1995). Analysis of rainfall-induced debris flows. *Journal of Geotechnical Engineering, ASCE*, 121(7), 544–552.

Basha, H. A. (2000). Multidimensional linearised nonsteady infiltration toward a shallow water table. *Water Resources Research*, 36(9), 2567–2573.

Beven, K., Horton, Robert, E. (2004). Horton's perceptual model of infiltration processes. *Hydrological Processes*, 18(17), 3447–3460.

Bishop, A. W. (1959). The principle of effective stress. *Teknisk Ukeblad*, 106(39), 859–63.

Blatz, J. A., Ferreira, N. J., and Graham, J. (2004). Effects of near-surface environmental conditions on instability of an unsaturated soil slope. *Canadian Geotechnical Journal*, 41(6), 1111–1126.

Borja, R. I., and White, J. A. (2010). Continuum deformation and stability analyses of a steep hillside slope under rainfall infiltration. *Acta Geotechnica*, 5(1), 1–14.

Brand, E. W. (1984). Landslides in south Asia: A state-of-art report. *Proceedings of the 4th International Symposium on Landslides*, Toronto, Canada, pp. 17–59.

Briggs, K. M., Smethurst, J. A., Powrie, W., and O'Brien, A. S. (2013). Wet winter pore pressures in railway embankments. *Proceedings of the ICE-Geotechnical Engineering*, 166(5), 451–465.

Cascini, L., Cuomo, S., and Guida, D. (2008). Typical source areas of May 1998 flow-like mass movements in the Campania region, Southern Italy. *Engineering Geology*, 96(3), 107–125.

Cascini, L., Cuomo, S., Pastor, M., and Sorbin, G. (2010). Modelling of rainfall-induced shallow landslides of the flow-type. *Journal of Geotechnical and Geoenvironmental Engineering*, 136(1), 85–98.

Chen, H., Lee, C. F., and Law, K. T. (2004). Causative mechanisms of rainfall-induced fill slope failures. *Journal of Geotechnical and Geoenvironmental Engineering, ASCE*, 130(6), 593–602.

Chen, J. M., Tan, Y. C., and Chen, C. H. (2001). Multidimensional infiltration with arbitrary surface fluxes. *Journal of Irrigation and Drainage Engineering*, 127(6), 370–377.

Cheuk, C. Y., Ng, C. W. W., and Sun, H. W. (2005). Numerical experiments of soil nails in loose fill slopes subjected to rainfall infiltration effects. *Computers and Geotechnics*, 32(4), 290–303.

Chiu, C. F., and Ng, C. W. W. (2003). A state-dependent elasto-plastic model for saturated and unsaturated soils. *Géotechnique*, 53(9), 809–829.

Cho, S. E. (2014). Probabilistic stability analysis of rainfall-induced landslides considering spatial variability of permeability. *Engineering Geology*, 171, 11–20.

Cho, S. E., and Lee, S. R. (2001). Instability of unsaturated soil slopes due to infiltration. *Computers and Geotechnics*, 28, 185–208.

Collins, B. D., and Znidarcic, D. (2004). Stability analyses of rainfall induced landslides. *Journal of Geotechnical and Geoenvironmental Engineering, ASCE*, 130(4), 362–372.

Crosta, G. B., and Frattini, P. (2003). Distributed modelling of shallow landslides triggered by intense rainfall. *Natural Hazards and Earth System Science*, 3(1/2), 81–93.

Cui, P., Chen, X., Waqng, Y., Hu, K., and Li, Y. (2005). Jiangjia Ravine debris flows in south-western China. *Debris-Flow Hazards and Related Phenomena*. Springer, Berlin, pp. 565–594.

Cui, Y. J., and Delage, P. (1996). Yielding and plastic behaviour of an unsaturated compacted silt. *Géotechnique*, 46(2), 291–311.

Da Costa Nunes, A. J., Costa Couto e Fonseca, A. M. M., and Hunt, R. E. (1979). Landslides of Brazil. In: Voight, B. (Ed.). *Rockslides and Avalanches, Vol. 2: Engineering Sites*. Elsevier, New York, pp. 419–446.

Dai, F. C., Lee, C. F., and Wang, S. J. (2003). Characterization of rainfall-induced landslides. *International Journal of Remote Sensing*, 24(23), 4817–4834.

Day, R. W., and Axten, G. W. (1989). Surficial stability of compacted clay slopes. *Journal of Geotechnical Engineering, ASCE*, 115(4), 577–580.

Dou, H. Q., Han, T. C., Gong, X. N., Qiu, Z. Y., and Li, Z. N. (2015). Effects of the spatial variability of permeability on rainfall-induced landslides. *Engineering Geology*, 192, 92–100.

Fourie, A. B. (1996). Predicting rainfall-induced slope instability. *Proceedings of the Institution of Civil Engineers Geotechnical Engineering*, 119(4), 211–218.

Fourie, A. B., Rowe, D., and Blight, G. E. (1999). The effect of infiltration on the stability of the slopes of a dry ash dump. *Géotechnique*, 49(1), 1–13.

Fredlund, D. G., and Morgenstern, N. R. (1977). Stress state variables for unsaturated soils. *Journal of Geotechnical and Geoenvironmental Engineering*, 103(GT5), 447–466.

Fredlund, D. G., Morgenstern, N. R., and Widger, R. A. (1978). The shear strength of unsaturated soils. *Canadian Geotechnical Journal*, 15(3), 313–321.

Fredlund, D. G., and Rahardjo, H. (1993). *Soil Mechanics for Unsaturated Soils*. Wiley, New York.

Fredlund, D. G., Rahardjo, H., and Fredlund, M. D. (2012). *Unsaturated Soil Mechanics in Engineering Practice*. Wiley, New York.

Gasmo, J. M., Rahardjo, H., and Leong, E.C. (2000). Infiltration effects on stability of a residual soil slope. *Computers and Geotechnics*, 26(2), 145–165.

Gens, A., Guimaraes, L., Sánchez, M., and Sheng, D. (2008). Developments in modelling the generalised behaviour of unsaturated soils. *Unsaturated Soils: Advances in Geo-Engineering*. Taylor & Francis Group, London, pp. 53–61.

Georgiadis, K. (2003). *Development, Implementation and Application of Partially Saturated Soil Models in Finite Element Analysis*. Ph.D. Thesis, Imperial College of Science, Technology and Medicine, University of London, London, UK.

Gerscovich, D.M.S., Vargas, E.A., and de Campos, T.M.P. (2006). On the evaluation of unsaturated flow in a natural slope in Rio de Janeiro, Brazil. *Engineering Geology*, 88(1–2), 23–40.

Glade, T., Anderson, M.G., and Crozier, M.J. (2006). *Landslide Hazard and Risk*. Wiley, New York.

Green, W. H., and Ampt, C. A. (1911). Studies on soil physics: Flow of air and water through soils. *Journal of Agricultural Science*, 4, 1–24.

Hamdhan, I. N., and Schweiger, H. F. (2013). Finite element method–based analysis of an unsaturated soil slope subjected to rainfall infiltration. *International Journal of Geomechanics*. 13(5), 653–658.

Holtan, H. N. (1961). *A Concept of Infiltration Estimates in Watershed Engineering* (ARS41-51). US Department of Agricultural Service, Washington, DC.

Hong, Y., and Adler, R. F. (2008). Predicting global landslide spatiotemporal distribution: Integrating landslide susceptibility zoning techniques and real-time satellite rainfall estimates. *International Journal of Sediment Research*, 23(3), 249–257.

Horton, R. E. (1938). The interpretation and application of runoff plot experiments with reference to soil erosion problems. *Soil Science Society of America Proceedings*, 3, 340–349.

Huang, R. Q. (2007). Large-scale landslides and their sliding mechanisms in China since the 20th century. *Chinese Journal of Rock Mechanics and Engineering*, 26(3), 433–454 (in Chinese).

Huang, R. Q., and Wu, L. Z. (2012). Analytical solutions to 1-D horizontal and vertical water infiltration in saturated/unsaturated soils considering time-varying rainfall. *Computers and Geotechnics*, 39, 66–72.

Huat, B. B. K., Ali, F. H., and Low, T. H. (2006). Water infiltration characteristics of unsaturated soil slope and its effect on suction and stability. *Geotechnical and Geological Engineering*, 24, 1293–1306.

Iverson, R. M. (2000). Landslide triggering by rain infiltration. *Water Resources Research*, 36(7), 1897–1910.

Iverson, R. M., George, D. L., Allstadt, K., Reid, M. E., Collins, B. D., Vallance, J. Schillinga S. P., Godtd J. W., Cannone C. M., Magirlf, C. S., Baumd R. L., Coed J. A., Schulzd W. H., and Bower, J. B. (2015). Landslide mobility and hazards: Implications of the 2014 Oso disaster. *Earth and Planetary Science Letters*, 412, 197–208.

Jamei, M., Guiras, H., and Olivella, S. (2015). Analysis of slope movement initiation induced by rainfall using the Elastoplastic Barcelona Basic Model. *European Journal of Environmental and Civil Engineering*, 19(9), 1033–1058.

Jibson, R. W. (2005). *Landslide Hazards at La Conchita, California*. Open-File Report 2005–1067, US Geological Survey, Reston, VA, p. 12.

Johnson, K. A., and Sitar, N. (1990). Hydrologic conditions leading to debris-flow initiation. *Canadian Geotechnical Journal*, 27(6), 789–801.

Kim, J., Jeong, S., and Regueiro, R. A. (2012). Instability of partially saturated soil slopes due to alteration of rainfall pattern. *Engineering Geology*, 147, 28–36.

Kitamura, R., and Sako, K. (2010). Contribution of "Soils and Foundations" to studies on rainfall-induced slope failure. *Soils and Foundations*, 50(6), 955–964.

Lan, H. X., Zhou, C. H., Lee, C. F., Wang, S. J., and Wu, F. Q. (2003). Rainfall-induced landslide stability analysis in response to transient pore pressure—A case study of natural terrain landslide in Hong Kong. *Science in China Series E: Engineering and Materials Science*, 46, 52–68.

Lee, L. M., Gofar, N., and Rahardjo, H. (2009). A simple model for preliminary evaluation of rainfall-induced slope instability. *Engineering Geology*, 108(3–4), 272–285.

Li, J. H., and Zhang, L. M. (2010). Geometric parameters and REV of a crack network in soil. *Computers and Geotechnics*, 37, 466–475.

Li, J. H., and Zhang, L. M. (2011). Study of desiccation crack initiation and development at ground surface. *Engineering Geology*, 123, 347–358.

Li, J. H., Zhang, L. M., Wang, Y., and Fredlund, D. G. (2009). Permeability tensor and representative elementary volume of saturated cracked soil. *Canadian Geotechnical Journal*, 46(8), 928–942.

Lloret, A., Gens, A., Batlle, F., and Alonso, E. E. (1987). Flow and deformation analysis of partial saturated soils. *Proceedings of the 9th European Conference on Soil Mechanics and Foundation Engineering*, Dublin, Ireland, pp. 565–568.

Loret, B., and Khalili, N. (2002). An effective stress elastic–plastic model for unsaturated porous media. *Mechanics of Materials*, 34, 97–116.

Lu, N., and Godt, J. W. (2013). *Hillslope Hydrology and Stability*. Cambridge University Press, Cambridge, UK, pp. 1–458.

Lu, X. B., Ye, T. L., Zhang, X. H., Cui, P., and Hu, K. H. (2012). Experimental and numerical analysis on the responses of slope under rainfall. *Natural Hazards*, 64(1), 887–902.

Lumb, P. (1962). Effect of rain storms on slope stability. *Proceedings of the Symposium on Hong Kong Soils*, Hong Kong, pp. 73–87.

Mein, R. G. and Larson, C. L. (1973). Modelling infiltration during a steady rain. *Water Resources Research*, 9(2), 384–394.

Morris, P. H., Graham, J., and Williams, D. J. (1992). Cracking in drying soils. *Canadian Geotechnical Journal*, 29(2), 263–277.

Ng, C. W. W., and Shi, Q. (1998). Numerical investigation of the stability of unsaturated soil slopes subjected to transient seepage. *Computers and Geotechnics*, 22(1), 1–28.

Novak, V., Simunek, J. and van Genuchten, M. T. (2000). Infiltration of water into soil with cracks. *Journal of Irrigation and Drainage Engineering-ASCE*, 126(1), 41–47.

Park, H. J., Lee, J. H., and Woo, I. (2013). Assessment of rainfall-induced shallow landslide susceptibility using a GIS-based probabilistic approach. *Engineering Geology*, 161, 1–15.

Penalba, R. F., Luo, Z., and Juang, C. H. (2009). Framework for probabilistic assessment of landslide: A case study of El Berrinche. *Environmental Earth Sciences*, 59(3), 489–499.

Peron, H., Delenne, J. Y., Laloui, L., and El Youssoufi, M. S. (2009). Discrete element modelling of drying shrinkage and cracking of soils. *Computers and Geotechnics*, 36(1–2), 61–69.

Pierson, T. C. (1983). Soil pipes and slope stability. *Quarterly Journal of Engineering Geology and Hydrogeology*, 16(1), 1–11.

Prokešová, R., Medveďová, A., Tábořík, P., and Snopková, Z. (2013). Towards hydrological triggering mechanisms of large deep-seated landslides. *Landslides*, 10(3), 239–254.

Rahardjo, H., Li, X. W., Toll, D. G., and Leong, E. C. (2001). The effect of antecedent rainfall on slope stability. *Geotechnical and Geological Engineering*, 19, 371–399.

Rahardjo, H., Ong, T. H., Rezaur, R. B., and Leong, E. C. (2007). Factors controlling instability of homogeneous soil slopes under rainfall. *Journal of Geotechnical and Geoenvironmental Engineering*, 133(12), 1532–1543.

Rahimi, A., Rahardjo, H., and Leong, E. C. (2010). Effect of hydraulic properties of soil on rainfall-induced slope failure. *Engineering Geology*, 114(3–4), 135–143.

Sako, K., Kitamura, R. and Fukagawa, R. (2006). Study of slope failure due to rainfall: A comparison between experiment and simulation. *Proceedings of the 4th International Conference on Unsaturated Soils*, April 2–6, Carefree, AZ, ASCE, pp. 2324–2335.

Santoso, A. M., Phoon, K. K., and Quek, S. T. (2011). Effects of soil spatial variability on rainfall-induced landslides. *Computers and Structures*, 89(11–12), 893–900.

Schuster, R. L., and Highland, L. M. (2007). The Third Hans Cloos Lecture. Urban landslides: Socioeconomic impacts and overview of mitigative strategies. *Bulletin of Engineering Geology and the Environment*, 66(1), 1–27.

Serrano, S. E. (2004). Modeling infiltration with approximate solutions to Richard's equation. *Journal of Hydrologic Engineering*, 9(5), 421–432.

Sheng, D., Gens, A., Fredlund, D. G., and Sloan, S. W. (2008). Unsaturated soils: From constitutive modelling to numerical algorithms. *Computers and Geotechnics*, 35(6), 810–824.

Sidle, R. C., and Ochiai, H. (2006). *Landslides: Processes, Prediction, and Land Use.* Water Resources Monograph Series, Vol. 18. American Geophysical Union, Washington, DC, pp. 1–312.

Šimůnek, J. (2006). Models of water flow and solute transport in the unsaturated zone. *Encyclopedia of Hydrological Sciences*, 6, 78.

Sivakumar Babu, G. L., and Murthy, D. S. N. (2005). Reliability analysis of unsaturated soil slopes. *Journal of Geotechnical and Geoenvironmental Engineering*, 131(11), 1423–1428.

Smith, P. G. C. (2003). *Numerical Analysis of Infiltration into Partially Saturated Soil Slopes.* Ph.D. Thesis, Imperial College of Science, Technology and Medicine, University of London, London, UK.

Srivastava, A., Sivakumar Babu, G. L., and Haldar, S. (2010). Influence of spatial variability of permeability property on steady state seepage flow and slope stability analysis. *Engineering Geology*, 110(3–4), 93–101.

Srivastava, R., and Yeh, T. C. J. (1991). Analytical solutions for one dimension transient infiltration toward the water table in homogeneous and layered soils. *Water Resources Research*, 27(5), 753–762.

Sun, D. A., Sheng, D., Sloan, S. W. (2007). Elastoplastic modelling of hydraulic and stress-strain behaviour of unsaturated soil. *Mechanics of Materials*, 39(3), 212–221.

Sun, H. W., Wong, H. N., and Ho, K. K. S. (1998). Analysis of infiltration in unsaturated ground. *Slope Engineering in Hong Kong: Proceedings of the Annual Seminar on Slope Engineering in Hong Kong*, Balkema, pp. 101–109.

Tan, X. H., Hu, N., Li, D., Shen, M. F., and Hou, X. L. (2013). Time-variant reliability analysis of unsaturated soil Slopes under rainfall. *Geotechnical and Geological Engineering*, 31(1), 319–327.

Thomas, H. R., and He, Y. (1998). Modelling the behaviour of unsaturated soil using an elastoplastic constitutive model. *Géotechnique*, 48(5), 589–603.

Trandafir, A. C., Sidle, R. C., Gomi, T., and Kamai, T. (2008). Monitored and simulated variations in matric suction during rainfall in a residual soil slope. *Environmental Geology*, 55(5), 951–961.

Tsaparas, I., Rahardjo, H., Toll, D. G., and Leong, E. C. (2002). Controlling parameters for rainfall-induced landslides. *Computers and Geotechnics*, 29(1), 1–27.

Wang, F. W., Sassa, K., and Wang, G. (2002). Mechanism of a long-runout landslide triggered by the August 1998 heavy rainfall in Fukushima Prefecture, Japan. *Engineering Geology*, 63(1), 169–185.

Wang, G. and Sassa, K. (2001). Factors affecting rainfall-induced flowslides in laboratory flume tests. *Géotechnique*, 51(7), 587–599.

Wang, G., Sassa, K., and Fukuoka, H. (2003). Downslope volume enlargement of a debris slide–debris flow in the 1999 Hiroshima, Japan, rainstorm. *Engineering Geology*, 69(3), 309–330.

Wheeler, S. J., Sharma, R. S., and Buisson, M. S. R. (2003). Coupling of hydraulic hysteresis and stress-strain behaviour in unsaturated soils. *Géotechnique*, 53(1), 41–54.

Wheeler, S. J., and Sivakumar, V. (1995). An elasto-plastic critical state framework for unsaturated soil. *Géotechnique*, 45(1), 35–53.

Wilkinson, P. L., Anderson, M. G., and Lloyd, D. M. (2002). An integrated hydrological model for rain-induced landslide prediction. *Earth Surface Processes and Landforms*, 27(12), 1285–1297.

Wu, L. Z., and Zhang, L. M. (2009). Analytical solution to 1D coupled water infiltration and deformation in unsaturated soils. *International Journal for Numerical and Analytical Methods in Geomechanics*, 33(6), 773–790.

Xiong, Y., Bao, X., Ye, B., and Zhang, F. (2014). Soil–water–air fully coupling finite element analysis of slope failure in unsaturated ground. *Soils and Foundations*, 54(3), 377–395.

Ye, G. L., Zhang, F., Yashima, A., Sumi, T., and Ikemura, T. (2005). Numerical analyses on progressive failure of slope due to heavy rain with 2D and 3D FEM. *Soils and Foundations*, 45(2), 1–15.

Yuan, F., and Lu, Z. (2005). Analytical solutions for vertical flow in unsaturated, rooted soils with variable surface fluxes. *Vadose Zone Journal*, 4(4), 1210–1218.

Zhan, L. T., and Ng, C. W. W. (2001). Analytical analysis of rainfall infiltration mechanism in unsaturated soils. *International Journal of Geomechanics*, 4, 273–284.

Zhang, F., and Ikariya, T. (2011). A new model for unsaturated soil using skeleton stress and degree of saturation as state variables. *Soils and Foundations*, 57, 67–81.

Zhang, J., Huang, H. W., Zhang, L. M., Zhu, H. H., and Shi, B. (2014a). Probabilistic prediction of rainfall-induced slope failure using a mechanics-based model. *Engineering Geology*, 168, 129–140.

Zhang, L. L., Fredlund, D. G., Zhang, L. M., and Tang, W. H. (2004). Numerical study of soil conditions under which matric suction can be maintained. *Canadian Geotechnical Journal*, 41(4), 569–582.

Zhang, L. L., Zhang, J., Zhang, L. M., and Tang, W. H. (2010). Back analysis of slope failure with Markov chain Monte Carlo simulation. *Computers and Geotechnics*, 37(7), 905–912.

Zhang, L. L., Zhang, J., Zhang, L. M., and Tang, W. H. (2011). Stability analysis of rainfall-induced slope failure: A review. *Proceedings of the ICE-Geotechnical Engineering*, 164(5), 299–316.

Zhang, L. L., Zhang, L. M., and Tang, W. H. (2005). Rainfall-induced slope failure considering variability of soil properties. *Géotechnique*, 55(2), 183–188.

Zhang, L. L., Zheng, Y. F., Zhang, L. M. (2014b). Probabilistic model calibration for soil slope under rainfall: Effects of measurement duration and frequency in field monitoring. *Géotechnique*, 64(5), 365–378.

Zhou, Y. F., Tham, L. G., Yan, R. W. M., and Xu, L. (2014). The mechanism of soil failures along cracks subjected to water infiltration. *Computers and Geotechnics*, 55, 330–341.

Zhu, H., Zhang, L. M., Zhang, L. L., Zhou, C. B. (2013). Two-dimensional probabilistic infiltration analysis with a spatially varying permeability function. *Computers and Geotechnics*, 48, 249–259.

Zhu, J. H., and Anderson, S. A. (1998). Determination of shear strength of Hawaiian residual soil subjected to rainfall-induced landslides. *Géotechnique*, 48(1), 73–82.

Chapter 2

Infiltration and seepage analysis in soil slopes

2.1 INTRODUCTION

Infiltration plays a significant role in the instability of slopes (Zhang et al., 2011). The effect of seepage and infiltration on slope stability is traditionally addressed by calculating the factor of safety or critical depth for an infinite slope subject to seepage parallel to the slope surface. This type of analysis assumes that saturated steady-state flow takes place over a given depth. To simplify the analysis as a worst-infiltration scenario, it is often assumed that the phreatic surface rises to coincide with the slope surface and that the slope is completely saturated (Collins and Znidarcic, 2004). For such fully saturated slopes, additional infiltration is not possible and rainfall will have no further effect on slope stability.

Field measurements of soil suction show that the soil suction will not necessarily be destroyed, even under long-term rainfall infiltration. A good example is a field monitoring study in Hong Kong, in which situ suction measurements in situ were made throughout the year of 1980 in a 30 m high completely weathered rhyolite slope (Sweeney, 1982). Matric suction at shallow depths showed a gradual reduction during the rainy season but, at 5–17 m depths, the soil suction remained constant throughout the year (Figure 2.1). The pore-water pressures remained negative even during the rainy season. For slopes that have unsaturated zones, the rainfall will have a different effect compared with full-saturated slopes. The pore-water pressure profile and the wetting front vary continuously in the unsaturated zone as a transient process as the infiltrating water moves downward into the soil. The shear strength of the soil and hence the slope stability are affected by the transient pore-water pressure profiles. In this chapter, the estimation methods based on the wetting front model for both level ground and sloping ground is introduced in Section 2.2. The governing equation of water flow in unsaturated soil, the Richards equation, is presented in Section 2.3. Analytical solutions to the Richards equation are described in Section 2.4. In Section 2.5, the important issues related to numerical solutions of the Richards equation are discussed. Typical pore-water pressure profiles under rainfall infiltration are described in Section 2.6. Finally, a numerical parametric study is conducted to illustrate under certain conditions soil suction can be maintained even under rainfall infiltration.

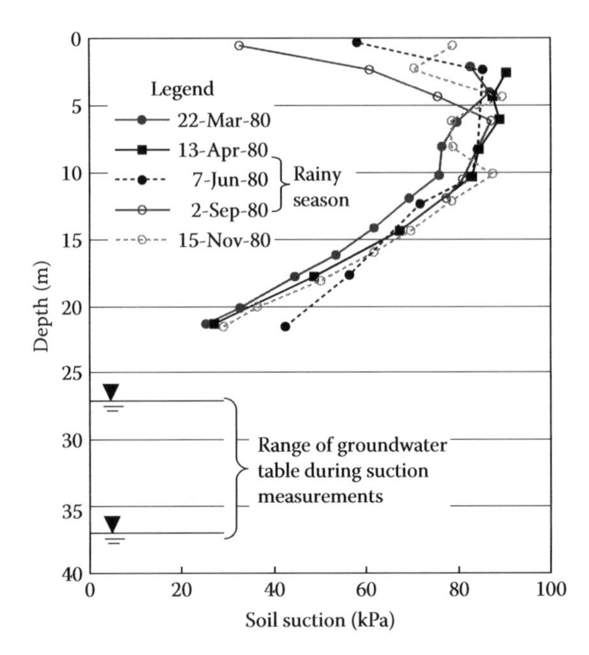

Figure 2.1 Suction measurements in a weathered rhyolite in Hong Kong. (From Sweeney, D. J., Some in situ soil suction measurements in Hong Kong's residual soil slopes, *Proceedings of the 7th Southeast Asian Geotechnical Conference*, Vol. 1, Southeast Asian Geotechnical Society, Thailand, pp. 91–106, 1982. With permission.)

2.2 ESTIMATION OF INFILTRATION RATE BASED ON CONCEPTUAL MODEL

2.2.1 Mechanism of infiltration in soils

In field situation, the actual infiltration rate into soils may not be equal to the rainfall intensity because the rainwater in excess of the infiltration capacity of a soil cannot infiltrate into soils. The amount of water that can infiltrate into the ground is governed by the infiltration capacity of the soil, which is defined as the maximum rate of the rain which can be absorbed by the soil at a given condition. Surface runoff in a sloping ground or surface ponding in a level ground occurs as the rainfall intensity is larger than the infiltration capacity of the soil.

Figure 2.2 illustrates a schematic plot of the relationship between the infiltration rate and the rainfall intensity. At the beginning of a rainstorm, the infiltration capacity of soil is very large (often assumed infinite), and all the rainwater is absorbed by the soil. During this period, the infiltration rate is equal to the rain intensity. As the surface soil becomes wet, the infiltration capacity of soil decreases rapidly. When the rainfall intensity is larger than the infiltration capacity, surface runoff takes place. In such a case, the actual infiltration rate is equal to the infiltration capacity of soil.

The infiltration capacity is closely related to but not equal to the saturated permeability of soil. The condition of the surface soil is an important factor affecting the infiltration capacity. For instance, surface cracking and holes may make the infiltration capacity much larger than the saturated permeability of the soil. On the other hand, surface sealing could substantially reduce the infiltration capacity.

Many theoretical and empirical models have been developed in the literature in fields of agriculture and hydrology, such as the Green–Ampt model (Green and Ampt, 1911), the Horton

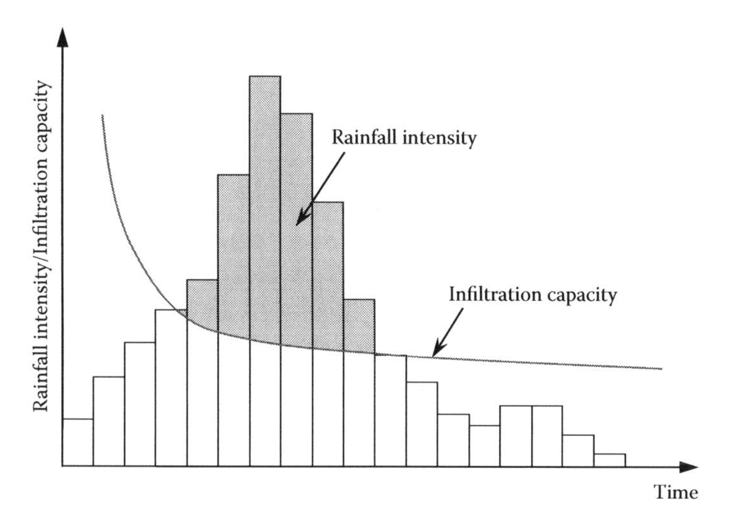

Figure 2.2 Rainfall intensity and infiltration capacity of soil.

model (Horton, 1938; Beven and Horton, 2004), and the Holton model (Holtan, 1961). A comprehensive comparison of different infiltration models can be found in Mishra et al. (2003). The Green–Ampt model is one of the most widely used infiltration models and is derived based on the Darcy's law with physically measurable parameters. In the following paragraphs, how to analyze the infiltration process based on the Green–Ampt model will be described.

It should be noted that the actual infiltration process into soil in field is much more complicated than the simplified estimation models, due to soil heterogeneity, spatial variability, and the potential interaction between surface flow and subsurface flow (Kwok and Tung, 2003). Field measurements (Li et al., 2005; Rahardjo et al., 2005) and numerical modeling results (Gasmo et al., 2000) have shown that the actual infiltration rate is not limited by the saturated permeability and surface ponding/runoff can occur even before the soil is fully saturated. Li et al. (2005) recorded the infiltration rate into a completely decomposed granite slope in Hong Kong during the rainy season in 2001. The saturated permeability of the soil is in the range of 1×10^{-6} to 1×10^{-5} m/s. The rainfall intensity during the test period ranged from 2.8×10^{-7} to 2.3×10^{-6} m/s. The study showed that runoff began even before the near surface soils became fully saturated. Rahardjo et al. (2005) applied an artificial rainfall of 13×10^{-6} m/s in intensity to an initially unsaturated soil slope with a saturated permeability k_s of 5.18×10^{-6} m/s and found that the infiltration capacity of the slope converged to 2.0×10^{-6} m/s ($\approx 0.4\ k_s$). Results from the numerical study of Gasmo et al. (2000) also showed that the initial infiltration rate can be larger than k_s and gradually decreases to a steady-state value which is less than k_s (Figure 2.3).

2.2.2 Green–Ampt model for infiltration of constant rainfall on a level ground

The Green–Ampt model was originally proposed for infiltration analysis on a level ground under ponded infiltration (Green and Ampt, 1911). The Green–Ampt model assumes that a distinct wetting front develops between the dry soil and the wetted soil (Figure 2.4). Assume the initial water content before the rainfall event starts is θ_0. Let θ_1 denote the water content of the soil in the wetted (saturation) zone. Supposing the depth of ponding is negligible (pressure head is zero at the ground surface), the difference between the total heads at the ground surface and wetting front is

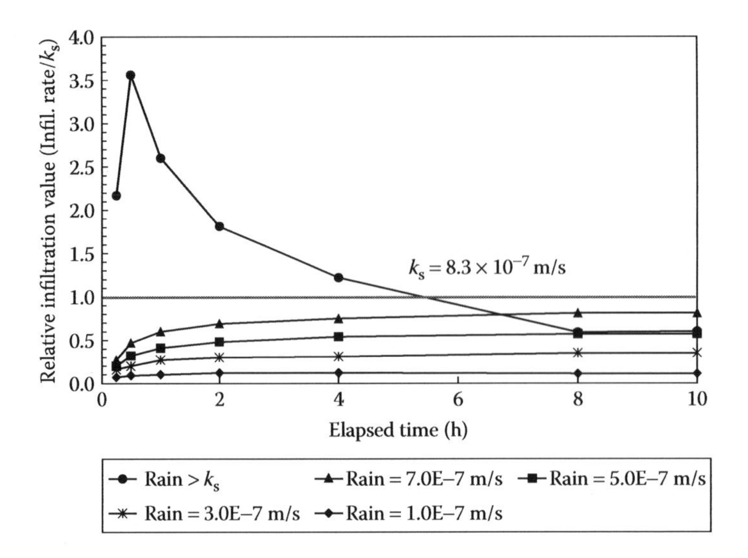

Figure 2.3 Infiltration rate at the crest of a slope. (From *Computers and Geotechnics*, 26, Gasmo, J. M. et al., Infiltration effects on stability of a residual soil slope, 145–165, Copyright 2000, with permission from Elsevier.)

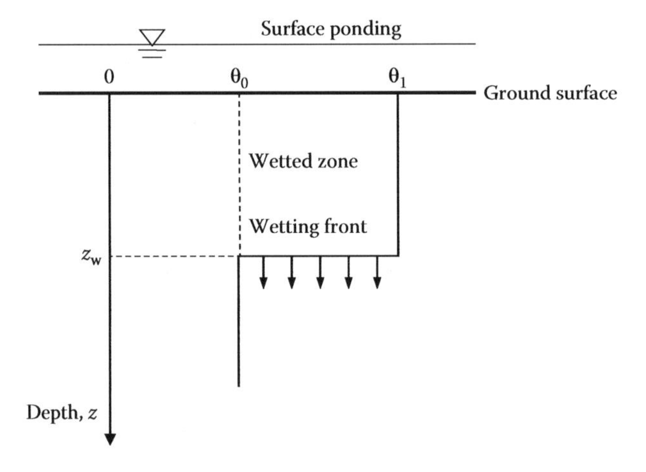

Figure 2.4 Illustration of the Green–Ampt (1911) infiltration model.

$$\Delta H = z_w + s_f \tag{2.1}$$

where:

s_f is the suction head at the wetting front (positive for suction corresponding to θ_0)
z_w is the depth of wetting front
H is the total head

Based on the principle of continuity and the Darcy's law, the infiltrate capacity under ponded infiltration is equal to the flow rate inside the soil:

$$i = k_w \frac{\Delta H}{\Delta z} = k_w \frac{z_w + s_f}{z_w} \tag{2.2}$$

where:

k_w is the coefficient of permeability of the soil in the wetted zone
i is the infiltration capacity

Note that the depth of the wetting front can be related to the cumulative infiltration I (unit: length) as follows:

$$z_w = \frac{I}{\theta_1 - \theta_0} \quad (2.3)$$

Substituting Equation 2.3 into Equation 2.2, the infiltration capacity can be written as

$$i = k_w\left[1 + \frac{s_f(\theta_1 - \theta_0)}{I}\right] \quad (2.4)$$

Based on the infiltration mechanism as described in the previous section, ponding occurs when the infiltration capacity is equal to the rainfall intensity q. Substituting i with q in Equation 2.4, the cumulative infiltration at the moment of ponding, I_p, can be solved as follows:

$$I_p = \frac{s_f k_w(\theta_1 - \theta_0)}{q - k_w} \quad (q > k_w) \quad (2.5)$$

As the cumulative infiltration cannot be negative, Equation 2.5 indicates that q must be larger than k_w to achieve ponding.

Hence, the time needed to achieve ponding can be calculated as

$$t_p = \frac{I_p}{q} = \frac{s_f k_w(\theta_1 - \theta_0)}{q(q - k_w)} \quad (q > k_w) \quad (2.6)$$

where t_p is the time of ponding.

2.2.3 Infiltration of constant rainfall on a sloping ground

For the slope stability problem, the infiltration model on the sloping ground should be adopted. The Green–Ampt model (Green and Ampt, 1911) was extended for a sloping surface by Chen and Young (2006). On a sloping surface, only rainfall normal to the surface causes infiltration; the down-slope component of rainfall and gravity cause flow, but do not change the water content profile along the normal direction on a planar slope (Philip, 1991). Suppose the rainfall direction is vertical with an intensity of q. The slope angle is α_s. The rainfall normal to the slope surface is thus $q \cos \alpha_s$. Assume that before ponding occurs all the rainwater infiltrates into the soil, that is, the infiltration capacity is equal to the rainfall intensity. After ponding occurs, the infiltration capacity i is (Chen and Young, 2006)

$$i = k_w\left[\cos\alpha_s + \frac{s_f(\theta_1 - \theta_0)}{I}\right] \quad (2.7)$$

when $\alpha_s = 0$, Equation 2.7 is reduced to Equation 2.4.

The depth of the wetting front, z_w, can be calculated based on the cumulative infiltration I as follows:

$$z_w = \frac{I}{(\theta_1 - \theta_0)\cos\alpha_s} \tag{2.8}$$

Consider the case of a constant rain intensity, q. At the moment ponding occurs, the rainfall intensity is equal to the infiltration capacity. Equating the rainfall intensity $q \cos \alpha_s$ and the infiltration capacity at the moment of ponding (Equation 2.7), the corresponding cumulative infiltration can be calculated as

$$I_p = \frac{s_f k_w (\theta_1 - \theta_0)}{(q - k_w)\cos\alpha_s} \quad (q > k_w) \tag{2.9}$$

The time of ponding, t_p, can thus be calculated as follows:

$$t_p = \frac{I_p}{q\cos\alpha_s} = \frac{s_f k_w (\theta_1 - \theta_0)}{q(q - k_w)\cos^2\alpha_s} \tag{2.10}$$

Note $i = dI/dt$. Equation 2.7 can also be written as

$$\frac{dt}{dI} = \frac{1}{k_w\left[\cos\alpha_s + \left(s_f(\theta_1 - \theta_0)/I\right)\right]} \tag{2.11}$$

Solving the above equation with the initial condition that at $t = t_p$ and I is equal to I_p, the time needed to achieve a certain cumulative infiltration I after ponding is

$$t = t_p + \frac{I - I_p}{k_{w2}} - \frac{\zeta}{k_{w2}} \ln\left(\frac{I + \zeta}{I_p + \zeta}\right) \quad (t > t_p) \tag{2.12}$$

where:

$k_{w2} = k_w \cos \alpha_s$
$\zeta = [s_f(\theta_1 - \theta_0)]/\cos \alpha_s$

Example 2.1

Considering an example with a sloping ground subjected to a rainfall of $q = 1.0 \times 10^{-5}$ m/s. The slope angle is $\alpha_s = 30°$. The soil properties of the ground are as follows: $\theta_0 = 0.25$, $\theta_1 = 0.45$, $k_w = 7.5 \times 10^{-6}$ m/s, $s_f = 25$ cm.

Based on Equation 2.10, the time to ponding can be calculated as follows:

$$t_p = \frac{s_f k_w (\theta_1 - \theta_0)}{q(q - k_w)\cos^2\alpha_s} = \frac{0.25 \times 7.5 \times 10^{-6} \times (0.45 - 0.25)}{1.0 \times 10^{-5} \times (1.0 \times 10^{-5} - 7.5 \times 10^{-6}) \times (\cos 30°)^2} = 2.0 \times 10^4 \text{ s} = 5.56 \text{ h}$$

Based on Equation 2.9, the cumulative infiltration (total infiltration depth) at the moment of ponding is

$$I_p = \frac{s_f k_w(\theta_1 - \theta_0)}{(q - k_w)\cos\alpha_s} = \frac{0.25 \times 7.5 \times 10^{-6} \times (0.45 - 0.25)}{(1.0 \times 10^{-5} - 7.5 \times 10^{-6}) \times \cos 30^\circ} = 0.173 \text{ m}$$

Based on Equation 2.8, the depth of wetting front at the moment of ponding is

$$z_w = \frac{I_p}{(\theta_1 - \theta_0)\cos\alpha_s} = \frac{0.173}{(0.45 - 0.25) \times \cos 30^\circ} = 1.0 \text{ m}$$

Before the ponding occurs, the infiltration rate is equal to the rainfall intensity, that is, $q \cos \alpha_s = 1.0 \times 10^{-5} \times \cos(30^\circ) = 8.66 \times 10^{-6}$ m/s. After the ponding, the infiltration rate is equal to the infiltration capacity as indicated by Equation 2.7. For instance, when the cumulative infiltration is $I = 0.2$ m, the infiltration rate is

$$i = k_w\left[\cos\alpha_s + \frac{s_f(\theta_1 - \theta_0)}{I}\right] = 7.5 \times 10^{-6} \times \left[\cos 30^\circ + \frac{0.25 \times (0.45 - 0.25)}{0.20}\right] = 8.37 \times 10^{-6} \text{ m/s}$$

The time needed to achieve a cumulative infiltration of $I = 0.2$ m can be calculated with Equation 2.12 as follows:

$$t = t_p + \frac{I - I_p}{k_{w2}} - \frac{\zeta}{k_{w2}} \ln\left(\frac{I + \zeta}{I_p + \zeta}\right)$$

$$= 2.0 \times 10^4 + \frac{0.2 - 0.17}{6.5 \times 10^{-6}} - \frac{0.058}{6.5 \times 10^{-6}} \ln\left(\frac{0.20 + 0.058}{0.17 + 0.058}\right) = 2.31 \times 10^4 \text{ s} = 6.43 \text{ h}$$

where $k_{w2} = k_w \cos \alpha_s = 7.5 \times 10^{-6} \times \cos(30^\circ) = 6.50 \times 10^{-6}$ m/s and $\zeta = [s_f(\theta_1 - \theta_0)]/\cos \alpha_s = [0.25 \times (0.45-0.25)]/\cos(30^\circ) = 0.058$ m. The corresponding depth of wetting front is

$$z_w = \frac{I}{(\theta_1 - \theta_0)\cos\alpha_s} = \frac{0.2}{(0.45 - 0.25) \times \cos 30^\circ} = 1.155 \text{ m}$$

Repeating the above procedure, the infiltration rate, the depth of wetting front, and the time to achieve different cumulative infiltration can be calculated. The relationship between time versus actual infiltration rate as well as the relationship between time versus depth of wetting front are shown in Figures 2.5 and 2.6, respectively. The above calculation can be calculated conveniently in a spreadsheet, as shown in Figure 2.7.

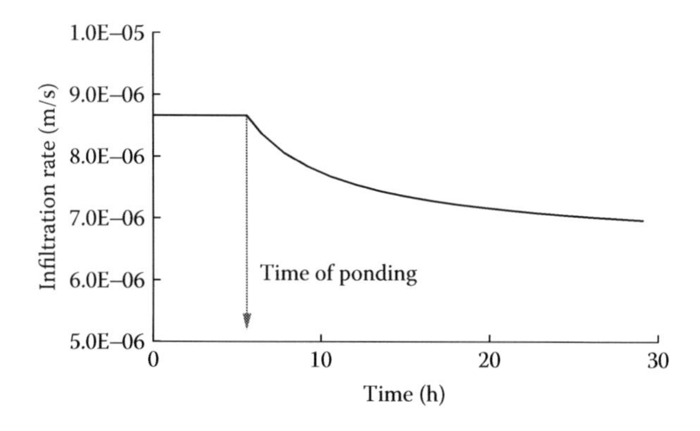

Figure 2.5 **Relationship between actual infiltration rate and time.**

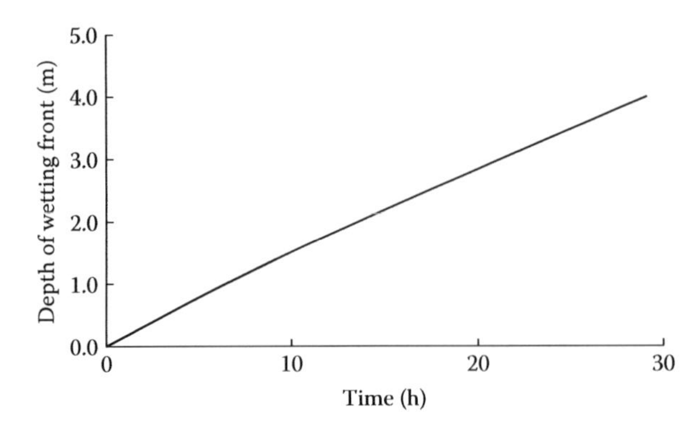

Figure 2.6 Depth of wetting front with time.

	B	C	D	E	F	G	H	I	J
2	Spreadsheet template for infiltration analysis under constant rainfall								
3		k_s (m/s)	7.50E–06		I (m)	t (h)	i (m/s)	z_w (m)	
4		s_f (m)	0.25		0.000	0.000	8.66E–06	0.000	
5		θ_1	0.45		0.173	5.556	8.66E–06	0.866	
6		θ_0	0.25		0.200	6.430	8.37E–06	1.000	
7		q (m/s)	1.00E–05		0.240	7.785	8.06E–06	1.200	
8		α_s (°)	30		0.280	9.184	7.83E–06	1.400	
9					0.320	10.619	7.67E–06	1.600	
10		I_p (m)	1.73E–01		0.360	12.081	7.54E–06	1.800	
11		t_p (h)	5.56E+00		0.400	13.566	7.43E–06	2.000	
12					0.440	15.069	7.35E–06	2.200	
13					0.480	16.589	7.28E–06	2.400	
14					0.520	18.123	7.22E–06	2.600	
15					0.560	19.668	7.16E–06	2.800	
16					0.600	21.224	7.12E–06	3.000	
17					0.640	22.789	7.08E–06	3.200	
18					0.680	24.362	7.05E–06	3.400	
19					0.720	25.942	7.02E–06	3.600	
20					0.760	27.529	6.99E–06	3.800	
21					0.800	29.122	6.96E–06	4.000	
22									

Figure 2.7 Spreadsheet template for infiltration analysis under constant rainfall.

2.2.4 Infiltration of time-varied rainfall on a sloping ground

A time-varied rainfall process can be modeled as a series of constant intensity rainfall events of short durations as shown in Figure 2.8. Suppose the rainfall intensity during the nth time interval is q. Let I_0 denote the cumulative infiltration when the nth time interval begins.

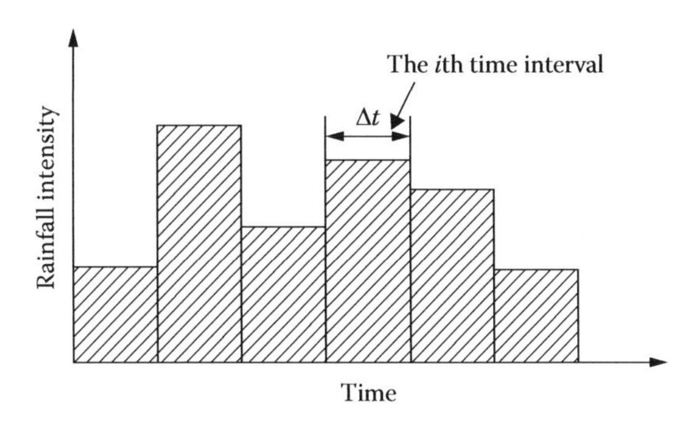

Figure 2.8 Illustration of a time-varied rainfall process.

According to Equation 2.4, the infiltration capacity of the soil can be determined by I_0 and reduces as the cumulative infiltration increases. Thus, the infiltration capacity of the soil at the end of the nth time interval is smaller than that at the beginning of the nth time interval. Comparing the rainfall intensity with the infiltration capacity at the beginning and the end of the nth time interval, there are three possibilities (Chow et al., 1988):

1. The rainfall intensity is always larger than the infiltration capacity.
2. The rainfall intensity is always smaller than the infiltration capacity.
3. The rainfall intensity is initially smaller than the infiltration capacity but becomes larger than the rainfall capacity at the end of the nth time interval.

To determine the evolution of cumulative infiltration with time in a time-varied rainfall process, one needs first to determine the increment of the cumulative infiltration in the nth time interval, ΔI. Consider first the case where the rainfall intensity is always larger than the infiltration capacity. In such a case, the infiltration capacity is determined by Equation 2.7. The increment in the cumulative infiltration is indeed an implicit function of time and must be solved numerically. Following the idea suggested by Li et al. (1976) for infiltration analysis on a level ground, a two-step procedure can be used to solve ΔI, that is, the increment of the cumulative infiltration during the time increment Δt, as follows (Zhang et al., 2014):

$$\Delta I_0 = \frac{1}{2}\left[k_{w2}\Delta t - 2I_0 + \sqrt{k_{w2}\Delta t\left(k_w\Delta t + 4I_0 + 8\zeta\right) + 4{I_0}^2}\right] \tag{2.13}$$

$$\Delta I = -I_0 - \frac{\left(I_0 + \Delta I_0\right)^2}{\zeta} + \frac{I_0 + \zeta + \Delta I_0}{\zeta}\sqrt{\left(I_0 + \Delta I_0\right)^2 + 2\zeta\left[\zeta \ln\left(1 + \frac{\Delta I_0}{I_0 + \zeta}\right) + k_{w2}\Delta t - \Delta I_0\right]} \tag{2.14}$$

where:
ΔI is the increment of the cumulative infiltration in the nth time interval
ΔI_0 is an intermediate variable to calculate ΔI
Δt is the duration of the nth time interval

For the rest two cases where the rainfall intensity is initially smaller than the infiltration capacity, one needs to first judge if at the end of the nth time interval the rainfall intensity is smaller than the rainfall capacity. Assuming that the rainfall intensity is always smaller than the infiltration capacity during the time interval, the increment of cumulative infiltration should be $\Delta I = \Delta t q \cos \alpha_s$, and the cumulative infiltration at the end of this time interval should be $I_1 = I_0 + \Delta I$. Substituting I_1 into Equation 2.7, the infiltration capacity of the soil at the end of the nth time interval is obtained, which is denoted as i_1 here. If $q \cos \alpha_s$ is indeed smaller than i_1, it indicates that the assumption that during the nth interval the rainfall intensity is always smaller than the infiltration capacity is valid. In such a case, ΔI is

$$\Delta I = q \cos \alpha_s \Delta t \tag{2.15}$$

If $q \cos \alpha_s$ is larger than i_1, it indicates that the rainfall becomes larger than the infiltration capacity at the end of the nth time interval. Equating the rainfall intensity $q \cos \alpha_s$ with the infiltration capacity as indicated by Equation 2.7, the corresponding cumulative infiltration on a sloping ground can be obtained, which is denoted as I_s here. During the period where the rainfall intensity is smaller than the infiltration capacity, the increment of the cumulative infiltration is

$$\Delta I_1 = I_s - I_0 \tag{2.16}$$

The durations of the periods where the rainfall intensity is smaller and larger than the infiltration capacity, which are respectively denoted as Δt_1 and Δt_2, can be calculated as follows:

$$\Delta t_1 = \frac{\Delta I_1}{q \cos \alpha_s} \tag{2.17}$$

$$\Delta t_2 = \Delta t - \Delta t_1 \tag{2.18}$$

Because the rainfall intensity is always larger than the infiltration capacity during the time interval Δt_2, the increment of cumulative infiltration in this time interval, which is denoted as ΔI_2 here, can be calculated using the two-step procedure as indicated by Equations 2.13 and 2.14 by substituting I_0 and Δt with I_s and Δt_2, respectively. Thus, the total increment of I is

$$\Delta I = \Delta I_1 + \Delta I_2 \tag{2.19}$$

Applying the above analysis to each time interval, the cumulative infiltration after each time interval can then be calculated, and the depth of the wetting front can then be calculated using Equation 2.8.

To facilitate the practical application, VBA codes that can be executed in Microsoft Excel are developed to calculate the incremental cumulative infiltration ΔI during a time interval Δt, which can judge the three conditions automatically, as shown in Figure 2.9.

Example 2.2

Solve Example 2.1 using the method for time-varied rainfall infiltration analysis and check its accuracy with the analytical solution.

The spreadsheet form for time-varied rainfall infiltration analysis is shown in Figure 2.10. Figure 2.11 compares the predicted cumulative infiltration from the time-varied rainfall method and the constant rainfall method. The predictions from the two methods are practically the same.

```
Function getDIc(Ks, sf, Qs, Qi, aerfa, Ic0, q, Dt)
'get the change in the cumulative infiltration when the initial cumulative infiltration is Ic0 and the rainfall
duration is Dt
Ii = getIi(Ks, sf, Qs, Qi, aerfa, Ic0)
aerfa_rad = aerfa * WorksheetFunction.Pi() / 180
q2 = q * Cos(aerfa_rad)
If q2 > Ii Then
getDIc = getDIc_GA(Ks, sf, Qs, Qi, aerfa, Dt, Ic0)
Else
Ic1 = Ic0 + q2 * Dt
newIi = getIi(Ks, sf, Qs, Qi, aerfa, Ic1)
   If newIi > q2 Then
   getDIc = q2 * Dt
   Else
   Ictemp = getIc(Ks, sf, Qs, Qi, aerfa, q)
   ts = (Ictemp - Ic0) / q2
   DIc1 = ts * q2
   Dt2 = Dt – ts
   DIc2 = getDIc_GA(Ks, sf, Qs, Qi, aerfa, Dt2, Ictemp)
   getDIc = DIc1 + DIc2
   End If
End If
End Function

Function getDIc_GA(Ks, sf, Qs, Qi, aerfa, Dt, Ic0)
'get the change in the cumulative infiltration depth when the rainfall intensity is larger than infiltration
capacity and when the initial cumulative infiltration depth is Ic0 and the rainfall duration is Dt
aerfa_rad = aerfa * WorksheetFunction.Pi() / 180
Ks2 = Ks * Cos(aerfa_rad)
kexi = sf * (Qs – Qi) / Cos(aerfa_rad)
temp0 = Ks2 * Dt – 2 * Ic0 + Sqr(Ks2 * Dt * (Ks2 * Dt + 4 * Ic0 + 8 * kexi) + 4 * Ic0 ^ 2)
DIc0 = 0.5 * temp0
temp1 = (Ic0 + DIc0) ^ 2 + 2 * kexi * (kexi * WorksheetFunction.Ln(1 + DIc0 / (Ic0 + kexi)) + Ks2
Dt–DIc0)
getDIc_GA = –Ic0 – (Ic0 + DIc0) ^ 2 / kexi + (Ic0 + kexi + DIc0) * Sqr(temp1) / kexi
End Function

Function getIc(Ks, sf, Qs, Qi, aerfa, q)
'get the corresponding cumulative infiltration depth if the rainfall capacity is q
aerfa_rad = aerfa * WorksheetFunction.Pi() / 180
temp1 = sf * (Qs – Qi)
temp2 = Ks / (q – Ks)
getIc = temp1 * temp2 / Cos(aerfa_rad)
End Function

Function getIi(Ks, sf, Qs, Qi, aerfa, Ic0)
'to get the GA infiltration capacity when the cumulative rainfall depth is Ic0
aerfa_rad = aerfa * WorksheetFunction.Pi() / 180
temp = sf * (Qs – Qi) / Ic0 + Cos(aerfa_rad)
getIi = Ks * temp
End Function
```

Figure 2.9 VBA functions for calculating ΔI during a time interval Δt.

Spreadsheet template for infiltration analysis under time-varied rainfall

k_s (m/s)	s_f (m)	θ_1	θ_0	α_s (°)
7.50E–06	0.25	0.45	0.25	30

I_0 (m)	q (m/s)	Δt (s)	ΔI (m)	t (h)	I (m)
1E–07	1.00E–05	3600	0.0312	1.0	0.0312
0.0312	1.00E–05	3600	0.0312	2.0	0.0624
0.0624	1.00E–05	3600	0.0312	3.0	0.0935
0.0935	1.00E–05	3600	0.0312	4.0	0.1247
0.1247	1.00E–05	3600	0.0312	5.0	0.1559
0.1559	1.00E–05	3600	0.0310	6.0	0.1869
0.1869	1.00E–05	3600	0.0301	7.0	0.2170
0.2170	1.00E–05	7200	0.0578	9.0	0.2748
0.2748	1.00E–05	7200	0.0557	11.0	0.3305
0.3305	1.00E–05	7200	0.0543	13.0	0.3848
0.3848	1.00E–05	7200	0.0533	15.0	0.4382
0.4382	1.00E–05	7200	0.0526	17.0	0.4907
0.4907	1.00E–05	7200	0.0520	19.0	0.5427
0.5427	1.00E–05	7200	0.0515	21.0	0.5943
0.5943	1.00E–05	7200	0.0511	23.0	0.6454
0.6454	1.00E–05	7200	0.0508	25.0	0.6962
0.6962	1.00E–05	7200	0.0505	27.0	0.7467

Figure 2.10 Spreadsheet template for infiltration analysis under time-varied rainfall.

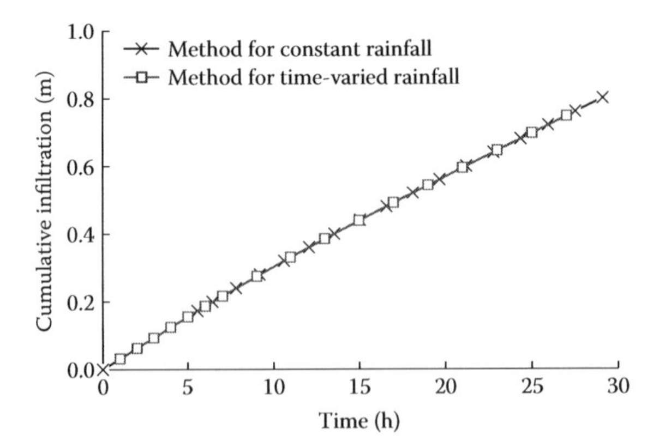

Figure 2.11 Comparison of methods for infiltration analysis.

2.3 SEEPAGE ANALYSIS IN UNSATURATED SOIL SLOPES BASED ON PHYSICAL GOVERNING EQUATION

2.3.1 Governing equation of water flow in unsaturated soil

In unsaturated soils, both water phase and air phase are classified as fluids that can flow. In this book, only the flow of water phase is discussed. The pore-air pressure remains at the atmospheric condition. The total suction or soil suction ψ of an unsaturated soil has two components, that is, the matric suction $(u_a - u_w)$, where u_a is the pore-air pressure and u_w is the pore-water pressure and the osmotic suction π which is related to the salt concentration in the free pore water of a soil. As the change in the total suction is essentially equivalent to a change in the matric suction and the effect of osmotic suction on the mechanical behavior is minor compared with the effect of matric suction for most of the rainfall-induced landslides (Fredlund et al., 2012), the total suction ψ and the matric suction $(u_a - u_w)$ are interchangeably used in this book. It should also be noted that matric suction, total suction or their potentials do not cause water flow. The driving potential for the flow of water is the hydraulic head (or total head), which is the sum of the gravitational head and the pressure head

$$H = z + h = z + \frac{u_w}{\rho_w g} \tag{2.20}$$

where:

H is the hydraulic head or the total head
z is the elevation head or the coordinate of elevation
$h = (u_w/\rho_w g)$ is the pore pressure head
ρ_w is the density of water
u_w is the pore-water pressure
g is the gravitational acceleration

The flow of water in a saturated/unsaturated soil system can be described by the Darcy's law (Childs and Collis-George, 1950), that is, the rate of water flow through a soil mass was is proportional to the hydraulic head gradient. Darcy's law written in a vector form is as follows:

$$\mathrm{v} = -\mathbf{k}\nabla(H) \tag{2.21}$$

where:

v is the vector of flow rate
k is the tensor of coefficient of permeability
$\nabla(H)$ is the hydraulic gradient

Based on mass balance and the Darcy's law, the governing equation for transient flow through an unsaturated soil can be formulated as follows (Richards, 1931):

$$\nabla\left(-\mathbf{k}\nabla H\right) = -\frac{\partial \theta_w}{\partial t} \tag{2.22}$$

where θ_w is the volumetric water content. This equation is also called the Richards equation in groundwater hydrology field.

Under the steady-state condition, the governing equation is simplified as

$$\nabla\left(-\mathbf{k}\nabla H\right) = 0 \tag{2.23}$$

2.3.2 Hydraulic properties of unsaturated soils

The saturated coefficient of permeability k_s, is a function of the void ratio or soil density (Lambe and Whitman, 1969). For an unsaturated soil, both the volumetric water content θ_w and the coefficient of permeability, k, are significantly affected by combined changes in the void ratio, e, and matric suction (Fredlund et al., 2012). In this chapter, the hydraulic property functions of unsaturated soil are discussed for rigid unsaturated soils, that is, the deformation of unsaturated soil is negligible. The coupling hydro-mechanical behavior of unsaturated soil will be discussed later in Chapter 4.

2.3.2.1 Soil–water characteristic curve

The relationship between volumetric water content θ_w and soil suction ψ for an unsaturated soil is defined as the soil–water characteristic curve (SWCC). This curve is referred to as water retention curve in soil sciences and groundwater hydrology. It can also be expressed as the relationships of the gravimetric water content w_w—soil suction ψ, or the degree of saturation S—soil suction ψ. The SWCC of a soil is usually obtained using a pressure plate device in the laboratory. Typical SWCCs for a sandy soil, a silty soil, and a clayey soil are shown in Figure 2.12. Several empirical equations have been proposed to describe SWCC (Table 2.1). A detailed review about equations of the SWCC was presented by Leong and Rahardjo (1997b) and Fredlund et al. (2012).

The water content of a soil decreases during a drying process and increases when the soil suction decreases following a wetting path. The drying and wetting curves in general are not identical. This phenomenon is called hysteresis. Figure 2.13 illustrates a drying curve and a wetting curve for a silty soil. Hysteresis in the SWCC may be caused by nonuniformity of the pore cross sections ("ink-bottle" effect), different contact angles during drying and wetting, entrapped air, swelling or shrinking, and the aging effect (Klausner, 1991; Fredlund et al., 2012).

2.3.2.2 Coefficient of permeability function

The shape of the coefficient of permeability function bears a relationship to the shape of the SWCC as shown in Figure 2.14. The coefficient of permeability remains relatively constant until the air-entry value of the soil is exceeded. At suctions greater than the air-entry value,

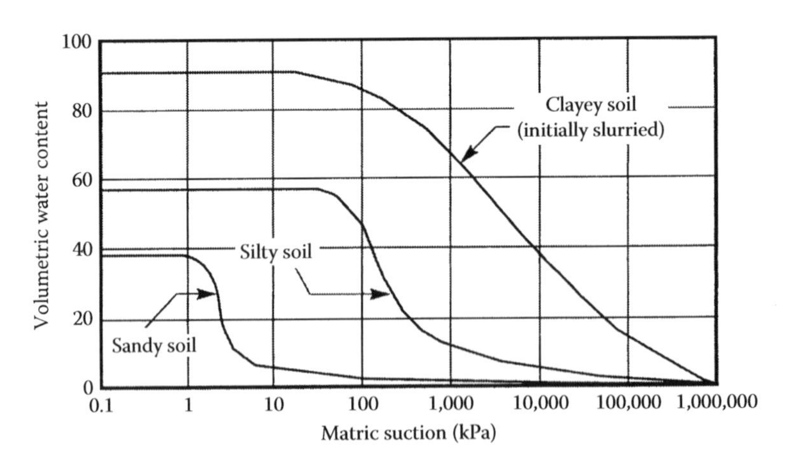

Figure 2.12 SWCCs for a sandy soil, silty soil, and clayey soil. (From Fredlund, D. G. et al., *Canadian Geotechnical Journal*, 31, 533–546, Copyright 1994 Canadian Science Publishing or its licensors. Reproduced with permission.)

Table 2.1 Empirical equations for soil–water characteristic curve

Equation	*Coefficients*	*References*
$\theta_w = \theta_r + (\theta_s - \theta_r)\left(\frac{\psi_{aev}}{\psi}\right)^{\lambda_{bc}}$	ψ_{aev}, λ_{bc}	Brooks and Corey (1964)
$\theta_w = \theta_r + \frac{\theta_s - \theta_r}{1 + a_g \psi^{b_g}}$	a_g, b_g	Gardner (1958)
$\theta_w = \theta_r + \frac{\theta_s - \theta_r}{(1 + (a_{vg} h_s)^{n_{vg}})^{m_{vg}}}$	a_{vg}, n_{vg}, m_{vg}	van Genuchten (1980)
$\theta_w = \theta_r + \frac{\theta_s - \theta_r}{(1 + (a_{vgm} h_s)^{n_{vgm}})^{m_{vgm}}}$ where $m_{vgm} = 1 - \frac{1}{n_{vgm}}$	$a_{vgm}, n_{vgm}, m_{vgm}$	van Genuchten (1980); Mualem (1976)
$\theta_w = \theta_r + (\theta_s - \theta_r)\exp\left(\frac{a_{mb1} - \psi}{b_{mb1}}\right)$	$\theta_r, a_{mb1}, b_{mb1}$	McKee and Bumb (1984)
$\theta_w = \theta_r + \frac{\theta_s - \theta_r}{1 + \exp\left(\psi - a_{mb2}/b_{mb2}\right)}$	$\theta_r, a_{mb2}, b_{mb2}$	McKee and Bumb (1987)
$\theta_w = \frac{\theta_s}{\left\{\ln\left[\exp(1) + (\psi/a_f)^{n_f}\right]\right\}^{m_f}}$	a_f, n_f, m_f	Fredlund and Xing (1994)
$\theta_w = \left[1 - \frac{\ln\left(1 + (\psi/\psi_r)\right)}{\ln\left(1 + (10^6/\psi_r)\right)}\right] \frac{\theta_s}{\left\{\ln\left[\exp(1) + (\psi/a_f)^{n_f}\right]\right\}^{m_f}}$	ψ_r, a_f, n_f, m_f	Fredlund and Xing (1994) with correction factor

Notes: θ_w is the volumetric water content; θ_s is the saturated volumetric water content; θ_r is the residual volumetric water content; ψ is the soil suction; h_s is the suction head (m); ψ_{aev} is the air-entry value (AEV) or air-entry suction; ψ_r is the residual suction.

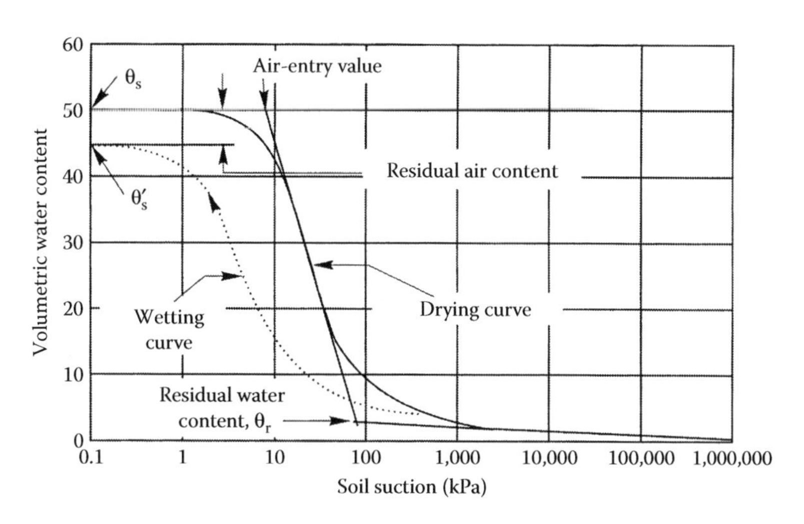

Figure 2.13 Typical drying and wetting SWCC curves for a silty soil. (From Fredlund, D. G. et al., *Canadian Geotechnical Journal*, 31, 533–546, Copyright 1994 Canadian Science Publishing or its licensors. Reproduced with permission.)

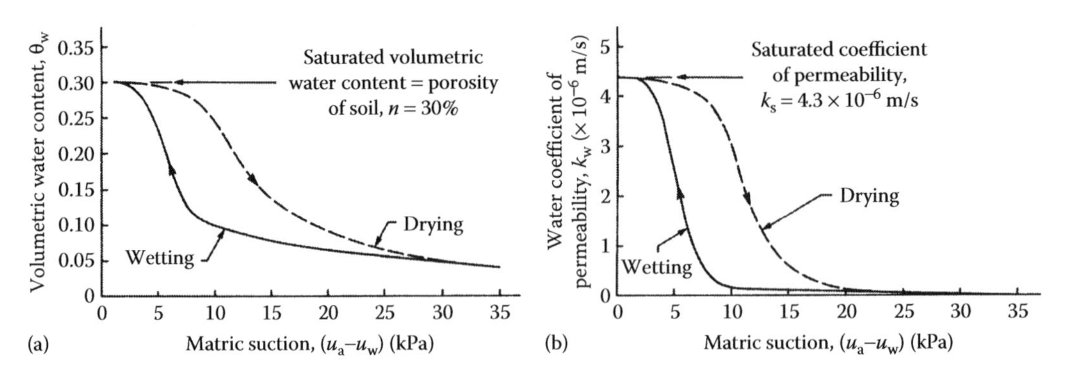

Figure 2.14 Hysteresis in (a) SWCCs and (b) permeability functions for a naturally deposited sand. (From Fredlund, D. G., and Rahardjo, H.: *Soil Mechanics for Unsaturated Soils*. 1993. Copyright Wiley-VCH Verlag GmbH & Co. KGaA. Reproduced with permission.)

Table 2.2 Equations for coefficient of permeability of unsaturated soils

	Equation	*Unknown coefficients*	*References*
Empirical	$k = k_s \quad (\psi \le \psi_{aev})$ $k = k_s\left(\psi_{aev}/\psi\right)^{2+3\lambda_{bc}} \quad (\psi > \psi_{aev})$	ψ_{aev}, λ_{bc}	Brooks and Corey (1964)
	$k = \dfrac{k_s}{1 + a_g\psi^{b_g}}$	a_g, b_g	Gardner (1958)
	$k = k_s\left[\Theta_d(\psi)\right]^{\delta_{lr}}$	δ_{lr}	Leong and Rahardjo (1997a)
Statistical	$k = k_s\Theta_n^{\delta_b}\dfrac{\int_{\theta_r}^{\theta_w}\left(d\theta_w/\psi^2(\theta_w)\right)}{\int_{\theta_r}^{\theta_s}\left(d\theta_w/\psi^2(\theta_w)\right)}$	δ_b	Burdine (1953)
	$k = k_s\Theta_n^{\delta_m}\left(\dfrac{\int_{\theta_r}^{\theta_w}\left(d\theta_w/\psi(\theta_w)\right)}{\int_{\theta_r}^{\theta_s}\left(d\theta_w/\psi(\theta_w)\right)}\right)^2$	δ_m	Mualem (1976)
	$k = k_s\dfrac{\int_{\theta_r}^{\theta_w}\left(\theta_w - x/\psi^2(x)\right)dx}{\int_{\theta_r}^{\theta_s}\left(\theta_w - x/\psi^2(x)\right)dx}$		Fredlund et al (1994)
	$k = k_s\dfrac{1-(a_{vb}h_s)^{(n_{vb}-2)}\left[1+(a_{vb}h_s)^{n_{vb}}\right]^{-m_{vb}}}{\left[1+(a_{vb}h_s)^{n_{vb}}\right]^{2n_{vb}}}$	a_{vb}, n_{vb}, m_{vb}	van Genuchten (1980); Burdine (1953)
	$k = k_s\dfrac{\left\{1-(a_{vgm}h_s)^{(n_{vgm}-1)}\left[1+(a_{vgm}h_s)^{n_{vgm}}\right]^{-m_{vgm}}\right\}^2}{\left[1+(a_{vgm}h_s)^{n_{vgm}}\right]^{0.5}}$	$a_{vgm}, n_{vgm}, m_{vgm}$	van Genuchten (1980); Mualem (1976)

Notes: k_s is the saturated coefficient of permeability; ψ is the soil suction; ψ_{aev} is the air-entry suction; h_s is the suction head (m); Θ_d is the dimensionless water content is the θ_w/θ_s; Θ_n is the $(\theta_w - \theta_r)/(\theta_s - \theta_r)$ is the normalized volumetric water content; x is the dummy variable of integration; δ_b is the a constant set to 2; δ_m is the a constant set to 0.5.

the coefficient of permeability decreases rapidly. Hysteresis has also been found in coefficient of permeability functions of unsaturated soils. Figure 2.14 demonstrates a similar hysteresis form in both SWCCs and permeability functions for a natural deposited sand.

The coefficient of permeability function can be directly measured in laboratory. However, the measurement is a very time-consuming process because the coefficient of permeability of unsaturated soils is extremely low, especially at high matric suction values. As an alternative, researchers have proposed empirical and statistical methods to estimate the coefficient of permeability of an unsaturated soil (Fredlund et al., 1994, 2012). Empirical models utilize the relationship between the character of the SWCC and the permeability function in an empirical manner. Statistical models are derived based on a physical model of the assemblage of pore channels and utilize an integration procedure along the SWCC. Table 2.2 lists some commonly used equations of coefficient of permeability function for unsaturated soils.

2.4 ANALYTICAL SOLUTIONS OF THE RICHARDS EQUATION

The solution of partial differential equation (2.22) is complicated because the SWCC ($\theta_w - \psi$) and the unsaturated permeability function ($k - \psi$) are strongly nonlinear. Analytical solutions, which represent a classical mathematical approach to solve differential equations by applying various transformations (e.g., Laplace or Fourier transformations) or green functions, usually result in an explicit equation for the pressure head or water content at particular time and location. Using analytical solutions one can more easily evaluate interrelationships among parameters, and get better insight into how various processes control the basic flow processes (Šimůnek, 2006).

Several analytical and quasi-analytical solutions to unsaturated flow problems have been developed. Srivastava and Yeh (1991) assumed that the SWCC and the unsaturated permeability function are both exponential and presented analytical solutions to the one-dimensional (1D) infiltration problem in a homogeneous soil layer and a two-layered soil system. With the same exponential equations for θ_w and k, Yuan and Lu (2005) further developed analytical solutions to transient flow in rooted, homogeneous soils with time-varied surface fluxes. Basha (1999, 2000) used Green's function to derive multidimensional transient solutions for domains with prescribed surface flux boundary conditions and bottom boundary conditions. Chen et al. (2001) employed a Fourier integral transformation to obtain a series solution that has the merit of easy calculation. Serrano (2004) developed an approximate analytical solution for the nonlinear Richards equation. Huang and Wu (2012) presented analytical solutions to 1D horizontal and vertical infiltration into the soils considering time-varied surface fluxes.

Many analytical solutions lead to relatively complicated formulations that include infinite series and/or integrals that need to be evaluated numerically, which often exceeds some ambiguity in the often-claimed advantage of exactness of analytical methods over numerical techniques (Šimůnek, 2006). Another limitation of analytical solution is that, it can usually only be derived for simplified systems involving linearization of governing equations, homogeneous soils, simplified geometries, and constant or highly simplified initial and boundary conditions (Šimůnek, 2006). For two-dimensional (2D) problems such as slope stability analyses involving unsaturated infiltration, analytical solutions are only available with various simplifications and assumptions (Griffiths and Lu, 2005). In the following sections, we present the analytical solutions for a homogenous soil layer with vertical infiltration by Srivastava and Yeh (1991) for constant rainfall and Yuan and Lu (2005) for time-varied rainfall.

2.4.1 Analytical solution of one-dimensional infiltration under constant rainfall

Based on Equation 2.22, the governing equation for 1D infiltration in unsaturated soils is given by

$$\frac{\partial}{\partial z}\left[k(h)\frac{\partial(h+z)}{\partial z}\right]=\frac{\partial\theta_w}{\partial t} \tag{2.24}$$

where z is the coordinate of elevation (see Figure 2.15, $z = 0$ at lower boundary).

Srivastava and Yeh (1991) assumed that the SWCC and the permeability function of the unsaturated soil are both exponential:

$$\theta_w=\begin{cases}\theta_r+(\theta_s-\theta_r)e^{\alpha_{sy}h} & h<0\\ \theta_s & h\geq 0\end{cases} \tag{2.25}$$

$$k=\begin{cases}k_s e^{\alpha_{sy}h} & h<0\\ k_s & h\geq 0\end{cases} \tag{2.26}$$

where:

θ_s is saturated volumetric water content
θ_r is residual volumetric water content
α_{sy} is the coefficient represents the desaturation rate of the SWCC

The lower boundary is located at the groundwater table, where the pore-water pressure head is equal to 0 (i.e., $h_l = 0$). The ground surface is subjected to a rainfall intensity q_1 (positive for infiltration), which is constant throughout the duration of the rainfall. The initial pore-water pressure distribution is obtained by performing a steady-state analysis, in which the upper boundary is subjected to antecedent infiltration rate q_0.

Through Laplace's transformation, the solution for a homogeneous unsaturated soil layer under constant infiltration q_1 is written as (Srivastava and Yeh, 1991)

$$k^*=q_1^*-\left[q_1^*-e^{\alpha_{sy}h1}\right]\cdot e^{-z^*} -4\left(q_1^*-q_0^*\right)\cdot e^{(L^*-z^*)/2}\sum_{n=1}^{\infty}\frac{\sin(\lambda_n z^*)\sin(\lambda_n L^*)\cdot e^{-\lambda_n^2 t^*}\cdot e^{-t^*}}{1+(L^*/2)+2\lambda_n^2 L^*} \tag{2.27}$$

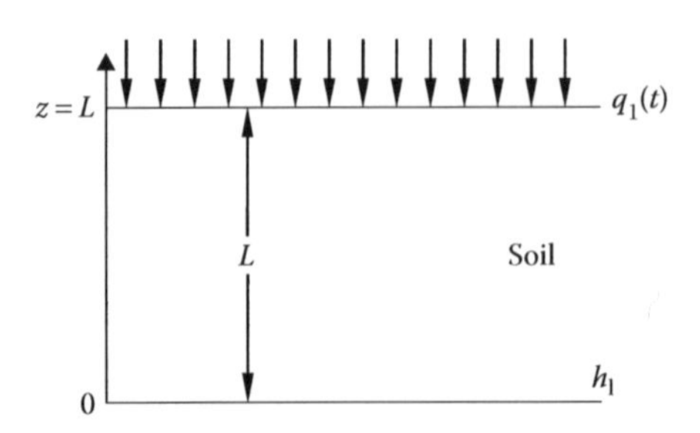

Figure 2.15 Schematic plot of the soil profile for the analytical solution of one-dimensional infiltration.

where $z^* = \alpha_{sy}z$, $L^* = \alpha_{sy}L$ with L being the total depth of the soil, $k^* = k/k_s$, $q_0^* = q_0/k_s$, $q_1^* = q_1/k_s$, $t^* = \alpha_{sy}k_s t/\theta_s - \theta_r$, and λ_n is the nth positive root of the following characteristic equation:

$$\tan(\lambda L^*) + 2\lambda = 0 \tag{2.28}$$

Using Equations 2.25 and 2.26, the pore-water pressure head and the volumetric water content can be obtained as follows:

$$h = \frac{\ln k^*}{\alpha_{sy}} \tag{2.29}$$

$$\theta_w = \theta_r + (\theta_s - \theta_r)e^{\alpha_{sy}h} \tag{2.30}$$

2.4.2 Analytical solution of one-dimensional infiltration under time-varied rainfall

Yuan and Lu (2005) assumed the same exponential functions for both SWCC and permeability functions (Equations 2.25 and 2.26) and further developed analytical solutions for transient flow in rooted, homogeneous soils with time-dependent varying surface fluxes. In their analytical solutions, the sign of infiltration rate is positive for evaporation and negative for precipitation. Define Φ as the matric flux potential:

$$\Phi(z,t) = \frac{k(h)}{\alpha_{sy}} \tag{2.31}$$

The steady-state matric flux potential Φ_s is written as

$$\Phi_s(z) = \frac{k_s \exp\left[\alpha_{sy}(h_l - z)\right]}{\alpha_{sy}} + \frac{q_0}{\alpha_{sy}}\left[\exp(-\alpha_{sy}z) - 1\right] \tag{2.32}$$

where:

q_0 is the surface flux at the time $t = 0$ (negative for infiltration)
h_l is the prescribed pressure head at the lower boundary

Considering time-dependent varying surface flux, the matric flux potential Φ for transient flow is

$$\Phi(z,t) = \Phi_s(z) + 8D\exp\left[\frac{\alpha_{sy}(L-z)}{2}\right]\sum_{n=1}^{\infty}\left[\frac{\left(\lambda_n^2 + \frac{\alpha_{sy}^2}{4}\right)\sin(\lambda_n L)\sin(\lambda_n z)}{2\alpha_{sy} + \alpha_{sy}^2 L + 4L\lambda_n^2} g(t)\right] \tag{2.33}$$

where:

$D = k_s / \left[\alpha_{sy}(\theta_s - \theta_r)\right]$
λ_n is the nth positive root of equation $\sin(\lambda L) + (2\lambda/\alpha_{sy})\cos(\lambda L) = 0$
$g(t)$ can be obtained using the following equation:

$$g(t)=\int_0^t [q_0-q_1(\tau)]\exp\left[-D\left(\lambda_n^2+\frac{\alpha_{sy}^2}{4}\right)(t-\tau)\right]d\tau \tag{2.34}$$

where $q_1(t)$ is the time-dependent flux at the upper boundary (negative for infiltration).

With the calculated matric flux potential in Equation 2.33, the pressure head can be computed based on Equations 2.26 and 2.31 as follows:

$$h=\frac{\ln(\alpha_{sy}\Phi/k_s)}{\alpha_{sy}} \tag{2.35}$$

2.4.3 Effect of soil properties, boundary conditions, and initial conditions

In this section, we investigate the effects of soil parameters, that is, the saturated permeability k_s, the desaturation coefficient of SWCC α_{sy}, the effective water content $(\theta_s-\theta_r)$, and boundary conditions, including, antecedent infiltration q_0 and main infiltration q_1, and the thickness of soil layer L on pore pressure profiles in an unsaturated soil layer during surface infiltration. The analytical solution by Yuan and Lu (2005) in Section 2.4.2 is adopted. The input parameters of the parametric study are shown in Table 2.3.

2.4.3.1 *Saturated permeability* k_s

Figure 2.16 shows the coefficient of permeability functions for the three soils with different values of k_s (i.e., 0.1 cm/h, 1 cm/h, and 10 cm/h) in Group 1. The SWCCs for the three

Table 2.3 Study schemes and parameters of the parametric study

Group no.	*Case no.*	k_s *(cm/h)*	α_{sy} *(cm⁻¹)*	$\theta_s-\theta_r$	q_0 *(cm/h)*	q_1 *(cm/h)*	*L (cm)*	h_i *(cm)*
1	1a	0.1						
	1b	1	0.01	0.3	0	−1.0	100	0
	1c	10						
2	2a		0.001					
	2b	1	0.01	0.3	0	−1.0	100	0
	2c		0.04					
3	3a			0.15				
	3b	1	0.01	0.30	0	−1.0	100	0
	3c			0.45				
4	4a				−0.1			
	4b	1	0.01	0.30	0	−1.0	100	0
	4c				0.1			
5	5a					−0.5		
	5b	1	0.01	0.30	0	−1.0	100	0
	5c					−2.0		
6	6a						50	
	6b	1	0.01	0.30	0	−1.0	100	0
	6c						200	

Note: q_0 and q_1 are positive for evaporation and negative for precipitation.

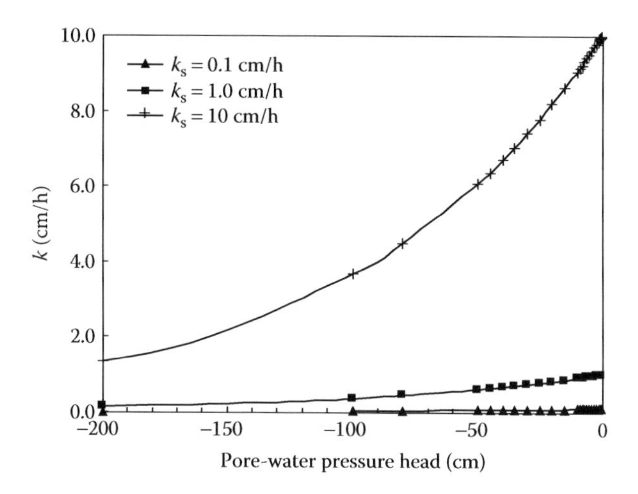

Figure 2.16 Effect of k_s on the coefficient of permeability functions (Group I).

soils are the same and not presented here. The intensity of infiltration is 1 cm/h. Figure 2.17 shows pore-water pressure profiles of the three soils. When k_s equals to 0.1 cm/h, rain water infiltrates only about 5 cm after 24 hours of rain. The remaining matric suction head at ground surface is about 30 cm. When k_s is 1 cm/h, the wetting front is vertical instead of horizontal. After 8 hours of rainfall, the matric suction near the ground surface is reduced to about 20 cm. After 24 hours, the unsaturated zone becomes almost fully saturated. The pore-water pressure head at the ground surface is positive, which means rainwater accumulated on the ground surface. When k_s is 10 cm/h, after 2 hours of rainfall the soils in shallow depth is completely saturated with a wetting front of about 20 cm deep. After 8 hours, the wetting front advances to 60 cm deep. After 24 hours, the soil layer is completely saturated and a significant surface ponding is accumulated.

These results of pore-water pressure profiles show that the greater the k_s value, the faster the wetting front advances. As shown in Figure 2.16, with the same initial pore-water pressure distribution, when k_s value is larger, the unsaturated permeability is greater. Hence, the infiltration and dissipation of matric suction is faster.

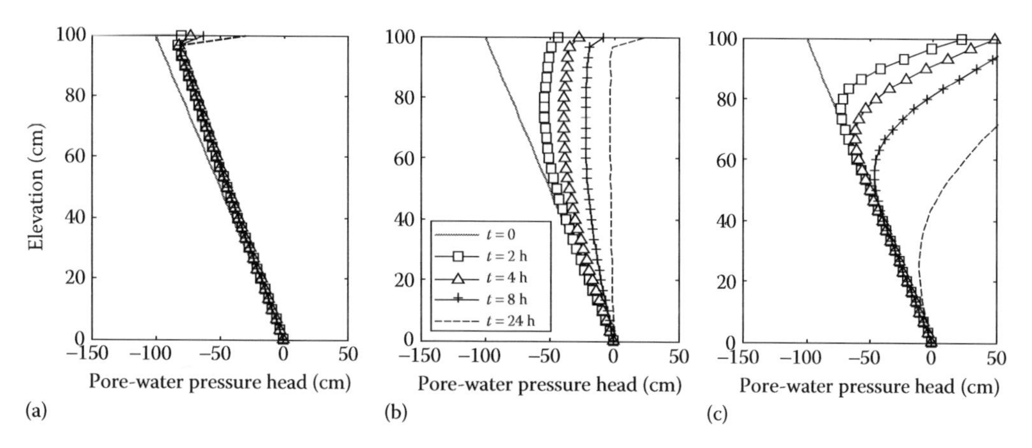

Figure 2.17 Variation of pore-water pressure profiles with different k_s (Group I): (a) k_s = 0.1 cm/h, (b) k_s = 1.0 cm/h, and (c) k_s = 10 cm/h.

2.4.3.2 Desaturation coefficient α_{sy}

Figure 2.18 illustrates the SWCC and coefficient of permeability functions for the soils with different values of α_{sy} in Group 2. With a larger value of α_{sy}, the rate of decrease of water content or permeability is greater. Figure 2.19 shows pore-water pressure profiles for the three soils with different values of α_{sy}. When α_{sy} is 0. 001 cm/h, matric suction in unsaturated zone dissipates very rapidly. After 2 hours, matric suction reduces to less than 20 cm. The soil layer is completely saturated after 4 hours with a ponding rainwater accumulating on the ground surface. When α_{sy} is 0. 01 cm/h, the wetting front is vertical. The matric suction dissipation rate is relatively slow. After 24 hours, unsaturated zone of the soil layer is completely saturated and there is water accumulation on the ground surface. When α_{sy} is 0.04 cm/h, the rainwater infiltrates to a depth of 40 cm after 2 hours. However, the soil layer is not fully saturated. After 8 hours, the average suction head is about 20 cm. The whole soil layer is fully saturated after 24 hours. The graphs show that when the α_{sy} value is greater, the velocity of infiltration and the rate of suction dissipation become slower.

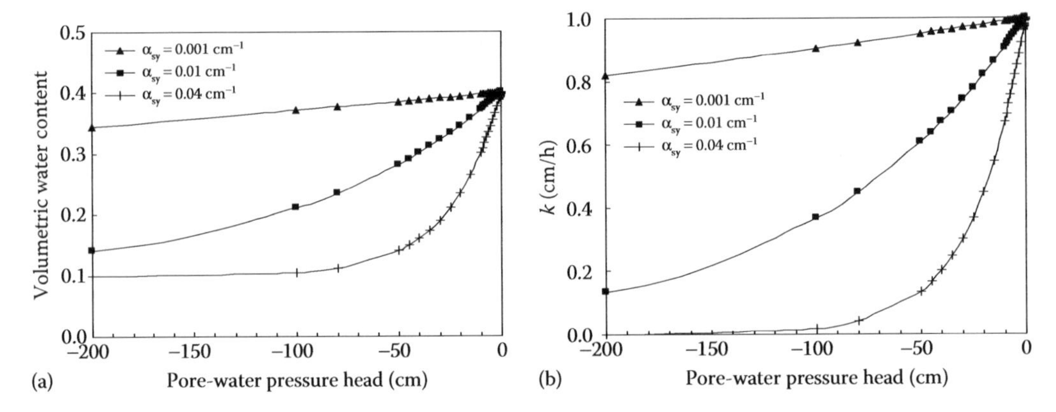

Figure 2.18 Effect of α_{sy} on (a) the SWCCs and (b) coefficient of permeability functions (Group 2).

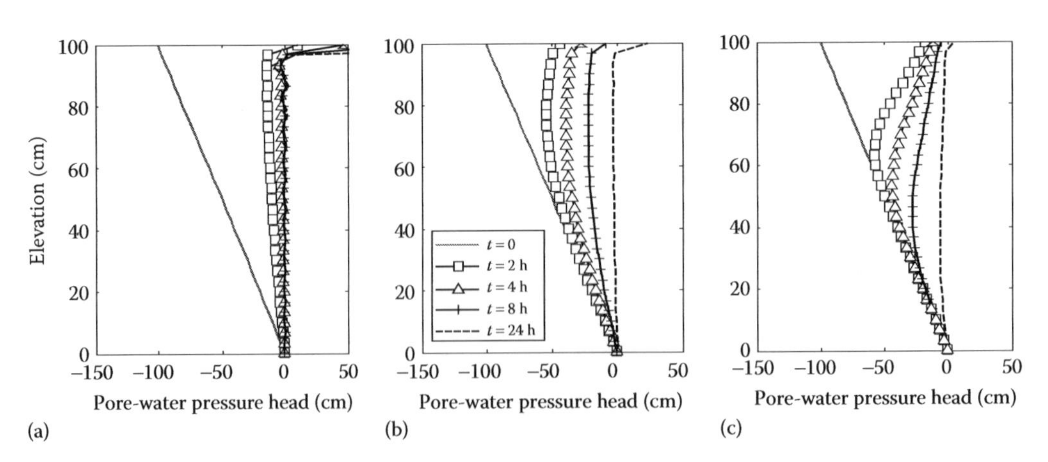

Figure 2.19 Variation of pore-water pressure profiles with different α_{sy} (Group 2): (a) α_{sy} = 0.001 cm^{-1}, (b) α_{sy} = 0.01 cm^{-1}, and (c) α_{sy} = 0.001 cm^{-1}.

2.4.3.3 *Effective water content* $\theta_s - \theta_r$

The SWCC curves of soils with different values of $\theta_s - \theta_r$ in Group 3 are shown in Figure 2.20. The coefficient of permeability functions are identical and not shown here. Figure 2.21 shows a series of pore-water pressure profiles with respect to three different values of $\theta_s - \theta_r$. When $\theta_s - \theta_r$ is 0.15, the matric suction in the unsaturated zone is reduced to about 40 cm after 2 hours of rainfall. After 8 hours, the matric suction decreases to about 5 cm in the unsaturated zone. When $\theta_s - \theta_r$ is 0.45, the matric suction in the unsaturated zone decreases to 60 cm and 30 cm after 2 hours and 8 hours of rainfall, respectively. For all the three soils, the unsaturated zone is almost fully saturated after 24 hours. Based on these results, a soil with a larger effective water content can store more infiltrated water and hence result in slower dissipation of matric suction in unsaturated zone. Compared to the results in the previous sections, the effect of $(\theta_s - \theta_r)$ is not as significant as the effects of α_{sy} and k_s on the dissipation of matric suction.

2.4.3.4 *Antecedent surface flux* q_0

As is shown in Figure 2.22, q_0 has a significant effect on the initial pore-water pressure distribution. If q_0 is –0.1 cm/h (positive means infiltration), the initial matric suction at ground surface is reduced to 85 cm. If q_0 is 0.1 cm/h (a positive value means evaporation), the initial matric suction at ground surface is increased to 120 cm. q_0 can affect the dissipation of matric suction especially at the beginning of the rainfall period. The differences of pore-water pressure profiles with different q_0 values are dismissed as time increases, which indicates that the effect of antecedent infiltration rate q_0 gradually reduces.

2.4.3.5 *Rainfall intensity* q_1

As shown in Figure 2.23, the rainfall intensity q_1 has a significant impact on the variation of pore-water pressure. When q_1 is –0.5 cm/h, the matric suction at ground surface is reduced to about 70 cm after 2 hours. After 24 hours of infiltration, the unsaturated zone is not completely saturated and the remained maximum matric suction head is 40 cm. When q_1 is –1 cm/h, the matric suction at ground surface is about 45 cm after 2 hours of rainfall.

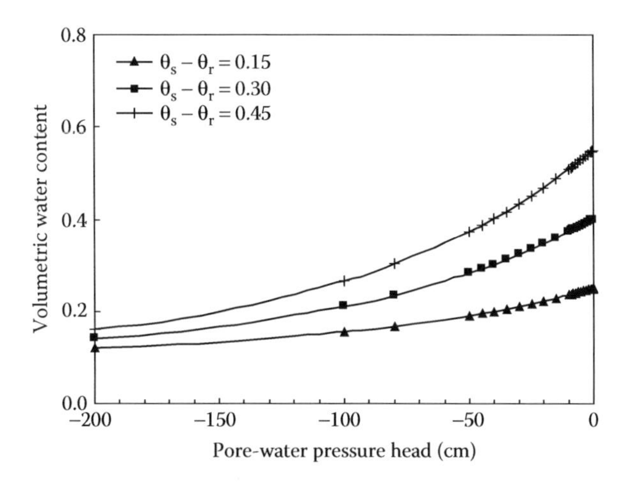

Figure 2.20 Effect of $\theta_s - \theta_r$ on the SWCCs ($\theta_r = 0.1$).

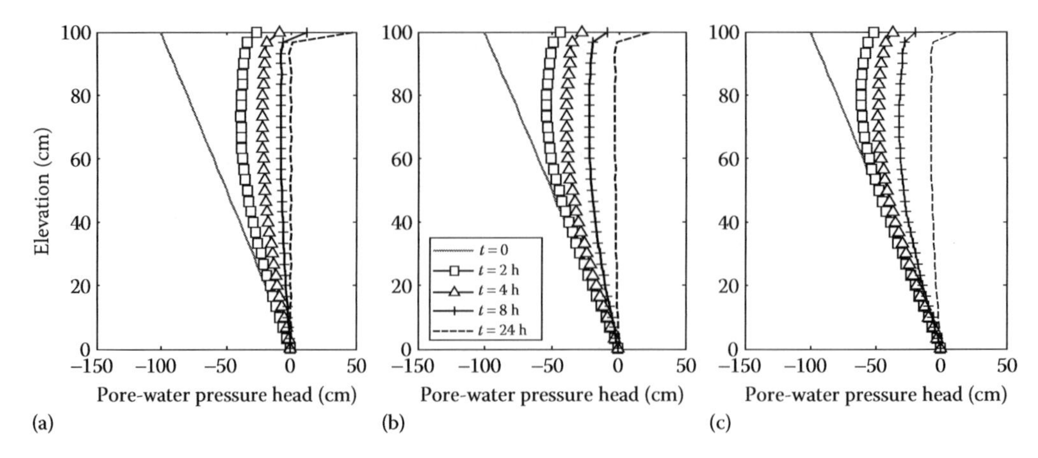

Figure 2.21 Variation of pore-water pressure profiles with different $\theta_s - \theta_r$ (Group 3): (a) $\theta_s - \theta_r = 0.15$, (b) $\theta_s - \theta_r = 0.30$, and (c) $\theta_s - \theta_r = 0.45$.

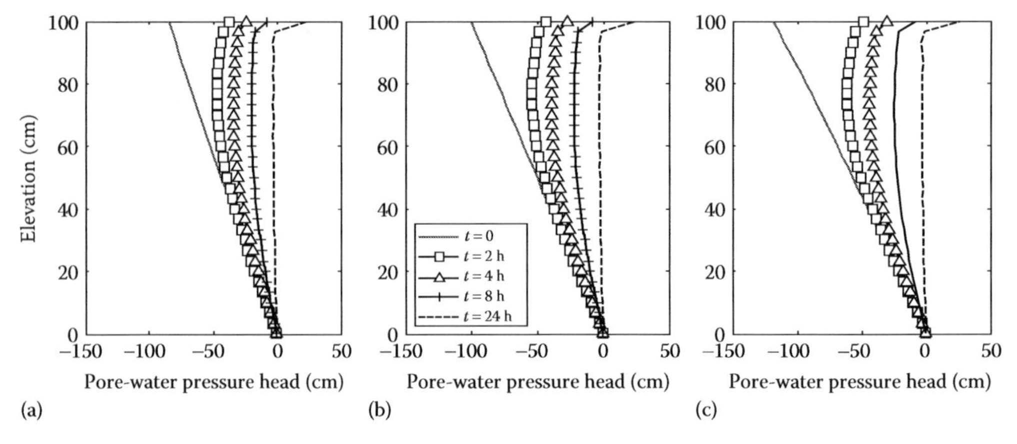

Figure 2.22 Variation of pore-water pressure profiles with different q_0 (Group 4): (a) $q_0 = -0.1$ cm/h, (b) $q_0 = 0$ cm/h, and (c) $q_0 = 0.1$ cm/h.

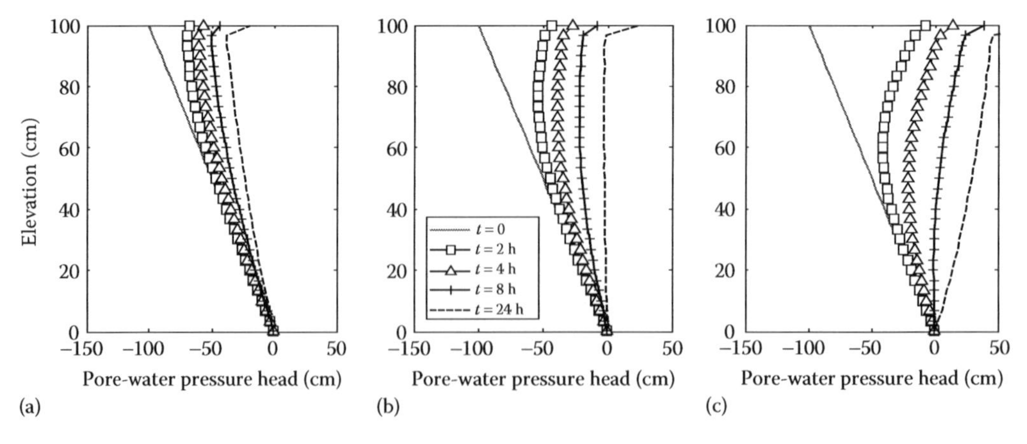

Figure 2.23 Variation of pore-water pressure profiles with different q_1 (Group 5): (a) $q_1 = -0.5$ cm/h, (b) $q_1 = -1.0$ cm/h, and (c) $q_1 = -2.0$ cm/h.

When q_1 is –2 cm/h, after 2 hours, the wetting front penetrates to a depth of 60 cm and the average matric suction in the unsaturated zone is 40 cm. The unsaturated zone is completely saturated after 8 hours. It can be seen that with larger rainfall intensity, the advance of wetting front and the dissipation of matric suction is faster.

2.4.3.6 *Thickness of soil layer* L

Figure 2.24 shows pore-water pressure profiles with respect to different soil thicknesses. It can be seen that soil layer thickness mainly affects the initial distribution of pore-water pressure. The initial matric suction at ground surface is greater and hence the water storage capacity of the whole unsaturated zone is larger when the soil layer is thicker. Therefore, the dissipation of matric suction in unsaturated zone is slower.

2.5 NUMERICAL ANALYSIS OF THE RICHARDS EQUATION

Because of the high nonlinearity of hydraulic parameters involved in the governing equation of unsaturated flow, analytical solutions for the infiltration problem can only be obtained by making some assumptions and under some given initial and boundary conditions (Zhan and Ng, 2001; Griffiths and Lu, 2005; Šimůnek, 2006). The advantage of numerical analysis is that it can incorporate more sophisticated and advanced models of soil hydraulic properties. Tremendous numerical studies (Ng and Shi, 1998a, b; Gasmo et al., 2000; Ng et al., 2001; Tsaparas et al., 2002; Blatz et al., 2004; Tami et al., 2004; Zhang et al., 2004; Chen and Zhang, 2006; Rahardjo et al., 2007; Rahimi et al., 2010; among others) have been carried out to simulate the seepage and infiltration in soil slopes under rainfall conditions. Computer programs that have been applied for numerical modeling of seepage and infiltration in unsaturated slopes include Seep/W (Geo-slope Ltd., 2012), SVFlux (SoilVision System Ltd., 2009), HYDRUS (Šimůnek et al., 1999), SWAP (van Dam et al., 1997), Flow3D (Gerscovich, 1994), and FEMWATER (Lin et al., 1997).

2.5.1 Standard formulations

The governing equation of infiltration in unsaturated soils may be written in different forms, with either pore pressure head h $[L]$ or water content θ_w $[L^3/L^3]$ as the dependent variable.

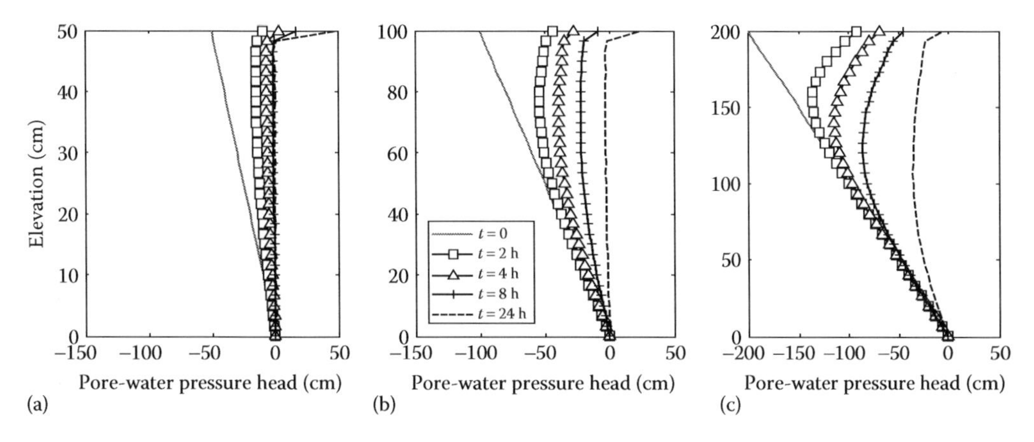

Figure 2.24 Variation of pore-water pressure profiles with different L (Group 6): (a) $L = 50$ cm, (b) $L = 100$ cm, and (c) $L = 200$ cm.

Three standard forms of the unsaturated flow equation are commonly adopted, that is, the h-based form, the θ_w-based form, and the mixed form as follows:

$$h\text{-based form: } C(h)\frac{\partial h}{\partial t} - \nabla\left(\mathbf{k}(h)\nabla(h+z)\right) = 0 \tag{2.36}$$

$$\theta\text{-based form: } \frac{\partial \theta_w}{\partial t} - \nabla\left(\mathbf{D}(\theta_w)\nabla\theta_w\right) = 0 \tag{2.37}$$

$$\text{Mixed form: } \frac{\partial \theta_w}{\partial t} - \nabla\left(\mathbf{k}(h)\nabla(h+z)\right) = 0 \tag{2.38}$$

where:

$C(h) = d\theta_w/dh$ is the specific storage capacity $[1/L]$

$\mathbf{k}(h)$ is the tensor of permeability

$\mathbf{D}(\theta_w) = \mathbf{k}(\theta_w)/C(\theta_w)$ is the tensor of unsaturated diffusivity $[L^2/T]$

The h-based form of the Richards equation is difficult to solve accurately using standard time integration methods. Mass balance errors grow as the integration progresses unless very small time steps are taken. The θ_w-based form allows for very efficient numerical solutions, even for infiltration into initially dry soils as it formulates discrete approximations that are perfectly mass conservative. However, because material discontinuities can produce discontinuous water content profiles, special provisions must be taken for heterogeneous soils and hence the θ_w based form is rarely used in numerical models (Rathfelder and Abriola, 1994; Šimůnek, 2006). Celia et al. (1990) proposed a numerical scheme that uses the mixed form to overcome the mass conservation difficulties and maintain the advantage of h-based formulation. After that, the mixed form is commonly used in computer codes of unsaturated flow.

2.5.2 Spatial approximation and time discretization

The spatial approximation and time discretization of the Richards equation is usually accomplished using finite element or finite difference methods. In finite element modeling, the Galerkin method is commonly used for space approximation. A weighted finite difference scheme is usually used for time discretization. Discretization in time yields the following system of nonlinear equations (Paniconi and Putti, 1994):

$$\mathbf{f}(\mathbf{h}^{n+1}) \equiv \mathbf{A}(\mathbf{h}^{n+\delta})\mathbf{h}^{n+\delta} + \mathbf{F}(\mathbf{h}^{n+\delta})\frac{\mathbf{h}^{n+1} - \mathbf{h}^{n}}{\Delta t^{n+1}} + \mathbf{b}(\mathbf{h}^{n+\delta}) - \mathbf{q}(t^{n+\delta}) = 0 \tag{2.39}$$

where:

$\mathbf{h}$ is the vector of nodal pressure heads

superscript n denotes time step

$\mathbf{f}(\mathbf{h})$ is a nonlinear function of $\mathbf{h}$, with $\mathbf{h}^{n+\delta} = \delta\mathbf{h}^{n+1} + (1-\delta)\,\mathbf{h}^{n}$

δ is a parameter that represents the type of approximation method ($\delta = 0$, explicit; $\delta = 0.5$, Crank–Nicolson; $\delta = 1$, backward implicit Euler)

$\mathbf{A}$ is the global finite element stiffness matrix

$\mathbf{F}$ is the global finite element storage or mass matrix

$\mathbf{b}$ is the gravitational gradient vector

$\mathbf{q}$ is the generalized forcing vector which contains the specified Darcy flux boundary conditions

For any time scheme other than the fully explicit forward method, linearization and/or iteration procedure must be used to solve the discrete nonlinear algebraic equations.

2.5.3 Nonlinear solution methods

Because the permeability function and the SWCC are nonlinear functions of matric suction, iterative schemes such as the Newton–Raphson method and the Picard iteration method are required for the solution process. The Picard method, also known as successive approximation or simple iteration, enjoys great popularity because it is computationally inexpensive and preserves symmetry of the discrete system of equations. However, the method may diverge under certain conditions (e.g., Huyakorn et al., 1984; Celia et al., 1990). The Newton–Raphson iteration method, also known as the Newton scheme, yields nonsymmetric system matrices and is more complex and expensive than the Picard linearization. However, the Newton method can achieves a higher rate of convergence and can be more robust than the Picard iteration for certain types of problems (Paniconi and Putti, 1994). Applied to Equation 2.39, the Newton scheme can be written as (Paniconi and Putti, 1994)

$$\mathbf{f}'(\mathbf{h}^{n+1,(m)})\Delta\mathbf{h} = -\mathbf{f}(\mathbf{h}^{n+1,(m)}) \tag{2.40}$$

where:

$\Delta\mathbf{h} = \mathbf{h}^{n+1,(m+1)} - \mathbf{h}^{n+1,(m)}$

superscript (m) is an iteration index

$\mathbf{f}'()$ is the Jacobian matrix

The Picard scheme may be written as (Paniconi and Putti, 1994)

$$\left[\delta\mathbf{A}(\mathbf{h}^{n+\delta,(m)}) + \frac{1}{\Delta t^{n+1}}\mathbf{F}(\mathbf{h}^{n+\delta,(m)})\right]\Delta\mathbf{h} = -\mathbf{f}(\mathbf{h}^{n+1,(m)}) \tag{2.41}$$

Comparing Equations 2.40 and 2.41, it shows that the Picard scheme can be viewed as an approximate Newton method. The calculation of derivative terms in the Jacobian matrix makes the Newton scheme more costly and algebraically complex than the Picard scheme. Figure 2.25 shows a schematic plot of convergence pattern for the Newton iteration method and the Picard iteration method.

As the slow convergence rate often occur in sharp wetting fronts with highly nonlinear soil hydraulic properties or initial dry conditions, various under-relaxation (or damping) techniques are developed to enhance convergence of a nonlinear iterative scheme (Paniconi and Putti, 1994; Tan et al., 2004). The transformation methods are also developed to overcome the numerical difficulties caused by the strong nonlinearity of the hydraulic properties. Transformation methods (Williams et al., 2000; Cheng et al., 2008) can reduce the nonlinearity of the solution profiles through the identification and application of an appropriate change of variable applied to the dependent variable in the governing equations. The solution of the original problem may then be retrieved by applying an inverse transformation. Brief review on different under-relaxation techniques and transformation methods can be found in Tan et al. (2004) and Williams et al. (2000).

2.5.4 Numerical oscillation

Numerical oscillation describes a phenomenon where the calculated solution oscillates around the correct value (Segerlind, 1984). In time-dependent field problems such as

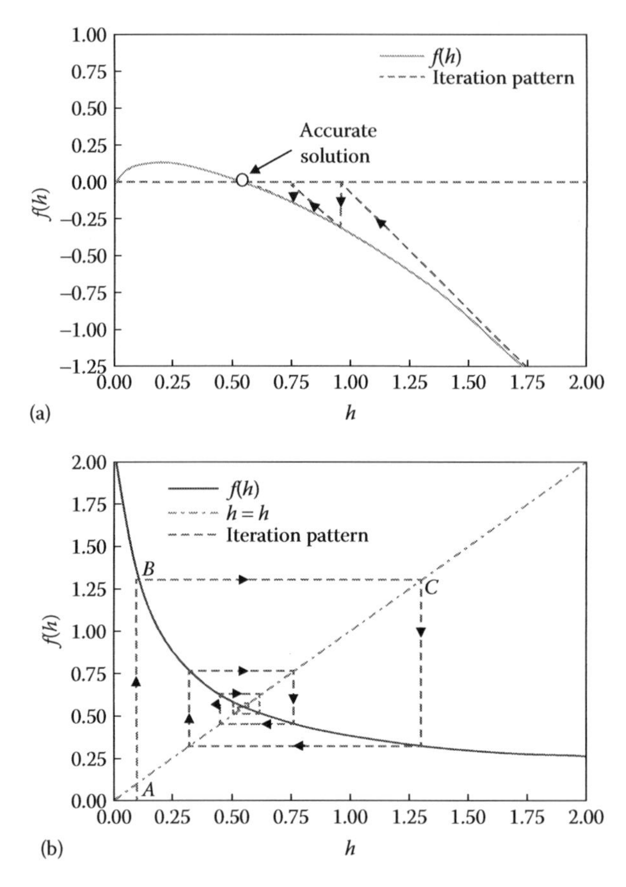

Figure 2.25 Schematic plot of convergence pattern for (a) the Newton iterative method and (b) the Picard method. (From Mehl, S.: Use of Picard and Newton iteration for solving nonlinear ground water flow equations. *Ground Water.* 2006. 44. 583–594. Copyright Wiley-VCH Verlag GmbH & Co. KGaA. Reproduced with permission.)

consolidation, heat diffusion, and seepage flow, oscillatory results in the finite element solution are quite common (Sandhu et al., 1977; Vermeer and Verruijt, 1981; Celia et al., 1990; Pan et al., 1996; Ju and Kung, 1997; Thomas and Zhou, 1997). Such oscillations cause the solution to deviate from its true value, which may lead to erroneous evaluation of slope stability under rainfall condition. For problems with sharp wetting fronts, lower initial water contents, and larger grid spacings, numerical oscillations become extremely serious (van Genuchten, 1982).

Some researchers attributed the numerical oscillation to the mass matrix scheme. In a finite element formulation, the mass matrix can be consistent (mass-distributed) or lumped (mass-lumped). Celia et al. (1990) observed numerical oscillation in finite element solutions which however does not appear in any finite difference solutions. As the only difference between the two solution procedures is the treatment of the time derivative term, the results implied that diagonalized time (mass) matrices are to be preferred in solving the oscillation problem. They explained that only in a diagonal matrix the numerical solution can satisfy the maximum principle (Bouloutas, 1989). This principle states that the maximum value of the numerical solution is dictated by either the boundary conditions or the initial data. The maximum value at current time-step level should be less than or equal to the maximum value at the previous one.

Similarly, the minimum value at current time-step should be greater than or equal to the minimum value at the previous time-step. The maximum principle cannot be guaranteed when the time (mass) matrix is not diagonal as in finite element solutions. Therefore, finite element models may produce oscillatory solutions. Pan et al. (1996) compared the consistent (mass-distributed) formulation with the lumped (mass-lumped) formulation for an unsaturated seepage flow analysis. It is found that the mass-distributed scheme generates numerical oscillation at the sharp wetting fronts due to the highly nonlinear properties of water flow when linear elements are used, whereas no oscillation was observed in mass-lumped approach. However, they also found that the mass-lumped scheme may cause smearing of the wetting front. Ju and Kung (1997) found that only the linear elements with the lumped mass scheme offered accurate transient solutions (Figure 2.26). The consistent mass scheme with any elements and the lumped mass scheme with quadratic/cubic elements would cause numerical oscillation.

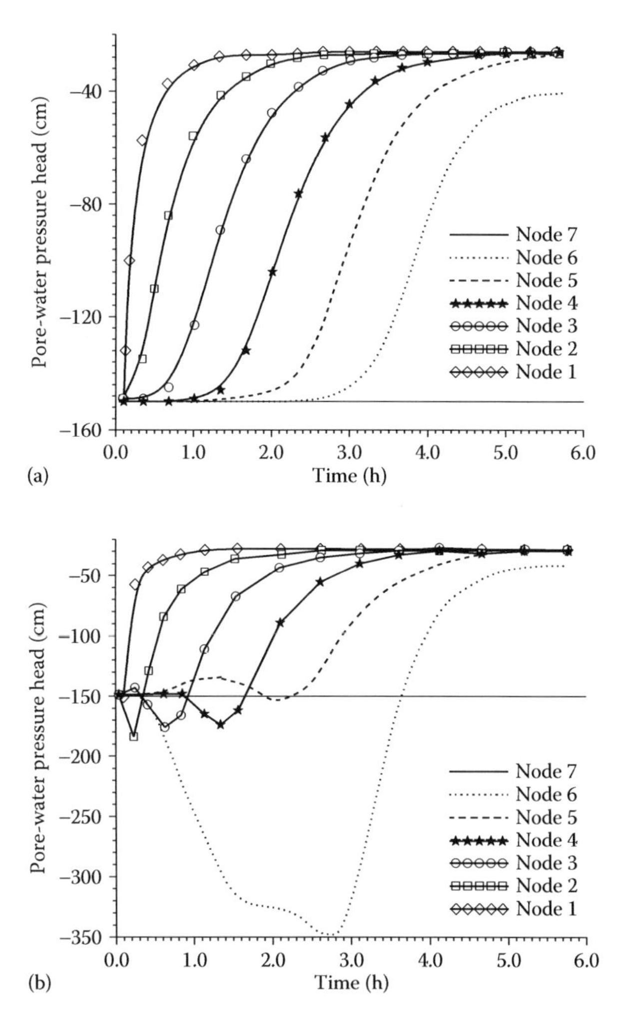

Figure 2.26 Finite-element results of one-dimensional infiltration with (a) lumped mass with linear element and (b) consistent mass with linear element. (From *Computers and Geosciences*, 23, Ju, S. H., and Kung, K. J. S., Mass types, element orders and solution schemes for the Richards equation, 175–187, Copyright 1997, with permission from Elsevier.)

In addition to choosing a lumped mass matrix scheme, some researchers solved the numerical oscillation problem by controlling element size and time step. The time step cannot be arbitrarily reduced to achieve the convergence and the mesh size must first be reduced to avoid numerical oscillations (Ju and Kung, 1997). Sandhu et al. (1977) and Vermeer and Verruijt (1981) investigated oscillations of pore-water pressures in consolidation problems. Vermeer and Verruijt (1981) suggested a minimum time-step size in terms of mesh size and the coefficient of consolidation:

$$\Delta t \geq \frac{1}{6}\frac{(\Delta L)^2}{\delta c} \tag{2.42}$$

where:

ΔL is the element length

δ is a parameter to represent the type of approximation in numerical modeling (same as δ in Equation 2.39)

$c = a_v C_v$, where C_v is the coefficient of consolidation

a_v is a dimensionless coefficient representing the relative influence of the compressibility of the pore fluid which can be expressed as

$$a_v = \frac{m_v}{m_v + n\beta_w} \tag{2.43}$$

where:

m_v is coefficient of volume compressibility

n is the porosity of the soil

β_w is the compressibility of water

Thomas and Zhou (1997) derived two minimum time-step criteria to avoid numerical oscillations in heat diffusion problems. The criteria are formulated in terms of constant thermal conductivity, and specific heat capacity. Based on an analogy between the h-based form of the Richards equation and the heat diffusion equation, Karthikeyan et al. (2001) reinterpreted the minimum time-step criteria proposed by Thomas and Zhou (1997) in terms of soil permeability and specific storage capacity for unsaturated seepage problems (Table 2.4). To account for material nonlinearity, the criteria are suggested to be calculated based on the most critical state, in which the material properties correspond to the highest negative pore-water pressure under the initial condition.

Table 2.4 Minimum time-step size for different elements

Oscillation type	*Minimum time-step size for one-dimensional element*		*Minimum time-step size for two-dimensional element*	
	Two-noded	*Three-noded*	*Four-noded*	*Eight-noded*
Type 1	$\Delta t \geq \Delta L^2 C/6k$	$\Delta t \geq \Delta L^2 C/40k$	$\Delta t \geq \Delta L^2 C/2k$	$\Delta t \geq \Delta L^2 C/40k$
Type 2	–	$\Delta t \geq \Delta L^2 C/20k$	–	$\Delta t \geq \Delta L^2 C/20k$

Source: Karthikeyan et al., *Canadian Geotechnical Journal*, 38, 639–651, 2001. With permission.

Notes: (1) Δt, time increment; C, the specific storage capacity ($=d\theta_w/dh$); k, permeability; ΔL, element length or width perpendicular to the direction of flow; γ_w, unit weight of water (=9.81 kN/m^3). (2) Type 1 oscillation means the calculated total head at some nodes are less than their initial values. Type 2 oscillation means the calculated distribution of total head is oscillatory even though the values at all nodes are greater than their initial values.

2.5.5 Rainfall infiltration boundary condition

In the analytical solutions of the Richards equation, it often assumed that the infiltration rate is equal to the rainfall intensity, which may lead to unrealistic surface ponding as shown in Figure 2.17(c). In the numerical modeling, there are several different approaches to handle the rainfall infiltration boundary conditions. Iverson (2000) assumed that the rainfall can totally infiltrate into the soil if the rainfall intensity is less than or equal to the saturated permeability. When the rainfall intensity is greater than the saturated permeability, the infiltration rate is equal to the saturated permeability and the surplus rainfall runs off the slope as surface flow. This assumption is different from those in some conceptual models such as the Green–Ampt infiltration model where the infiltration rate depends on the infiltration capacity. Tsai and Yang (2006) showed that unrealistically high pressure heads obtained using Iverson's solution are mainly due to the overestimation of the infiltration rate.

A more realistic approach to simulate rainfall infiltration boundary condition is to switch between Dirichlet and Neumann boundary conditions depending on the pore pressure at the soil surface. For example, a Neumann boundary condition along the soil surface can be defined as follows (Chui and Freyberg, 2009):

$$\mathbf{n} \times \boldsymbol{k} \cdot \nabla(H) = m_N q + m_b R_b (H_b - H) \tag{2.44}$$

where:

- $\mathbf{n}$ is the normal vector to the boundary
- $\boldsymbol{k}$ is the tensor of permeability
- m_N and m_b are the complementary smoothing functions
- q is the rainfall intensity
- R_b is the external resistance
- H_b is the external total head at surface

The above-mentioned general boundary condition reduces to a Neumann condition when the second term on the right-hand side is zero and to a Dirichlet condition when the first term is zero. To minimize the number of switches during iterative solution, the boundary conditions are changed gradually over a small range of boundary pressure using complementary smoothing functions, m_N and m_b (Figure 2.27). R_b is usually set to a very large value to simulate a Dirichlet condition when the pore-water pressure at soil surface is greater

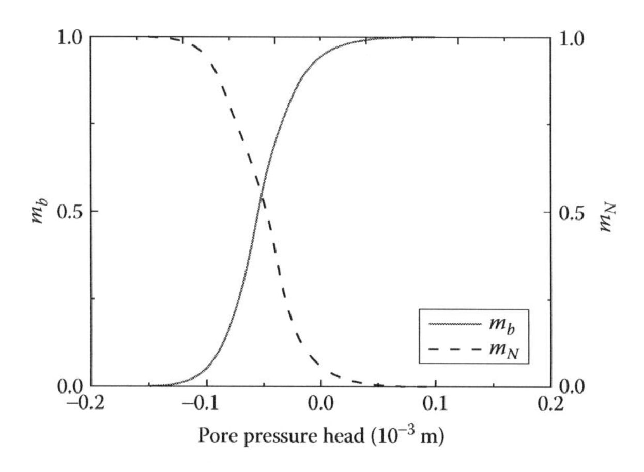

Figure 2.27 Smoothing functions in the surface boundary condition.

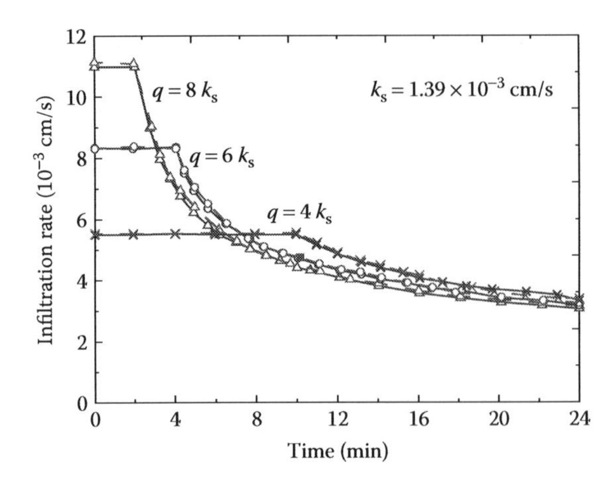

Figure 2.28 Comparison of infiltration rates by Mein and Larson (1973) (solid lines) and the numerical solution (dotted lines) at different constant rainfall intensity. (From Chui, T. F. M., and Freyberg, D. L., *Journal of Hydrologic Engineering*, 14, 12, 1374–1377, 2009. Reproduced with permission from ASCE.)

than zero. In addition, the external total head H_b can be adjusted to simulate a ponded surface with a specified value.

In Seep/W and SVFlux, the review boundary condition for seepage face that is used to simulate the rainfall boundary condition is similar to the approach in Equation 2.44. Chui and Freyberg (2009) compared the numerical results for an unsaturated soil column with the results by Mein and Larson (1973) and found very good agreement (Figure 2.28).

Besides precipitation and runoff along soil surface, evaporation and transpiration may also affect the net infiltration into soils. A comprehensive description on the quantification of moisture flux boundary conditions focusing on the estimation of the actual evaporation rate can be found in Fredlund et al. (2012).

2.6 TYPICAL PORE-WATER PRESSURE PROFILES UNDER RAINFALL CONDITION

The wetting front concept provides a simplified methodology for considering changes in pore pressure (or matric suction) under a change in rainfall conditions. However, there is not always a distinct difference between the infiltration zone and the zone where the negative pore-water pressures have been maintained. Typical pore-water pressure distributions under various ground surface fluxes are shown in Figure 2.29. Figure 2.29(a) illustrates the pore-water pressure distributions for the steady-state condition. According to Kisch (1959), the gradient of pore-water pressure under the steady-state condition can be written as

$$\frac{d(u_w/\gamma_w)}{dz} = (q/k - 1) \tag{2.45}$$

Under hydrostatic conditions, there is no ground flux. According to Equation 2.45, the gradient of the pore-water pressure head is –1 as shown in Figure 2.29(a). Whenever the magnitude of the ground surface flux, q, approaches the coefficient of permeability of the unsaturated soil, k, at a particular value of matric suction, the pressure gradient is zero as shown in Figure 2.29(a). Figures 2.29(b) and (c) show the pore-water pressure profiles under transient seepage condition for the cases where $q/k_s < 1$ and $q/k_s \geq 1$, respectively. Infiltration

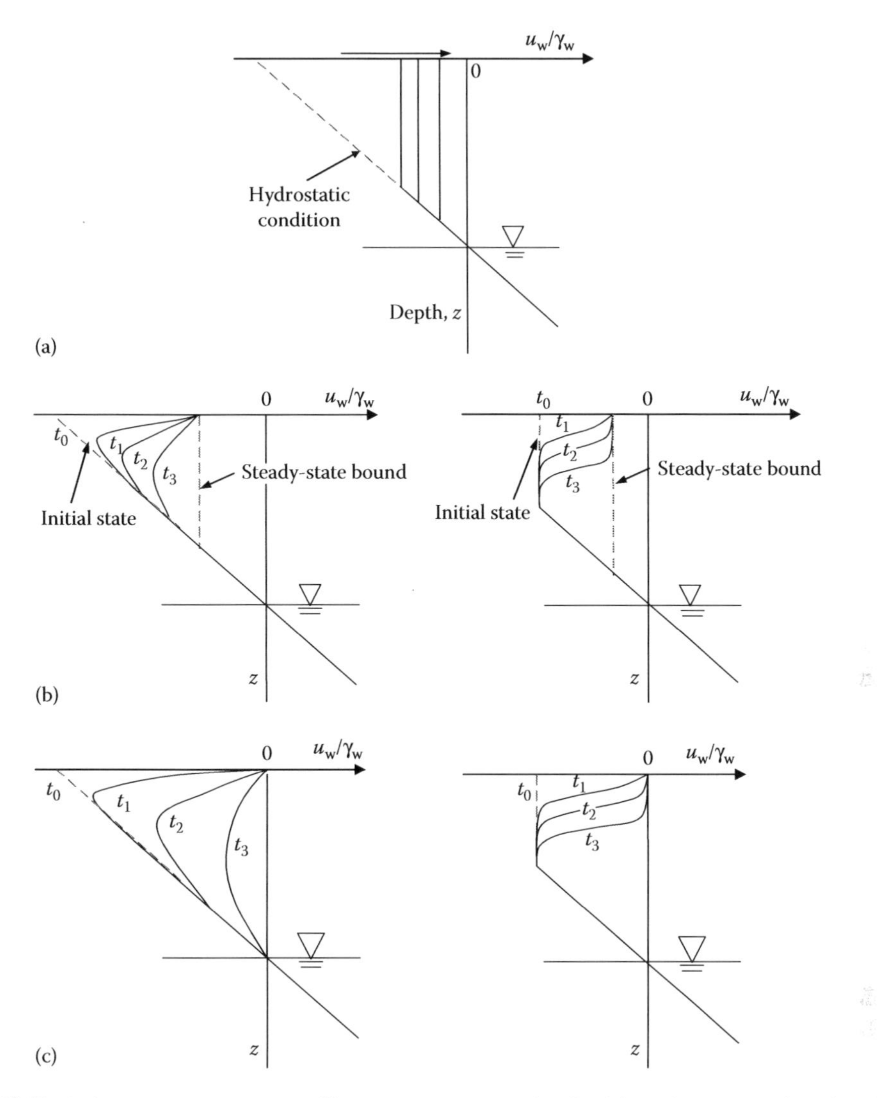

Figure 2.29 Typical pore-water pressure files in an unsaturated soil with various ground surface fluxes: (a) steady-state condition; (b) transient condition, $q/k_s < 1$; and (c) transient condition, $q/k_s \geq 1$.

under transient seepage conditions can be considered as a transitional state between initial state and the final steady states. The initial states of the left diagrams in Figures 2.29(b) and (c) are the hydrostatic condition ($q = 0$). The initial states of the right diagrams in Figures 2.29(b) and (c) are a steady-state condition with the surface flux q greater than zero. The time to reach the steady state is a function of the surface flux, the coefficient of permeability of the soil and the water storage of the soil. When the ground flux is less than the saturated coefficient of permeability (Figure 2.29b), the matric suction in the unsaturated soil can decrease but does not disappear. Only when the ground surface flux is equal to or greater than the saturated coefficient of permeability (Figure 2.29c) can the matric suction be eliminated if the duration of the rainfall is long enough.

Rahardjo et al. (1995) suggested three idealized pore-water pressure profiles in unsaturated soil slopes for the nonhydrostatic condition (Figure 2.30). Profile *a* represents the situation where matric suction is reduced to zero at the ground surface. Profile *b* represents the condition with

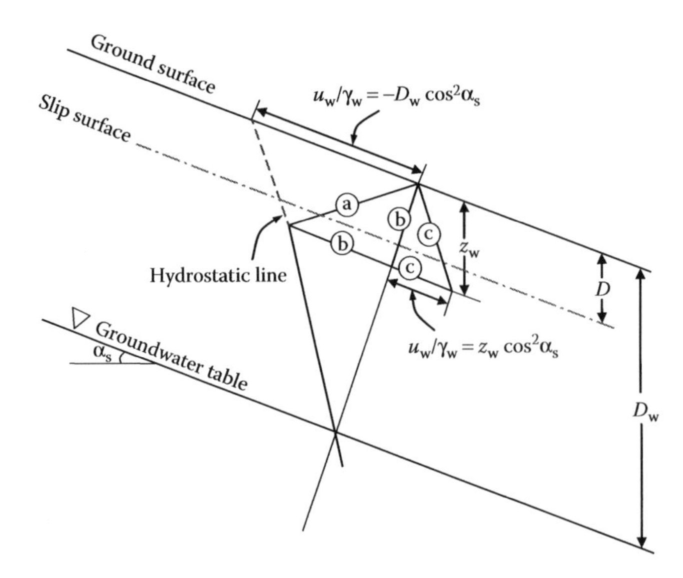

Figure 2.30 Idealized pore-water pressure profiles in a residual soil slope. (From Rahardjo, H. et al., *Canadian Geotechnical Journal*, 32, 1, 60–77, Copyright 1995 Canadian Science Publishing or its licensors. Reproduced with permission.)

sharp wetting front with a depth of z_w. Profile *c* corresponds to the condition with a perched water table at depth z_w. The potential pore-water pressure profile in a coarse-grained soil is profile *b*. The sharp wetting front in profile *b* is mainly due to steep slopes of the SWCC and the permeability function of the unsaturated soil. For fine-grained soils, the slopes of the SWCC and permeability function are generally gentler. Thus, profile *a* can be considered as the potential pore-water pressure profile in a fine-grained soil. The perched water table (profile *c*) commonly occurs in layered soils (Ng and Bruce, 2007; Cho, 2009). Lee et al. (2009) proposed a rational approach to determine critical pore-water pressure profiles considering both major and antecedent rainfalls. They defined a rainfall as a major rainfall if the duration is less than 24 hours. The antecedent rainfall is defined as a rainfall with duration greater than 1 day. Suction profiles (Figure 2.31) from nine rainfall patterns (i.e., 1-day, 2-day, 3-day, 5-day, 7-day, 14-day, and 30-day antecedent rainfalls, plus two critical combinations of antecedent and major rainfalls) as well as the suction redistribution patterns can be used to determine the worst scenario of suction distribution for the assessment of slope stability.

2.7 SOIL CONDITIONS UNDER WHICH MATRIC SUCTION CAN BE MAINTAINED

There is a perception among geotechnical engineers that negative pore-water pressures will dissipate with rainfall infiltration and cannot be relied upon in design considerations. Therefore, the effect of negative pore-water pressure is often ignored in slope stability studies. Here a parametric study is conducted to illustrate the conditions under which soil suction can be maintained.

In this section, analyses for both steady-state and transient seepage conditions were conducted on a 20 m high slope inclined at 30 degree. The slope is composed of a homogenous, isotropic soil. Seep/W is used for numerical simulation. The finite element mesh, along with the boundary conditions, is shown in Figure 2.32. Along the left and right boundaries beneath the groundwater table, a constant head was applied. A zero flux boundary was applied along the left and

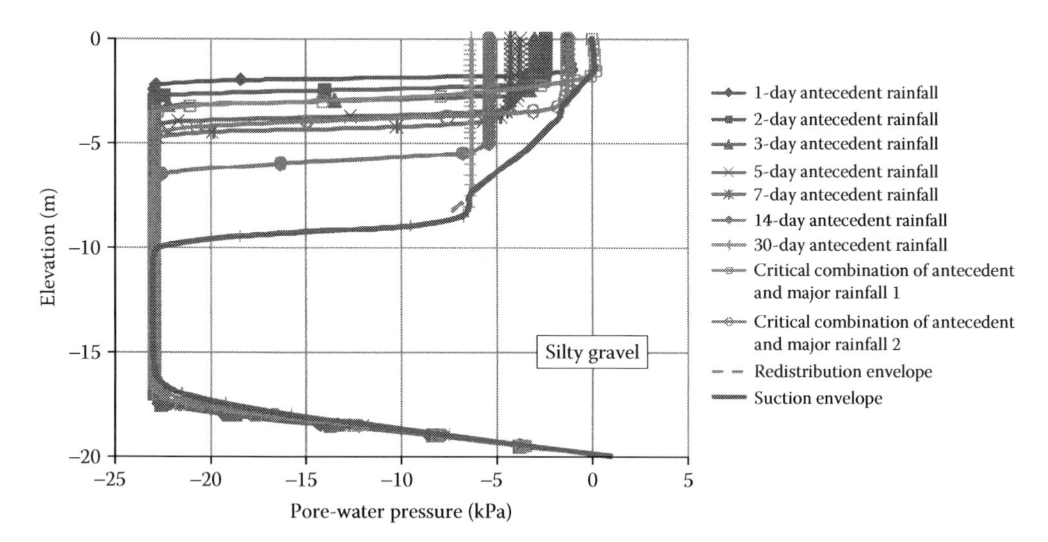

Figure 2.31 Suction envelopes computed from the PERISI model. (From *Engineering Geology*, 108, Lee, L. M. et al., A simple model for preliminary evaluation of rainfall-induced slope instability, 272–285, Copyright 2009, with permission from Elsevier.)

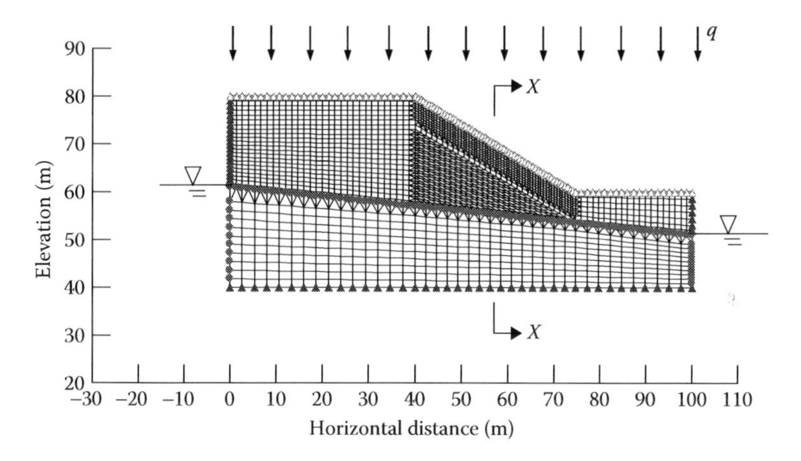

Figure 2.32 Finite element mesh and boundary conditions of the slope. (From Zhang, L. L. et al., *Canadian Geotechnical Journal*, 41, 569–582, Copyright 2004 Canadian Science Publishing or its licensors. Reproduced with permission.)

right boundaries above the groundwater table. To illustrate the pore-water pressure profiles more clearly, the groundwater table was fixed by applying a constant pressure head equal to zero at the groundwater table, on the nodes of the mesh. However, the groundwater table in a soil slope may rise during rainfall in real situations. An example with a free groundwater table will be presented later to illustrate the effect of different groundwater conditions.

It was assumed that the base of the finite element mesh was impermeable. Precipitation was modeled as a flux boundary, q, applied along the slope surface. Section X–X is in the middle of the slope. The groundwater table at the selected section is 15.28 m under the sloping ground surface. The pore-water pressure profiles at section X–X are presented to illustrate conditions under which soil suction can be maintained. However, it should be pointed out that it is the infiltration behavior in the entire slope that controls the stability of the slope.

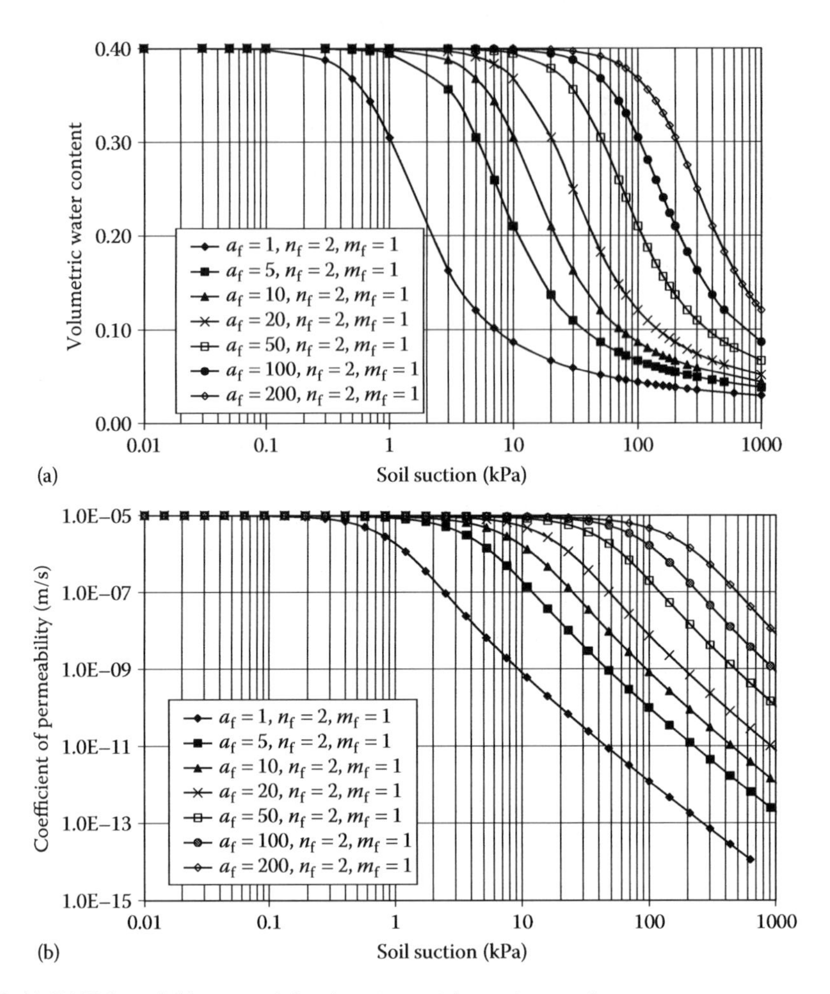

Figure 2.33 (a) SWCCs and (b) permeability functions with varying a_f values.

The Fredlund and Xing (1994) model (Table 2.1) is used for SWCC. Soils with various a_f values (e.g., 1, 5, 10, 20, 50, 100, and 200 kPa) and the same n_f, m_f, and saturated coefficient of permeability k_s values (i.e., $n_f = 2$, $m_f = 1$, and $k_s = 1 \times 10^{-5}$ m/s) were studied in detail for the parametric study. The permeability function is estimated using the Fredlund et al. (1994) method. The SWCCs and the corresponding coefficient of permeability functions for the soils with different a_f values are shown in Figures 2.33. The effect of n_f and k_s will be presented later.

2.7.1 Pore-water pressure profiles under steady-state conditions

The results in Figure 2.34 were obtained for soils having the same saturated coefficients of permeability of 1×10^{-5} m/s, $n_f = 2$, and $m_f = 1$. The pore-water pressure profiles shown from Figure 2.34(a) to (f) correspond to soils with a_f equal to 5, 10, 20, 50, 100, and 200 kPa, respectively. Different ratios of ground surface flux to saturated coefficient of permeability, q/k_s (i.e., 0.001, 0.01, 0.1, 0.2, 0.5), were applied to the surface of the slope in the numerical analysis.

The results illustrate that as the rainfall flux approaches the saturated coefficient of permeability of the soil, the matric suction at the surface of the slope approaches zero. There is essentially

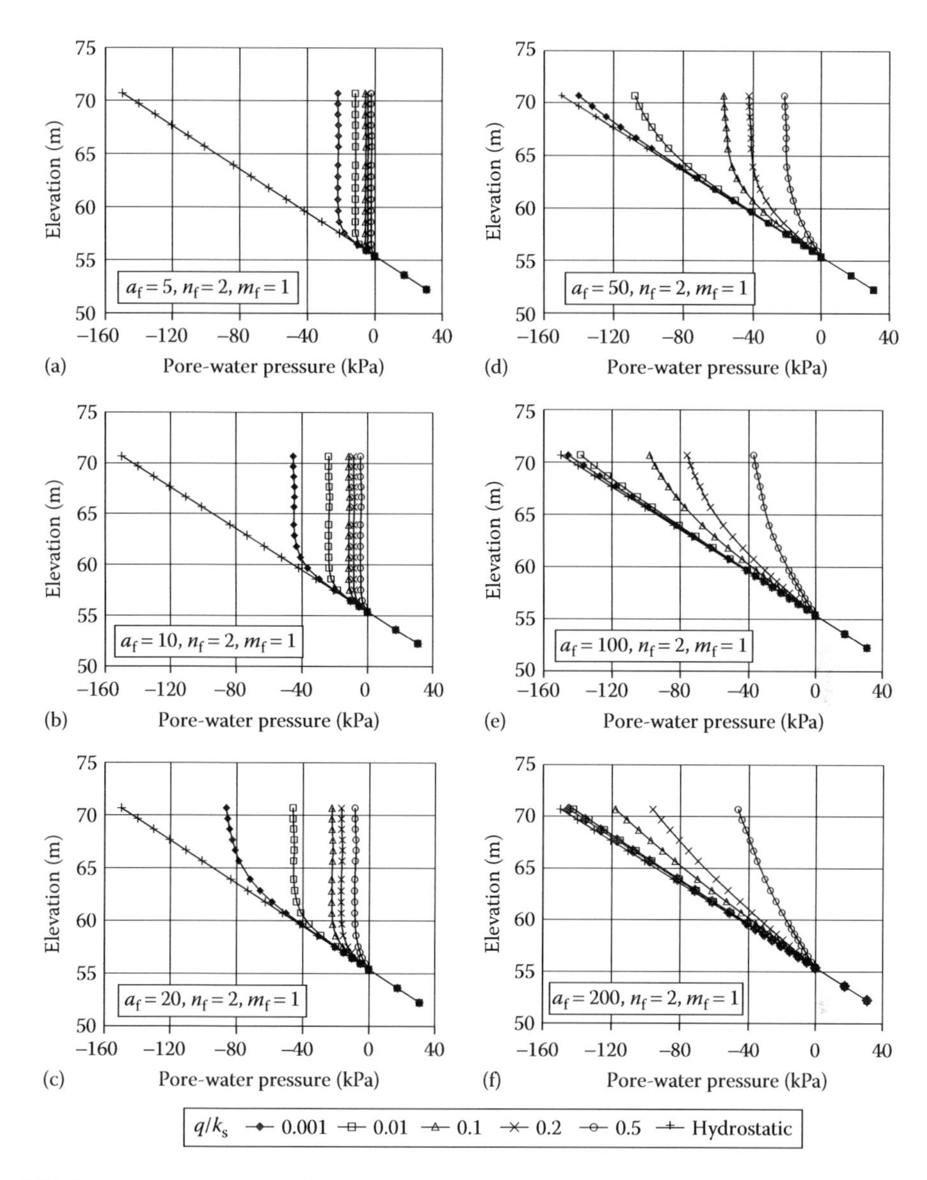

Figure 2.34 Pore-water pressure profiles in slopes with various a values subjected to various rainfall fluxes under steady-state conditions: (a) $a_f = 5$ kPa, (b) $a_f = 10$ kPa, (c) $a_f = 20$ kPa, (d) $a_f = 50$ kPa, (e) $a_f = 100$ kPa, and (f) $a_f = 200$ kPa. (From Zhang, L. L. et al., *Canadian Geotechnical Journal*, 41, 569–582, Copyright 2004 Canadian Science Publishing or its licensors. Reproduced with permission.)

a vertical matric suction profile (i.e., $d(u_w/\gamma_w)/dz$ is zero), established under steady-state conditions when the ground surface flux approaches the saturated coefficient of permeability. The depth of the constant matric suction profile increases when increasing, q/k_s. It should be noted that these results are based on the boundary condition of a fixed groundwater table. The pore-water pressure profiles can be significantly influenced by varying boundary conditions.

The values of matric suction on the vertical portion of the pressure profiles decrease with decreasing a_f values. The long-term matric suction does not disappear but remains essentially close to or unchanged from the hydrostatic profile when the steady-state rainfall flux is two or more orders of magnitude less than the saturated coefficient of permeability and a_f is greater than 100 (Figure 2.34f).

Figure 2.35 illustrates that the long-term matric suction profiles in the slope for the soils with different n_f values. It shows that the values of matric suction at the same value of q/k_s do not decrease or increase monotonically with n_f values.

As shown conceptually in Figure 2.29 and further illustrated by Figures 2.34 and 2.35 from the numerical modeling, the matric suction profiles under steady-state conditions consist of a section of constant pore-water pressure, a transition section, and a section of the hydrostatic condition. Kasim (1997) proposed a method to estimate the value of matric suction $(u_a - u_w)_1$ at the constant pore-water pressure section, associated with the steady-state rainfall flux q_1, using the coefficient of permeability function for the case of a horizontal ground surface (Figure 2.36). Figure 2.37 presents the SWCCs and the corresponding

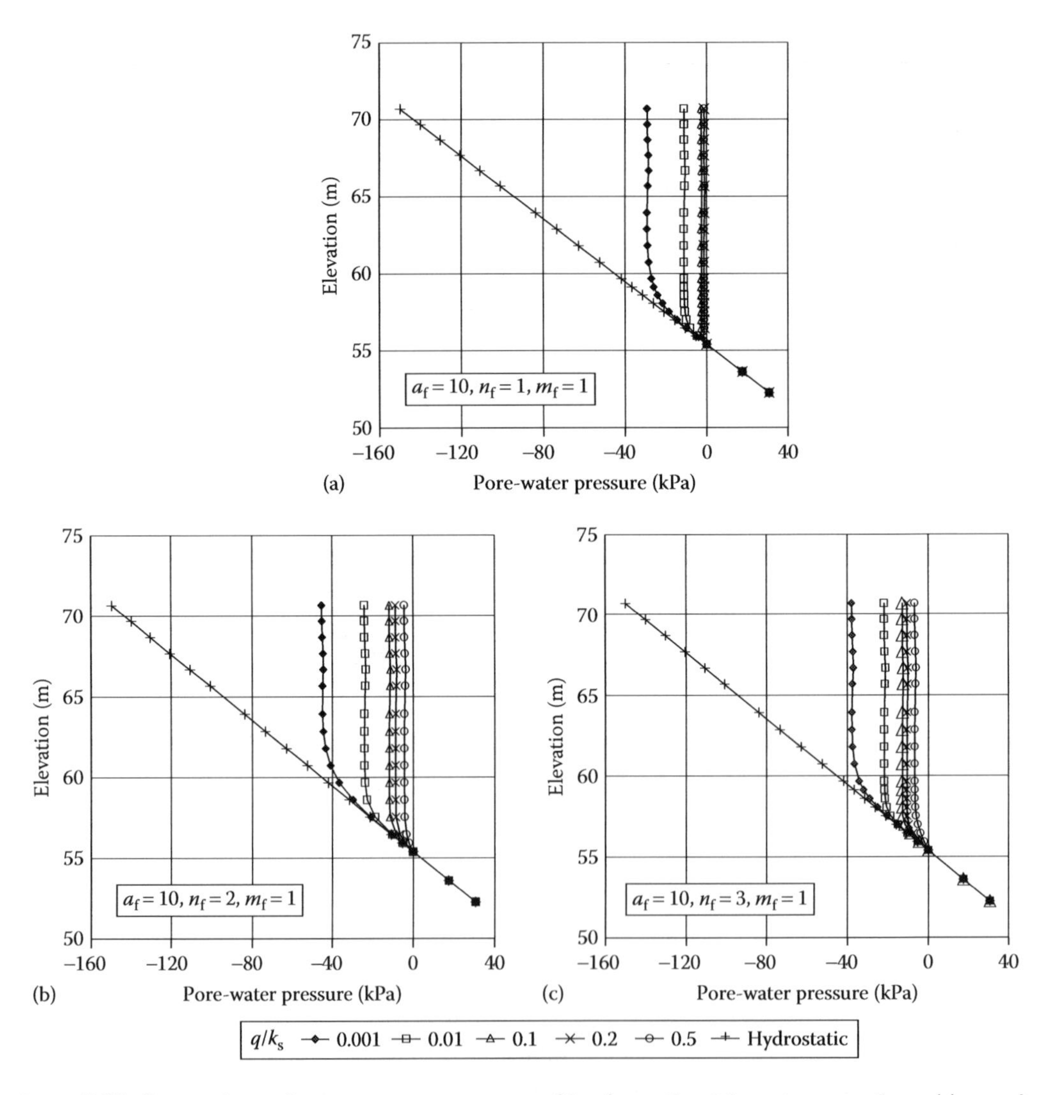

Figure 2.35 Comparison of pore-water pressure profiles for soils with various n_f values: (a) $n_f = 1$, (b) $n_f = 2$, and (c) $n_f = 3$. (From Zhang, L. L. et al., *Canadian Geotechnical Journal*, 41, 569–582, Copyright 2004 Canadian Science Publishing or its licensors. Reproduced with permission.)

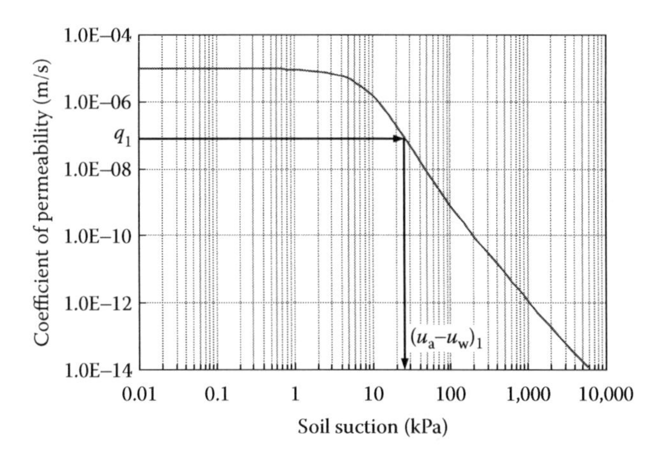

Figure 2.36 Determination of the matric suction values at a horizontal ground surface for a soil subjected to a steady-state rainfall flux, q_1.

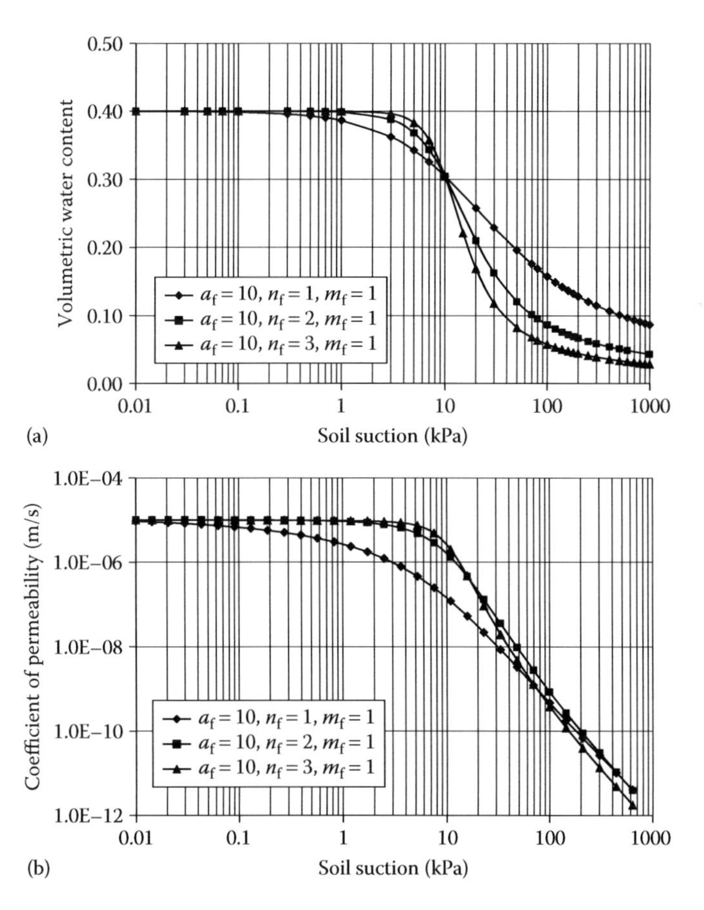

Figure 2.37 (a) SWCCs and (b) permeability functions with varying n_f values of the soils.

permeability functions for soils with varying n_f values. Applying the estimation method proposed by Kasim (1997), the theoretical matric suction $(u_a - u_w)_1$ for soils with varying n_f values does not decrease or increase monotonically at the same value of q_1. Therefore, the pore-water pressure profiles shown in Figure 2.35 are reasonable.

2.7.2 Pore-water pressure profiles under transient seepage conditions

Effects of air-entry value, water storage capacity, and saturated coefficient of permeability and groundwater boundary conditions on pore-water pressure profiles are illustrated in this section. The initial state is taken to be the hydrostatic condition for all cases. It should be noted that varying initial soil moisture conditions could significantly influence the rainfall infiltration and pore-water pressure profiles in soil slopes, as has been studied in the previous section and Freeze (1969) and Tsaparas et al. (2002).

2.7.2.1 Effect of air-entry value on the wetting front

Figures 2.38(a)–(f) show the pore-water pressure profiles under transient seepage condition with various a_f values. The rainfall flux is equal to the saturated coefficient of permeability.

The results illustrate the different patterns created throughout the pore-water pressure profile. For the soil with a_f equal to 1 kPa, the wetting front is sharp and distinct. The matric suction near the ground surface decreases with time, but the rate of downward movement is small. After 50 days of rainfall with a flux equal to the saturated coefficient permeability, the depth of the wetting front is only about 1 meter below the ground surface.

For a soil with a_f equal to 5 kPa, the transition zone between the infiltration zone and the unaffected zone is still quite sharp and distinct. At shallow depths into the slope, the matric suction decreases but remains essentially constant for deeper soils. It takes only one day for the wetting front to move to a depth of about one meter. After 3 days the infiltration depth is about 4 meters. The negative pore-water pressure almost disappears after 6 days of rainfall.

The pore-water pressure profiles of the soil with a_f equal to 10 kPa resembles the profiles of the soil with a_f equal to 5 kPa. However, the negative pore-water pressures almost disappear after 4 days of rainfall infiltration. For a_f values greater than 10 kPa, the transition between the infiltration zone and the affected zone becomes less distinct. The remaining matric suction in the soil decreases more rapidly with an increase in the a_f value.

The results show that as the a_f value increases, the wetting front becomes less distinct. According to Equation 2.45, the gradient of the pore-water pressure depends on the ratio of flux and the coefficient of permeability of the unsaturated soil. Given the same initial matric suctions (e.g., all the cases are in hydrostatic conditions initially in this study), the soil with a smaller air-entry value corresponds to a smaller initial coefficient of permeability while the soil with a larger air-entry value has a larger coefficient of permeability (see Figure 2.33). If the moisture flux q is the same for both cases, then the value of q/k is larger for the soil with a smaller air-entry value. Consequently, the gradient of the pore-water pressure is greater for the soil with the smaller air-entry value.

The above observations can be further illustrated by the examples of pore-water pressure profiles when the rainfall flux is 10^{-6} m/s, which is 10% of the saturated coefficient of permeability of the soils (Figure 2.39). Comparing the two graphs, the most significant difference is the shape of the wetting front. For the soil with a equal to 10 kPa (Figure 2.39a), the wetting front is approximately horizontal, which means that the infiltration rate is much greater than the unsaturated coefficient of permeability according to Equation 2.45. However, for the soil with a_f equal to 100 (Figure 2.39b), the initial coefficient of

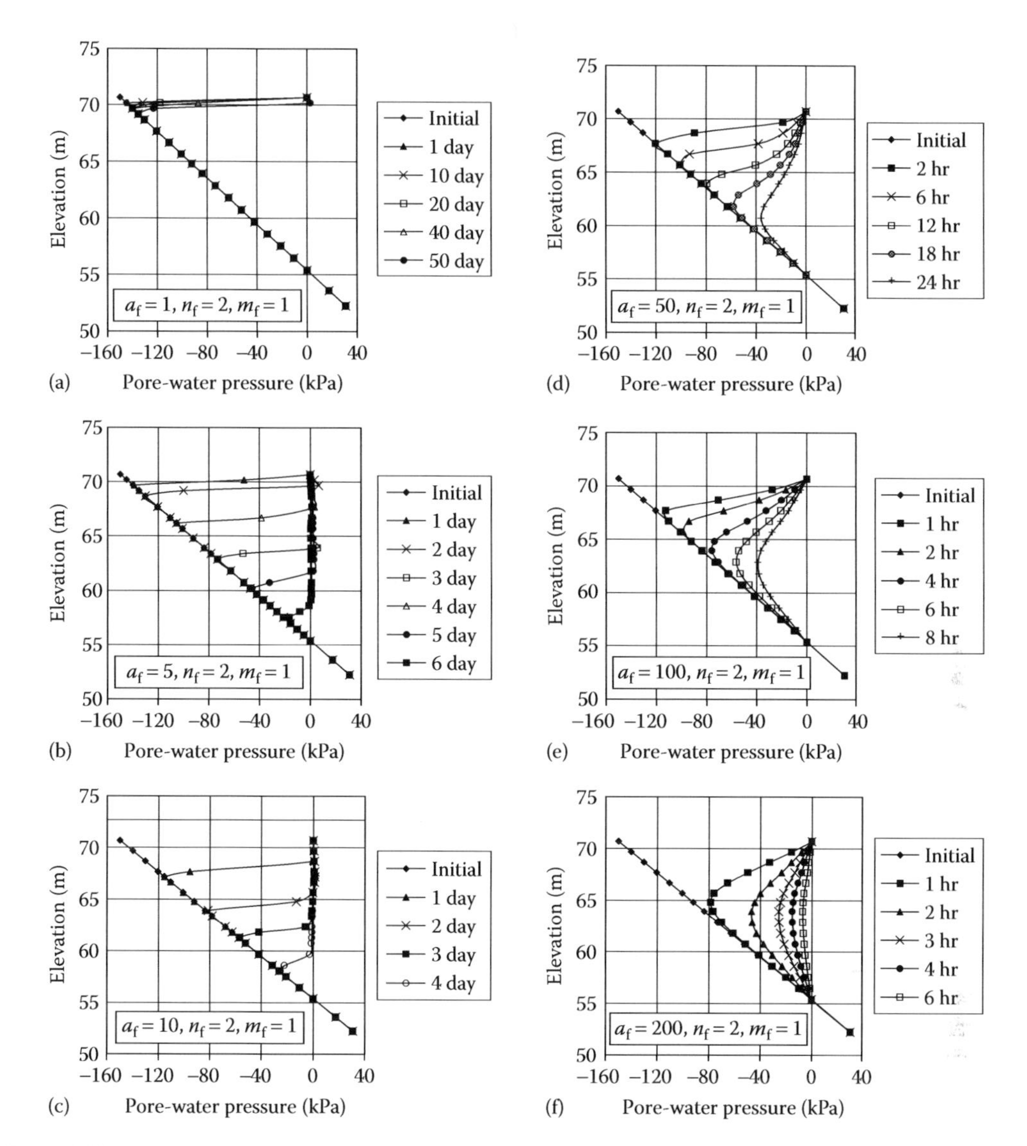

Figure 2.38 Pore-water pressure profiles for soils with various air-entry values and $k_s = 10^{-5}$ m/s subjected to a flux of $q = 10^{-5}$ m/s under transient seepage conditions: (a) $a_f = 1$ kPa, (b) $a_f = 5$ kPa, (c) $a_f = 10$ kPa, (d) $a_f = 50$ kPa, (e) $a_f = 100$ kPa, and (f) $a_f = 200$ kPa. (From Zhang, L. L. et al., *Canadian Geotechnical Journal*, 41, 569–582, Copyright 2004 Canadian Science Publishing or its licensors. Reproduced with permission.)

permeability in the soil is comparable to the flux rate and the pore-water pressure gradient approaches zero.

2.7.2.2 Effect of the saturated coefficient of permeability

Figures 2.40(a)–(c) show the pore-water pressure profiles for soils with the same SWCC (i.e., $a_f = 100, n_f = 2, m_f = 1$) but different saturated coefficients of permeability (i.e., 10^{-7} m/s, 10^{-5} m/s, and 10^{-3} m/s, respectively). The rainfall fluxes are equal to the saturated coefficient of permeability for all the three cases (i.e., the ratios of flux versus the saturated coefficient of

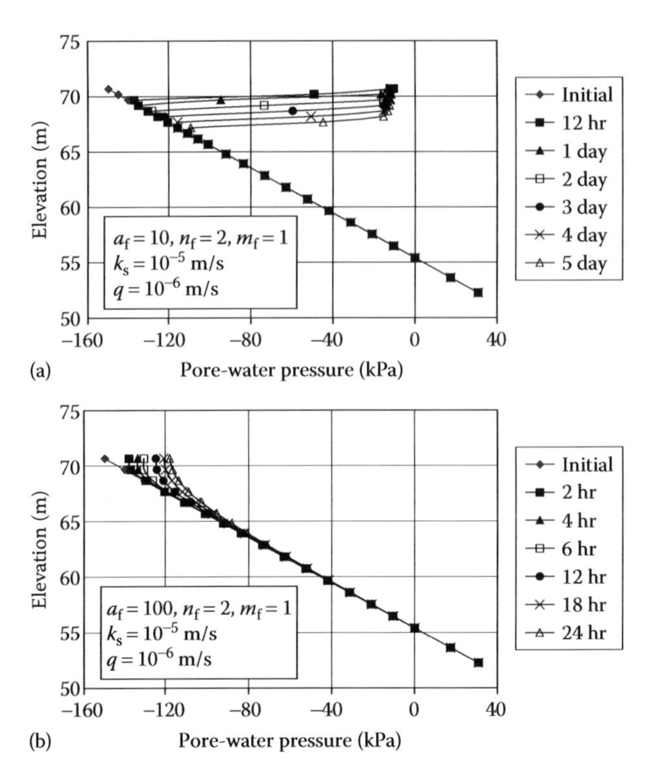

Figure 2.39 Examples of pore-water pressure profiles in a soil subjected to a surface flux, $q = 10^{-6}$ m/s: (a) $a_f = 10$ kPa and (b) $a_f = 100$ kPa. (From Zhang, L. L. et al., *Canadian Geotechnical Journal*, 41, 569–582, Copyright 2004 Canadian Science Publishing or its licensors. Reproduced with permission.)

permeability are unity for the three cases). Comparing the patterns of the pore-water pressure profiles, it can be seen that the shapes of the profiles are similar but the rates of downward movement of the wetting fronts are different, which indicates that the behavior of rainfall infiltration under transient seepage conditions should be related to the absolute intensity of the rainfall and the soil properties.

2.7.2.3 *Influence of water storage coefficient*

The change of volumetric water content can be related to a change in pore-water pressure based on the SWCC:

$$d\theta_w = m_2^w d(u_a - u_w) \tag{2.46}$$

As water content decreases with the increase of matric suction, m_2^w is negative. For convenience, a positive value $|m_2^w|$ instead of m_2^w is defined as the water storage coefficient, which represents the water storage capacity of the unsaturated soil at certain suction. Note that the specific storage capacity C in Equation 2.36 equals to $|m_2^w|\gamma_w$.

Figure 2.41 can be used to explain the behavior of infiltration for soils with varying air-entry values. The SWCCs and the corresponding water storage functions are shown on the graph. The desaturation rate ($n_f = 2$) is the same for both soils but the air-entry values of the two SWCCs are different. The maximum water storage coefficient for the soil with a_f equal to 1 kPa is 0.1445, while for the soil with a_f equal to 10 kPa is only 0.0145. These values are reasonable because the desaturation rate n_f represents the slope

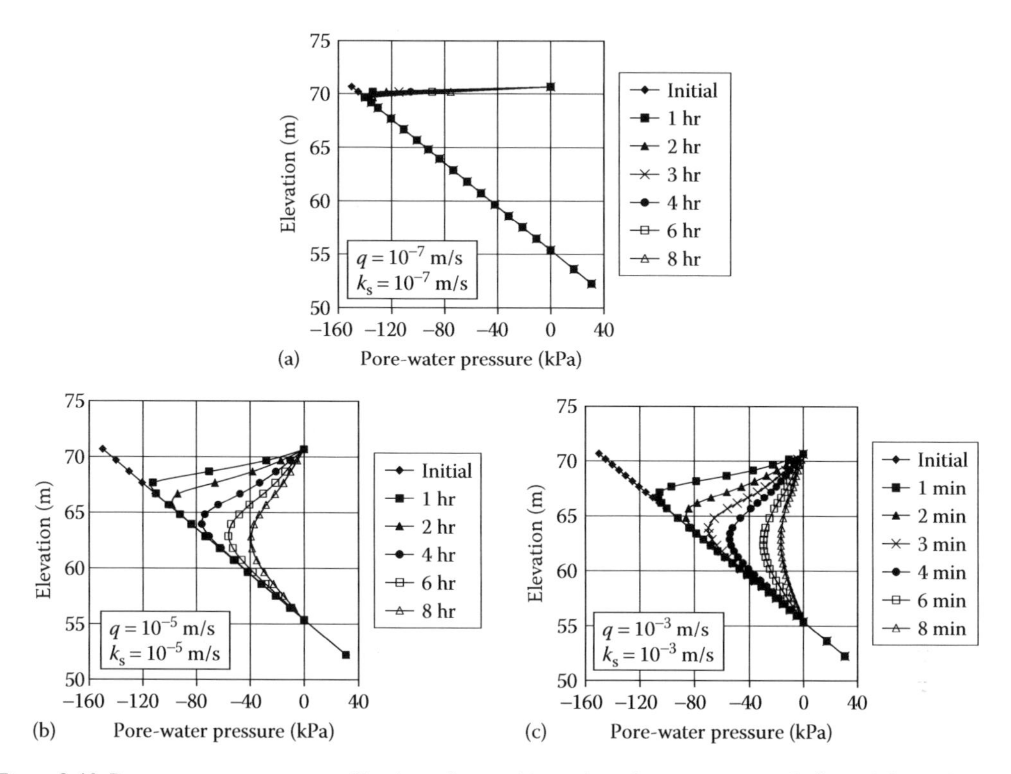

Figure 2.40 Pore-water pressure profiles in a slope subjected to the same ratio of $q/k_s = 1$ for soils with $a_f = 100$, $n_f = 2$, $m_f = 1$ and different values of k_s: (a) $k_s = 10^{-7}$ m/s, (b) $k_s = 10^{-5}$ m/s, and (c) $k_s = 10^{-3}$ m/s. (From Zhang, L. L. et al., *Canadian Geotechnical Journal*, 41, 569–582, Copyright 2004 Canadian Science Publishing or its licensors. Reproduced with permission.)

at the inflection point of the SWCC, which is expressed on a logarithm scale of matric suction, whereas the water storage coefficient, $|m_2^w|$, is the arithmetic slope of the SWCC. As the n_f values of the two SWCCs are the same, the same change of volumetric water content, $\Delta\theta_w$, is associated with the same change of matric suction on a logarithm scale (i.e., $\Delta\ln(u_a - u_w)$). However, the change of matric suction, $\Delta(u_a - u_w)$, is much smaller for the case of a_f equal to 1 than for the case of a_f equal to 10 because the desaturation part is in the low matric suction portion for a_f equal to 1 and consequently yields a larger water storage coefficient.

The difference in the water storage functions can help explain why the rate of the movement of the wetting front for soils with different air-entry values is distinctly different as shown in Figures 2.38(a)–(f). For the soils with the same coefficient of permeability and the same desaturation rate, subjected to the same magnitude of rainfall flux, the one with a lower air-entry value has a greater water storage capacity than the one with a larger air-entry value. Thus, the movement of the wetting front for the previous soil is much slower than that for the latter one.

On the other hand, if the soils have the same SWCC (i.e., the same water storage function), the one with a smaller saturated coefficient of permeability allows less infiltration than the one with a larger coefficient of permeability although both have the same water storage capacity. That is why the rate of downward movement of the wetting front in a soil with a k_s equal to 10^{-7} m/s is much smaller than that in a soil with a k_s equal to 10^{-3} m/s as shown in Figures 2.40(a) and (c).

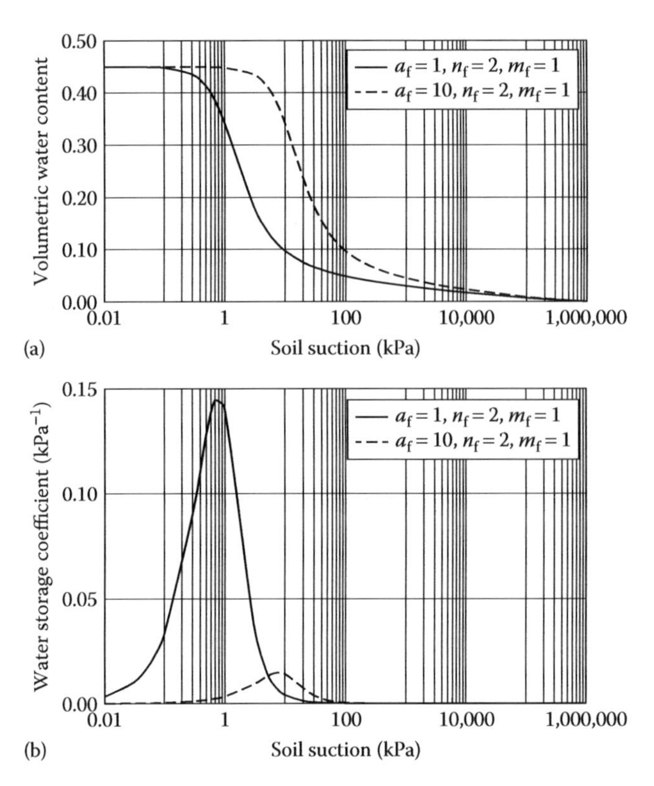

Figure 2.41 Comparison of water storage functions for soils with (a) $a_f = 1$ and (b) $a_f = 10$. (From Zhang, L. L. et al., *Canadian Geotechnical Journal*, 41, 569–582, Copyright 2004 Canadian Science Publishing or its licensors. Reproduced with permission.)

2.7.2.4 Effect of the groundwater boundary conditions

In all the previous cases, it was assumed that the groundwater table in the slope did not vary during the rainstorms. However, in reality, the pore-water pressure profiles and the groundwater tables in soil slopes can be significantly influenced by varying boundary conditions.

Figures 2.42(a) and (b) show the pore-water pressure profiles when the groundwater table is allowed to rise (i.e., the constant pressure boundary condition for the groundwater table inside the slope in Figure 2.32 is removed). Comparing Figure 2.42(a) with Figure 2.38(c) and comparing Figure 2.42(b) with Figure 2.38(e), it can be observed that the rates of the downward movement of the wetting front are comparable. The rising of the groundwater table is more marked for the soils with a larger air-entry value because the soil has a smaller water storage capacity.

All the pore-water pressure profiles presented earlier are at section X–X in the middle part of the slope (Figure 2.32). However, the pore-water pressure distribution in the entire slope needs to be analyzed to better understand the overall permanency of matric suction. Figure 2.43 illustrates an example of pore-water pressure contours in a soil slope subjected to a rainfall flux q equal to 10^{-5} m/s. While absolute pore-water pressure values vary from one cross section to another, the change of pore-water pressure with time in the entire slope is similar to that presented for the X–X section.

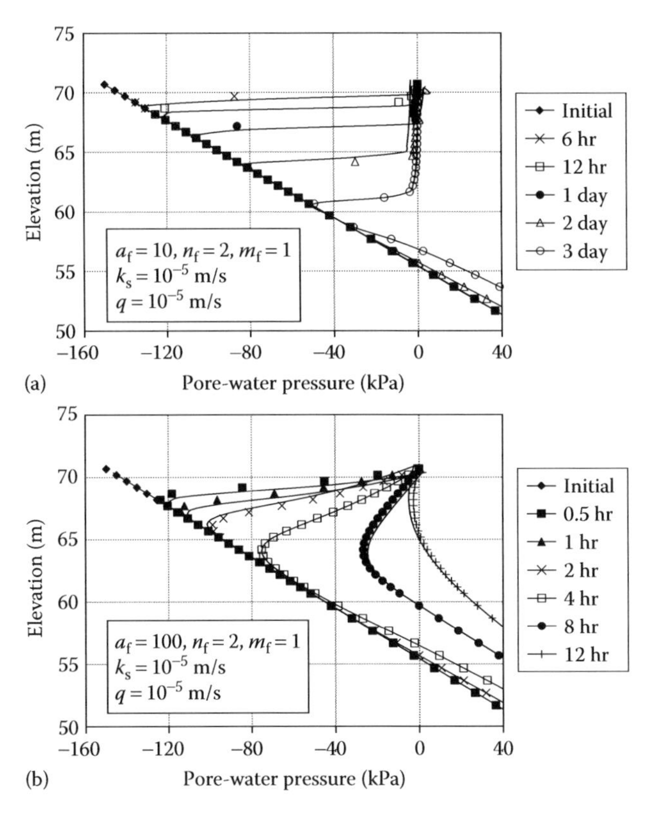

Figure 2.42 Pore-water pressure profiles in a slope with free water table subjected to a rainfall flux of $q = 10^{-5}$ m/s: (a) $a_f = 10$ kPa and (b) $a_f = 100$ kPa. (From Zhang, L. L. et al., *Canadian Geotechnical Journal*, 41, 569–582, Copyright 2004 Canadian Science Publishing or its licensors. Reproduced with permission.)

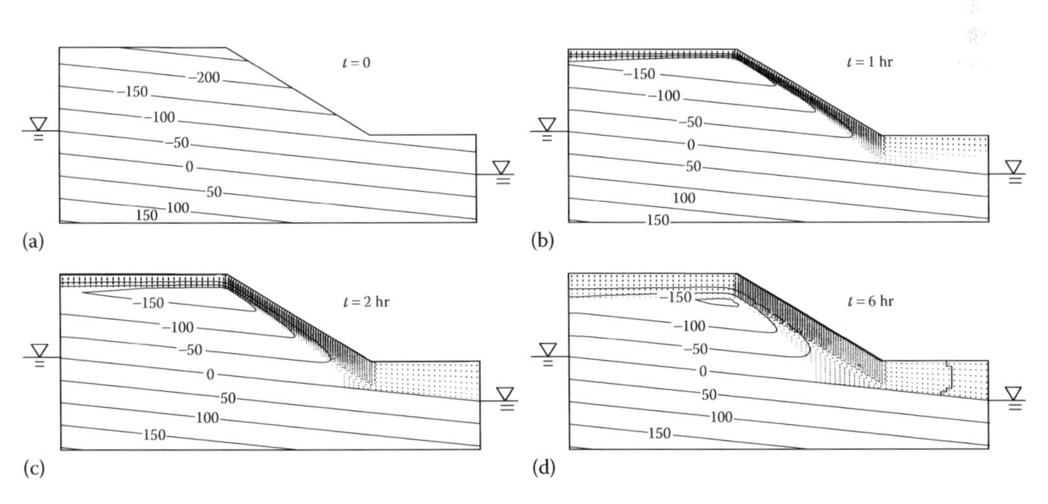

Figure 2.43 Pore-water pressure contours (in kPa) in a slope subjected to a surface flux of $q = 10^{-5}$ m/s with a soil of $k_s = 10^{-5}$ m/s, $a_f = 100$, $n_f = 2$, and $m_f = 1$: (a) $t = 0$, (b) $t = 1$ hr, (c) $t = 2$ hr, and (d) t = 6 hr.

2.7.3 Geotechnical engineering implications

The long-term matric suction changes in a slope are controlled by factors related to the ground surface moisture flux as well as the hydraulic properties of the soils near the ground surface. The ratio of the ground surface moisture flux to the saturated coefficient of permeability of the soil near the ground surface is the primary variable to be considered when assessing the potential permanence of matric suction. To maintain negative pore-water pressures in a slope, one may reduce the infiltration flux through the use of a suitable cover system at the ground surface. A common practice in Hong Kong is to provide a layer of soil–cement–lime plaster cover called "chunam" on the soil slopes. Lim et al. (1996) carried out a field instrumentation program to monitor negative pore-water pressures in residual soil slopes in Singapore that were protected by different types of surface covers. The changes in matric suction due to ground surface moisture flux were found to be least significant under a canvas-covered slope and most significant in a bare slope. Other relatively impermeable surface covers can be adopted depending on the saturated coefficients of permeability of the surface materials on the slope.

It is also effective to maintain negative pore-water pressures in a slope through reducing the saturated permeability of the surface soil. After the 1976 Sau Mau Ping failure, the investigation panel (Government of Hong Kong, 1976) recommended that a minimum stabilization requirement of a loose slope should consist of removing the loose surface soil by excavating a vertical depth of not less than 3 meters, and re-compacting at 95% of standard compaction density. For the Sau Mau Ping fill material, the average saturated coefficient of permeability decreases by three times when the relative compaction of the fill was increased from 82% to 95%.

It should be noted that variability of in situ hydraulic properties of a soil may influence the distribution of matric suction in a soil slope. For example, the average annual rainfall in Hong Kong is about 2200 mm. If the rainfall is averaged over one year, the rainfall flux intensity is approximately 7×10^{-8} m/s. For a typical soil slope comprising completely decomposed rhyolites, this averaged rainfall flux intensity would correspond to 0.7% to 70% of the saturated permeability that might vary from 10^{-5} to 10^{-7} m/s due to spatial variability of soil properties.

REFERENCES

Basha, H. A. (1999). Multidimensional linearised nonsteady infiltration with prescribed boundary conditions at the soil surface. *Water Resources Research*, 35(1), 75–83.

Basha, H. A. (2000). Multidimensional linearised nonsteady infiltration toward a shallow water table. *Water Resources Research*, 36(9), 2567–2573.

Beven, K., and Horton, R. E. (2004). Robert E. Horton's perceptual model of infiltration processes. *Hydrological Processes*, 18(17), 3447–3460.

Blatz, J. A., Ferreira, N. J., and Graham, J. (2004). Effects of near-surface environmental conditions on instability of an unsaturated soil slope. *Canadian Geotechnical Journal*, 41(6), 1111–1126.

Bouloutas, E. T. (1989). *Improved Numerical Approximations for Flow and Transport in the Unsaturated Zone*. PhD Thesis, Massachusetts Institute of Technology, Cambridge.

Brooks, R. H., and Corey, A. T. (1964). *Hydraulic Properties of Porous Medium*. Colorado State University, Fort Collins, CO.

Burdine, N. T. (1953). Relative permeability calculations from pore size distribution data. *Transactions of the American Institute of Mining, Metallurgical, and Petroleum Engineers*, 5(3), 71–78.

Celia, M. A., Bououtas, E. T., and Zarba, R. L. (1990). A general mass-conservative numerical solution for the unsaturated flow equation. *Water Resources Research*, 26(7), 1483–1496.

Chen, J. M., Tan, Y. C., and Chen, C. H. (2001). Multidimensional infiltration with arbitrary surface fluxes. *Journal of Irrigation and Drainage Engineering*, 127(6), 370–377.
Chen, L., and Young, M. H. (2006). Green–Ampt infiltration model for sloping surfaces. *Water Resources Research*, 42, 1–9.
Chen, Q., and Zhang, L. M. (2006). Three-dimensional analysis of water infiltration into the Gouhou rock-fill dam using saturated-unsaturated seepage theory. *Canadian Geotechnical Journal*, 43(5), 449–461.
Cheng, Y. G., Phoon, K. K., and Tan, T. S. (2008). Unsaturated soil seepage analysis using a rational transformation method with under-relaxation. *International Journal of Geomechanics*, 8(3), 207–212.
Childs, E. C., and Collis-George, N. (1950). The permeability of porous material. *Proceedings of the Royal Society of London A: Mathematical, Physical and Engineering Sciences*, 201(1066), 392–405.
Cho, S. E. (2009). Infiltration analysis to evaluate the surficial stability of two-layered slopes considering rainfall characteristics. *Engineering Geology*, 105(1–2), 32–43.
Chow, V. T., Maidment, D. R., and Mays, L. W. (1988). *Applied Hydrology*. McGraw-Hill, New York.
Chui, T. F. M., and Freyberg, D. L. (2009). Implementing hydrologic boundary conditions in multiphysics model. *Journal of Hydrologic Engineering*, 14(12), 1374–1377.
Collins, B. D., and Znidarcic, D. (2004). Stability analyses of rainfall induced landslides. *Journal of Geotechnical and Geoenvironmental Engineering*, 130(4), 362–372.
Fredlund, D. G., and Rahardjo, H. (1993). *Soil Mechanics for Unsaturated Soils*. Wiley, New York.
Fredlund, D. G., Rahardjo, H., and Fredlund, M. D. (2012). *Unsaturated Soil Mechanics in Engineering Practice*. Wiley, Hoboken, NJ.
Fredlund, D. G., and Xing, A. Q. (1994). Equations for the soil-water characteristic curve. *Canadian Geotechnical Journal*, 31(4), 521–532.
Fredlund, D. G., Xing, A. Q., and Huang, S. Y. (1994). Predicting the permeability function for unsaturated soils using the soil-water characteristic curve. *Canadian Geotechnical Journal*, 31(4), 533–546.
Freeze, R. A. (1969). The mechanism of natural ground-water recharge and discharge 1. One-dimensional, vertical, unsteady, unsaturated flow above a recharging or discharging ground-water flow system. *Water Resources Research*, 5(1), 153–171.
Gardner, W. R. (1958). Some steady state solutions of the unsaturated moisture flow equation with application to evaporation from a water table. *Soil Science*, 85(4), 228–232.
Gasmo, J. M., Rahardjo, H., and Leong, E. C. (2000). Infiltration effects on stability of a residual soil slope. *Computers and Geotechnics*, 26(2), 145–165.
Geo-slope Ltd. (2012). *Seep/W for Finite Element Seepage Analysis, User's Guide*. Geo-slope Ltd., Calgary, Alberta, Canada.
Gerscovich, D. M. S. (1994). *Flow Through Saturated-unsaturated Porous Media: Numerical Modelling and Slope Stability Studies of Rio de Janeiro Natural Slopes*. Ph.D Thesis, Catholic University of Rio de Janeiro, Rio de Janeiro, Brazil.
Government of Hong Kong (1976). *Report on the Slope Failures at Sau Mau Ping 25th August 1976*. Hong Kong Government Printer, Hong Kong.
Green, W. H., and Ampt, C. A. (1911). Studies on soil physics, 1: Flow of air and water through soils. *Journal of Agricultural Science*, 4, 1–24
Griffiths, D. V., and Lu, N. (2005). Unsaturated slope stability analysis with steady infiltration or evaporation using elasto-plastic finite elements. *International Journal of Numerical and Analytical Methods in Geomechanics*, 29(3), 249–267.
Holtan, H. N. (1961). *A Concept of Infiltration Estimates in Watershed Engineering* (ARS41-51). US Department of Agricultural Service, Washington, DC.
Horton, R. E. (1938). The interpretation and application of runoff plot experiments with reference to soil erosion problems. *Proceedings of Soil Science Society of America*, 3, 340–349.
Huang, R. Q., and Wu, L. Z. (2012). Analytical solutions to 1-D horizontal and vertical water infiltration in saturated/unsaturated soils considering time-varying rainfall. *Computers and Geotechnics*, 39, 66–72.
Huyakorn, P. S., Thomas, S. D., and Thompson, B. M. (1984). Techniques for making finite elements competitive in modeling flow in variably saturated porous media. *Water Resources Research*, 20(8), 1099–1115.

Iverson, R. M. (2000). Landslide triggering by rain infiltration. *Water Resources Research*, 36(7), 1897–1910.

Ju, S. H., and Kung, K. J. S. (1997). Mass types, element orders and solution schemes for the Richards equation. *Computers and Geosciences*, 23(2), 175–187.

Karthikeyan, M., Tan, T. S., and Phoon, K. K. (2001). Numerical oscillation in seepage analysis of unsaturated soils. *Canadian Geotechnical Journal*, 38(3), 639–651.

Kasim, F. B. (1997). *Effects of Steady State Rainfall on Long Term Matric Suction Conditions in Slopes*. Unsaturated Soils Group, University of Saskatchewan, Saskatoon, Saskatchewan, Canada.

Kisch, M. (1959). The theory of seepage from clay-blanked reservoirs. *Géotechnique*, 9, 9–21.

Klausner, Y. (1991). *Fundamentals of Continuum Mechanics of Soils*. Springer-Verlag, Berlin, Germany.

Kwok, S. Y. F., and Tung, Y. K. (2003). Uncertainty and sensitivity analysis of coupled surface and subsurface flows. *Groundwater Quality Modeling And Management Under Uncertainty: Proceedings of Symposium on Probabilistic Approaches and Groundwater Modeling, 2003 EWRI World Congress*. Philadelphia, PA, pp. 58–71.

Lambe, T. W., and Whitman, R. V. (1969). *Soil Mechanics*. Wiley, Hoboken, NJ.

Lee, L. M., Gofar, N., and Rahardjo, H. (2009). A simple model for preliminary evaluation of rainfall-induced slope instability. *Engineering Geology*, 108(3–4), 272–285.

Leong, E. C., and Rahardjo, H. (1997a). Permeability functions for unsaturated soils. *Journal of Geotechnical and Geoenvironmental Engineering*, 123(12), 1118–1126.

Leong, E. C., and Rahardjo, H. (1997b). Review of soil-water characteristic curve equations. *Journal of Geotechnical and Geoenvironmental Engineering*, 123(12), 1106–1117.

Li, A. G., Yue, Z. Q., Tham, L. G., Lee, C. F., and Law, K. T. (2005). Field-monitored variations of soil moisture and matric suction in a saprolite slope. *Canadian Geotechnical Journal*, 42(1), 13–26.

Li, R. H., Simons, D. B., and Stevens, M. A. (1976). Solution to Green–Ampt infiltration equation. *Journal of the Irrigation and Drainage Division*, 102 (IR2), 239–248.

Lim, T. T., Rahardjo, H., Chang, M. F., and Fredlund, D. G. (1996). Effect of rainfall on matric suctions in a residual soil slope. *Canadian Geotechnical Journal*, 33(4), 618–628.

Lin, H. C., Richards, D. R., Talbot, C. A., Yeh, G. T., Cheng, J. R., Cheng, H. P., and Jones, N. L. (1997). *FEMWATER: A Three-dimensional Finite Element Computer Model for Simulation of Density-dependent Flow and Transport in Variably Saturated Media* (Technical Report CHL-97–12), United States Waterways Experiment Station, Coastal and Hydraulics Laboratory, Vicksburg, MS.

McKee, C. R., and Bumb, A. C. (1984). The importance of unsaturated flow parameters in designing a monitoring system for hazardous wastes and environmental emergencies. *Proceedings of Hazardous Materials Control Research Institute National Conference*. Houston, TX, pp. 50–58.

McKee, C. R., and Bumb, A. C. (1987). Flow-testing coalbed methane production wells in the presence of water and gas. *SPE Formation Evaluation*, 2(04), 599–608.

Mehl, S. (2006). Use of Picard and Newton iteration for solving nonlinear ground water flow equations. *Ground Water*, 44(4), 583–594.

Mein, R. G., and Larson, C. L. (1973). Modelling infiltration during a steady rain. *Water Resources Research*, 9(2), 384–394.

Mishra, S. K., Tyagi, J. V., and Singh, V. P. (2003). Comparison of infiltration models. *Hydrological Processes*, 17(13), 2629–2652.

Mualem, Y. (1976). A new model for predicting the hydraulic conductivity of unsaturated porous media. *Water Resources Research*, 12(3), 593–622.

Ng, C. W. W., and Bruce, K. M. (2007). *Advanced Unsaturated Soil Mechanics and Engineering*. Taylor and Francis, Routledge, London.

Ng, C. W. W., and Shi, Q. (1998a). Influence of rainfall intensity and duration on slope stability in unsaturated soils. *Quarterly Journal of Engineering Geology*, 31(2), 105–113.

Ng, C. W. W., and Shi, Q. (1998b). Numerical investigation of the stability of unsaturated soil slopes subjected to transient seepage. *Computers and Geotechnics*, 22(1), 1–28.

Ng, C. W. W., Wang, B., and Tung, Y. K. (2001). Three-dimensional numerical investigations of groundwater responses in an unsaturated slope subjected to various rainfall patterns. *Canadian Geotechnical Journal*, 38(5), 1049–1062.

Pan, L., Warrick, A. W., and Wierenga, P. J. (1996). Finite element methods for modeling water flow in variably saturated porous media: Numerical oscillation and mass-distributed schemes. *Water Resources Research*, 32(6), 1883–1889.

Paniconi, C., and Putti, M. (1994). A comparison of Picard and Newton iteration in the numerical solution of multidimensional variably saturated flow problems. *Water Resources Research*, 30(12), 335–3374.

Philip, J. R. (1991). Hillslope infiltration: Planar slopes. *Water Resources Research*. 27(1), 109–117.

Rahardjo, H., Lee, T., Leong, E., and Rezaur, R. (2005). Response of a residual soil slope to rainfall. *Canadian Geotechnical Journal*, 42(2), 340–351.

Rahardjo, H., Lim, T. T., Chang, M. F., and Fredlund, D. G. (1995). Shear strength characteristics of a residual soil. *Canadian Geotechnical Journal*, 32(1), 60–77.

Rahardjo, H., Ong, T. H., Rezaur, R. B., and Leong, E. C. (2007). Factors controlling instability of homogeneous soil slopes under rainfall. *Journal of Geotechnical and Geoenvironmental Engineering*, 133(12), 1532–1543.

Rahimi, A., Rahardjo, H., and Leong, E. C. (2010). Effect of hydraulic properties of soil on rainfall-induced slope failure. *Engineering Geology*, 114(3–4), 135–143.

Rathfelder, K., and Abriola, L. M. (1994). Mass conservative numerical solutions of the head-based Richards equation. *Water Resources Research*, 30(9), 2579–2586.

Richards, L. A. (1931). Capillary conduction of liquids through porous mediums. *Physics*, 1(5), 318–333.

Sandhu, R. S., Liu, H., and Singh, K. J. (1977). Numerical performance of some finite element schemes for analysis of seepage in porous elastic media. *International Journal of Numerical and Analytical Methods in Geomechanics*, 1(2), 177–194.

Segerlind, L. J. (1984). *Applied Finite Element Analysis*. Wiley, New York.

Serrano, S. E. (2004). Modeling infiltration with approximate solutions to Richard's equation. *Journal of Hydrologic Engineering*, 9(5):421–432.

Šimůnek, J. (2006). Models of water flow and solute transport in the unsaturated zone. In: Anderson, M. G. (Ed.). *Encyclopedia of Hydrological Sciences, Vol. 6*. Wiley, Hoboken, NJ, p. 78.

Šimůnek, J., Šejna, M., and van Genuchten, M. Th. (1999). *The HYDRUS-2D Software Package for Simulating Two-dimensional Movement of Water, Heat, and Multiple Solutes in Variably Saturated Media*. International Ground Water Modeling Center, Colorado School of Mines, Golden, CO.

SoilVision System Ltd. (2009). *SVFlux User's Manual*. SoilVision System Ltd., Saskatoon, Canada.

Srivastava, R., and Yeh, T. C. J. (1991). Analytical solutions for one-dimensional, transient infiltration toward the water table in homogenous and layered soils. *Water Resources Research*, 27(5): 753–762.

Sweeney, D. J. (1982). Some in situ soil suction measurements in Hong Kong's residual soil slopes, *Proceedings of the 7th Southeast Asian Geotechnical Conference, Vol. 1*, Southeast Asian Geotechnical Society, Thailand, pp. 91–106.

Tami, D., Rahardjo, H., and Leong, E. C. (2004). Effects of hysteresis on steady-state infiltration in unsaturated slopes. *Journal of Geotechnical and Geoenvironmental Engineering*, 130(9), 956–967.

Tan, T. S., Phoon, K. K., and Chong, P. C. (2004). Numerical study of finite element method based solutions for propagation of wetting fronts. *Journal of Geotechnical and Geoenvironmental Engineering*, 130(3), 254–263.

Thomas, H. R., and Zhou, Z. (1997). Minimum time-step size for diffusion problem in FEM analysis. *International Journal for Numerical Methods in Engineering*, 40(20), 3865–3880.

Tsai, T. L., and Yang, J. C. (2006). Modelling of rainfall-triggered shallow landslide. *Environmental Geology*, 50(4), 525–534.

Tsaparas, I., Rahardjo, H., Toll, D. G., and Leong, E. C. (2002). Controlling parameters for rainfall-induced landslides. *Computers and Geotechnics*, 29(1), 1–27.

van Dam, J. C., Huygen, J., Wesseling, J. G., Feddes, R. A., Kabat, P., van Valsum, P. E. V., Groenendijk, P., and van Diepen, C. A. (1997). *Theory of SWAP, Version 2.0. Simulation of Water Flow, Solute Transport and Plant Growth in the Soil- Water-Atmosphere-Plant Environment*. Department Water Resources, WAU, Report 71, DLO Winand Staring Centre, Technical Document 45, Wageningen.

van Genuchten, M. Th. (1980). A close-form equation for predicting the hydraulic conductivity of unsaturated soils. *Soil Science Society of America Journal*, 44(5), 892–898.

van Genuchten, M. Th. (1982). A comparison of numerical solutions of the one dimensional unsaturated–saturated flow and mass transport equations. *Advances in Water Resources*, 5(1), 47–55.

Vermeer, P. A., and Verruijt, A. (1981). An accuracy conditions for consolidation by finite elements. *International Journal of Numerical and Analytical Methods in Geomechanics*, 5(1), 1–14.

Williams, G. A., Miller, C. T., and Kelley, C. T. (2000). Transformation approaches for simulating flow in variably saturated porous media. *Water Resources Research*, 36(4), 923–934.

Yuan F, and Lu Z. (2005). Analytical solutions for vertical flow in unsaturated, rooted soils with variable surface fluxes. *Vadose Zone Journal*, 4(4), 1210–1218.

Zhan, L. T., and Ng, C. W. W. (2001). Analytical analysis of rainfall infiltration mechanism in unsaturated soils. *International Journal of Geomechanics*, 4(4), 273–284.

Zhang, J., Huang, H. W., Zhang, L. M., Zhu, H. H., and Shi, B. (2014). Probabilistic prediction of rainfall-induced slope failure using a mechanics-based model. *Engineering Geology*, 168, 129–140.

Zhang, L. L., Fredlund, D. G., Zhang, L. M., and Tang, W. H. (2004). Numerical study of soil conditions under which matric suction can be maintained. *Canadian Geotechnical Journal*, 41(4), 569–582.

Zhang, L. L., Zhang, J., Zhang, L. M., and Tang, W. H. (2011). Stability analysis of rainfall-induced slope failure: A review. *ICE Proceedings: Geotechnical Engineering*, 164(GE5), 299–316.

Chapter 3

Stability analysis of slope under rainfall infiltration based on limit equilibrium

3.1 INTRODUCTION

Traditional slope stability analyses incorporate rainfall influences through changing the groundwater flow patterns with increasing pressure heads or a rising groundwater table. However, there is not much evidence of a rise of groundwater table in many shallow failures (Fourie et al., 1999). The failures are mainly attributed to the advance of a wetting front and the reduction of shear strength due to the decrease of matric suction in the unsaturated soils (Fredlund and Rahardjo, 1993; Rahardjo et al., 1995). Hence, such failures would not be properly analyzed using the traditional approaches.

Because initial failures due to rainfall infiltration often have small depth-to-length ratios and form failure planes parallel to the slope surface, the use of infinite-slope stability analysis for evaluation of rainfall-induced landslides is justified and often preferred for its simplicity. The methods used in traditional infinite-slope analysis (Skempton and Delory, 1957; Duncan and Wright, 1995) can be modified to take into account the variation of the pore-water pressure profile. Usually the pore-water pressure profile for the infinite-slope stability analysis comes from conceptual models of infiltration, calculated pore-water pressure distribution of numerical or analytical infiltration analyses, and field measured profiles. Besides the infinite-slope stability analysis method, the limit equilibrium method of slope stability, that is, methods of slices are also widely adopted. In this chapter, the method for the infinite-slope stability analysis and the two-dimensional limit equilibrium methods of slope stability for unsaturated soil slopes under rainfall infiltration together with two examples are presented in Sections 3.2 and 3.3. The controlling factors including soil shear strength parameters, hydraulic parameters, and rainfall characteristics are discussed in detail in Section 3.4. In Section 3.5, the spatially distributed models for landslide hazard assessment are introduced.

3.2 INFINITE-SLOPE STABILITY ANALYSIS BASED ON ONE-DIMENSIONAL INFILTRATION PROFILE

3.2.1 General equation of factor of safety for infinite unsaturated slope

For an infinite slope with unsaturated zone, the shear strength of unsaturated soil can be obtained based on the extended Mohr–Coulomb failure criterion (Fredlund et al., 1978):

$$\tau_f = c' + (\sigma_n - u_a)\tan\phi' + (u_a - u_w)\tan\phi^b \quad (3.1)$$

where:

τ_f is the shear strength
c' is the effective cohesion of a saturated soil
ϕ' is the effective angle of internal friction
ϕ^b is the angle indicating the rate of increase in shear strength related to matric suction
$(u_a - u_w)$ is matric suction with u_a = pore air pressure and u_w = pore-water pressure.
$(\sigma_n - u_a)$ is net normal stress on the slip surface normal with σ_n = total stress on the slip surface

The safety factor for the slip surface at depth D can be expressed as (Cho and Lee, 2001) follows:

$$F_s = \frac{c' + (\sigma_n - u_a)\tan\phi' + (u_a - u_w)\tan\phi^b}{\gamma_t D \sin\alpha_s \cos\alpha_s} \quad (3.2)$$

where:

γ_t = total unit weight of the soil
D = the depth of slip surface
α_s = slope angle

Assume that the normal stress on the slip surface can be computed from the weight of the soil:

$$(\sigma_n - u_a) = \gamma_t D \cos^2\alpha_s \quad (3.3)$$

The factor of safety, F_s, can therefore be written as

$$F_s = \frac{c'}{\gamma_t D \sin\alpha_s \cos\alpha_s} + \frac{\tan\phi'}{\tan\alpha_s} + \frac{(u_a - u_w)\tan\phi^b}{\gamma_t D \sin\alpha_s \cos\alpha_s} \quad (3.4)$$

Some researchers presented Equation 3.4 in slightly different forms, for example, using the suction stress σ^s (Lu and Likos, 2004) instead of suction (Lu and Griffiths, 2004; Lu and Godt, 2008), or considering ψ and ϕ^b as functions of H (Travis et al., 2010) or expressed the equation in terms of pressure head (Iverson, 2000; Cho and Lee, 2002; Muntohar and Liao, 2009).

3.2.2 Simplified scenarios of various seepage conditions

3.2.2.1 Steady-state condition with seepage parallel slope surface

For steady-state water flow condition, water flows through both saturated and unsaturated zones and is parallel to the phreatic line. Hence, the hydraulic head gradient is equal to zero in a direction perpendicular to the phreatic line. The matric suction along the slip surface at depth D (e.g., points A and B, as shown in Figure 3.1) can be expressed as follows (Fredlund et al., 2012):

$$\psi = \gamma_w z \cos^2\alpha_s = \gamma_w (D_w - D)\cos^2\alpha_s \quad (3.5)$$

where:

γ_w is the unit weight of water
z is the vertical coordinate with the origin from the phreatic line
D_w is the vertical depth of water table, as shown in Figure 3.1

Hence, the factor of safety for the slip surface at depth D is

$$F_s = \frac{c'}{\gamma_t D \sin\alpha_s \cos\alpha_s} + \frac{\tan\phi'}{\tan\alpha_s} + \left(\frac{D_w}{D} - 1\right)\left(\frac{\gamma_w}{\gamma_t}\right)\left(\frac{\tan\phi^b}{\tan\alpha_s}\right) \tag{3.6}$$

For the slip surface that is below the phreatic line as shown in Figure 3.2, the safety factor is as follows (Skempton and Delory, 1957; Duncan and Wright, 1995):

$$F_s = \frac{c'}{\gamma_t D \sin\alpha_s \cos\alpha_s} + \frac{\tan\phi'}{\tan\alpha_s} - \frac{(D - D_w)\gamma_w \tan\phi'}{\gamma_t D \tan\alpha_s} \tag{3.7}$$

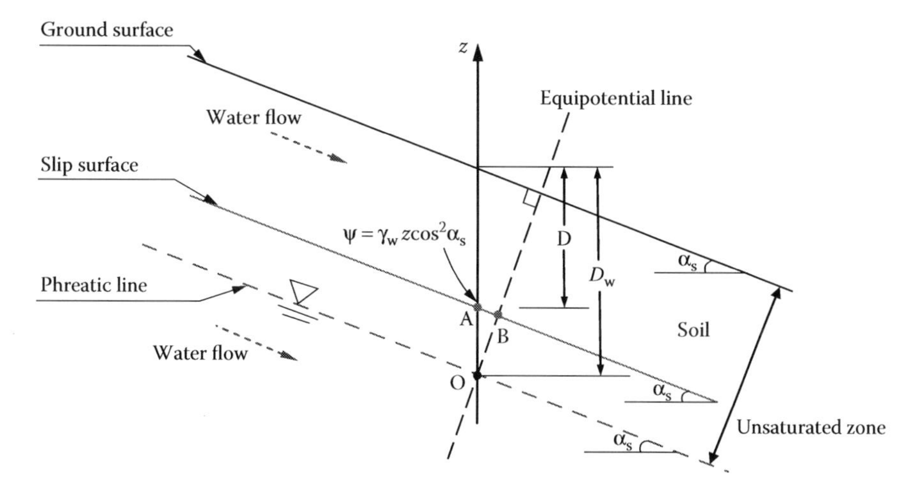

Figure 3.1 Schematic of an infinite slope with steady-state flow parallel to slope surface and a slip surface above the phreatic line.

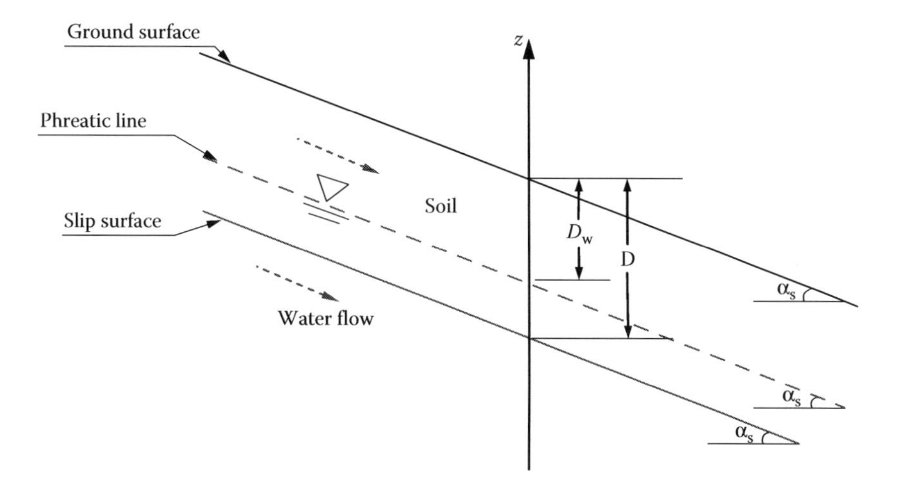

Figure 3.2 Schematic of an infinite slope with a slip surface below the phreatic line.

3.2.2.2 Transient condition with different shapes of wetting front

Rahardjo et al. (1995) presented the equations of factor of safety for three typical pore pressure profiles under rainfall infiltration (Figure 2.30). For a smooth wetting front like profile *a* in Figure 2.30:

$$F_s = \frac{c'}{\gamma_t D \sin\alpha_s \cos\alpha_s} + \frac{\tan\phi'}{\tan\alpha_s} + \left(\frac{D_w}{z_w} - 1\right)\left(\frac{\gamma_w}{\gamma_t}\right)\left(\frac{\tan\phi^b}{\tan\alpha_s}\right) \tag{3.8}$$

where z_w = depth of wetting front.

For a sharp wetting front (profile *b* in Figure 2.30):

$$F_s = \frac{c'}{\gamma_t D \sin\alpha_s \cos\alpha_s} + \frac{\tan\phi'}{\tan\alpha_s} \tag{3.9}$$

For a perched water table with positive pore pressure (profile *c* in Figure 2.30):

$$F_s = \frac{c'}{\gamma_t D \sin\alpha_s \cos\alpha_s} + \frac{\tan\phi'}{\tan\alpha_s} - \left(\frac{\gamma_w}{\gamma_t}\right)\left(\frac{\tan\phi'}{\tan\alpha_s}\right) \tag{3.10}$$

As illustrated in Chapter 2, smooth wetting fronts (profile *a*) are usually observed in slopes with fine materials while sharp wetting front (profile *b*) often occurs in soil slopes with coarse materials. The perched water table in unsaturated zone is sometimes observed in layered soil slopes where a less permeable soil layer exists.

3.2.3 Slope stability based on estimated nonlinear unsaturated shear strength

The linear increase in unsaturated shear strength in accordance with the angle ϕ^b had been based on a limited number of datasets that were available in the research literature in the 1970s. Over time it was found that the shear strength of an unsaturated soil had a nonlinear form for soils tested over a wide range of soil suctions (Donald, 1956; Escario and Saez, 1986; Fredlund et al., 1987). The nonlinearity of the shear strength curve was noticed to bear a clear relationship with the soil–water characteristic curve (SWCC) (Fredlund et al., 1996; Vanapalli et al., 1996). Figure 3.3 shows a schematic of the general anticipated unsaturated shear strength envelopes for a typical soil. The shear strength envelope for all soil types appears to respond as a saturated soil when the matric suction is less than the air-entry value of the soil. The shear strength function begins to curve once the air-entry value is exceeded. In most cases there is an increase in shear strength with an increase in soil suction beyond the air-entry value. The unsaturated shear strength envelope bends toward a near horizontal line at the residual suction for soils with considerable silt or clay content. Sandy soils generally show a leveling off in strength even prior to the residual suction being reached and can tend to decrease in strength at higher soil suctions.

Several empirical estimation equations have been proposed relating the unsaturated shear strength to the SWCC (Fredlund et al., 1996; Vanapalli et al., 1996; Oberg and Sallfours, 1997).

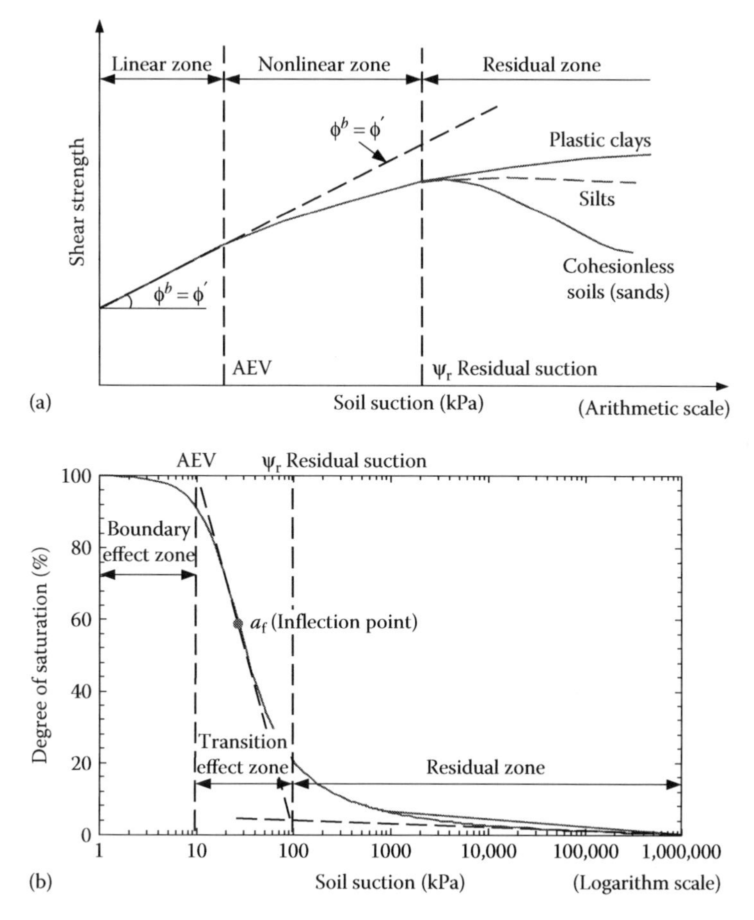

Figure 3.3 Relationship between the unsaturated shear strength envelope and the SWCC; (a) typical unsaturated shear strength envelopes and (b) SWCC for a typical soil.

Other unsaturated shear strength equations such as those proposed by Khalili and Khabbaz (1998), Bao et al. (1998), and Sheng et al. (2008) make use of the air-entry value of the soil but not the entire SWCC.

3.2.3.1 Fredlund et al. nonlinear shear strength equation

Fredlund et al. (1996) proposed a nonlinear shear strength equation, which incorporated the SWCC written in terms of dimensionless water content, Θ_d, as follows:

$$\tau_f = c' + (\sigma_n - u_a)\tan\phi' + (u_a - u_w)\Theta_d^{\kappa_f}\tan\phi' \tag{3.11}$$

where:

Θ_d is the dimensionless water content defined as (θ_w/θ_s), where θ_w is the volumetric water content and θ_s is the saturated volumetric water content

κ_f is the fitting parameter

Garven and Vanapalli (2006) established the correlation between the fitting parameter, κ_f, and the plastic index, I_p, of a soil based on 10 datasets of statically compacted soils:

$$\kappa_f = -0.0016 I_p^2 + 0.0975 I_p + 1 \tag{3.12}$$

where I_p is the plasticity index (%). Once the κ_f parameter is estimated, the Fredlund et al. (1996) equation can be used to describe the unsaturated shear strength.

3.2.3.2 Vanapalli et al. nonlinear shear strength equation

Vanapalli et al. (1996) proposed a nonlinear shear strength equation using a normalization of the SWCC between the saturated and residual soil conditions:

$$\tau_f = c' + (\sigma_n - u_a)\tan\phi' + (u_a - u_w)\left[(\tan\phi')\left(\frac{\theta_w - \theta_r}{\theta_s - \theta_r}\right)\right] \tag{3.13}$$

or

$$\tau_f = c' + (\sigma_n - u_a)\tan\phi' + (u_a - u_w)\left[(\tan\phi')\left(\frac{S - S_r}{100 - S_r}\right)\right] \tag{3.14}$$

where:

θ_r is the residual volumetric water content
S is the degree of saturation
S_r is the residual degree of saturation

The shear strength equation shows a nonlinear reduction from the saturated to residual soil conditions. According to Equations 3.13 and 3.14, the shear strength contributed by soil suction becomes zero once soil suction is greater than the residual suction.

3.2.3.3 Vilar nonlinear shear strength equation

Vilar (2006) proposed the use of an empirical hyperbolic function to fit experimental data of saturated and unsaturated shear strength. The total cohesion of the soil (i.e., effective cohesion plus apparent cohesion due to matric suction) was assumed to be a hyperbolic equation of matric suction as follows:

$$c_{tot} = c' + \frac{u_a - u_w}{a_{vilar} + b_{vilar}(u_a - u_w)} \tag{3.15}$$

where:

c_{tot} is the total cohesion
a_{vilar} and b_{vilar} are the fitting parameters

Let us assume that the slope of the shear strength envelope is equal to $\tan\phi'$ when matric suction approaches zero and that there is no significant change in the shear strength when the soil suction is greater than the residual suction. The fitting parameters (a_{vilar} and b_{vilar}) can then be obtained as follows:

$$\left.\frac{dc_{tot}}{d\psi}\right|_{\psi\to 0} = \frac{1}{a_{vilar}} = \tan\phi' \tag{3.16}$$

$$\lim_{\psi\to\infty} c_{\text{tot}} = c_{\text{ult}} = c' + \frac{1}{b_{\text{vilar}}} \tag{3.17}$$

where c_{ult} is the ultimate undrained shear strength of air-dried soil sample.

It should be noted that soil parameters from the SWCC are not directly used in the Vilar (2006) model. One shear strength test on a soil specimen that is at or above the residual suction state together with saturated shear strength parameters is required for the Vilar (2006) model. The Vilar (2006) model is initially presented as a fitting model, but with a measured or reasonably assumed c_{ult} value, the model can be used for estimation of unsaturated shear strength.

3.2.3.4 *Khalili and Khabbaz nonlinear shear strength equation*

Khalili and Khabbaz (1998) assumed that the suction component of shear strength was reduced by multiplying soil suction by the variable λ_{kk}, and the shear strength equation was written as

$$\tau = c' + (\sigma_{\text{n}} - u_{\text{a}})\tan\phi' + (u_{\text{a}} - u_{\text{w}})\lambda_{kk}\tan\phi' \tag{3.18}$$

The parameter λ_{kk} was defined as

$$\lambda_{kk} = \begin{cases} 1.0 & (u_{\text{a}} - u_{\text{w}}) \le \psi_{\text{aev}} \\ \left[\dfrac{u_{\text{a}} - u_{\text{w}}}{\psi_{\text{aev}}}\right]^{-0.55} & (u_{\text{a}} - u_{\text{w}}) > \psi_{\text{aev}} \end{cases} \tag{3.19}$$

where ψ_{aev} is the air-entry value of the soil, which is a suction value beyond which the soil starts to desaturate. According to Equation 3.19, the λ_{kk} value is 1.0 for shear strengths up to the air-entry value and then decreases without any influence from residual suction. Therefore, the soil behaved as a saturated soil as long as the matric suction was less than the air-entry value. Once the air-entry value was exceeded, soil suction always provides a positive increase in strength as λ_{kk} is always greater than zero. The rate of increasing of shear strength due to soil suction is simply influenced by the air-entry value of the soil.

3.2.3.5 *Bao et al. nonlinear shear strength equation*

The Bao et al. (1998) nonlinear shear strength equation can be expressed as follows:

$$\tau = c' + (\sigma_{\text{n}} - u_{\text{a}})\tan\phi' + (u_{\text{a}} - u_{\text{w}})\zeta_{\text{bao}}\tan\phi' \tag{3.20}$$

The parameter ζ_{bao} was defined based on the air-entry value and residual suction of an unsaturated soil:

$$\zeta_{\text{bao}} = \begin{cases} 1.0 & (u_{\text{a}} - u_{\text{w}}) \le \psi_{\text{aev}} \\ \dfrac{\log\psi_{\text{r}} - \log(u_{\text{a}} - u_{\text{w}})}{\log\psi_{\text{r}} - \log(\psi_{\text{aev}})} & \psi_{\text{aev}} < (u_a - u_{\text{w}}) < \psi_{\text{r}} \\ 0 & (u_{\text{a}} - u_{\text{w}}) \ge \psi_{\text{r}} \end{cases} \tag{3.21}$$

where ψ_{r} is the residual suction.

According to Equation 3.21, the soil behaved as a saturated soil as long as the matric suction was less than the air-entry value of the soil. The influence of soil suction on the shear strength of an unsaturated soil was normalized between the air-entry value and residual suction.

Table 3.1 Apparent cohesion ($c(u_a - u_w)$ or $c(\psi)$) contributed by soil suction

Unsaturated shear strength equation	*Apparent cohesion due to soil suction*
Fredlund et al. (1978)	$c(u_a - u_w) = (u_a - u_w)\tan\phi^b$
Fredlund et al. (1996)	$c(u_a - u_w) = (u_a - u_w)\Theta_d^{\kappa f}\tan\phi'$
Vanapalli et al. (1996)	$c(u_a - u_w) = (u_a - u_w)\left[(\tan\phi')\left(\frac{\theta_w - \theta_r}{\theta_s - \theta_r}\right)\right]$
	$c(u_a - u_w) = (u_a - u_w)\left[(\tan\phi')\left(\frac{S - S_r}{100 - S_r}\right)\right]$
Vilar (2006)	$c(u_a - u_w) = \frac{u_a - u_w}{a_{vilar} + b_{vilar}(u_a - u_w)}$ where $a_{vilar} = \frac{1}{\tan\phi'}$, $b_{vilar} = \frac{1}{(c_{ult} - c')}$
Khalili and Khabbaz (1998)	$c(u_a - u_w) = (u_a - u_w)\lambda_{kk}\tan\phi'$ where $\lambda_{kk} = \begin{cases} 1.0 & (u_a - u_w) \le \psi_{aev} \\ \left[\frac{u_a - u_w}{\psi_{aev}}\right]^{-0.55} & (u_a - u_w) > \psi_{aev} \end{cases}$
Bao et al. (1998)	$c(u_a - u_w) = (u_a - u_w)\left[\zeta_{bao}\right]\tan\phi'$ where $\zeta_{bao} = \begin{cases} 1.0 & (u_a - u_w) \le \psi_{aev} \\ \frac{\log\psi_r - \log(u_a - u_w)}{\log\psi_r - \log(\psi_{aev})} & \psi_{aev} < (u_a - u_w) < \psi_r \\ 0 & (u_a - u_w) \ge \psi_r \end{cases}$

Notes: θ_w is the volumetric water content; θ_r is the residual volumetric water content; θ_s is the saturated volumetric water content; S is the degree of saturation, and S_r is the residual degree of saturation; Θ_d is the dimensionless water content defined as (θ_w/θ_s); C_{ult} is the ultimate undrained shear strength of air-dried soil sample; ψ_{aev} is the air-entry value of the soil; ψ_r is the residual suction.

To consider the nonlinearity of unsaturated shear strength, the factor of safety F_s of an infinite slope with unsaturated zone can be written as in a more general form:

$$F_s = \frac{c'}{\gamma_t H \sin\alpha_s \cos\alpha_s} + \frac{\tan\phi'}{\tan\alpha_s} + \frac{c(\psi)}{\gamma_t H \sin\alpha_s \cos\alpha_s} \tag{3.22}$$

where $c(\psi)$ or $c(u_a - u_w)$ is the function of apparent cohesion due to soil suction. The equations of apparent cohesion due to soil suction corresponding to the different shear strength models are presented in Table 3.1.

Example 3.1

An example of infinite-slope stability analysis based on 1D infiltration analysis.

Let us consider an infinite slope with a soil layer of 6 m thick. The slope angle is 35°. The groundwater table is 6 m below the slope surface and the initial pore-water pressure distribution in the slope is hydrostatic. The unit weight of the soil in the slope is assumed to be 20 kN/m^3. The effective angle of internal friction, ϕ' and the effective cohesion, c' are 34° and 2 kPa, respectively.

1. Hydrostatic condition
   Figure 3.4 shows the hydrostatic pore-water pressure profile and the factor of safety with respect to depth of soil using Equation 3.6. Three different values of ϕ^b are assumed. The difference of factor of safety between the cases of ignoring matric suction (i.e., $\phi^b = 0$) and the case assuming $\phi^b = 15°$ is significant, especially for shallow slip surfaces. For example, when the slip surface is at a depth of 2 m (elevation = 4 m), the factor of safety when ignoring matric suction (i.e., $\phi^b = 0$) is only 1.07 while the assumption of ϕ^b equal to 15° yields a factor of safety of 1.44.

2. Transient condition under rainfall infiltration
   Assume that the van Genuchten (1980) and Mualem (1976) (VGM) model is adopted for the SWCC and permeability function of unsaturated soil. The input soil parameters are listed in Table 3.2. The rainfall intensity q is 2×10^{-5} m/s. Figures 3.5

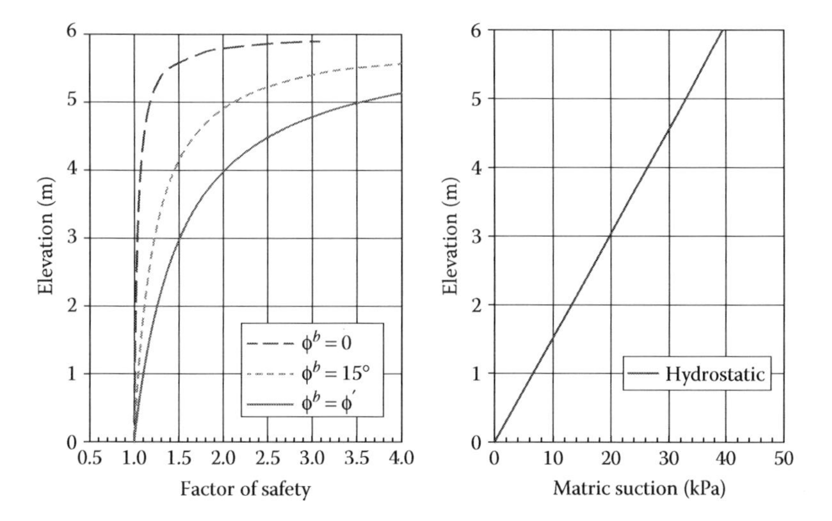

Figure 3.4 Pore-water pressure profile of hydrostatic condition and factor of safety of an infinite slope ($\alpha_s = 35°$, $\gamma_t = 20$ kN/m^3, $\phi' = 34°$, $c' = 2$ kPa) (Example 3.1).

Table 3.2 Soil parameters for the infinite-slope stability analysis under transient rainfall condition (Example 3.1)

Parameters (Unit)	*Values*
ϕ' (°)	34
c' (kPa)	2
ϕ^b (°)	15
θ_r	0.10
θ_s	0.40
a_{vgm}	0.06, 0.08, 0.10
n_{vgm}	2
k_s (m/s)	5×10^{-6}, 2×10^{-5}, 8×10^{-5}
q (m/s)	2×10^{-5}
γ_t (kN/m^3)	20
Slope angle α_s (°)	35

and 3.6 present the pore-water pressure profiles and the corresponding factor of safety for the infinite slope for different values of k_s and a_{vgm}. The infiltration analysis is conducted using a finite element model established on the FE platform COMSOL. As shown in Figures 3.5(a) and (b), when k_s is less than or equal to rainfall intensity q, the unsaturated zone can be fully saturated as long as the rain duration is long enough. The penetration of wetting front is faster in a soil slope with higher value of k_s. When k_s is greater than the rainfall intensity q, the matric suction can be slightly maintained and the factor of safety of slope stability may be kept to a level above the dangerous level (Figure 3.5c). In Figure 3.6, the effects of a_{vgm} on wetting front and factor of safety are presented. With a lower value of a_{vgm} (higher AEV, finer soil), the moving of wetting front is faster and the factor of safety is reduced faster.

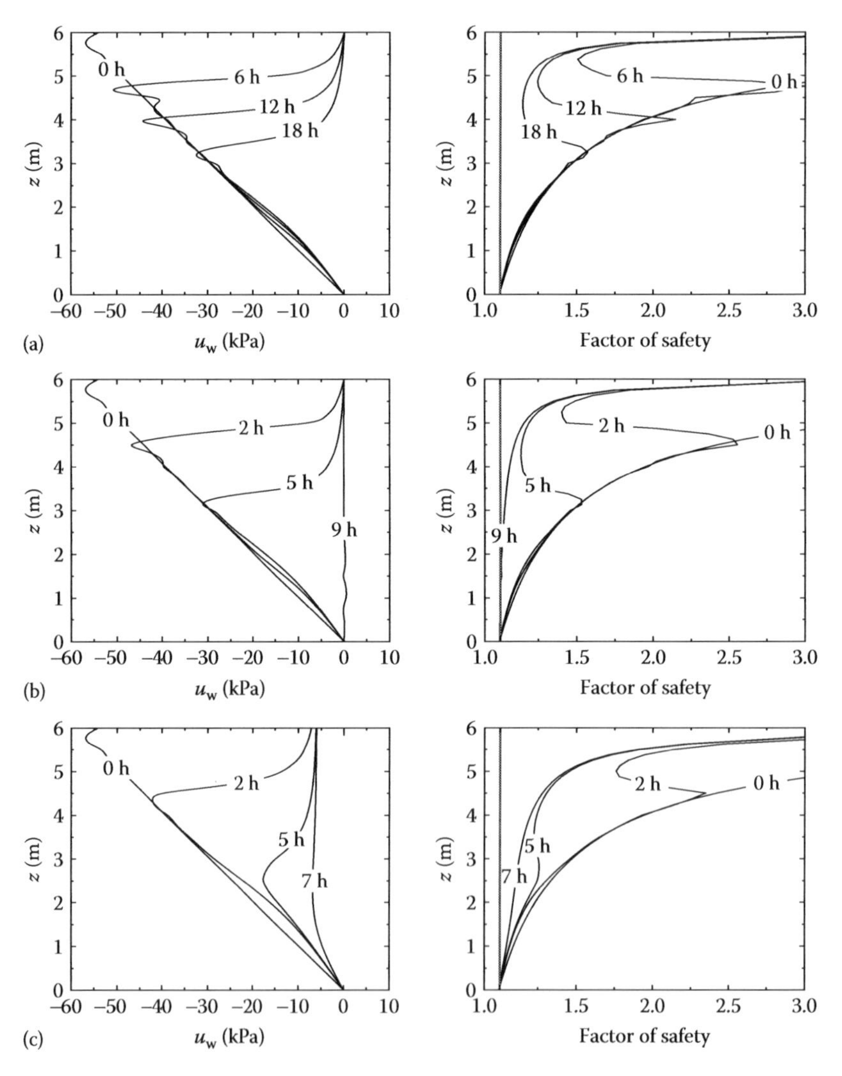

Figure 3.5 Transient pore-water pressure profile and factor of safety of a 35° infinite slope ($q = 2 \times 10^{-5}$ m/s, $a_{vgm} = 0.08$): (a) $k_s = 5 \times 10^{-6}$ m/s; (b) $k_s = 2 \times 10^{-5}$ m/s; and (c) $k_s = 8 \times 10^{-5}$ m/s (Example 3.1).

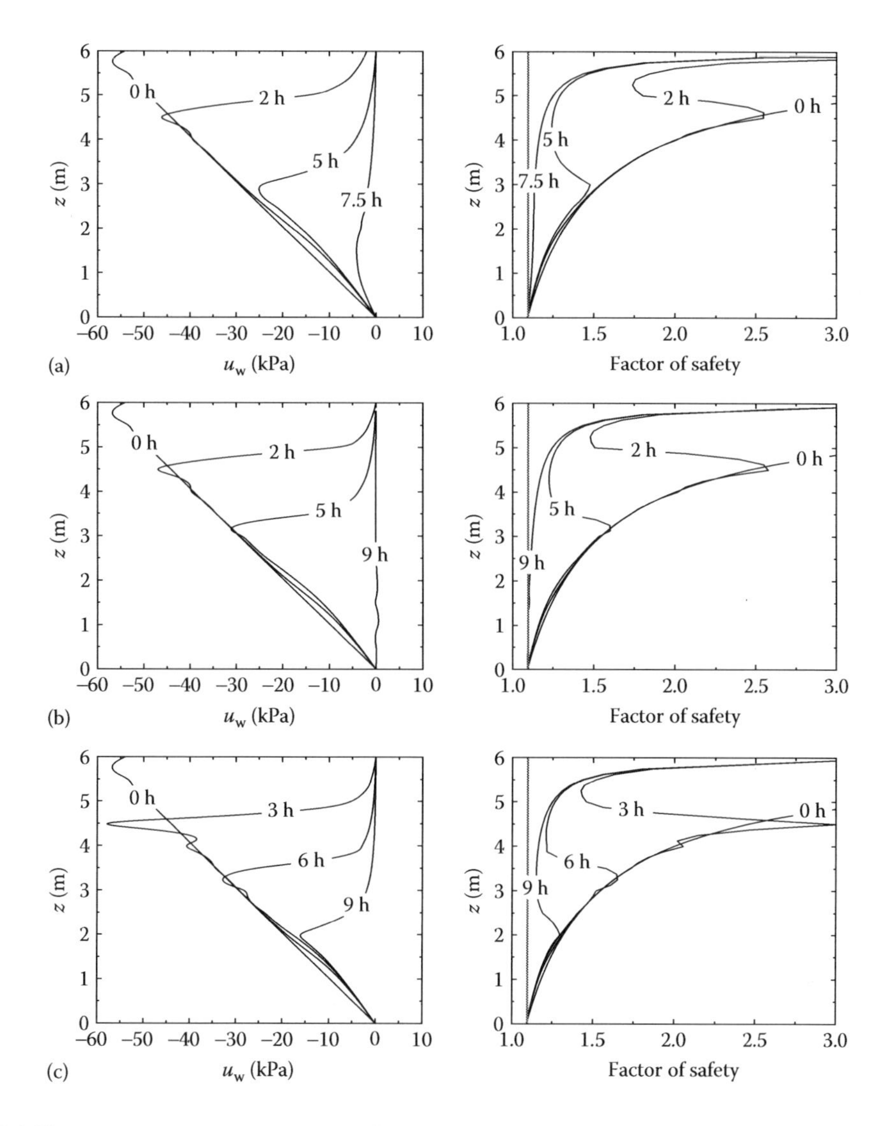

Figure 3.6 Transient pore-water pressure profile and factor of safety of a 35° infinite slope ($k_s = 2 \times 10^{-5}$ m/s, $q = 2 \times 10^{-5}$ m/s): (a) $a_{vgm} = 0.06$; (b) $a_{vgm} = 0.08$; and (c) $a_{vgm} = 0.10$ (Example 3.1).

3.3 SLOPE STABILITY ANALYSIS BASED ON LIMIT EQUILIBRIUM METHODS

The pore-water pressure distributions used as input data in the limit equilibrium slope stability analysis can be classified into three types: calculated pore-water pressure distribution from numerical seepage analyses (Tsaparas et al., 2002), assumed pore-water pressure distribution based on the wetting front concept (Chen et al., 2009), and actual field measured pore-water pressures (Gasmo et al., 2000). Various methods of slices were used to calculate the safety factor of the slope, for example, Bishop's simplified method with circular slip surfaces (Ng and Shi, 1998; Rahardjo et al., 2001, 2007; Tsaparas et al., 2002), Morgenstern and Price's method (Blatz et al., 2004; Cascini et al., 2010), and Janbu's method with non-circular slip surfaces (Wilkinson et al., 2002). The extended Mohr–Coulomb failure criterion

(Fredlund et al., 1978) is usually adopted. Commercial software packages such as SVflux and SVSlope in the SVoffice (SoilVision System Ltd, 2010a, b) and Seep/W and Slope/W in GeoStudio (Geo-slope International Ltd, 2012a, b) are available for performing integrated infiltration analysis and slope stability analysis for unsaturated soil slopes. Other adopted computer programs for slope stability analysis include STABL (Chen et al., 2009), etc.

The following formulations summarize the general limit equilibrium (GLE) method (Fredlund and Krahn, 1977), which incorporates the shear strength contribution from matric suction based on the extended Mohr–Coulomb failure criterion and satisfy both force and moment equilibrium (Fredlund et al., 2012).

The soil mass above an assumed slip surface is divided into vertical slices. The forces acting on a slice within the sliding soil mass are shown in Figure 3.7 for a composite slip surface. The forces are designated for a unit width (i.e., direction perpendicular to movement) of the slope. The variables in Figure 3.7 are defined as follows:

W = total weight of the slice
N = total normal force on the base of the slice
S_m = shear force mobilized on the base of each slice
E = horizontal interslice normal forces (the L and R subscripts designate the left- and right-hand sides of slice, respectively)
X = vertical interslice shear forces (the L and R subscripts designate the left- and right-hand sides of slice, respectively)
R = radius of a circular slip surface or the moment arm associated with the mobilized shear force
f_{slice} = perpendicular offset of the normal force from the center of moments
x_{slice} = horizontal distance from the center line of each slice to the center of moments
h_{slice} = vertical distance from the center of the base of each slice to the uppermost line in the geometry (i.e., generally ground surface), or height of the slice
b_{slice} = width of the slice
α_{slice} = the inclination of the slice base
β_{slice} = the base length of the slice

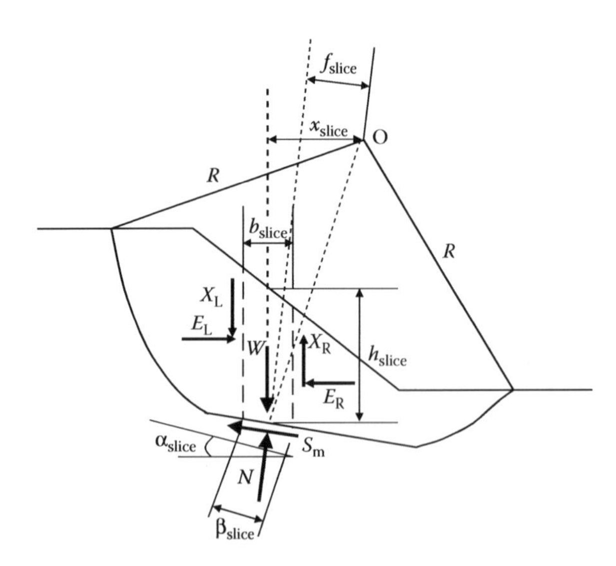

Figure 3.7 Geometry and forces acting on one slice of a sliding mass with a composite slip surface.

In accordance with conventional limit equilibrium methods of slices, the factors of safety are assumed to be equal for all parameters and for all slices. The mobilized shear force, S_m, at the base of a slice can be written as follows:

$$S_m = \frac{\beta_{slice}}{F_s}\left\{c' + (\sigma_n - u_a)\tan\phi' + (u_a - u_w)\tan\phi^b\right\} \tag{3.23}$$

The total normal force on the base of the slice, N, can be written as follows (Fredlund et al. 2012):

$$N = \frac{W - (X_R - X_L) + \left[-c' + u_a \tan\phi' - (u_a - u_w)\tan\phi^b\right]\dfrac{\beta_{slice}\sin\alpha_{slice}}{F_s}}{m_\alpha} \tag{3.24}$$

where $m_\alpha = \cos\alpha_{slice} + (\sin\alpha_{slice}\tan\phi')/F_s$. The interslice shear force, X, is computed by the interslice function (Morgenstern and Price 1965):

$$X = \lambda f_{IS}(x_{slice})E \tag{3.25}$$

where:

$f_{IS}(x)$ is the interslice function that describes the ratio of X/E varies across the slip surface

λ is the scaling constant that represents the percentage of the interslice function for solving the factor of the safety equations.

The interslice normal forces are then computed from the summation of the horizontal forces on each slice starting from left to right.

The factors of safety with respect to moment and force equilibriums can then be formulated as follows (Fredlund et al., 2012):

$$(F_s)_m = \Sigma \frac{\left\{c'\beta_{slice}R + \left[N - u_a\beta_{slice} + (u_a - u_w)\beta_{slice}\dfrac{\tan\phi^b}{\tan\phi'}\right]R\tan\phi'\right\}}{\left(\Sigma W x_{slice} - \Sigma N f_{slice}\right)} \tag{3.26}$$

where $(F_s)_m$ is the factor of safety with respect to moment equilibrium.

$$(F_s)_f = \Sigma \frac{\left\{c'\beta_{slice}\cos\alpha + \left[(N - u_a\beta_{slice}) + (u_a - u_w)\dfrac{\tan\phi^b}{\tan\phi'}\right]\tan\phi'\cos\alpha_{slice}\right\}}{\Sigma N\sin\alpha_{slice}} \tag{3.27}$$

where $(F_s)_f$ is the factor of safety with respect to force equilibrium.

Example 3.2

An example of 2D slope stability analysis based on 2D infiltration analysis. Assume a steep simple geometry homogenous slope that is 20 m high at a slope angle of 40° (Figure 3.8). The unit weight of the soil is 18.0 kN/m^3. The cohesion is 10 kPa and the effective angle of internal friction is 35°. The soil in the slope is a silty soil. The VGM model is adopted for the SWCC and unsaturated permeability function. The SWCC and permeability function and corresponding model parameters are presented in Figure 3.9.

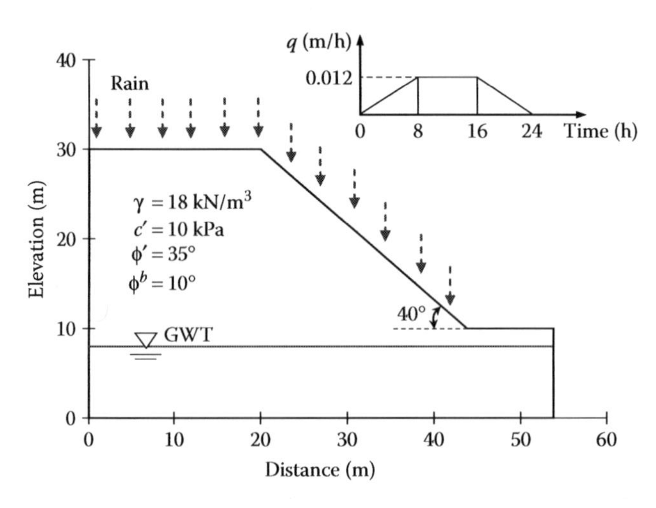

Figure 3.8 Geometry of a simple geometry slope (Example 3.2).

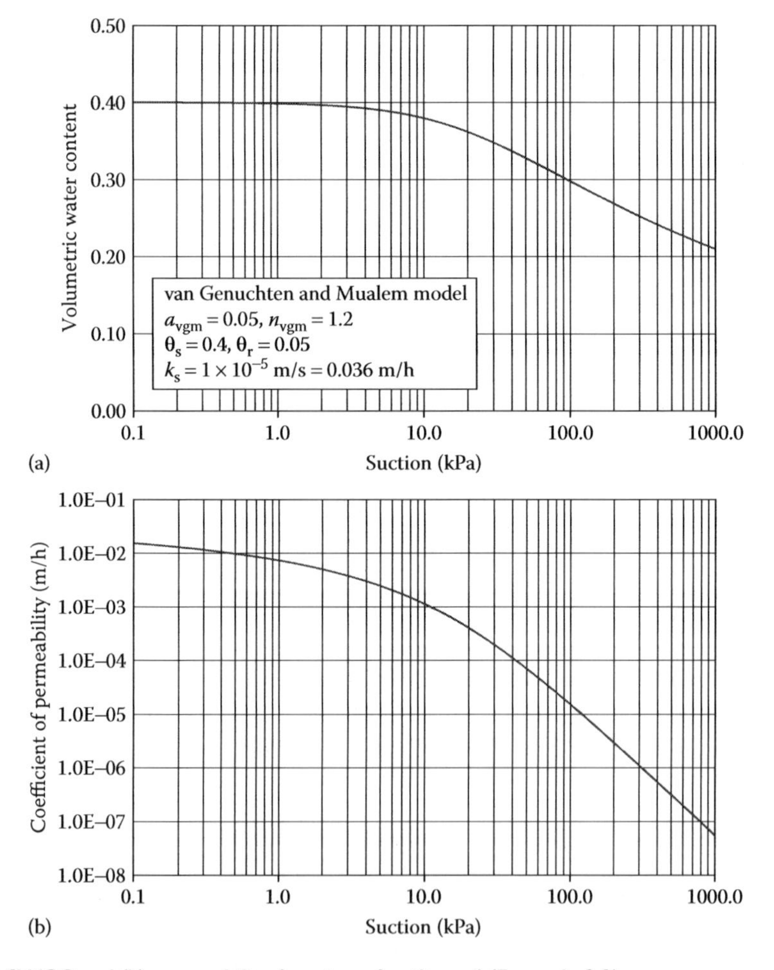

Figure 3.9 (a) SWCC and (b) permeability function of a silty soil (Example 3.2).

The groundwater table is on average 20 m below the surface of the slope. Assume that the initial pore pressure is hydrostatic with a maximum suction in the unsaturated zone of 50 kPa. The rainfall lasts 24 h with the varied intensity q shown in Figure 3.8. The 2D unsaturated seepage analysis and slope stability analysis for the soil slope are conducted using SVflux and SVSlope.

Figure 3.10 illustrates the contours of pore-water pressure distributions in the slope during the rainfall process. Figure 3.11 shows the variation of critical slip surfaces and the reduction of factor of safety due to rainfall infiltration. The critical slip surface becomes shallower when the soils near the slope surface are saturated. If the rain intensity is increased to three times of the one in Figure 3.8, the slope stability will be further deteriorated. Figure 3.12 shows that the safety factor of the slope is reduced to 1.184 and the critical slip surface is much shallower than that in Figure 3.11.

3.4 CONTROLLING FACTORS FOR RAINFALL-INDUCED LANDSLIDES

The results of seepage and infiltration in an unsaturated soil and slope stability analyses are sometimes difficult to interpret because the analysis involves many parameters related with soil properties, slope geometry, groundwater condition, initial conditions, and rainfall characteristics. To separate the influence of different factors and find out the controlling parameters has been the major objective of many research studies.

Slopes with a large slope angle and slope height and a shallow initial water table constitute the worst combination of factors for failure and are more likely to fail due to rainfall (Ng and Shi, 1998; Rahardjo et al., 2007). According to Rahardjo et al. (2007), the slope geometry and the initial water table determine the initial safety factor, and the actual failure conditions are much affected by rainfall characteristics and properties of the soils in the slope. Therefore, attention should be paid to soil properties and rainfall characteristics when dealing with rainfall-induced slope failures.

3.4.1 Soil shear strength properties

Based on either the extended Mohr–Coulomb failure criterion (Fredlund et al., 1978) or the suction stress concept (Lu and Likos, 2004), it is generally accepted that the presence of soil suction will increase the shear strength of the soil and hence the safety factor of the slope (Fredlund and Rahardjo, 1993; Griffiths and Lu, 2005). As shown in Figure 3.13 based on the study by Rahardjo and Fredlund (1995), the safety factor increases approximately linearly with the ratio of ϕ^b/ϕ'. The safety factor continues to decrease until the rainfall stops and the decrease in the safety factor is more substantial as the ratio of ϕ^b/ϕ' increases.

Zhang et al. (2014) illustrated the effect of estimated unsaturated shear strength on stability of homogenous slopes with different types of soils (sand, silt, clay). Figure 3.14 clearly shows that the difference of various nonlinear shear strength models exists. For a steep simple geometry slope, which is 30 m high at a slope angle of 50° (Figure 3.15), the critical slip surfaces obtained when using the Fredlund et al. (1996) model, the Vanapalli et al. (1996) model, and the Vilar (2006) model are overlapped. The Bao et al. (1998) model yields the shallowest critical slip surface. The Khalili and Khabbaz (1998) model yields the highest factor of safety and the deepest critical slip surface.

Figure 3.16 summarizes the relationship between the minimum safety factor and the air-entry values (AEV) of unsaturated soils for the steep slope. The factor of safety obtained

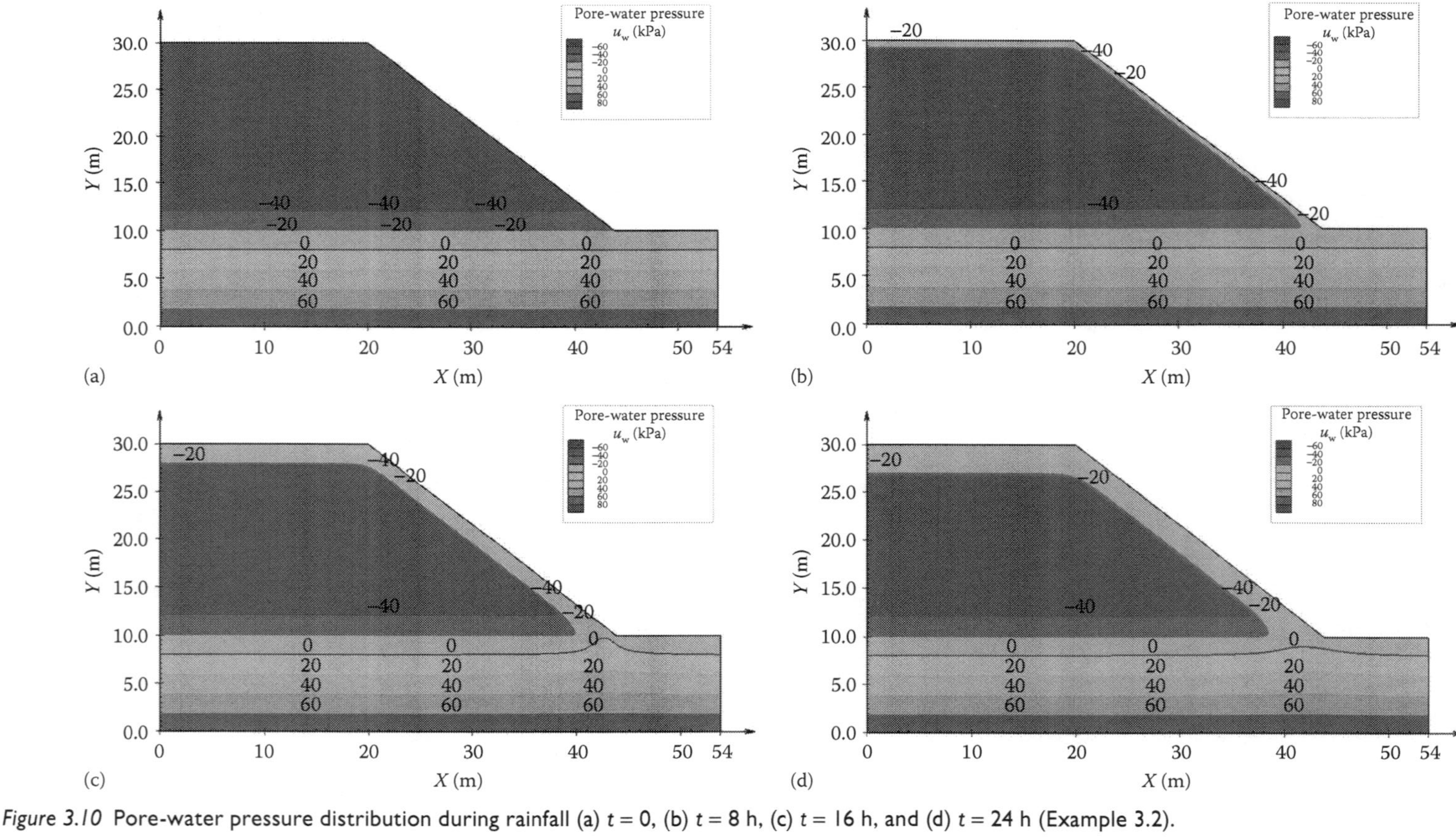

Figure 3.10 Pore-water pressure distribution during rainfall (a) $t = 0$, (b) $t = 8$ h, (c) $t = 16$ h, and (d) $t = 24$ h (Example 3.2).

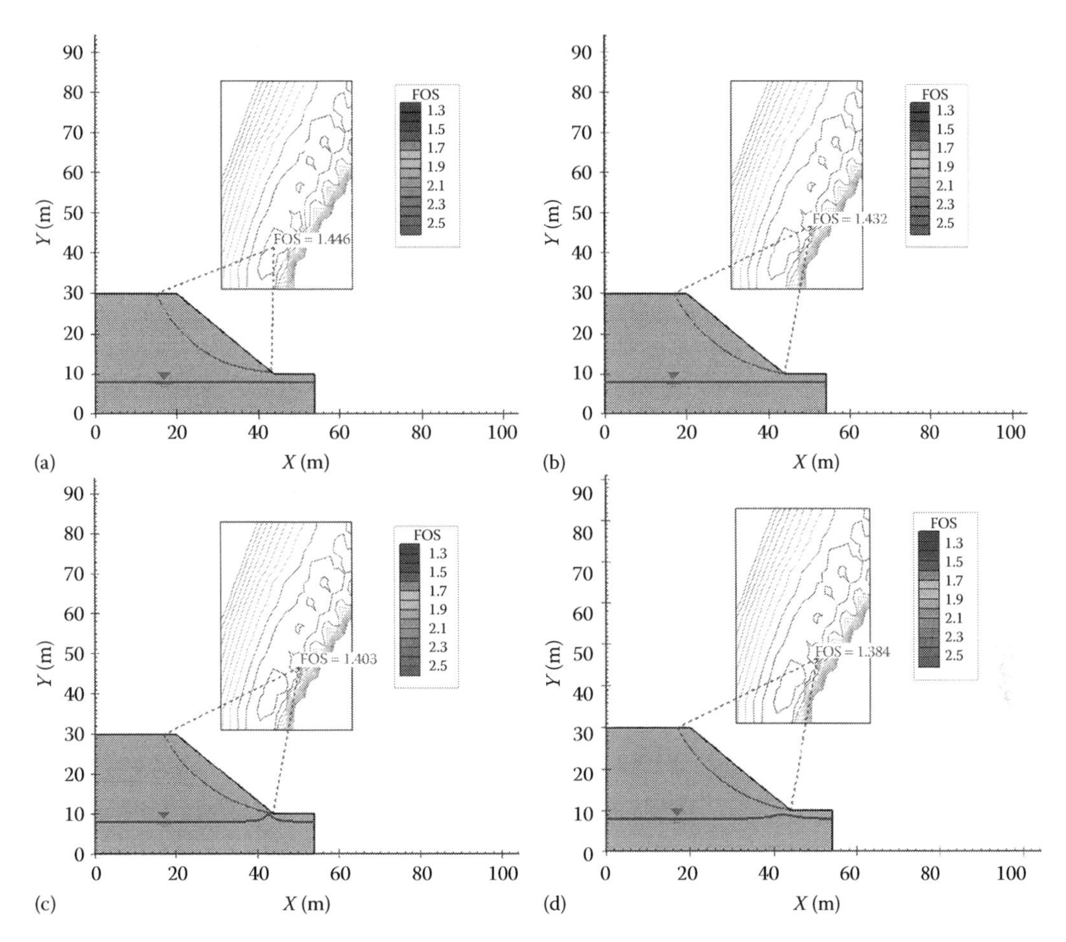

Figure 3.11 Critical slip surfaces and factor of safety during rainfall by 2D GLE method (a) $t = 0$, (b) $t = 8$ h, (c) $t = 16$ h, and (d) $t = 24$ h (Example 3.2).

using various n_f values are also presented in the graph. When the AEV of a soil is smaller than 1 kPa, the effect of matric suction on slope stability is trivial and the ϕ^b value can be assumed to be zero. When the AEV of a soil is between 1 and 20 kPa, the assumption of ϕ^b equals to 15° will yield a nonconservative factor of safety. Hence, the nonlinear estimation equations of unsaturated shear strength should be adopted. Otherwise, the effect of matric suction on shear strength can be ignored as a conservative assumption. For soils with an AEV value between 20 and 200 kPa, a ϕ^b value of 15° can generally be assumed for unsaturated shear strength to represent the average factor of safety among different nonlinear models. For soils with an AEV greater than 200 kPa, ϕ^b can be assumed to be ϕ'.

3.4.2 Soil hydraulic properties

The soil hydraulic properties related to rainfall-induced landslide include SWCC and unsaturated permeability function. Research studies show that the saturated permeability of soil has been considered as one of the most important soil properties in rainwater infiltration. Tsaparas et al. (2002) (Figure 3.17) showed that for the same rainfall, the higher the value of k_s, the greater the increment of pore-water pressures from the initial conditions and

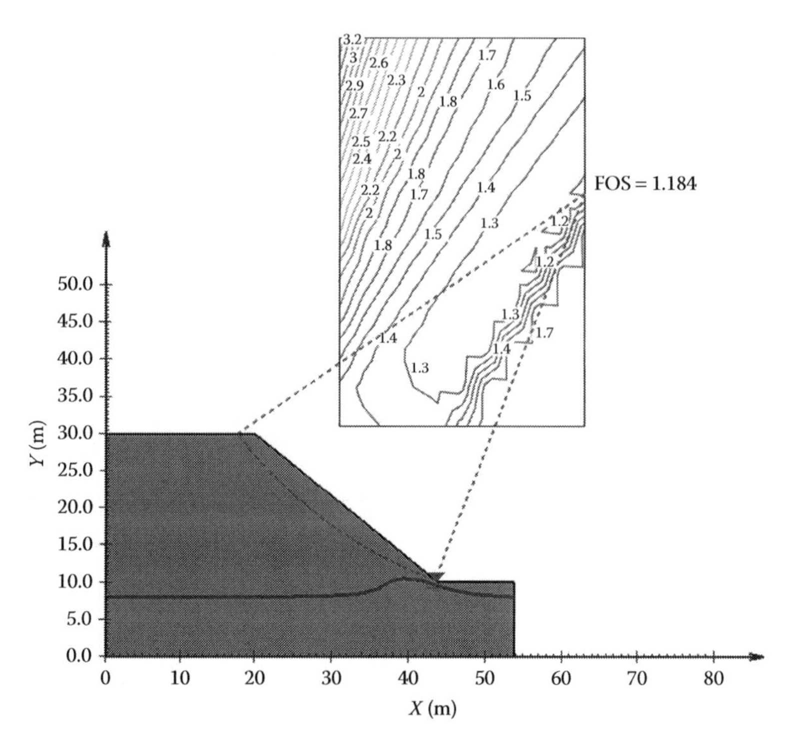

Figure 3.12 Critical slip surfaces and factor of safety by 2D GLE method after $t = 24$ h of rainfall with three times of the rainfall intensity in Figure 3.8 (Example 3.2).

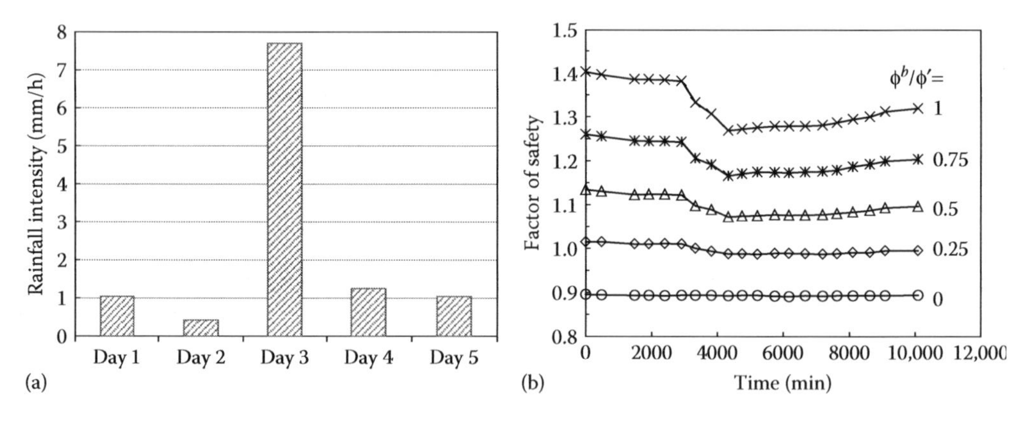

Figure 3.13 (a) Variation intensity of a rainfall event and (b) factor of safety of a steep weathered granite slope covered with colluvium. (Replotted based on Rahardjo, H., and Fredlund, D. G., Procedures for slope stability analyses involving unsaturated soils, *Proceedings of Asian Institute of Technology, Lecture Series on Slope Failures and Remedial Measures*, Bangkok, Thailand. Asian Institute of Technology, Klong Luang, Thailand, 1995.)

the deeper the wetting front will advance. Hence, the safety factor of the slope with a higher value of k_s is smaller. This observation is in agreement with the findings by Zhang et al. (2004) and Collins and Znidarcic (2004). The studies by Ng and Shi (1998) however illustrated that for soils with lower k_s, the reduction of matric suction and the rise of groundwater table are more significant (Figure 3.18). Hence, the decrease of safety factor

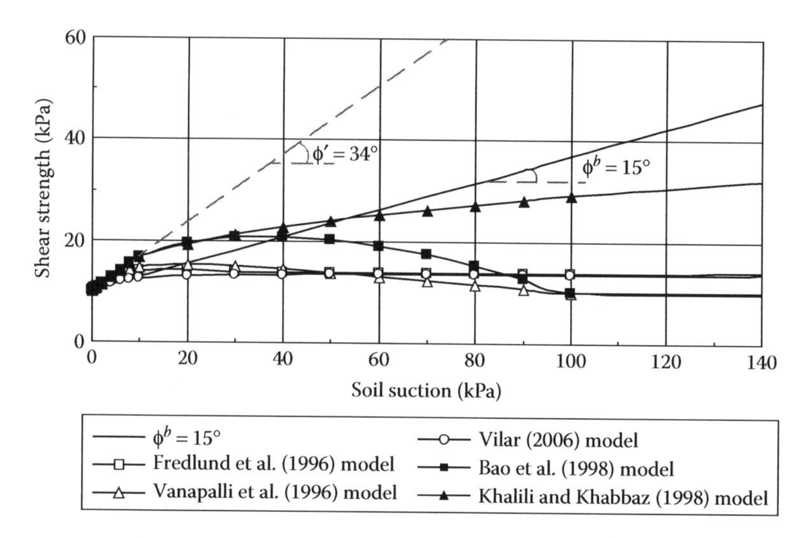

Figure 3.14 Shear strength envelopes for a silty soil ($\phi' = 34°$, $c' = 10$ kPa, Fredlund and Xing SWCC model with $\theta_s = 0.4$, $a_f = 10$, $n_f = 2$, $m_f = 1$, and $\psi_r = 100$). (From Zhang, L. L. et al., *Canadian Geotechnical Journal*, 51, 1–15, Copyright 2014 Canadian Science Publishing or its licensors. Reproduced with permission.)

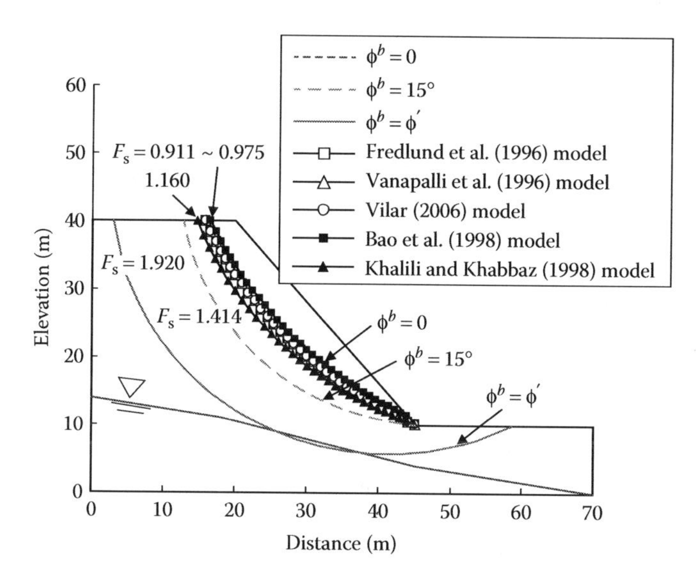

Figure 3.15 Comparison of critical slip surfaces for a steep slope ($\alpha_s = 50°$, $\gamma_t = 18$ kN/m^3, $\phi' = 34°$, $c' = 10$ kPa, Fredlund and Xing SWCC model with $\theta_s = 0.4$, $a_f = 10$, $n_f = 2$, $m_f = 1$, and $\psi_r = 100$). (From Zhang, L. L. et al., *Canadian Geotechnical Journal*, 51, 1–15, Copyright 2014 Canadian Science Publishing or its licensors. Reproduced with permission.)

is more significant for soil slopes with lower k_s. Results in Rahardjo et al. (2007) may be used to explain these contradictory conclusions about the influence of k_s on slope stability. Figure 3.19(a) shows the relationship between the rainfall intensity and minimum safety factor for a homogenous slope (10 m high, slope angle of 45°) subject to rainfall for 24 h. For the same rainfall intensity, the effect of k_s on the reduction of safety factor can be different. When the rainfall intensity is less than 10 mm/h, the minimum safety factor for the slope with soil type $f_{100,-6}$ ($k_s = 10^{-6}$ m/s) is the lowest, followed by that of $f_{50,-5}$ ($k_s = 10^{-5}$ m/s) and $f_{10,-4}$ ($k_s = 10^{-4}$ m/s). When the rainfall intensity is greater than 10 mm/h and less than

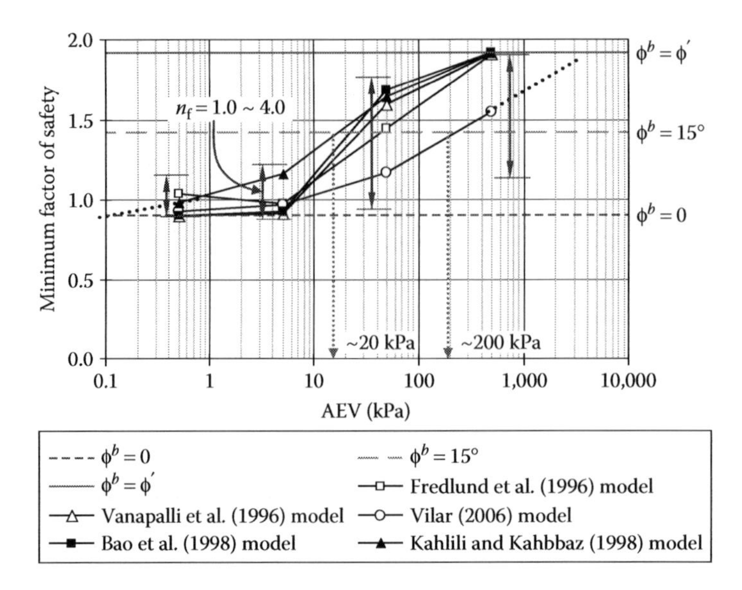

Figure 3.16 Minimum factor of safety versus air-entry value of unsaturated soils for the steep slope with low water table. (From Zhang, L. L. et al., *Canadian Geotechnical Journal*, 51, 1–15, Copyright 2014 Canadian Science Publishing or its licensors. Reproduced with permission.)

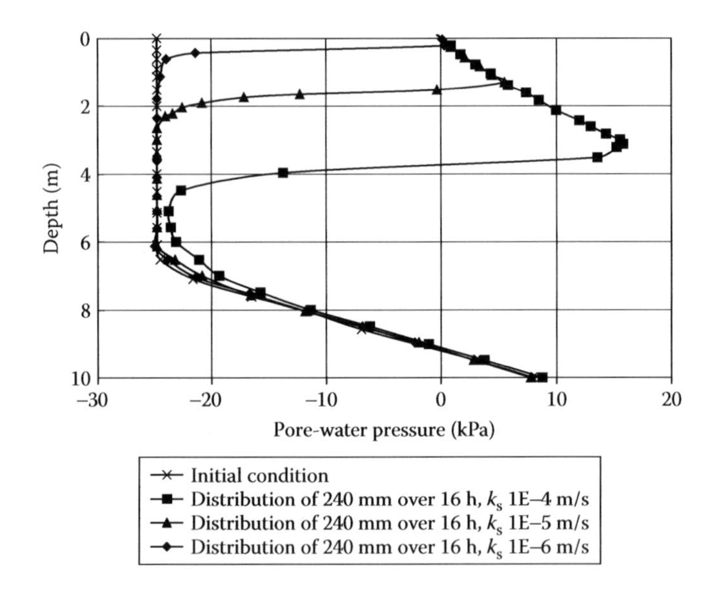

Figure 3.17 Comparison of the pore-water pressure profiles at the crest of a slope, at the end of the rainfall for different saturated coefficients of permeability. (Modified from *Computers and Geotechnics*, 29, Tsaparas, I. et al., Controlling parameters for rainfall-induced landslides, 1–27, Copyright 2002, with permission from Elsevier.)

200 mm/h, the minimum safety factor for the slope with soil type $f_{50,\,-5}$ ($k_s = 10^{-5}$ m/s) is the lowest. Figure 3.19(b) is plotted with q/k_s as the x-axis. It can be seen that with the same q/k_s, a higher value of k_s corresponds to a lower safety factor. Therefore, the ratio of q/k_s instead of q should be used to investigate the effect of soil hydraulic properties, which is consistent with the suggestions by Kasim (1997), Zhang et al. (2004), and Lee et al. (2009).

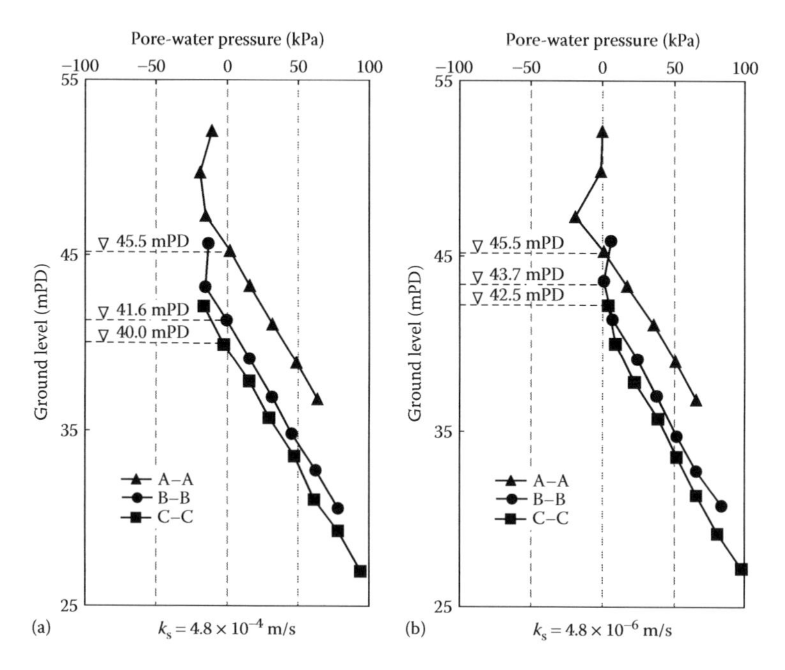

Figure 3.18 Pore-water pressure profiles for various soil permeability: (a) $k_s = 4.8 \times 10^{-4}$ m/s and (b) $k_s = 4.8 \times 10^{-6}$ m/s. A–A, B–B, and C–C represent different cross sections in the slope. (From *Computers and Geotechnics*, 22, Ng, C. W. W., and Shi, Q., Numerical investigation of the stability of unsaturated soil slopes subjected to transient seepage, 1–28, Copyright 1998, with permission from Elsevier.)

3.4.3 Rainfall characteristics

Critical rainfall conditions to initiate slope failures can be determined based on empirical (historical, statistical) methods by studying historical rainfall events that have resulted in landslides (Corominas, 2000; Crosta and Frattini, 2001; Aleotti, 2004; Wieczorek and Glade, 2005) or physical (process-based, conceptual) methods. Rainfall thresholds, which means once given rainfall conditions are reached or exceeded, landslides will occur with a given probability, are usually obtained by drawing lower bound lines visually to the rainfall conditions that resulted in landslides in Cartesian, semilogarithmic, or logarithmic coordinates (Lumb, 1975; Brand et al., 1984; Premchitt et al., 1994; Finlay et al., 1997; Pun et al., 1999; Dai and Lee, 2001). Sometimes, other types of parameters such as soil moisture or field hydrological conditions are combined with the rainfall parameters to form a threshold value.

Empirical thresholds can be classified as global, regional, or local thresholds. A comprehensive review of various thresholds can be found in Guzzetti et al. (2007). A global threshold attempts to establish a general ("universal") minimum level below which landslides do not occur, independently of local morphological, lithological, and land-use conditions (Caine, 1980; Innes, 1983; Jibson, 1989; Crosta and Frattini, 2001; Cannon and Gartner, 2005). Regional thresholds are defined for areas extending from a few to several thousand square kilometers of similar meteorological, climatic, and physiographic characteristics, and are potentially suited for landslide warning systems based on quantitative spatial rainfall forecasts or measurements. For example, Brand et al. (1984) suggested that an intensity of about 70 mm/h appeared to be the threshold rainfall intensity above which landslides would occur in Hong Kong and a 24 h rainfall of less than 100 mm is very unlikely to result in a large number of landslides occurring in a short time. Kay and Chen (1995) proposed that

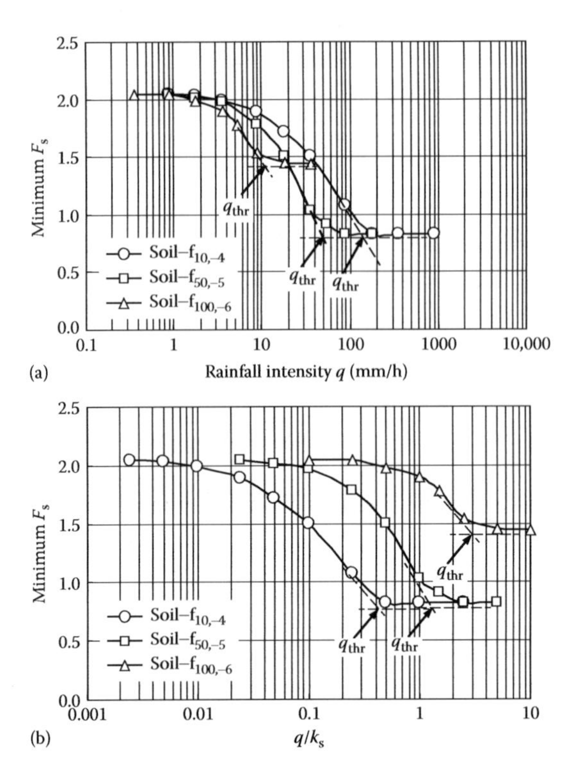

Figure 3.19 Relationship between rainfall intensity and minimum factor of safety for homogeneous soil slope subjected to rainfall for 24 h: (a) in terms of absolute rainfall intensity (Modified and reproduced from Rahardjo, H. et al., *Journal of Geotechnical and Geoenvironmental Engineering*, 133(12), 1532–1543, 2007. With permission from ASCE) and (b) in terms of q/k_s ratio.

the combination of maximum hourly rainfall and 24 h rainfall be used as an indicator of landslide activity. Local thresholds consider the local climatic regime and geomorphological setting, and are applicable to single landslides or to group of landslides in areas extending from a few to hundreds of square kilometers. Global thresholds are useful where local or regional thresholds are not available, but may result in false-positive predictions. Regional and local thresholds perform reasonably well in the area where they were developed, but cannot be easily exported to other areas.

The rainfall measurements to be considered in the rainfall threshold models can be classified into three groups: (i) precipitation measurements for a specific rainfall event, (ii) the antecedent conditions, and (iii) other types of meteorological conditions. Intensity–duration (I-D) thresholds are the most common type of thresholds proposed in the literature (Caine, 1980; Larsen and Simon, 1993; Guzzetti et al., 2007). Caine (1980) was the first to propose the global threshold curve defining the relationship between rainfall duration and intensity based on data from 73 rainfall events worldwide that resulted in shallow landslides and debris flows. The Caine (1980) I-D threshold curve equation is as follows:

$$I = 14.82D^{-0.39} \tag{3.28}$$

where:

D is the rainfall duration in hours
I is the rainfall intensity in millimeters per hour

More generally, I-D thresholds have the form as follows (Guzzetti et al., 2007):

$$I = c + aD^b \tag{3.29}$$

where $c \geq 0$, a, and b are the model parameters. Figure 3.20 shows the rainfall I-D thresholds summarized by Guzzetti et al. (2007). Most of the thresholds cover the range of durations between 1 and 100 h, and the range of intensities from 1 to 200 mm/h.

The main limitations of the empirical threshold models are obvious. These empirical thresholds are generally based on the assumption that there exists a direct relationship between the occurrence of landslides and the rainfall characteristics. However, the influence of initial geological and hydrological conditions and the physical process of rainfall seepage and infiltration on a slope are often not considered (Rahardjo et al., 2007). Long-period records of landslide and climate are required to establish empirical relationships. When including antecedent rainfall measurements, a significant scatter (3 days to several months) is observed for defining antecedent rainfall in the literature. In addition, different definitions have been used for the same or similar rainfall and climate variables in various threshold models. This language inconsistencies and disagreement on the rainfall and landslide variables of the models make it difficult to compare different models.

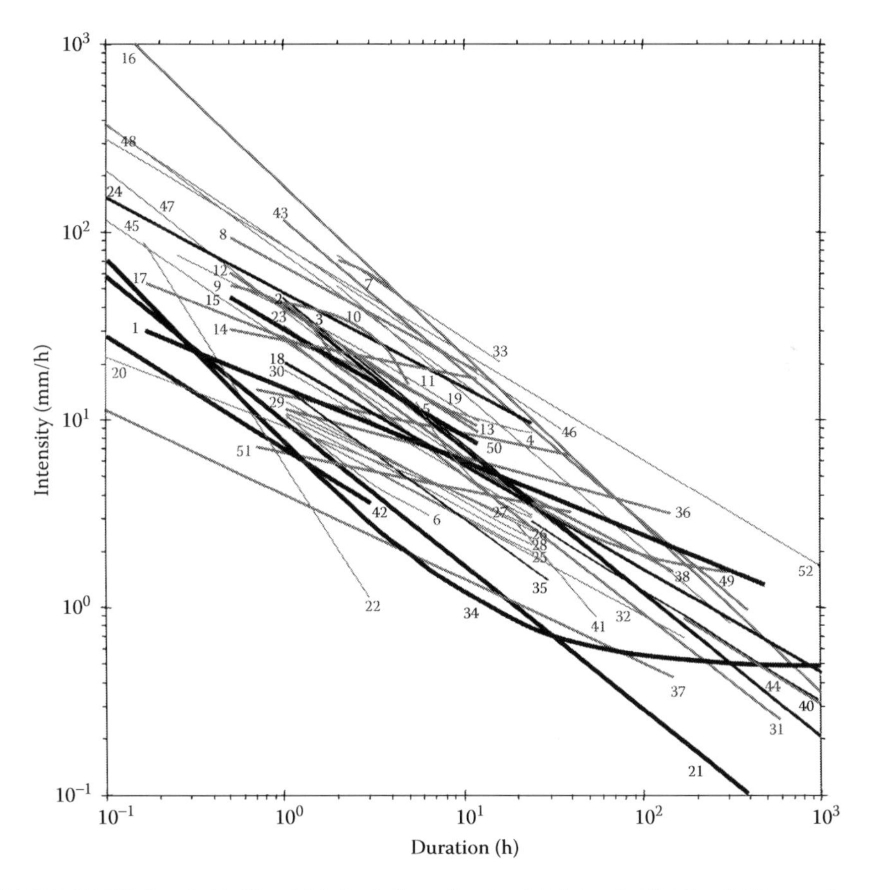

Figure 3.20 Rainfall I-D thresholds. The thick black lines denote global thresholds. The thick gray lines denote regional thresholds. The thin gray lines denote local thresholds. The numbers of the lines refer to the number of threshold equations in table 2 of Guzzetti et al. (2007). (Adopted from Guzzetti, F. et al., *Meteorology and Atmospheric Physics*, 98, 239–267, 2007. With permission.)

The effects of rainfall intensity, rainfall duration, rainfall pattern, and antecedent rainfall on slope stability have also been studied systematically through physical models based on infiltration analysis and slope stability analysis. It is generally accepted that more intense rainfall can lead to more reduction of negative pore-water pressure in the slope and a more significant rise of groundwater table (Ng and Shi, 1998; Rahardjo et al., 2007). Rahardjo et al. (2007) suggested that the threshold rainfall intensity can be determined by the maximum reduction of the factor of safety (q_{thr} in Figure 3.19). It was found that the threshold rainfall intensity is larger for soils with a higher k_s and the rainfall intensity equal to k_s will not necessarily produce the lowest safety factor. Based on their study, soil slopes with a high k_s are more likely affected by short-duration high-intensity rainfalls.

As rainfall intensity and duration are dependent parameters and depend on a certain return period, some researchers integrated the infiltration models and the intensity–duration–frequency (IDF) curve (i.e., relationship between the rainfall intensity and duration for various return periods) to determine the critical rainfall conditions. For example, Pradel and Raad (1993) proposed an approximate method based on the Green–Ampt model to determine the critical rainfall intensity and duration for rainfall-induced slope failure. Figure 3.21 illustrates the schematic plot of critical rainfall intensity and duration for coarse-grained soils is higher than that of the fine-grained soils. As shown in the graph, a more severe storm with a longer return period will be needed to trigger failure in coarse-grained soil slopes than in fine-grained soil slopes. This may probably explain why field observations indicate that soil slopes with high saturated permeability may be less likely to fail (Lumb, 1962; Pradel and Raad, 1993).

Although contradictory conclusions are made about the relative role of antecedent rainfall to landslides in different regions (Brand et al., 1984; Pitts, 1985), antecedent rainfall has usually been considered as an important factor to influence slope stability under rainfall as the negative pore-water pressure can be reduced by antecedent rain flux and the slope can be marginally safe before the major rainfall (Ng and Shi, 1998; Rahardjo et al., 2001; Tsaparas et al., 2002). Rahardjo et al. (2007) demonstrated that (Figure 3.22) for the same rainfall intensity and the same slope angle, the rate of reduction in safety factor is the fastest for soil type $f_{10,-4}$ ($k_s = 10^{-4}$ m/s) followed by soil types $f_{50,-5}$ ($k_s = 10^{-5}$ m/s) and $f_{100,-6}$ ($k_s = 10^{-6}$ m/s). After the rainfall ceases, the factor of safety recovered fastest for the soil type $f_{10,-4}$ followed by soil types $f_{50,-5}$ and $f_{100,-6}$. They

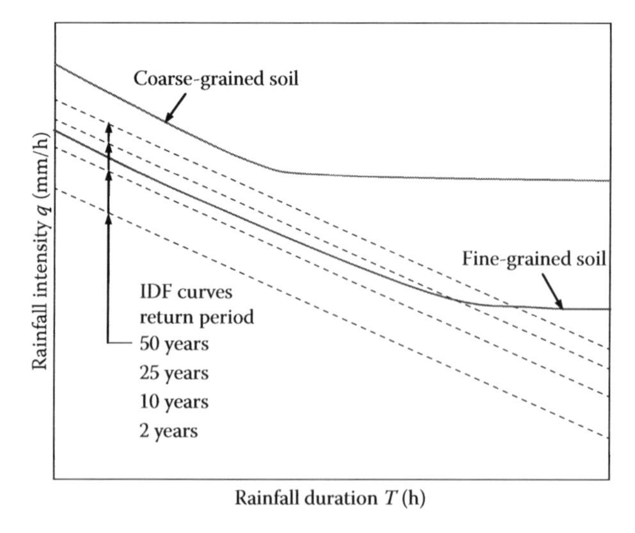

Figure 3.21 Rainfall thresholds of different soil types based on integration of IDF curve and the infiltration model.

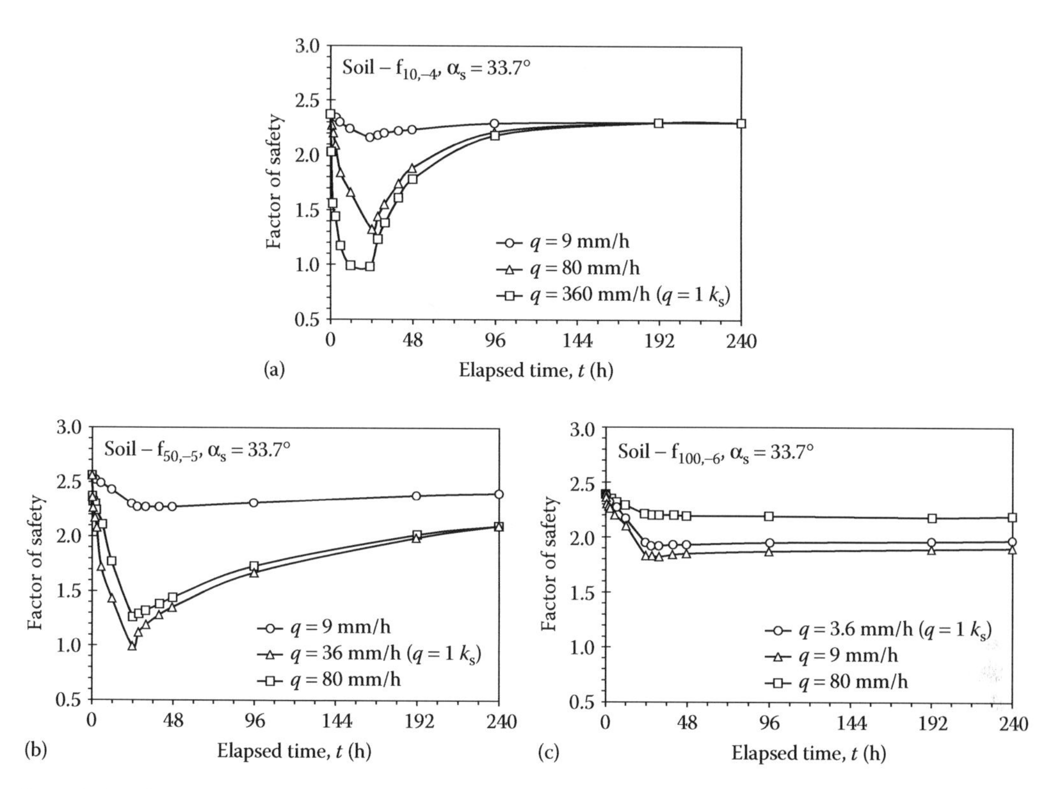

Figure 3.22 Effect of soil properties on variation of factor of safety with time for a homogenous soil slope of constant slope height: (a) $k_s = 10^{-4}$ m/s, (b) $k_s = 10^{-5}$ m/s, and (c) $k_s = 10^{-6}$ m/s. (Reproduced from Rahardjo, H. et al., *Journal of Geotechnical and Geoenvironmental Engineering*, 133, 1532–1543, 2007. With permission from ASCE.)

deduced that the effect of antecedent rainfalls is more significant in affecting the stability of soil slopes with low k_s ($k_s = 10^{-6}$ m/s) than those with high k_s ($k_s = 10^{-5}$ m/s). The conclusions are in agreement with the findings from previous numerical studies by Tsaparas et al. (2002), Cai and Ugai (2004), and field observations (Rahardjo et al. 2001).

Ng et al. (2001) and Tsaparas et al. (2002) investigated the effect of different rainfall patterns on the pore-water pressure response in slopes. According to Ng et al. (2001), for a 24 h rainfall with the same total rainfall amount, the pore-water pressure increases most rapidly and significantly in response to the advance rainfall pattern, followed by the central pattern, and then the delayed pattern. This observation is consistent with the results in Tsai (2008). Ng et al. (2001) also found that the effect of rainfall pattern is less significant with the increase in the depth of slip surface and under conditions of low rainfall intensity and long duration. Tsaparas et al. (2002) found that the changes of pore-water pressure in the scenarios of short rain events followed by a period of no rain are more limited than that in the scenario of continuous rainfall, because the soils are able to drain during the period of no rain.

3.5 SPATIALLY DISTRIBUTED MODEL OF HAZARD ASSESSMENT OF RAINFALL-INDUCED LANDSLIDES

For a large area where widespread slope failures may be triggered by rainfall, a distributed model for predicting rainfall-induced slope failures at regional scale is necessary. Both empirical models (e.g., Au, 1993; Aleotti, 2004; Lee et al., 2013) and physically-based models (e.g., Montgomery and Dietrich, 1994; Wu and Sidle, 1995; Dietrich and Montgomery, 1998; Crosta and Frattini, 2003; Zhou et al., 2003; Frattini et al., 2004; Baum et al., 2008; Godt et al., 2008; Arnone et al., 2011; Shuin et al., 2012; Park et al., 2013; Chen and Zhang, 2014) have been proposed to predict space and time for rainfall-induced slope failures in regional scale.

Empirical models predict the occurrence of landslide based on empirical relationships between historical landslides and various influencing factors of geomorphological, hydrological, and geological conditions and do not need detailed information regarding the physical and mechanical properties of the soil or rock mass and the geometry of the slopes (Lee et al., 2013). However, a careful verification is necessary when being adopted in another area. In Section 3.4, we have discussed the thresholds of rainfall-induced landslides. Hence, in this section, only the physically-based distributed model will be discussed.

3.5.1 General methodology

The main components of a physically-based distributed model typically consist of a hydrological module and a geotechnical module. The hydrological module is used to analyze the infiltration process and changes in the groundwater table. The geotechnical module is used to compute the factor of safety of a slope. One-dimensional infinite-slope stability analysis is most widely used due to its simplicity (Montgomery and Dietrich, 1994; Wu and Sidle, 1995; Borga et al., 1998; Pack et al., 1998; Iverson, 2000; Crosta and Frattini, 2001; Morrissey et al., 2001). Physically-based models can account for the mechanisms and process of the rainfall-induced slope failures. However, detailed investigations to obtain accurate site-specific conditions are required.

The commonly adopted physically based models include distributed Shallow Landslide Analysis Model (dSLAM) (Wu and Sidle, 1995; Dhakal and Sidle, 2003), Stability Index MAPping (SINMAP) (Pack et al., 1998; Morrissey et al., 2001), Shallow Slope Stability Model (SHALSTAB) (Dietrich et al., 1993, 1995; Montgomery and Dietrich, 1994; Montgomery et al., 1998), and Transient Rainfall Infiltration and Grid-Based Regional Slope-Stability (TRIGRS) (Iverson, 2000; Baum et al., 2002). Here, a brief introduction of the TRIGRS model will be provided to illustrate the basic methodology of a spatially distributed model. Theoretical details of the model can be found in TRIGRS open file reports (Baum et al., 2002, 2008).

The TRIGRS model is a FORTRAN program combining models for infiltration and subsurface flow, surface runoff and flow routing, and slope stability to investigate the effects of storms on the stability of slopes over large areas (Baum et al., 2002, 2008). The program operates on a gridded elevation model of a map area. Infiltration, hydraulic properties, and slope stability input parameters are allowed to vary over the grid area thus making it possible to analyze complex storm sequences over geologically complex terrain. Infiltration and slope stability analysis are conducted for each grid cell.

In the first version of TRIGRS (Baum et al., 2002), the infiltration model is based on Iverson's (2000) linearized solution of one-dimensional Richards equation in isotropic, homogeneous materials for wet (saturated or tension-saturated) initial conditions where the flow is in the linear range for Darcy's law and the hydraulic diffusivity is approximately constant. Baum et al. (2002) extended the Iverson's (2000) original solution for rainfall of constant intensity by using a series of Heaviside step functions to represent a general time-varying sequence of surface fluxes of variable intensities and durations. A solution for an impermeable basal boundary at a finite depth is added. In the second version of TRIGRS (Baum et al., 2008), the analytical solution of the Richards equation for one-dimensional, vertical infiltration by Srivastava and Yeh (1991) is added to consider the infiltration into an unsaturated soil layer above the water table. The TRIGRS program uses a simple infinite-slope model to compute factor of safety of slope stability for each grid cell.

Runoff routing schemes are adopted in TRIGRS to achieve mass balance between rainfall, infiltration, and runoff over the entire domain by allowing excess water of one cell to flow to downslope cells. Several runoff routing methods are available in TRIGRS. The simplest method for divert runoff flow is the eight-direction (D8) method (O'Callaghan and Mark 1984). The D8 method distributes flow from one cell to its eight neighbors through the steepest downslope path (Figure 3.23). The shortcoming of the D8 method is that the resulting flow distribution is irregular and sometimes unrealistic. The D-infinity method (Tarboton, 1997) assumes that water flows down one or two cells by partitioning the flow between the two cells nearest to the steepest slope direction. The other methods, in which the weighting factors are proportional to the slope raised to an exponent, are provided to allow the user to achieve greater dispersion or smoothing of the runoff distribution. Compared with the other methods, the D-infinity method was preferred because it is physically more realistic.

The limitations of the TRIGRS model are summarized as follows. The TRIGRS model assumes homogeneous, isotropic soil in both infiltration and stability analysis. Application to areas with soil anisotropy or heterogeneity in hydrologic properties may subject to significant errors. Applying grid cell-based infinite-slope stability analysis for three-dimensional topography may result in errors where slope, thickness, physical properties, or pore pressure changes abruptly. During long period of rainstorms, lateral flow has a significant influence on the magnitude and distribution of pore pressure. Hence, TRIGRS is not suitable for modeling long-term response because lateral flow is not considered (Baum et al., 2008). Furthermore, the runoff routing schemes do not carry runoff over from one time step to the next time step.

3.5.2 Performance index

In practice, the performance of a spatially distributed model can be evaluated through comparison of model evaluation of slope stability with actual observations. Most studies (Crosta and Frattini, 2003; Salciarini et al., 2006; Kim et al., 2010; Vieira et al., 2010) used agreement of parts (cells or areas) between the predicted and the actual landslides to evaluate the performance of their models. Some studies used the number of landslides between the observations and prediction. An ideal model can simultaneously maximize the agreement between known and predicted locations of landslides, and minimize overprediction. Various indices or estimators for measuring performance of model have been proposed in the literature (Frattini et al., 2010).

The success rate (SR) is one of the simplest indices that is defined as (Huang and Kao 2006):

$$\mathrm{SR} = \frac{\text{number of successfully predicted landslides}}{\text{total number of actual landslides}} \tag{3.30}$$

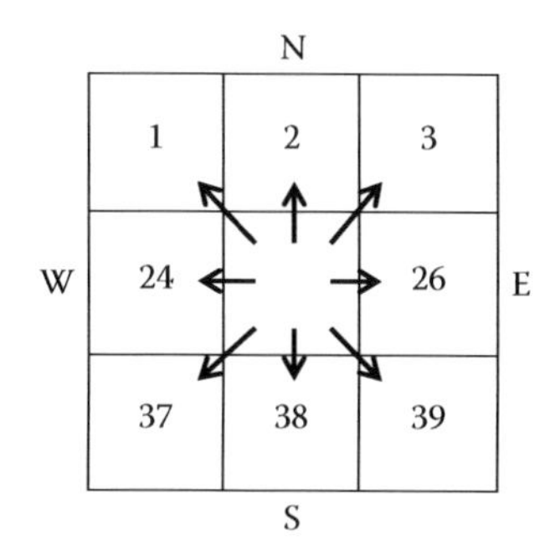

Figure 3.23 The D8 method for surface runoff distribution.

Noted that in this equation the number of landslide instead of number of cells is used while a landslide usually contains more than one cell.

Huang and Kao (2006) considered the successfully predicted stable cell and proposed the modified success rate (MSR):

$$\text{MSR} = 0.5 \cdot \text{SR} + 0.5 \cdot \frac{\text{successfully predicted stable cells}}{\text{total number of actual stable cells}} \tag{3.31}$$

In this equation, the SR and performance of stable cell prediction are equally weighted. The first component (SR) is still calculated based on landslide number rather than cell number to keep the conventional concept of SR. The values of SR and MSR range from 0.0 to 1.0. The MSR index substantially reduces the potential of landslide overprediction.

Crosta and Frattini (2003) defined a quality index for the performance of spatially distributed models as the ratio between the success rate and error rate:

$$\text{Quality index} = \frac{P(\text{predicted landslide within unstable area})}{P(\text{predicted landslide within stable area})} \tag{3.32}$$

Some other indices including the success index (SI) and the error index (EI) based on the area of stable and unstable zones defined by Sorbino et al. (2010); scars concentration (SC) and landslide potential (LP) defined by Vieira et al. (2010); probability of detection (POD), false alarm ratio (FAR), and critical success index (CSI) defined by Liao et al. (2011); LR_{class} defined by Park et al. (2013) and the D index defined by Liu and Wu (2008).

In the following paragraphs, we will briefly introduce the system of performance indices defined based on the confusion matrix. When comparing the actual landslides with the predicted ones, four types of outcomes are possible: (1) actual landslide cells are predicted as unstable cells (true positive, TP), (2) actual stable cells are predicted as unstable cells (false positive, FP), (3) actual landslide cells are predicted as stable cells (false negative, FN), and (4) actual stable cells are predicted as stable cells (true negative, TN). A confusion matrix (also known as contingency table, Figure 3.24) can be used to present the number of FP, FN, TP, and TN observations (e.g., Van Den Eeckhaut et al., 2006).

TP and TN are considered as success in model prediction, while FP and FN are regarded as error in prediction. FN is an important error because it indicates that the model is not able to gather the actual triggering condition (Crosta and Frattini, 2003). This can be due to incorrect rainfall input, wrong parameters, violation of model assumptions, oversimplification of hydrological processes, insufficient or erroneous description of slope morphology, and so forth. FP is a less important error because people are usually conservative when using

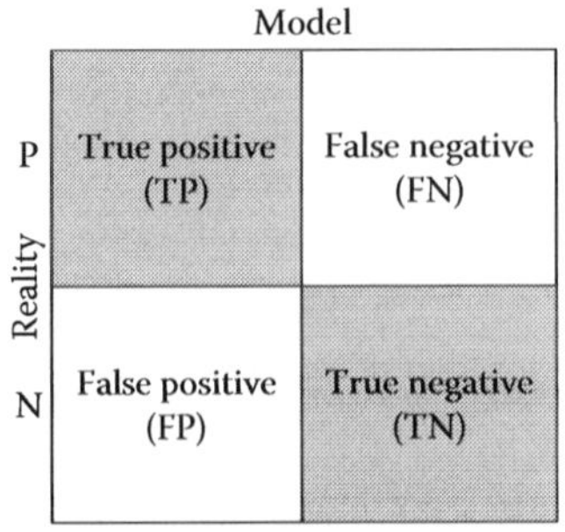

Reality \ Model		
P	True positive (TP)	False negative (FN)
N	False positive (FP)	True negative (TN)

Figure 3.24 A confusion matrix for a two-class problem.

a spatial distributed model and it is commonly accepted that the absence of a landslide in prediction does not mean that a certain slope cannot experience failure under slightly different conditions due to model uncertainties (Crosta and Frattini, 2003).

Some performances indices can be obtained by the confusion matrix:

$$\text{Accurancy} = \frac{\text{TP} + \text{TN}}{P + N} \tag{3.33}$$

where:

P is the total number of positives (landslide) observed in reality
N is the total number of negatives (no landslide) observed in reality

The precision of a model is defined as

$$\text{Precison} = \frac{\text{TP}}{\text{TP} + \text{FP}} \tag{3.34}$$

The sensitivity (also called the true positive rate) is the number of correctly predicted landslide cells (TP) over the total number of landslide cells (TP + FN):

$$\text{sensitivity} = \frac{\text{TP}}{\text{TP} + \text{FN}} \tag{3.35}$$

The specificity (also called the true negative rate) is the number of correctly predicted nonlandslides cells (TN) over the total number of nonlandslides cells (TN + FP):

$$\text{specifity} = \frac{\text{TN}}{\text{TN} + \text{FP}} \tag{3.36}$$

The performance of the model can be evaluated graphically by receiver operating characteristic (ROC) curve as shown in Figure 3.25 (Swets, 1988; Fawcett, 2006; Van Den Eeckhaut et al., 2006) in a plot of 'sensitivity' versus '1—specificity'. The origin (0, 0) represents a model predicts no instability. The converse, in which the entire area is

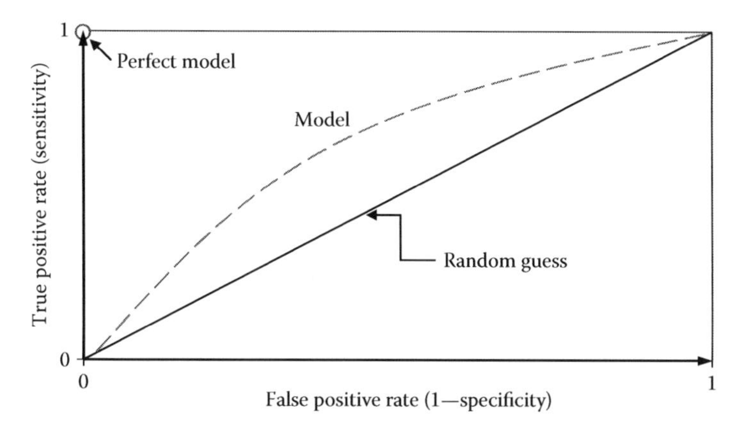

Figure 3.25 ROC curves.

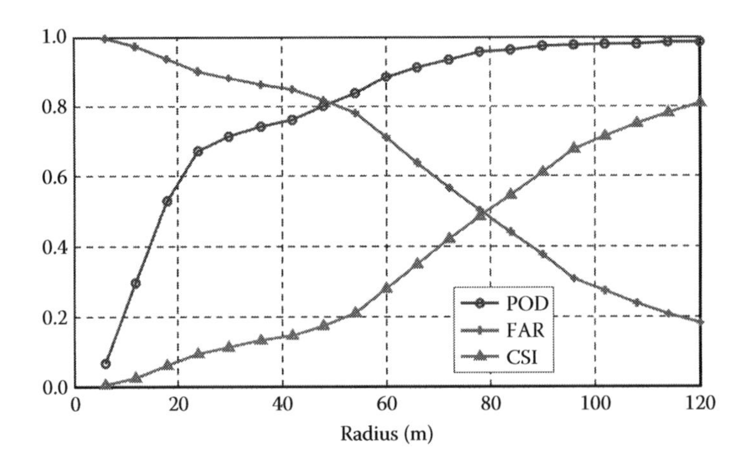

Figure 3.26 Schematic of POD, FAR, and CSI for comparison between landslide observations and predictions as a function of detection radius from center of the landslides. (With kind permission from Springer Science+Business Media: *Nature Hazards*, Evaluation of TRIGRS [transient rainfall infiltration and grid-based regional slope-stability analysis]'s predictive skill for hurricane-triggered landslides: A case study in Macon County, North Carolina, 58, 2011, 325–339, Liao, Z. et al.)

predicted to be unstable would be located at the upper right (1, 1). A perfect prediction would be located at the upper left (0, 1). Usually, sensitivity and specificity are determined for a prediction model with various conditions. For example, the factor of safety represents that a slop failure can vary in a range instead of a certain value 1.0 (Chen and Zhang, 2014). A model of random guessing would run diagonally from (0, 0) to (1, 1) (Van Den Eeckhaut et al., 2006). Model results that plot toward the upper left of the graph are generally considered superior. The area under the curve of ROC, AUC, is another index to test the performance of the model (e.g., Van Den Eeckhaut et al., 2006). A higher value of AUC means better performance.

In general, it is common for landslide to range in size from square meters to kilometers. The spatial uncertainty related with landslide mapping can influence the model performance. Crosta and Frattini (2003) adopted a reasonable buffer of 1200 m^2 for each landslide to calculate the total unstable area in a region. Liao et al. (2011) showed that the prediction performance varies with the specified detection radius from the center of landslide (Figure 3.26).

REFERENCES

Aleotti, P. (2004). A warning system for rainfall-induced shallow failures. *Engineering Geology*, 73(3–4), 247–265.

Arnone, E., Noto, L. V., Lepore, C., and Bras, R. L. (2011). Physically-based and distributed approach to analyze rainfall-triggered landslides at watershed scale. *Geomorphology*, 133(3), 121–131.

Au, S. W. C. (1993). Rainfall and slope failure in Hong Kong. *Engineering Geology*, 36(1–2), 141–147.

Bao, C., Gong, B., and Zhan, L. (1998). Properties of unsaturated soils and slope stability of expansive soils. *Proceedings of the Second International Conference on Unsaturated Soils (UNSAT 98)*, Beijing. Vol. 1, pp. 71–98.

Baum, R. L., Savage, W. Z., and Godt, J. W. (2002). *TRIGRS—A FORTRAN Program for Transient Rainfall Infiltration and Grid-based Regional Slope Stability Analysis*. US Geological Survey Open-File Report 02-424. US Department of the Interior, US Geological Survey, 38pp.

Baum, R. L., Savage, W. Z., and Godt, J. W. (2008). *TRIGRS – A FORTRAN Program for Transient Rainfall Infiltration and Grid-based Regional Slope Stability Analysis, Version 2.0.* US Geological Survey Open-File Report 2008–1159. US Department of the Interior, US Geological Survey, 75pp.

Blatz, J. A., Ferreira, N. J., and Graham, J. (2004). Effects of near-surface environmental conditions on instability of an unsaturated soil slope. *Canadian Geotechnical Journal*, 41(6), 1111–1126.

Borga, M., Dalla Fontana, G., De Ros, D., and Marchi, L. (1998). Shallow landslide hazard assessment using a physically based model and digital elevation data. *Environmental Geology*, 35(2–3), 81–88.

Brand, E. W. (1984). Landslides in south Asia: A state-of-art report. *Proceedings of the 4th International Symposium on Landslide*. Ontario BiTech, Vancouver, Canada, pp. 17–59.

Brand, E. W., Premchitt, J., and Phillipson, H. B. (1984). Relationship between rainfall and landslides. *Proceedings of the fourth International Symposium on Landslides*. Ontario BiTech, Vancouver, Canada, pp. 377–384.

Cai, F., and Ugai, K. (2004). Numerical analysis of rainfall effects on slope stability. *International Journal of Geomechanics*, 4(2), 69–78.

Caine, N. (1980). The rainfall intensity-duration control of shallow landslides and debris flows. *Geografiska Annaler. Series A, Physical Geography*, 62(1–2), 23–27.

Cannon, S. H., and Gartner, J. E. (2005). Wildfire-related debris flow from a hazards perspective. *Debris Flow Hazards and Related Phenomena*, Jakob, M. and Hungr, O. (eds). Springer, Berlin, pp. 363–385.

Cascini, L., Cuomo, S., Pastor, M., and Sorbin, G. (2010). Modelling of rainfall-induced shallow landslides of the flow-type. *Journal of Geotechnical and Geoenvironmental Engineering*, 136(1), 85–98.

Chen, H. X., and Zhang, L. M. (2014). A physically-based distributed cell model for predicting regional rainfall-induced shallow slope failures. *Engineering Geology*, 176, 79–92.

Chen, R. H., Chen, H. P., Chen, K. S., and Zhung, H. B. (2009). Simulation of a slope failure induced by rainfall infiltration. *Environmental Geology*, 58(5), 943–952.

Cho, S. E., and Lee, S. R. (2001). Instability of unsaturated soil slopes due to infiltration. *Computers and Geotechnics*, 28, 185–208.

Cho, S. E., and Lee, S. R. (2002). Evaluation of surficial stability for homogeneous slopes considering rainfall characteristics. *Journal of Geotechnical and Geoenvironmental Engineering*, 128(9), 756–763.

Collins, B. D., and Znidarcic, D. (2004). Stability analyses of rainfall induced landslides. *Journal of Geotechnical and Geoenvironmental Engineering*, 130(4), 362–372.

Corominas, J. (2000). Landslides and climate. *Proceedings of the 8th International Symposium on Landslides, Vol. 4*. Thomas Telford, Cardiff, pp. 1–33.

Crosta, G. B., and Frattini, P. (2001). Rainfall thresholds for triggering soil slips and debris flow. *Proceedings of the 2nd EGS Plinius Conference on Mediterranean Storms*. Siena, Italy, pp. 463–487.

Crosta, G. B., and Frattini, P. (2003). Distributed modelling of shallow landslides triggered by intense rainfall. *Natural Hazards and Earth System Science*, 3, 81–93.

Dai, F. C., and Lee, C. F. (2001). Frequency-volume relation and prediction of rainfall-induced landslides. *Engineering Geology*, 59(3–4), 253–266.

Dhakal, A. S., and Sidle, R. C. (2003). Long-term modelling of landslides for different forest management practices. *Earth Surface Processes and Landforms*, 28(8), 853–868.

Dietrich, W. E., and Montgomery, D. R., (1998). Hillslopes, channels and landscape scale. *Scale Dependence and Scale Invariance in Hydrology*, Sposito, G. (ed). Cambridge University Press, Cambridge, pp. 30–60.

Dietrich, W. E., Reiss, R., Hsu, M. L., and Montgomery, D. R. (1995). A process-based model for colluvial soil depth and shallow landsliding using digital elevation data. *Hydrological Processes*, 9(3), 383–400.

Dietrich, W. E., Wilson, C. J., Montgomery, D. R., and McKean, J. (1993). Analysis of erosion thresholds, channel networks, and landscape morphology using a digital terrain model. *The Journal of Geology*, 101(2), 259–278.

Donald, I. B. (1956). Shear strength measurements in unsaturated non-cohesive soils with negative pore pressures. *Proceedings of the Second Australia-New Zealand Conference on Soil Mechanics and Foundation Engineering*, Christchurch, New Zealand, pp. 200–205.

Duncan, J. M., and Wright, S. G. (1995). *Soil Strength and Slope Stability*. Wiley, Hoboken, NJ.

Escario, V., and Saez, J. (1986). The shear strength of partly saturated soils. *Géotechnique*, 36(3), 453–456.

Fawcett, T. (2006). An introduction to ROC analysis. *Pattern Recognition Letters*, 27(8), 861–874.

Finlay, P. J., Fell, R., and Maguire, P. K. (1997). The relationship between the probability of landslide occurrence and rainfall. *Canadian Geotechnical Journal*, 34, 811–824.

Fourie, A. B., Rowe, D., and Blight, G. E. (1999). The effect of infiltration on the stability of the slopes of a dry ash dump. *Géotechnique*, 49(1), 1–13.

Frattini, P., Crosta, G., and Carrara, A. (2010). Techniques for evaluating the performance of landslide susceptibility models. *Engineering Geology*, 111(1), 62–72.

Frattini, P., Crosta, G. B., Fusi, N., and Dal Negro, P., (2004). Shallow landslides in pyroclastic soils: A distributed modelling approach for hazard assessment. *Engineering Geology*, 73(3), 277–295.

Fredlund, D. G., and Krahn, J. (1977). Comparison of slope stability methods analysis. *Canadian Geotechnical Journal*, 14(3), 429–439.

Fredlund, D. G., Morgenstern, N. R., and Widger, R. A. (1978). The shear strength of unsaturated soils. *Canadian Geotechnical Journal*, 15(3), 313–321.

Fredlund, D. G., and Rahardjo, H. (1993). *Soil Mechanics for Unsaturated Soils*. Wiley, New York, 544 pp.

Fredlund, D. G., Rahardjo, H., and Fredlund, M. D. (2012). *Unsaturated Soil Mechanics in Engineering Practice*. John Wiley & Sons, Inc., Hoboken, NJ, 944 pp.

Fredlund, D. G., Rahardjo, H., and Gan, J. K. M. (1987). Nonlinearity of strength envelope for unsaturated soils. *Proceedings of the Sixth International Conference on Expansive Soils, Vol. 1*. New Delhi, pp. 49–54.

Fredlund, D. G., Xing, A., Fredlund, M. D., and Barbour, S.L. (1996). The relationship of the unsaturated soil shear strength to the soil water characteristic curve. *Canadian Geotechnical Journal*, 32(3), 440–448.

Garven, E. A., and Vanapalli, S. K. (2006). Evaluation of empirical procedures for predicting the shear strength of unsaturated soils. In: Miller G. A., Zapata C. E., Houston S. L, and Fredlund D. G. (Eds.). *Proceedings of the Fourth International Conference on Unsaturated Soils, Vol. 2*, GSP 147. American Society of Civil Engineers, Reston, VA, pp. 2570–2581.

Gasmo, J. M., Rahardjo, H., and Leong, E. C. (2000). Infiltration effects on stability of a residual soil slope. *Computers and Geotechnics*, 26(2), 145–165.

Geo-slope International Ltd (2012a). *Seep/W for Finite Element Seepage Analysis. User's Guide*, Geo-slope Ltd, Calgary, Canada.

Geo-slope International Ltd (2012b). *Slope/W for Slope Stability Analysis. User's Guide*, Geo-slope Ltd, Calgary, Canada.

Godt, J. W., Baum, R. L., Savage,W. Z., Salciarini, D., Schulz,W. H., and Harp, E. L. (2008). Transient deterministic shallow landslide modeling: Requirements for susceptibility and hazard assessments in a GIS framework. *Engineering Geology*, 102(3), 214–226.

Griffiths, D. V., and Lu, N. (2005). Unsaturated slope stability analysis with steady infiltration or evaporation using elasto-plastic finite elements. *International Journal for Numerical and Analytical Methods in Geomechanics*, 29(3), 249–267.

Guzzetti, F., Peruccacci, S., Rossi, M., and Stark, C. P. (2007). Rainfall thresholds for the initiation of landslides in central and southern Europe. *Meteorology and Atmospheric Physics*, 98(3–4), 239–267.

Huang, J. C., and Kao, S. J. (2006). Optimal estimator for assessing landslide model performance. *Hydrology and Earth System Sciences Discussions*, 10(6), 957–965.

Innes, J. L. (1983). Debris flows. *Progress in Physical Geography*, 7(4), 469–501.

Iverson, R. M. (2000). Landslide triggering by rain infiltration. *Water Resources Research*, 36(7), 1897–1910.

Jibson, R. W. (1989). Debris flow in southern Porto Rico. *Geological Society of America Special Paper*, 236, 29–55.

Kasim, F. B. (1997). *Effects of Steady State Rainfall on Long Term Matric Suction Conditions in Slopes.* Internal Engineering Research Report, Unsaturated Soils Group, University of Saskatchewan, Saskatoon, Saskatchewan, Canada.

Kay, J. N., and Chen, T. (1995). Rainfall–landslide relationship for Hong Kong. *Proceedings of the Institution of Civil Engineers-Geotechnical Engineering*, 113(2), 117–118.

Khalili, N., and Khabbaz, M. H. (1998). A unique relationship for the determination of the shear strength of unsaturated soils. *Géotechnique*, 48(5), 681–687.

Kim, D., Im, S., Lee, S. H., Hong, Y., and Cha, K. S. (2010). Predicting the rainfall-triggered landslides in a forested mountain region using TRIGRS model. *Journal of Mountain Science*, 7(1), 83–91.

Larsen, M. C., and Simon, A. (1993). A rainfall intensity-duration threshold for landslides in a humid-tropical environment, Puerto Rico. *Geografiska Annaler. Series A. Physical Geography*, 75(1–2), 13–23.

Lee, D. H., Lai, M. H., Wu, J. H., Chi, Y. Y., Ko, W. T., and Lee, B. L. (2013). Slope management criteria for Alishan Highway based on database of heavy rainfall-induced slope failures. *Engineering Geology*, 162, 97–107.

Lee, L. M., Gofar, N., and Rahardjo, H. (2009). A simple model for preliminary evaluation of rainfall-induced slope instability. *Engineering Geology*, 108(3–4), 272–285.

Liao, Z., Hong, Y., Kirschbaum, D., Adler, R. F., Gourley, J. J., and Wooten, R. (2011). Evaluation of TRIGRS (transient rainfall infiltration and grid-based regional slope-stability analysis)'s predictive skill for hurricane-triggered landslides: A case study in Macon County, North Carolina. *Nature Hazards*, 58(1), 325–339.

Liu, C. N., and Wu, C. C. (2008). Mapping susceptibility of rainfall triggered shallow landslides using a probabilistic approach. *Environment Geology*, 55(4), 907–915.

Lu, N., and Godt, J. (2008). Infinite slope stability under steady unsaturated seepage conditions. *Water Resources Research*, 44(11), W11404.

Lu, N., and Griffiths, D. V. (2004). Profiles of steady-state suction stress in unsaturated soils. *Journal of Geotechnical and Geoenvironmental Engineering*, 130(10), 1063–1076.

Lu, N., and Likos, W. J. (2004). *Unsaturated Soil Mechanics*. Wiley, Hoboken, NJ, p. 556.

Lumb, P. (1962). Effect of rain storms on slope stability. *Proceedings of the Symposium on Hong Kong Soils*. pp. 73–87.

Lumb, P. (1975). Slope failures in Hong Kong. *Quarterly Journal of Engineering Geology*, 8, 31–65.

Montgomery, D. R., and Dietrich, W. E. (1994). A physically based model for the topographic control on shallow landsliding. *Water Resources Research*, 30(4), 1153–1171.

Montgomery, D. R., Sullivan, K., and Greenberg, H. M. (1998). Regional test of a model for shallow landsliding. *Hydrological Processes*, 12(6), 943–955.

Morgenstern, N. R., and Price, V. E. (1965). The analysis of the stability of general slip surfaces. *Géotechnique*, 15(1), 79–93.

Morrissey, M. M., Wieczorek, G. F., and Morgan, B. A. (2001). *A Comparative Analysis of Hazard Models for Predicting Debris Flows in Madison County, Virginia* (US Geological Survey Open File Report 01-67). US Department of the Interior, US Geological Survey, p. 7.

Mualem, Y. (1976). A new model for predicting the hydraulic conductivity of unsaturated porous media. *Water Resources Research*, 12(3), 593–622.

Muntohar, A. S., and Liao, H. J. (2009). Analysis of rainfall-induced infinite slope failure during typhoon using a hydrological-geotechnical model. *Environmental Geology*, 56(6), 1145–1159.

Ng, C. W. W., and Shi, Q. (1998). Numerical investigation of the stability of unsaturated soil slopes subjected to transient seepage. *Computers and Geotechnics*, 22(1), 1–28.

Ng, C. W. W., Wang, B., and Tung, Y. K. (2001). Three-dimensional numerical investigations of groundwater responses in an unsaturated slope subjected to various rainfall patterns. *Canadian Geotechnical Journal*, 38(5), 1049–1062.

Oberg, A., and Sallfours, G. (1997). Determination of shear strength parameter of unsaturated silts and sands based on the water retention curve. *Geotechnical Testing Journal*, 20(1), 40–48.

O'Callaghan, J. F., and Mark, D. M. (1984). The extraction of drainage networks from digital elevation data. *Computer Vision, Graphics, and Image Process*, 28(3), 328–344.

Pack, R. T., Tarboton, D. G., and Goodwin, C. N. (1998). *Terrain Stability Mapping with SINMAP, Technical Description and Users Guide for Version 1.00* (Report No. 4114-0). Terratech Consulting Ltd, Salmon Arm, BC, Canada, p. 68.

Park, H. J., Lee, J. H., and Woo, I. (2013). Assessment of rainfall-induced shallow landslide susceptibility using a GIS-based probabilistic approach. *Engineering Geology*, 161, 1–15.

Pitts, J. (1985). *An Investigation of Slope Stability on the NTI Campus, Singapore, Applied Research Project RPI/83*. Nanyang Technological Institute, Singapore.

Pradel, D., and Raad, G. (1993). Effect of permeability on surficial stability of homogeneous slopes. *Journal of Geotechnical Engineering*, 109(1), 62–70.

Premchitt, J., Brand, E. W., and Chen, P. Y. M. (1994). Rain-induced landslides in Hong Kong. *Asia Engineer, Journal of the Hong Kong Institution of Engineers*, 43–51.

Pun, W. K., Wong, A. C. W., and Pang, P. L. R. (1999). *Review of Landslip Warning Criteria* (Special Project Report SPR 4/99). Geotechnical Engineering Office, The Government of the Hong Kong Special Administrative Region, 77pp.

Rahardjo, H., and Fredlund, D. G. (1995). Procedures for slope stability analyses involving unsaturated soils. *Proceedings of Asian Institute of Technology, Lecture Series on Slope Failures and Remedial Measures*, Bangkok, Thailand. Asian Institute of Technology, Klong Luang, Thailand.

Rahardjo, H., Li, X. W., Toll, D. G., and Leong, E. C. (2001). The effect of antecedent rainfall on slope stability. *Geotechnical and Geological Engineering*, 19, 371–399.

Rahardjo, H., Lim, T. T., Chang, M. F., and Fredlund, D. G. (1995). Shear strength characteristics of a residual soil. *Canadian Geotechnical Journal*, 32(1), 60–77.

Rahardjo, H., Ong, T. H., Rezaur, R. B., and Leong, E. C. (2007). Factors controlling instability of homogeneous soil slopes under rainfall. *Journal of Geotechnical and Geoenvironmental Engineering*, 133(12), 1532–1543.

Salciarini, D., Godt, J. W., Savage, W. Z., Conversini, P., Baum, R. L., and Michael, J. A. (2006). Modeling regional initiation of rainfallinduced shallow landslides in the eastern Umbria Region of central Italy. *Landslides*, 3(3), 181–194.

Sheng, D., Fredlund, D. G., and Gens, A. (2008). A new modelling approach for unsaturated soils using independent stress variables. *Canadian Geotechnical Journal*, 45(4), 511–534.

Shuin, Y., Hotta, N., Suzuki, M., and Ogawa, K. I. (2012). Estimating the effects of heavy rainfall conditions on shallow landslides using a distributed landslide conceptual model. *Physics and Chemistry of the Earth, Parts A/B/C*, 49, 44–51.

Skempton, A. W., and Delory, F. A. (1957). Stability of natural slopes in London clay. *Proceedings of the 4th International Conference of Soil Mechanics and Foundation Engineering*, London. Butterworths, London, pp. 378–381.

SoilVision System Ltd (2010a). *SVFlux User's Manual*. SoilVision System Ltd, Saskatoon, Canada.

SoilVision System Ltd (2010b). *SVSlope User's Manual*. SoilVision System Ltd, Saskatoon, Canada.

Sorbino, G., Sica, C., and Cascini, L. (2010). Susceptibility analysis of shallow landslides source areas using physically based models. *Nature Hazards*, 53(2), 313–332.

Srivastava, R., and Yeh, T. C. J. (1991). Analytical solutions for one-dimensional, transient infiltration toward the water table in homogeneous and layered soils. *Water Resources Research*, 27(5), 753–762.

Swets, J. (1988). Measuring the accuracy of diagnostic systems. *Science*, 240(4857), 1285–1293.

Tarboton, D. G. (1997). A new method for the determination of flow directions and contributing areas in grid digital elevation models. *Water Resources Research*, 33(2), 309–319.

Travis, Q. B., Houston, S. L., Marinho, F. A. M., and Schmeeckle, M. (2010). Unsaturated infinite slope stability considering surface flux conditions. *Journal of Geotechnical and Geoenvironmental Engineering*, 136(7), 963–974.

Tsai, T. L. (2008). The influence of rainstorm pattern on shallow landslide. *Environmental Geology*, 53(7), 1563–1569.

Tsaparas, I., Rahardjo, H., Toll, D. G., and Leong, E. C. (2002). Controlling parameters for rainfall-induced landslides. *Computers and Geotechnics*, 29(1), 1–27.

Vanapalli, S. K., Fredlund, D. G., Pufahl, D. E., and Clifton, A. W. (1996). Model for the prediction of shear strength with respect to soil suction. *Canadian Geotechnical Journal*, 33(3), 379–392.

Van Den Eeckhaut, M., Vanwalleghem, T., Poesen, J., Govers, G., Verstraeten, G., and Vandekerckhove, L. (2006). Prediction of landslide susceptibility using rare events logistic regression: A case-study in the Flemish Ardennes (Belgium). *Geomorphology*, 76(3), 392–410.

van Genuchten, M.Th. (1980). A closed-form equation for predicting the hydraulic conductivity of unsaturated soils. *Soil Science Society of America Journal*, 44(5), 892–898.

Vieira, B. C., Fernandes, N. F., and Filho, O. A. (2010). Shallow landslide prediction in the Serra do Mar, São Paulo, Brazil. *Natural Hazards and Earth System Science*, 10(9), 1829–1837.

Vilar, O. M. (2006). A simplified procedure to estimate the shear strength envelope of unsaturated soil. *Canadian Geotechnical Journal*, 43(10), 1088–1095.

Wieczorek, G. F., and Glade, T. (2005). Climatic factors influencing occurrence of debris flows. *Debris flow Hazards and Related Phenomena*, Jakob, M. and Hungr, O. (eds). Springer, Berlin, pp. 325–362.

Wilkinson, P. L., Anderson, M. G., and Lloyd, D. M. (2002). An integrated hydrological model for rain-induced landslide prediction. *Earth Surface Processes and Landforms*, 27(12), 1285–1297.

Wu, W., and Sidle, R. C. (1995). A distributed slope stability model for steep forested basins. *Water Resources Research*, 31(8), 2097–2110.

Zhang, L. L., Fredlund, D. G., Fredlund, M., and Wilson, G. W. (2014). Modeling the unsaturated soil zone in slope stability analysis. *Canadian Geotechnical Journal*, 51, 1–15.

Zhang, L. L., Fredlund, D. G., Zhang, L. M., and Tang, W. H. (2004). Numerical study of soil conditions under which matric suction can be maintained. *Canadian Geotechnical Journal*, 41(4), 569–582.

Zhou, G., Esaki, T., Mitani, Y., Xie, M., and Mori, J. (2003). Spatial probabilistic modeling of slope failure using an integrated GIS Monte Carlo simulation approach. *Engineering Geology*, 68(3), 373–386.

Chapter 4

Coupled hydro-mechanical analysis for unsaturated soil slope

4.1 INTRODUCTION

The behavior of soil slope under rainfall conditions is closely related not only to the pore-water pressure but also to the stress state during infiltration. Pore-water pressure changes due to rainfall infiltration will lead to changes in stresses and hence, deformation of a soil. Conversely, stress changes will modify the seepage process because soil hydraulic properties such as porosity, permeability, and water storage capacity are affected by the changes in stresses. Therefore, the seepage and stress-deformation problems are strongly linked in unsaturated soils under rainfall condition. More rigorous solution of the coupled governing equations for deformation and seepage should be obtained when the soils in the slope are considered deformable.

The seepage and deformation of an unsaturated soil slope under isothermal conditions require the coupled solution of the governing equations describing the equilibrium of the soil structure and the mass conservation of the water phase. Constitutive relationships for the solid and water are also required. A number of different formulations have been presented in the literature (Sheng et al., 2008b; Fredlund et al., 2012). In this chapter, both elastic-constitutive-model-based formulation and elastoplastic-model-based formulation will be introduced in Section 4.2. Because of the complexity of the governing equations and nonlinearity of soil behavior, numerical approaches are usually adopted for the coupled hydro-mechanical modeling of soil slope under rainfall. In this chapter, methods of slope stability analysis based on the coupled hydro-mechanical modeling are discussed in Section 4.3. An illustrative example of slope stability analysis for a hypothetical slope based on the effective stress concept of unsaturated soil and an elastoplastic model is presented in Section 4.4. Two case histories (1976 Sau Mau Ping landslide and a fully instrumented field test of rainfall-induced landslide) based on the fully coupled analysis are also presented in Section 4.4.

4.2 MODELING OF UNSATURATED SOIL BASED ON CONTINUITY MECHANICS

4.2.1 State variables

The state variables to describe the mechanical behavior of unsaturated soil under isothermal conditions include stress state variables and deformation state variables (Fredlund et al., 2012). Once stress state variables and deformation state variables are defined, constitutive relations can be proposed to represent relations between changes of stress state variables and deformation state variables. Ideally the definition of stresses should be independent of the behavior or the states of the material, so that the stress space does not change with the material state. The commonly used stress state variable for saturated soil is the effective stress $\sigma' = \sigma - u_w$. For soils that are partially saturated, the choice of the stress state variable becomes more complicated. Bishop (1959) proposed the single stress variable that is commonly been referred to as Bishop's effective stress for unsaturated soils:

$$\sigma' = (\sigma - u_a) + \chi(u_a - u_w) \tag{4.1}$$

where:

σ is the total stress
σ' is the effective stress
u_a is the pore air pressure
u_w is the pore-water pressure
χ is the parameter with the value between zero and unity

As the parameter χ usually depends on soil type, material states (e.g., the degree of saturation), and even on stress path, the Bishop's effective stress of unsaturated soils depends on the material behavior and changes with material states. Therefore, the constitutive behavior of the material is embodied in both the constitutive relation and the stress space (Fredlund and Rahardjo, 1993; Sheng et al., 2008b; Fredlund et al., 2012).

In 1960s and 1970s, researchers gradually turned to the approach to use two independent stress variables to model unsaturated soil behavior (Coleman, 1962; Bishop and Blight, 1963). Fredlund and Morgenstern (1977) provided a theoretical basis and justification for the use of two independent stress state variables. Three possible combinations of independent stress state variables, (1) $(\sigma - u_a)$ and $(u_a - u_w)$ (2) $(\sigma - u_w)$ and $(u_a - u_w)$; and (3) $(\sigma - u_a)$ and $(\sigma - u_w)$, were shown to be justifiable from the theoretical continuum mechanics analysis. The combination of the net normal stress $(\sigma - u_a)$ and the matric suction $(u_a - u_w)$ has been proved to be the easiest to apply in engineering practice.

The deformation state variables for unsaturated soils must satisfy continuity, compatibility, and conservation of mass (Fredlund et al., 2012). Both volume change and distortion should be considered. Based on the continuum mechanics, the strain tensor can be expressed as

$$\varepsilon = \begin{bmatrix} \varepsilon_x & \gamma_{xy} & \gamma_{xz} \\ \gamma_{yx} & \varepsilon_y & \gamma_{yz} \\ \gamma_{zx} & \gamma_{zy} & \varepsilon_z \end{bmatrix} \tag{4.2}$$

where:

ε is the strain tensor
ε_x, ε_y, and ε_z are the normal strain components in x-, y-, and z-directions, respectively
γ_{xy}, γ_{yx}, γ_{yz}, γ_{zy}, γ_{xz}, and γ_{zx} are the shear strain components

Denoting u, v, and w as the displacement components in x-, y-, and z-directions, respectively, and assuming the strains to be small, the strain components can be expressed as

$$\begin{Bmatrix} \varepsilon_x = \dfrac{\partial u}{\partial x}, \gamma_{xy} = \gamma_{yx} = \dfrac{\partial u}{\partial y} + \dfrac{\partial v}{\partial x} \\ \varepsilon_y = \dfrac{\partial v}{\partial y}, \gamma_{yz} = \gamma_{zy} = \dfrac{\partial v}{\partial z} + \dfrac{\partial w}{\partial y} \\ \varepsilon_z = \dfrac{\partial w}{\partial z}, \gamma_{xz} = \gamma_{zx} = \dfrac{\partial u}{\partial z} + \dfrac{\partial w}{\partial x} \end{Bmatrix} \tag{4.3}$$

The volumetric strain, ε_v is expressed as

$$\varepsilon_v = \varepsilon_x + \varepsilon_y + \varepsilon_z \tag{4.4}$$

4.2.2 Governing equations

4.2.2.1 Force equilibrium

Consider a small cubic element of the soil (Figure 4.1), whose sides are parallel with the coordinate axes. The element is large enough compared to the size of the pores and is considered as an infinitesimal in the mathematical treatment (Biot, 1941). The cubic element is referred to as a referential elemental volume (Fredlund et al., 2012). The corresponding total stress tensor is denoted by

$$\sigma = \begin{bmatrix} \sigma_x & \tau_{xy} & \tau_{xz} \\ \tau_{yx} & \sigma_y & \tau_{yz} \\ \tau_{zx} & \tau_{zy} & \sigma_z \end{bmatrix} \tag{4.5}$$

where:

σ is the total stress tensor

σ_x, σ_y, and σ_z are the normal stresses in x-, y-, and z-directions, respectively

τ_{xy} is the shear stress on the x-plane (normal direction along x-axis) in the y-direction ($\tau_{xy} = \tau_{yx}$)

τ_{yz} is the shear stress on the y-plane in the z-direction ($\tau_{yz} = \tau_{zy}$)

τ_{zx} is shear stress on the z-plane in the x-direction ($\tau_{zx} = \tau_{xz}$)

The force equilibrium equation for soil structure of a saturated/unsaturated soil is

$$\begin{Bmatrix} \dfrac{\partial \sigma_x}{\partial x} + \dfrac{\partial \tau_{xy}}{\partial y} + \dfrac{\partial \tau_{xz}}{\partial z} + b_x = 0 \\ \dfrac{\partial \tau_{yx}}{\partial x} + \dfrac{\partial \sigma_y}{\partial y} + \dfrac{\partial \tau_{yz}}{\partial z} + b_y = 0 \\ \dfrac{\partial \tau_{zx}}{\partial x} + \dfrac{\partial \tau_{zy}}{\partial y} + \dfrac{\partial \sigma_z}{\partial z} + b_z = 0 \end{Bmatrix} \tag{4.6}$$

where b_x, b_y, and b_z are the body forces in x-, y-, and z-directions, respectively.

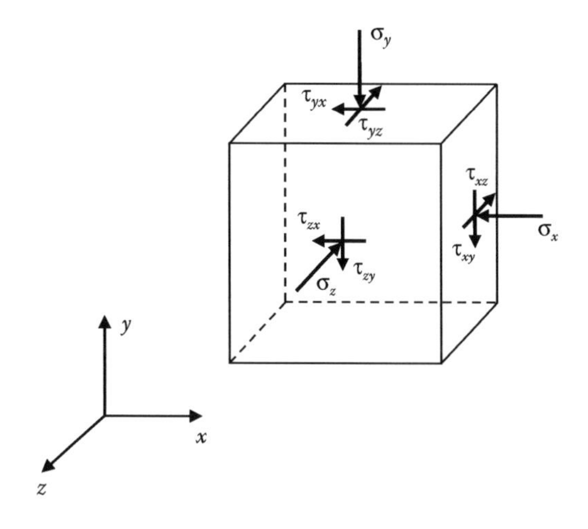

Figure 4.1 Soil element and total stress components.

4.2.2.2 *Mass continuity of water phase*

The governing equation of water flow in the pore of unsaturated soil based on mass conservation is

$$\nabla(\mathbf{v}) = -\frac{\partial\theta_w}{\partial t} \tag{4.7}$$

where:

$\mathbf{v}$ is the flow rate vector
θ_w is the volumetric water content
t is time

The flow of water in saturated and unsaturated soil system can be described by the Darcy's law (Childs and Collis-George, 1950):

$$\mathbf{v} = -\mathbf{k}\cdot\nabla H \tag{4.8}$$

where:

$\mathbf{k}$ is the tensor of coefficient of permeability
H is the total head

Therefore, the governing equation of water flow in saturated/unsaturated soil can be expressed as

$$\nabla(-\mathbf{k}\nabla H) = -\frac{\partial\theta_w}{\partial t} \tag{4.9}$$

4.2.3 Formulations based on elastic constitutive model

4.2.3.1 *Effective stress approach*

For simplicity of the formulations, the incremental forms of the constitutive models, that is, increments of strains, stresses, and the change of volumetric water content are simply expressed as absolute strains, stresses, and volumetric water content in the equations.

Based on the Hooke's law for an isotropic elastic medium, we have

$$\left\{\begin{array}{l}
\varepsilon_x = \dfrac{\sigma_x}{E} - \dfrac{\nu}{E}\left(\sigma_y + \sigma_z\right) \\
\varepsilon_y = \dfrac{\sigma_y}{E} - \dfrac{\nu}{E}\left(\sigma_z + \sigma_x\right) \\
\varepsilon_z = \dfrac{\sigma_z}{E} - \dfrac{\nu}{E}\left(\sigma_x + \sigma_y\right) \\
\gamma_{xy} = \gamma_{yx} = \tau_{xy}/G \\
\gamma_{yz} = \gamma_{zy} = \tau_{yz}/G \\
\gamma_{xz} = \gamma_{zx} = \tau_{zx}/G
\end{array}\right\} \tag{4.10}$$

where:

E is the Young's modulus
G is the shear modulus
ν is the Poisson's ratio of the solid skeleton

There are only two independent constants because of the relation

$$G = \frac{E}{2(1+\nu)} \tag{4.11}$$

Biot (1941) assumed that the pore-water pressure u_w cannot produce any shearing strain and its effect is the same on all three components of strain because of the assumed isotropy of the soil. The normal strain components of the soil are expressed as a function of the stresses in the soil and the pressure of the water in the pores:

$$\begin{Bmatrix} \varepsilon_x = \dfrac{\sigma_x}{E} - \dfrac{\nu}{E}(\sigma_y + \sigma_z) + \dfrac{u_w}{3H_B} \\ \varepsilon_y = \dfrac{\sigma_y}{E} - \dfrac{\nu}{E}(\sigma_z + \sigma_x) + \dfrac{u_w}{3H_B} \\ \varepsilon_z = \dfrac{\sigma_z}{E} - \dfrac{\nu}{E}(\sigma_x + \sigma_y) + \dfrac{u_w}{3H_B} \end{Bmatrix} \tag{4.12}$$

where H_B is the elastic modulus for soil structure with respect to pore-water pressure. It should be noted that the above equation implies the effective stress equation expressed as $\sigma' = \sigma + u_w$. The sign of stress of this formulation follows the rules of continuity mechanics, that is, positive sign is used for tension and negative sign represents compression for the stresses. However, for pore pressure, the positive sign means compression.

Assuming a change in shear stress cannot affect the water content and stresses in all three directions x, y, z have equivalent effect on the change of water content, the increment of water content:

$$\theta_w = \frac{1}{3H_B'}(\sigma_x + \sigma_y + \sigma_z) + \frac{u_w}{R_B} \tag{4.13}$$

where:

- θ_w is increment of water content
- H_B' is the coefficient of pore-water content change due to normal stresses
- R_B is the coefficient of pore-water content change due to pore-water pressure

Biot (1941) did not mention whether this water content θ_w is volumetric or gravimetric. Here, we use the same symbol as the volumetric water content. Biot (1941) proved that $H_B = H_B'$ under the assumption of the existence of a potential energy of the soil. This assumption means that if the changes occur at an infinitely slow rate, the work done to bring the soil from the initial condition to its final state of strain and water content, is independent of the way by which the final state is reached and is a definite function of the six strain components and the water content.

Based on Equations 4.12 and 4.4, the stresses can then be expressed as functions of the strain and the pore-water pressure:

$$\begin{Bmatrix} \sigma_x = 2G(\varepsilon_x + \dfrac{\nu\varepsilon_v}{1-2\nu}) - \alpha_B u_w \\ \sigma_y = 2G(\varepsilon_y + \dfrac{\nu\varepsilon_v}{1-2\nu}) - \alpha_B u_w \\ \sigma_z = 2G(\varepsilon_z + \dfrac{\nu\varepsilon_v}{1-2\nu}) - \alpha_B u_w \end{Bmatrix} \tag{4.14}$$

where

$$\alpha_B = \frac{2(1+\nu)}{3(1-2\nu)}\frac{G}{H_B} \tag{4.15}$$

Based on Equations 4.13 and 4.14 and the relation $H_B = H_B'$, the change of water content can be expressed as function of volumetric strain and pore-water pressure:

$$\theta_w = \alpha_B \varepsilon_v + u_w / Q_B \tag{4.16}$$

where

$$\frac{1}{Q_B} = \frac{1}{R_B} - \frac{\alpha_B}{H_B}$$

Substituting Equation 4.14 for the stresses into the equilibrium conditions Equation 4.6, we can obtain:

$$\begin{Bmatrix} G\nabla^2 u + \dfrac{G}{1-2\nu}\dfrac{\partial \varepsilon_v}{\partial x} - \alpha_B \dfrac{\partial u_w}{\partial x} + b_x = 0 \\ G\nabla^2 v + \dfrac{G}{1-2\nu}\dfrac{\partial \varepsilon_v}{\partial y} - \alpha_B \dfrac{\partial u_w}{\partial y} + b_y = 0 \\ G\nabla^2 w + \dfrac{G}{1-2\nu}\dfrac{\partial \varepsilon_v}{\partial z} - \alpha_B \dfrac{\partial u_w}{\partial z} + b_z = 0 \end{Bmatrix} \tag{4.17}$$

where ∇^2 is the Laplace operator or Laplacian, which is usually denoted by the symbols $\nabla \cdot \nabla$, ∇^2 or Δ. For the three-dimensional (3D) problem, $\nabla^2 = \partial^2 / \partial x^2 + \partial^2 / \partial y^2 + \partial^2 / \partial z^2$.

Assume the water to be incompressible, based on Equations 4.9 and 4.16 we obtain

$$\nabla \cdot [-\mathbf{k}\nabla H] = -\alpha_B \frac{\partial \varepsilon_v}{\partial t} - \frac{1}{Q_B}\frac{\partial u_w}{\partial t} \tag{4.18}$$

Equations 4.17 and 4.18 are the basic governing equations with the four unknowns u, v, w, and u_w. The coefficient $1/Q_B$ is a measure of the amount of water that can be forced into the soil under pore pressure. α_B is a coefficient that measures the ratio of the water volume squeezed out to the volume change of the soil. Biot (1941) pointed out that the coefficients α_B and $1/Q_B$ will be of significance for an unsaturated soil. For a completely saturated soil, $\alpha_B = 1$ and $1/Q_B = 0$. If the soil is unsaturated, $\alpha_B \neq 1$ and Q_B is finite, and the two coefficients depend on the degree of saturation of the soil.

Kim (1996, 2000, 2004) extended the Biot's consolidation theory for unsaturated soil by adopting a modified form of the Bishop's effective stress for unsaturated soil:

$$\sigma' = \sigma + \alpha_c \cdot S \cdot u_w \tag{4.19}$$

where:

S is the degree of saturation ($0 \leq S \leq 1$)
α_c is the Biot's hydro-mechanical coupling coefficient ($0 \leq \alpha_c \leq 1$)

Considering the change of water mass in a unit volume of unsaturated porous media was composed of three components: (1) compressibility of water; (2) deformation of soil skeleton;

and (3) variation of degree of saturation (Narasimhan & Witherspoon, 1977; Noorishad et al., 1982; Kim, 1996), and the governing equations for the coupled flow and deformation in unsaturated soils are expressed as follows (Kim, 2000):

$$\begin{cases} G\nabla^2 u + \dfrac{G}{1-2\nu}\dfrac{\partial \varepsilon_v}{\partial x} - \alpha_c S \gamma_w \dfrac{\partial h}{\partial x} = 0 \\ G\nabla^2 v + \dfrac{G}{1-2\nu}\dfrac{\partial \varepsilon_v}{\partial y} - \alpha_c S \gamma_w \dfrac{\partial h}{\partial y} = 0 \\ G\nabla^2 w + \dfrac{G}{1-2\nu}\dfrac{\partial \varepsilon_v}{\partial z} - \alpha_c S \gamma_w \dfrac{\partial h}{\partial z} = 0 \end{cases} \quad (4.20)$$

$$\nabla(-\mathbf{k}\nabla H) = -\alpha_c S_w \frac{\partial \varepsilon_v}{\partial t} - \left(n\frac{\mathrm{d}S}{\mathrm{d}h}\right)\frac{\partial h}{\partial t} - \left(nS\beta_w\gamma_w\right)\frac{\partial h}{\partial t} \quad (4.21)$$

where:

- $\mathbf{k}$ is the effective permeability tensor
- $H = h + z$ is the total hydraulic head
- z is the vertical coordinate and elevation head
- h is the pore-water pressure head
- n is the porosity
- γ_w is the unit weight of water
- β_w is the compressibility of water (5.0×10^{-10} m^2/N)

In the right-hand side of Equation 4.21, the first term represents the change of water content due to the deformation of soil skeleton. The second term represents the change of water content due to variation of degree of saturation. The third term represents change of water content due to the compressibility of water.

4.2.3.2 Two stress state variables approach

Some features of the volumetric behavior of unsaturated soils that cannot be explained using the effective stress concept have been observed through experimental work. Figure 4.2 shows the variation of void ratio in cycles of swell–shrink for an expansive clay from northern Karnataka, a state of India. The expansive behavior of unsaturated soils due to wetting can be explained by the effective stress concept as the effect stress of unsaturated soil is reduced after wetting. Figure 4.3 presents the void ratio state surface of a mixture of flint and kaolin. The soil structure of the mixture decreased in volume as the suction ($u_a - u_w$) was reduced. This collapse behavior caused by wetting cannot be explained the principle of effective stress and is often attributed to a metastable structure (Alonso et al., 1990; Fredlund et al., 2012). Figure 4.4 illustrates the results of odometer tests on Boom clay. The soil is initially dry and wetted up to a suction of 10 kPa and applied with cycles of suction between 10 and 600 kPa. The measured vertical strains show a progressive transition from swelling to collapse as the vertical applied load increases.

Fredlund and Rahardjo (1993) and Fredlund et al. (2012) described the volume change of the soil skeleton and the water content using two stress state variables, the net normal stress ($\sigma - u_a$) and the matric suction ($u_a - u_w$). The effects of net normal stress and matric suction on volume changes of soil structure and water are considered. The expansive and collapsible volume change behavior of unsaturated soil can be described.

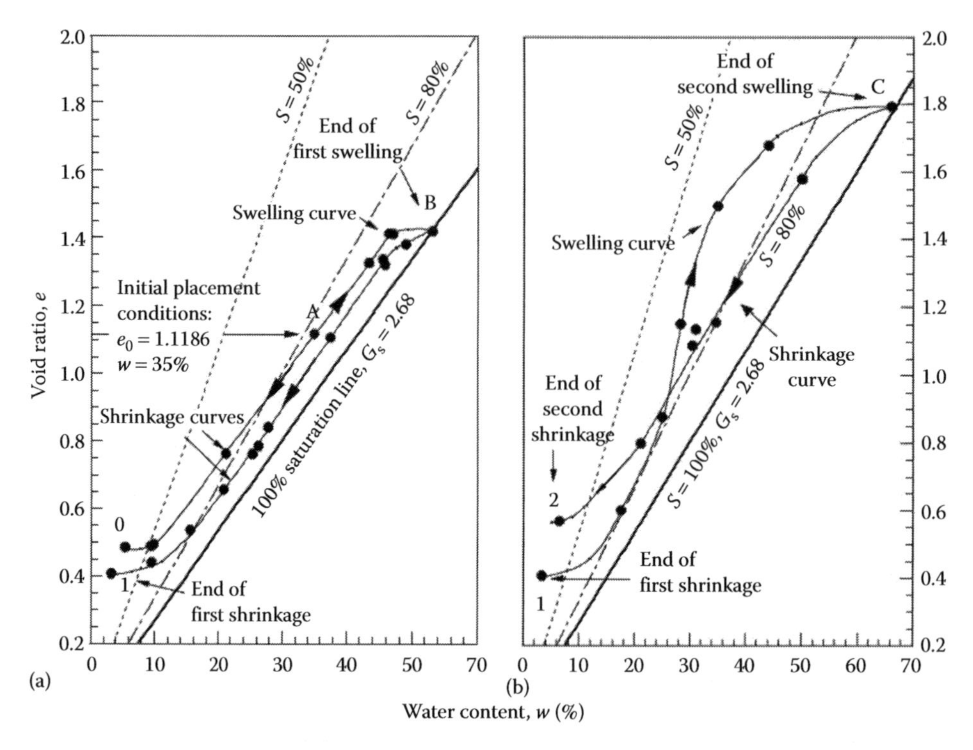

Figure 4.2 Water content—void ratio swell–shrink paths of an expansive clay for swelling—full shrinkage cycles: (a) cycle 1 and (b) cycle 2. (From Tripathy, S. et al., *Canadian Geotechnical Journal*, 39, 938–959, Copyright 2002 Canadian Science Publishing or its licensors. Reproduced with permission.)

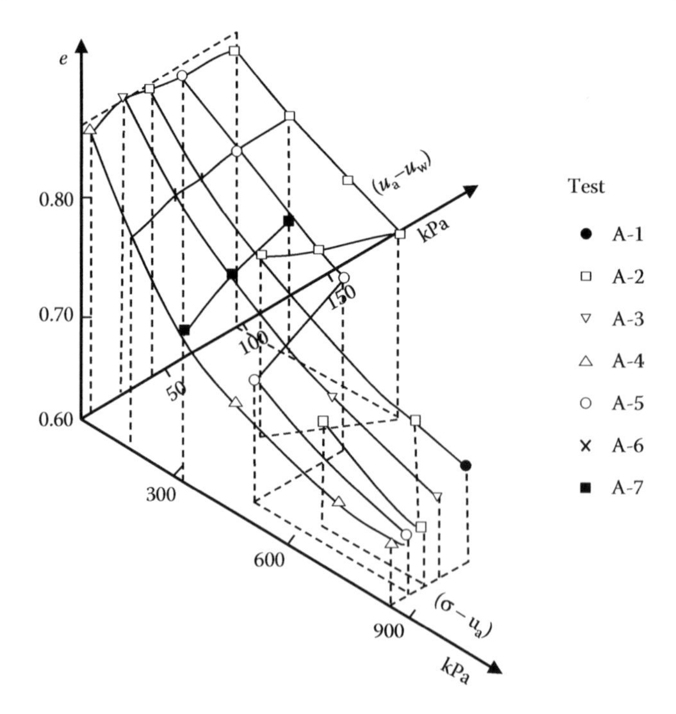

Figure 4.3 Void ratio surface for a mixture of flint and kaolin under isotropic loading condition. (After Matyas, E. L., and Radhakrishna, H. S., *Géotechnique*, 18, 432–448, 1968. Republished with permission of Thomas Telford.)

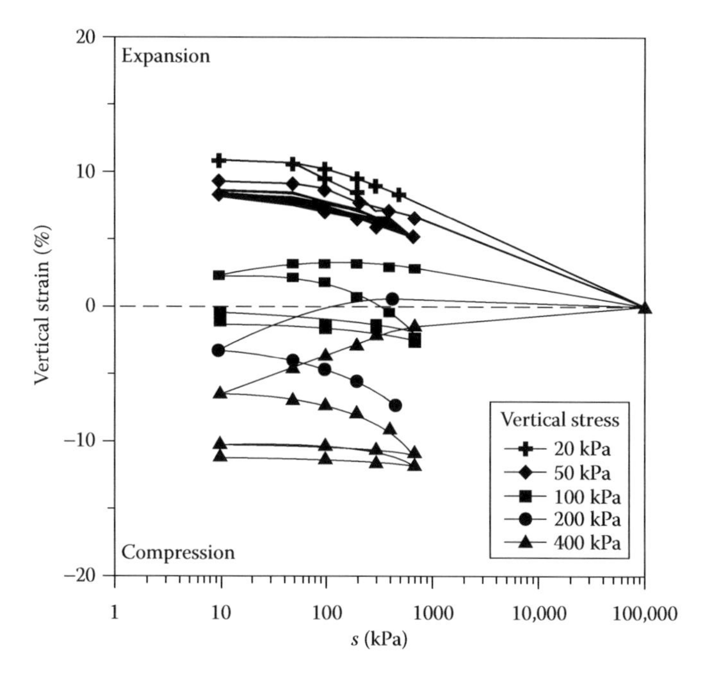

Figure 4.4 Suction-controlled odometer tests on compacted pellets of Boom clay. (From Alonso, E. E. et al., *Engineering Geology*, 54, 173–83, 1999. With permission.)

Assuming the soil behaves as an isotropic incrementally elastic material, the soil structure stress–strain relationships associated with the normal strains can be written as follows:

$$\begin{aligned}
\varepsilon_x &= \frac{(\sigma_x - u_a)}{E} - \frac{\nu}{E}\left(\sigma_y + \sigma_z - 2u_a\right) + \frac{(u_a - u_w)}{H_F} \\
\varepsilon_y &= \frac{(\sigma_y - u_a)}{E} - \frac{\nu}{E}\left(\sigma_x + \sigma_z - 2u_a\right) + \frac{(u_a - u_w)}{H_F} \\
\varepsilon_z &= \frac{(\sigma_z - u_a)}{E} - \frac{\nu}{E}\left(\sigma_y + \sigma_x - 2u_a\right) + \frac{(u_a - u_w)}{H_F}
\end{aligned} \tag{4.22}$$

where:

E is the elastic modulus for the soil structure with respect to changes in net normal stresses
H_F is the elastic modulus for soil structure with respect to a change in matric suction

It should be noted that the incremental form of constitutive modeling is also not used in this section for simplicity.

The volumetric strain is expressed as a function of the net normal stress and the matric suction under general three dimensional loading:

$$\varepsilon_v = \frac{3(1-2\nu)}{E}(\sigma_{mean} - u_a) + \frac{3}{H_F}(u_a - u_w) \tag{4.23}$$

where:

σ_{mean} is the mean total normal stress [i.e., $(\sigma_x + \sigma_y + \sigma_z)/3$]
$\sigma_{mean} - u_a$ is the mean net normal stress

In this formulation, the stresses are positive for compression. However, the normal strains are positive for expansion and negative for compression. The shear strain is positive if the angle between two positive directions of axes decreases. Therefore, the elastic modulus E has a negative sign because an increase of normal stress leads to a compression of a soil (Fredlund et al., 2012) as shown in Figure 4.5(a). The signs of H_F are related with the volume change behavior of an unsaturated soil. If the soil is an expansive soil (reduction of matric suction leads to an increase of soil volume), H_F has a negative sign. If the soil is a collapsible soil (reduction of matric suction leads to a decrease of soil volume), H_F has a positive sign. Figure 4.5(b) presents a schematic plot for the signs of H_F for collapsible and expansive soils.

The above equation can also be expressed in the form of coefficients of volume change:

$$\varepsilon_v = m_1^s(\sigma_{mean} - u_a) + m_2^s(u_a - u_w) \tag{4.24}$$

where:

m_1^s is the coefficient of volume change of soil skeleton with respect to a change in $(\sigma_{mean} - u_a)$, $m_1^s = 3(1-2\nu)/E$

m_2^s is the coefficient of volume change of soil skeleton with respect to a change in $(u_a - u_w)$, $m_2^s = 3/H_F$

Similarly, the change of volumetric water content can be written in a semiempirical linear combination of the stress state variables:

$$\theta_w = m_1^w(\sigma_{mean} - u_a) + m_2^w(u_a - u_w) \tag{4.25}$$

where:

m_1^w = coefficient of volume change of pore water with respect to a change in $(\sigma_{mean} - u_a)$

m_2^w = coefficient of volume change of pore water with respect to a change in $(u_a - u_w)$

The coefficients of volume changes for an expansive soil are illustrated in the schematic plot (Figure 4.6). Note that Equations 4.24 and 4.25, ε_v, θ_w, $(\sigma_{mean} - u_a)$, and $(u_a - u_w)$ all represent increment variables.

Usually, the state surfaces of void ratio can be obtained through laboratory tests on unsaturated soils. It is more convenient to express the volume change in term of these parameters. The change of void ratio can be written as

$$e = a_t(\sigma_{mean} - u_a) + a_m(u_a - u_w) \tag{4.26}$$

where:

e is the void ratio

a_t is the coefficient of compressibility with respect to a change in $(\sigma_{mean} - u_a)$

a_m is coefficient of compressibility with respect to a change in $(u_a - u_w)$

The coefficients of volume change can be calculated from the constitutive surface for the void ratio e of the soil:

$$m_1^s = \frac{a_t}{1+e_0} \tag{4.27}$$

$$m_2^s = \frac{a_m}{1+e_0} \tag{4.28}$$

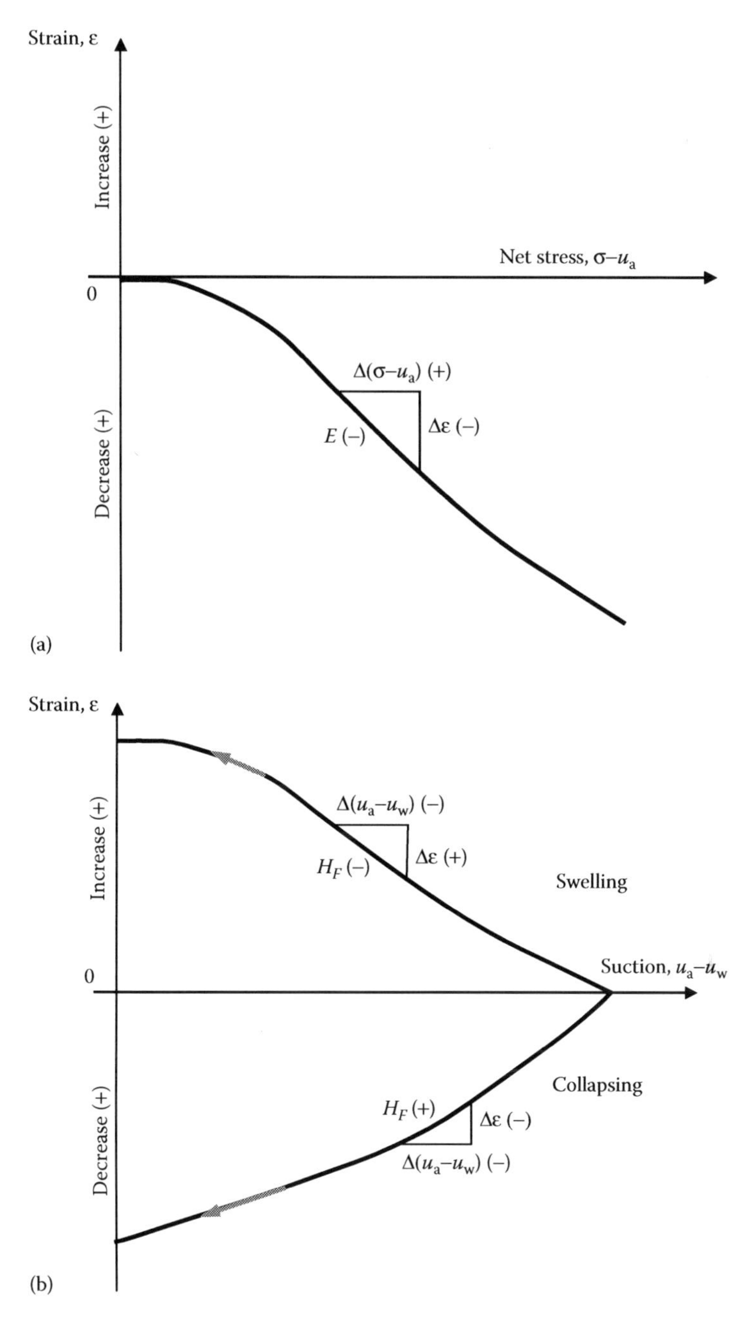

Figure 4.5 Definitions of elastic modulus (a) E and (b) H_F in Fredlund and Rahardjo (1993) formulation with two stress state variables. (Modified from Fredlund, D. G. et al., *Unsaturated Soil Mechanics in Engineering Practice*. John Wiley & Sons, New York, 2012. With permission.)

where e_o = initial void ratio.

If we substitute $(\sigma_{mean} - u_a)$ in Equation 4.24 into Equation 4.25, the change of volumetric water content can be written as a function of volumetric strain and the soil suction:

$$\theta_w = \beta_{w1}\varepsilon_v + \beta_{w2}(u_a - u_w) \tag{4.29}$$

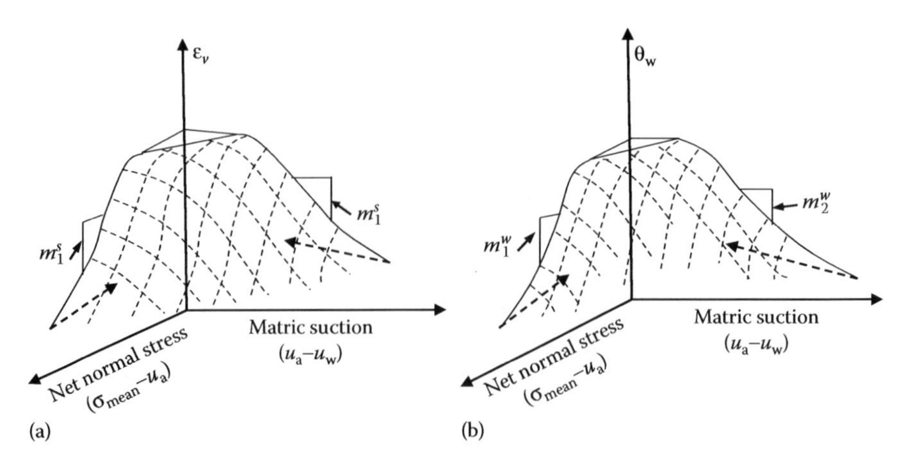

Figure 4.6 Three-dimensional constitutive surfaces for volume change of unsaturated soils (a) soil structure and (b) water phase. (Modified from Fredlund, D. G. et al., *Unsaturated Soil Mechanics in Engineering Practice*. John Wiley & Sons, New York, 2012. With permission.)

where:

$$\beta_{w1} = \frac{m_1^w}{m_1^s}$$

$$\beta_{w2} = \left(m_2^w - \frac{m_1^w m_2^s}{m_1^s} \right)$$

Substituting the strain displacement relations and the stress–strain relations into the equilibrium equation, the governing equations for general 3D condition are (Fredlund et al., 2012)

$$G\nabla^2 u_i + \frac{G}{1-2\nu}\frac{\partial \varepsilon_v}{\partial x_i} - \beta_F \frac{\partial(u_a - u_w)}{\partial x_i} + \frac{\partial u_a}{\partial x_i} + b_i = 0 \quad (i = x, y, z) \tag{4.30}$$

$$\nabla(-\mathbf{k}\nabla H) + \beta_{w1}\frac{\partial \varepsilon_v}{\partial t} + \beta_{w2}\frac{\partial(u_a - u_w)}{\partial t} = 0 \tag{4.31}$$

where $\beta_F = \dfrac{m_2^s}{m_1^s} = \dfrac{E/H_F}{(1-2\nu)}$

For the plane strain condition, the governing equations for coupled flow and deformation in unsaturated deformable soil are (Fredlund et al., 2012)

$$\frac{\partial}{\partial x}\left\{\frac{E}{(1-2\nu)(1+\nu)}\left[(1-\nu)\frac{\partial u}{\partial x} + \nu\frac{\partial v}{\partial y}\right]\right\} + \frac{\partial}{\partial y}\left\{G\left(\frac{\partial v}{\partial x} + \frac{\partial u}{\partial y}\right)\right\} - \beta_F\frac{\partial(u_a - u_w)}{\partial x} + \frac{\partial u_a}{\partial x} + b_x = 0 \tag{4.32}$$

$$\frac{\partial}{\partial x}\left\{G\left(\frac{\partial v}{\partial x} + \frac{\partial u}{\partial y}\right)\right\} + \frac{\partial}{\partial y}\left\{\frac{E}{(1-2\nu)(1+\nu)}\left[\nu\frac{\partial u}{\partial x} + (1-\nu)\frac{\partial v}{\partial y}\right]\right\} - \beta_F\frac{\partial(u_a - u_w)}{\partial y} + \frac{\partial u_a}{\partial y} + b_y = 0 \tag{4.33}$$

$$\frac{\partial}{\partial x}\left[k_x\frac{\partial}{\partial x}\left(\frac{u_w}{\gamma_w} + y\right)\right] + \frac{\partial}{\partial y}\left[k_y\frac{\partial}{\partial y}\left(\frac{u_w}{\gamma_w} + y\right)\right] = \frac{\partial}{\partial t}\left[\beta_{w1}(\frac{\partial u}{\partial x} + \frac{\partial v}{\partial y}) + \beta_{w2}(u_a - u_w)\right] \tag{4.34}$$

Comparing the above two equations with Equations 4.20 and 4.21, we can see that in Fredlund and Rahardjo's formulation, the volume change of water phase includes both deformation of soil skeleton and the variation of degree of saturation in pores. The compressibility of water phase is not considered. The deformation of soil skeleton and the variation of degree of saturation are affected by the net normal stress ($\sigma - u_a$) and the matric suction ($u_a - u_w$), separately.

4.2.4 Illustrative examples for elastic model

4.2.4.1 *Consolidation of unsaturated soil*

As is shown in Figure 4.7, a sandy soil column is used to simulate a soil layer of 10 m thick. The initial groundwater table is located at a certain depth below the ground surface. The initial pore pressure is hydrostatic pressure with the soil above the water table unsaturated. A surcharge load of 100 kPa is applied instantaneously on the ground surface. The soil column then deforms and consolidates under the surcharge load.

The boundary conditions for stress deformation and water flow are defined as follows. For stress and deformation of the soil skeleton, AB and CD are fixed horizontally but free vertically; the bottom surface BC is fixed both vertically and horizontally. A load of 100 kPa is applied on the top surface AD. For water flow, the left and right boundaries, AB and CD, and the bottom surface, BC, are impermeable. The top surface AD is a drained boundary. The hypothetical two-dimensional (2D) model is assumed to simulate a land with a very large area and the surcharge load is entirely applied on the ground surface. The flow of water is mainly along the vertical direction. Hence, the lateral boundaries are assumed to be impermeable to flow. The soil column is discretized into 250 rectangular elements. The average size of the element is 0.2 m. In this example, the formulation in Equations 4.20 and 4.21 is adopted. The degree of saturation and the permeability of the unsaturated soil are defined as follows:

$$S = \begin{cases} S_r + \dfrac{1 - S_r}{\left\{1 + \left[\beta_k (-h)^{\gamma_k}\right]\right\}^{\alpha_k}} & h < 0 \\ 1 & h \geq 0 \end{cases} \tag{4.35}$$

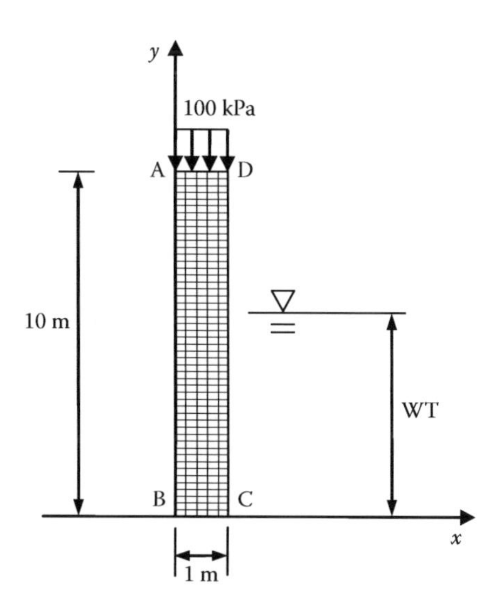

Figure 4.7 An illustrative example of unsaturated soil consolidation.

$$k = \begin{cases} k_s\left\{1+\left[a_k(-h)^{b_k}\right]\right\}^{-\alpha_k} & h<0 \\ k_s & h \geq 0 \end{cases} \tag{4.36}$$

where:

S_r is the residual degree of saturation

α_k, β_k, γ_k, a_k, and b_k are parameters of the hydraulic property functions

The values of the soil parameters are summarized in Table 4.1.

Three initial water table levels (i.e., 0 m, 6 m, and 10 m above the bottom surface) are considered. Figure 4.8 shows the variation of pore-water pressure and settlement along the vertical direction with time for the case with the initial water table at 6 m. As shown in Figure 4.8(a), in the unsaturated zone, the excess pore-water pressure almost dissipates in 1s. In the zone near the original water table, about 16 kPa of excess pore-water pressure is generated, which implies that the water table raises about 1.6 m. In Figure 4.8(b), it is clearly demonstrated that the initial surface settlement is mainly due to the deformation in the unsaturated zone. At the time of 1 s, the vertical displacement occurs mostly in the unsaturated zone. After that, the saturated soil below the water table deforms, which leads to the increase of surface settlement.

The results of pore-water pressure dissipation and surface settlement are shown in Figure 4.9. According to Figure 4.9, when the water level is at 10 m above the bottom surface, the soil column is fully saturated, and hence the initial ground settlement is almost zero and the excess pore-water pressure at the top surface is 100 kPa. During consolidation, the ground settlement gradually increases with the dissipation of excess pore-water pressure. After about 100 min, the excess pore-water pressure decreases to nearly zero and the final settlement of the soil column is about 39 mm. As shown in Figure 4.9, the dissipation of pore-water pressure at the bottom and the variation of the surface settlement with time from this study agree well with the results in Kim (2000).

When the water level is at 6 m above the bottom surface, the soil column is partially saturated with unsaturated soils above the water table. When the surface load is applied suddenly on the top surface, there is an initial settlement of 16 mm, which is mainly due to the deformation in the unsaturated zone. The consolidation of the partially saturated soil column is faster than that of a fully saturated soil column as the unsaturated soil zone absorbs a large amount of excess pore water in a short time after the surcharge load is applied. The final settlement does not change with the initial water table because the deformation due

Table 4.1 Soil properties of sandy soil in the example of unsaturated consolidation

Parameters (unit)	*Value*	*Parameters (unit)*	*Value*
ρ_w (kg/m^3)	1000	ν	0.3
g (m/s^2)	9.801	S_r	0.07
ρ_s (kg/m^3)	2.65×10^3	α_k	1.0
β_w (m^2/kN)	5×10^{-7}	β_k (m^{-1})	1.74
α_c	1.0	γ_k	2.5
n	0.45	a_k (m^{-1})	6.67
E (kPa)	1.9×10^4	b_k	5.00
k_s (m/s)	3.47×10^{-5}		

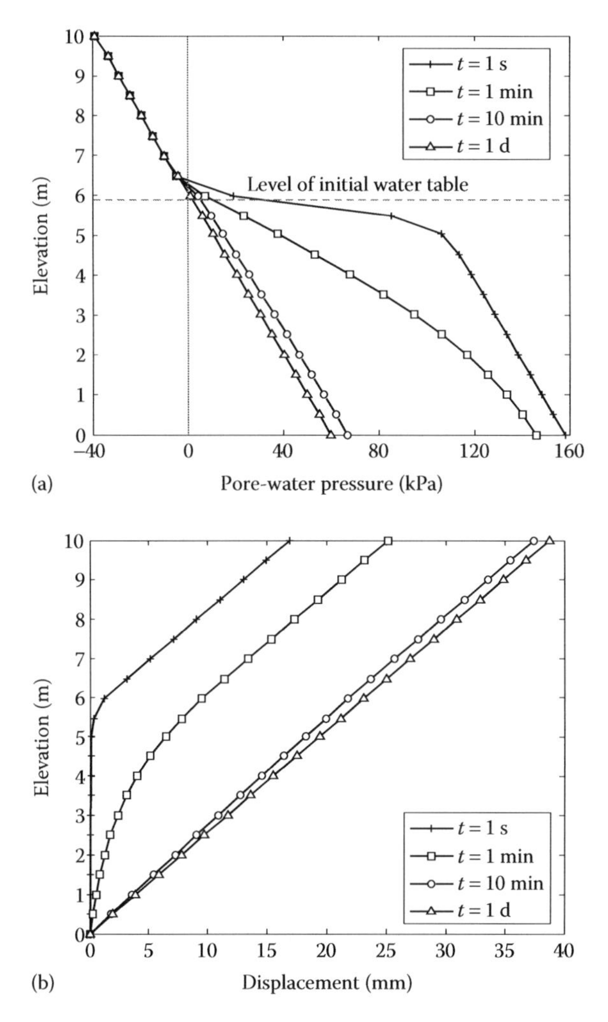

Figure 4.8 (a) Pore-water pressure and (b) vertical displacement profiles along elevation (WT = 6 m).

to consolidation induced by surcharge loading depends mostly on the amount of generated excess pore-water pressure, which is created by the applied load. If the water level is at 0 m, the soil column is completely composed of unsaturated soils. The generated excess pore-water pressure is absorbed by the unsaturated soil very quickly. The consolidation of the unsaturated soil column is completed immediately after being loaded, and hence the initial settlement is almost the same as the final settlement.

Here, we denote the soil in Table 4.1 as Soil A, which is a sandy soil. Assume a clayey soil with a set of unsaturated hydraulic parameters (α_k = 2.0, β_k = 0.01 m^{-1}, γ_k = 2.0, a_k = 0.01 m^{-1}, and b_k = 2.0) and denote this soil as Soil B. The air-entry value (AEV) of Soil B is larger than that of Soil A, and the slope of the hydraulic functions of Soil B is smaller than that of Soil A, which implies Soil B is finer than Soil A. Figure 4.10 illustrates the pore-water pressure and vertical displacement along elevation for Soil B. Compared with Figure 4.8, the excess pore-water pressure can be cumulated in the unsaturated zone for the clayey soil. Hence, the dissipation of pore-water pressure and deformation of Soil B is much slower than that in soil A.

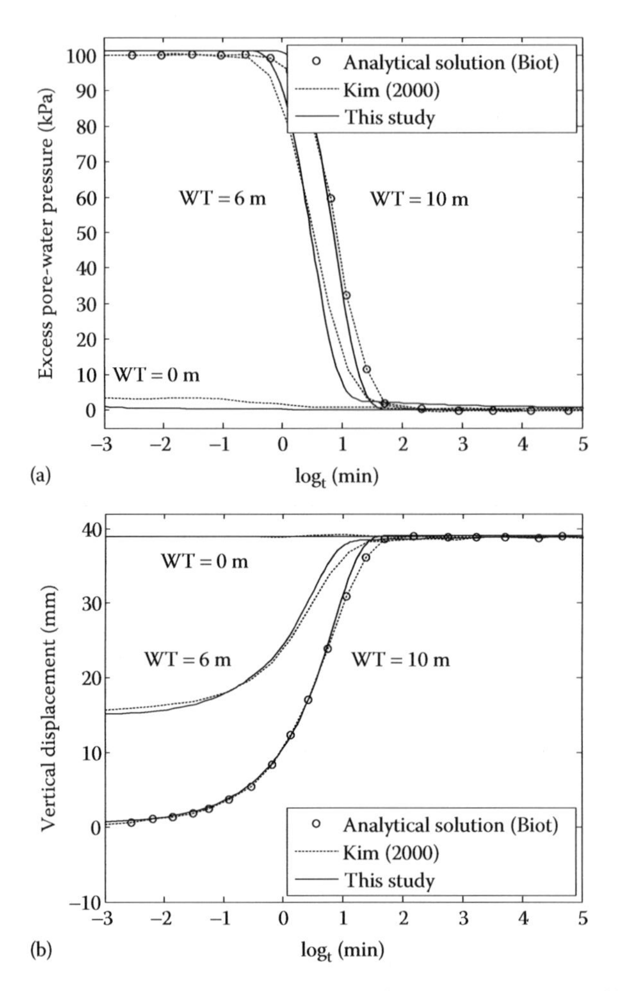

Figure 4.9 Verification of the coupled consolidation model for unsaturated soils: (a) excess pore-water pressure at bottom surface ($y = 0$) with time and (b) settlement with time at top surface ($y = 10$). WT is the groundwater table.

4.2.4.2 Heave of ground under evaporation

A problem considering expansion of swelling clay under a slab due to infiltration is solved based on the coupled formulation of Fredlund and Rahardjo (1993) and Fredlund et al. (2012). A 5 m deep deposit of Regina clay is partially covered with a flexible cover slab (Figure 4.11). The initial matric suction in the soil mass is assumed to be uniform and equal to 400 kPa. The transient wetting process is introduced by imposing a water infiltration rate equal to 2×10^{-8} m/s at the uncovered portion of the ground surface. Such a wetting condition simulates the water infiltration into the soil mass due to the watering of a lawn or a light rain. The analysis is performed to track both the swelling soil behavior and matric suction changes as the transient wetting front advances in the soil mass.

Experimental data obtained from tests on compacted specimens of Regina clay are used to determine the constitutive surfaces of the soil. The void ratio state surface and the degree of saturation are defined by a unique equation with different sets of fitting parameters (Vu, 2003):

$$e = a_{vu} + b_{vu} \log\left[\frac{1 + c_{vu}(\sigma - u_a) + d_{vu}(u_a - u_w)}{1 + f_{vu}(\sigma - u_a) + g_{vu}(u_a - u_w)}\right] \tag{4.37}$$

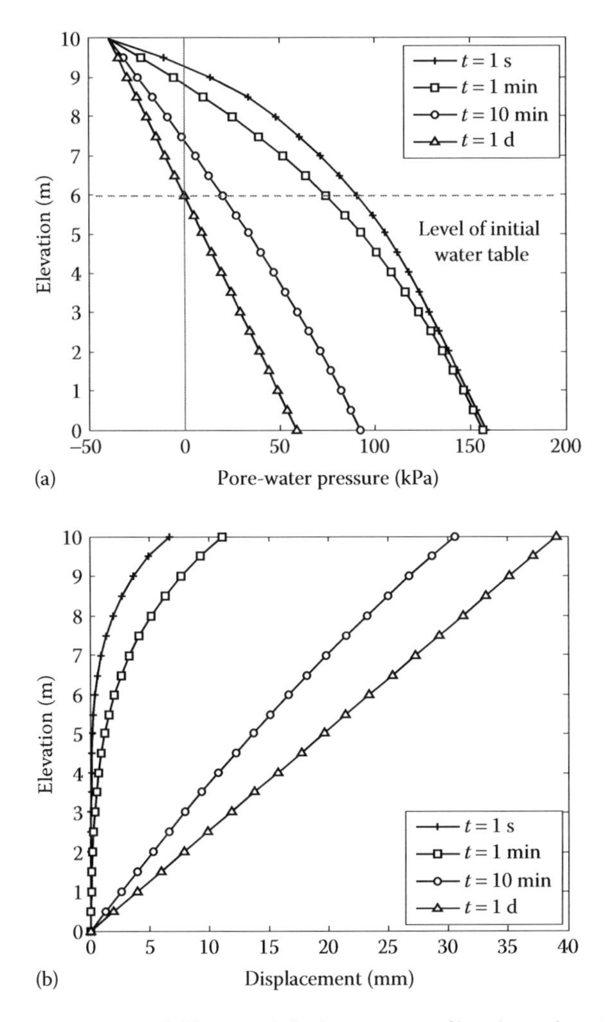

Figure 4.10 (a) Pore-water pressure and (b) vertical displacement profiles along elevation (soil B, WT = 6 m).

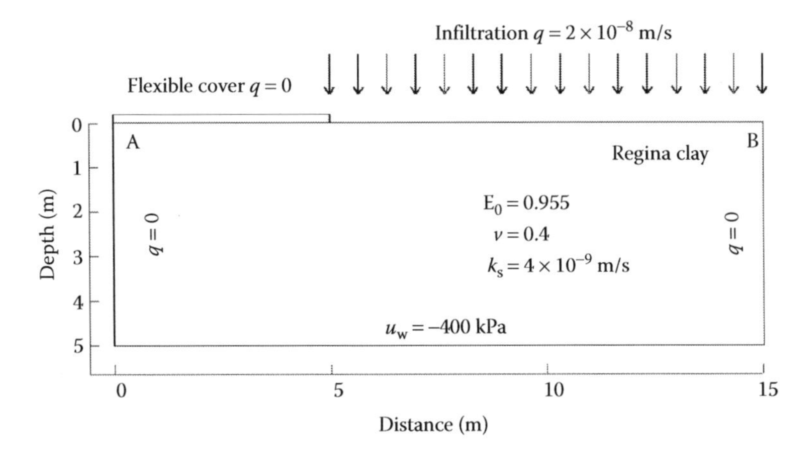

Figure 4.11 Illustration of the geometry and boundary conditions. (After Vu, H. Q., *Uncoupled and Coupled Solutions of Volume Change Problems in Expansive Soils.* Ph.D. Dissertation, University of Saskatchewan, Saskatoon, Canada, 2003. With permission.)

$$S = a_{vu} + b_{vu} \log\left[\frac{1 + c_{vu}(\sigma - u_a) + d_{vu}(u_a - u_w)}{1 + f_{vu}(\sigma - u_a) + g_{vu}(u_a - u_w)}\right] \quad (4.38)$$

where:
e is the void ratio
a_{vu}–g_{vu} are fitting parameters

Table 4.2 summarized the fitting parameters for the Regina clay. Figures 4.12 and 4.13 illustrate the constitutive surfaces for void ratio and degree of saturation, respectively.

The unsaturated permeability of the Regina clay is fitted using the Gardner (1958) equation and the change of void ratio is considered:

$$k = \frac{k_{s0} e^{18.5}}{1 + a_g \left(\frac{(u_a - u_w)}{\rho_w g}\right)^{b_g}} \quad (4.39)$$

where:
$k_{s0} = 0.4 \times 10^{-8}$ m/s
$a_g = 0.01$
$b_g = 1.1$

Table 4.2 Fitting parameters of constitutive surfaces for Regina clay

		Fitting parameters					
Surfaces	*Equations*	a_{vu}	b_{vu}	c_{vu}	d_{vu}	f_{vu}	g_{vu}
Void ratio e	Equation 4.37	1.2492	–0.0979	4.8240	3.3330	0.0009	0.0012
Degree of saturation S	Equation 4.38	1.0000	–0.0725	0.0125	11.7265	0.0125	0.0071

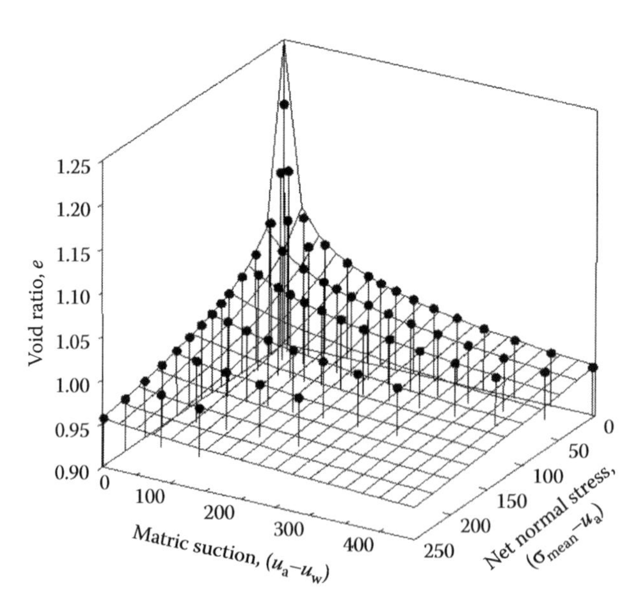

Figure 4.12 Void ratio state surface for Regina clay. (After Vu, H. Q., *Uncoupled and Coupled Solutions of Volume Change Problems in Expansive Soils*. Ph.D. Dissertation, University of Saskatchewan, Saskatoon, Canada, 2003. With permission.)

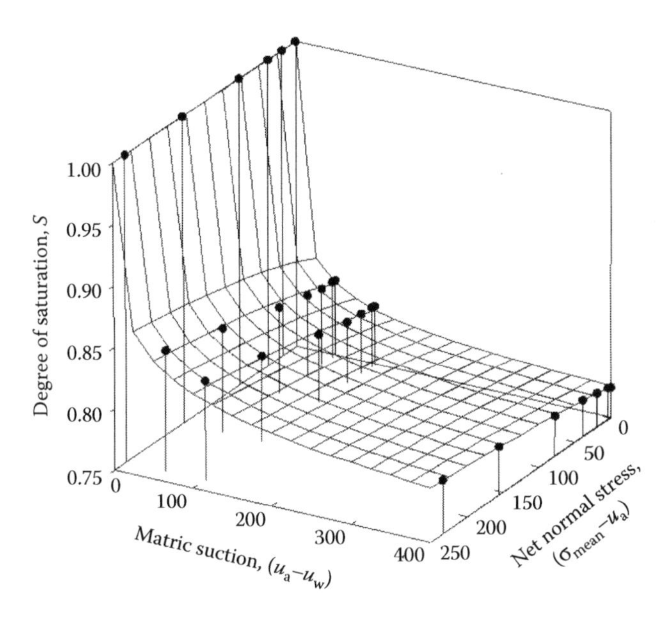

Figure 4.13 Degree of saturation constitutive surface for Regina clay. (After Vu, H. Q., *Uncoupled and Coupled Solutions of Volume Change Problems in Expansive Soils*. Ph.D. Dissertation, University of Saskatchewan, Saskatoon, Canada, 2003.)

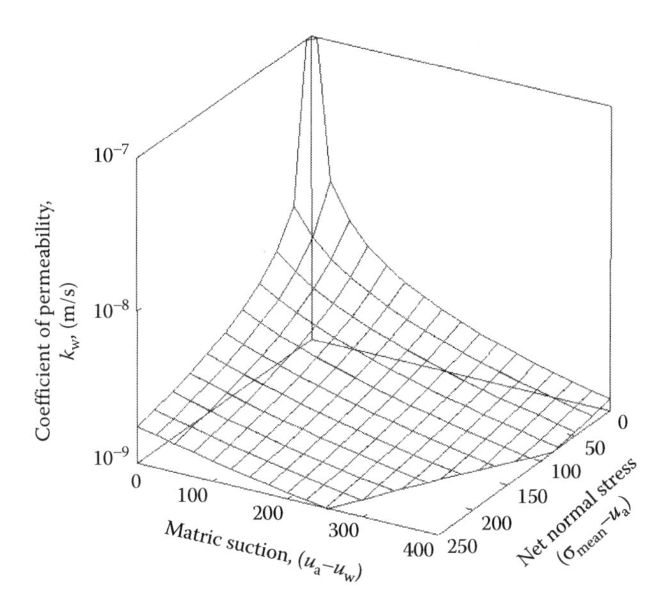

Figure 4.14 Coefficient of permeability constitutive surface for Regina clay. (After Vu, H. Q., *Uncoupled and Coupled Solutions of Volume Change Problems in Expansive Soils*. Ph.D. Dissertation, University of Saskatchewan, Saskatoon, Canada, 2003. With permission.)

The coefficient of permeability of the soil is presented graphically in Figure 4.14 for the void ratio surface shown in Figure 4.12. The Poisson's ratio is equal to 0.40.

Vu (2003), and Vu and Fredlund (2004) presented the coupled solutions of this problem, which was solved by a coupled hydro-mechanical numerical modeling program COUPSO developed by Pereira (1996). In this example, we adopt a finite element partial differential equation solver, FlexPDE (PDE Solutions Inc, 2015), to develop a coupled numerical model. The soil is assumed to be linear incremental elastic material. In the coupled analysis, the coefficients of deformation, the void ratio, soil-water characteristic curve (SWCC), and coefficient of permeability are updated according to the stresses and pore-water pressures. Figures 4.15 present the matric suction distributions after the wetting at day 53 and day 175, respectively. The patterns of contours are in good agreement with Vu (2003). Figure 4.16 illustrate the horizontal displacement and vertical displacement, respectively, at day 175. The displacements by this study are slightly greater than the result obtained by Vu (2003) because the change of matric suction is greater than that in the latter. Figure 4.17 shows the change of vertical displacement at the ground surface from point A to point B with time. The vertical displacements are greater than those obtained by Vu (2003). The difference between the results by this study and those by Vu (2003) are around 10%.

4.2.5 Formulations based on plastic constitutive models

An elastoplastic constitutive model based on the critical state concept for unsaturated soils was first presented in a qualitative form by Alonso et al. (1987). This was subsequently developed into a mathematical model by Alonso et al. (1990). The elastoplastic model was named Barcelona Basic Model (BBM). The model was intended for slightly or moderately expansive unsaturated soils. It was formulated based on the modified Cam-Clay model for saturated soils and was extended for unsaturated soils through the introduction of the concept of the loading-collapse (LC) yield surface using two independent stress variables. A large number of elastoplastic models for unsaturated soils (Pastor et al., 1990; Toll, 1990; Gens and Alonso, 1992; Kohgo et al., 1993; Wheeler and Sivakumar, 1995; Bolzon et al., 1996; Cui and Delage, 1996; Alonso et al., 1999; Loret and Khalili, 2002; Tang and Graham, 2002; Chiu and Ng, 2003; Gallipoli et al., 2003; Sheng et al., 2004, 2008a; Tamagnini, 2004; Sánchez et al., 2005; Tarantino and Tombolato, 2005; Russell and Khalili, 2006; Li, 2007; Sun et al., 2007; Thu and et al., 2007; Kohler and Hofstetter, 2008; Mašın and Khalili, 2008; Zhang

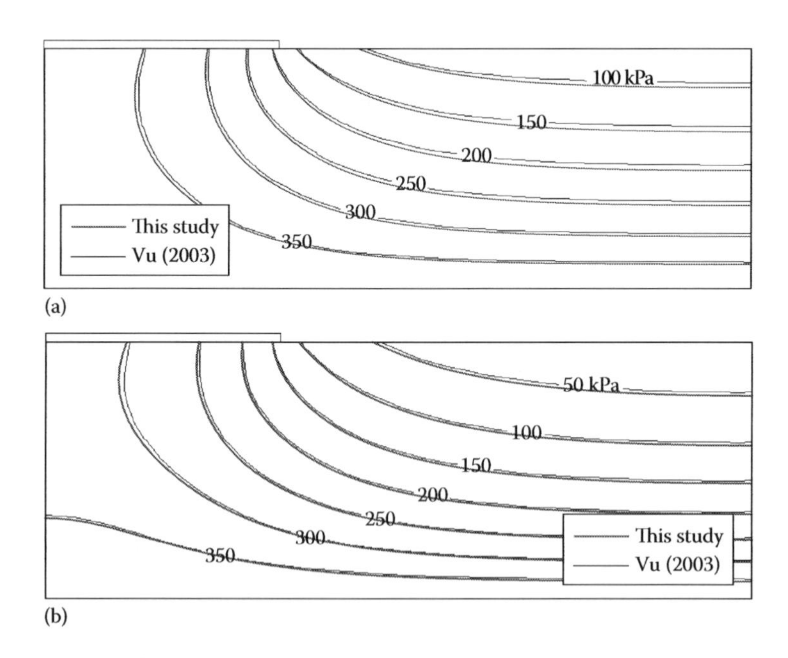

Figure 4.15 Comparison of distributions of matric suction (kPa) at (a) day 53 and (b) day 175.

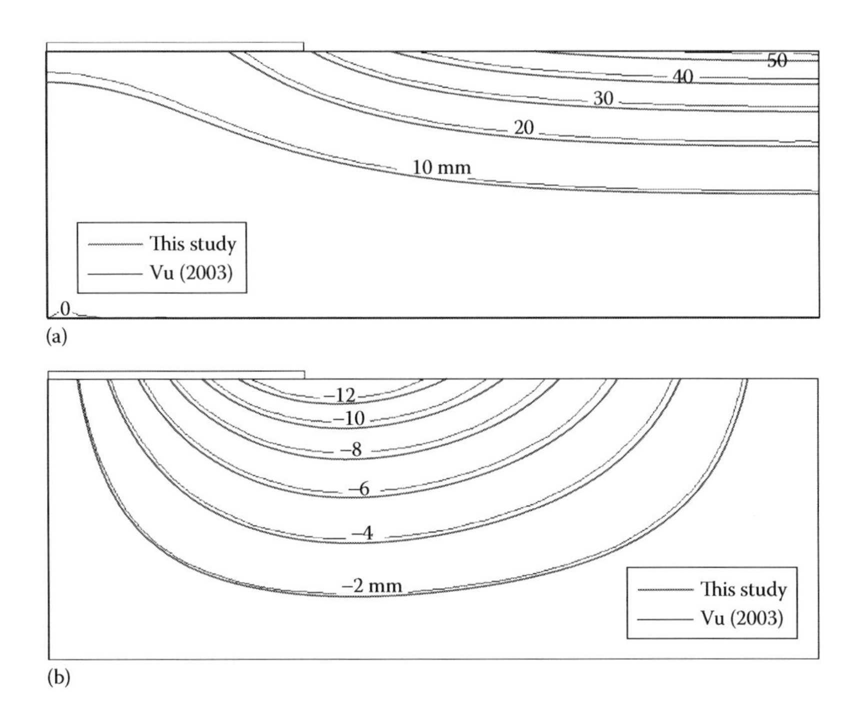

Figure 4.16 Comparison of (a) vertical displacement and (b) horizontal displacement (mm) at day 175.

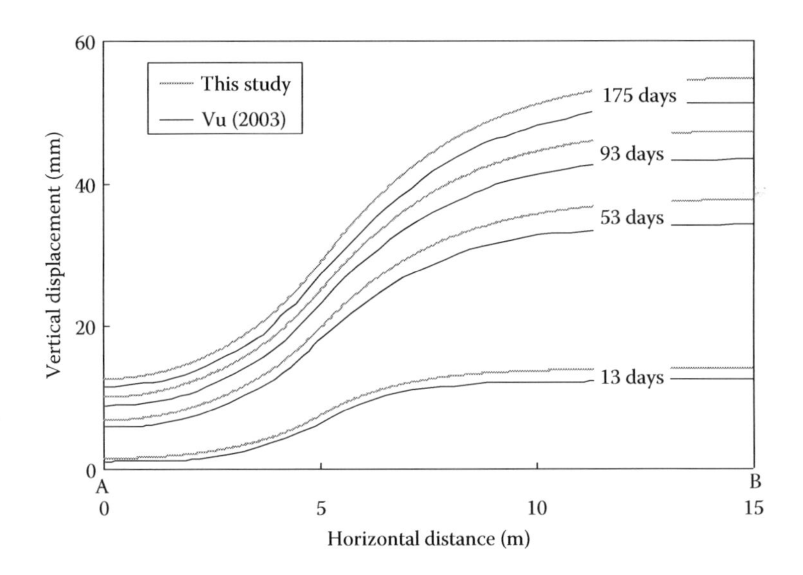

Figure 4.17 Change of vertical displacement at ground surface from point A to point B with time.

and Lytton, 2009; Zhang and Ikariya, 2011; Arairo et al., 2013; Vecchia and Romero, 2013; Gamnitzer and Hofstetter, 2015; Hu et al., 2015; among others) are developed following the work by Alonso et al. (1990). A comprehensive review of constitutive modeling of unsaturated soils and numerical algorithms can be found in Wheeler et al. (2003), Gens et al. (2008), Sheng et al. (2008b) and Sheng (2011).

In this section, we will briefly introduce the BBM model. To be consistent with the convention of symbols, in the following part of this section, we adopted the same system of symbols of the elastoplastic models based on critical state soil mechanics The stress variables in the BBM model are the mean net stress $\bar{p}$ and matric suction s, defined as follows:

$$\bar{p} = p - u_a, \quad s = u_a - u_w \tag{4.40}$$

where p is the mean total stress.

The isotropic normal compression lines (NCL) in the v:$\bar{p}$ plane for unsaturated soil is

$$v = N(s) - \lambda(s)\ln\left(\frac{\bar{p}}{p^c}\right) \tag{4.41}$$

where:

- v is the specific volume ($v = 1 + e$)
- $N(s)$ and $\lambda(s)$ are the intercept and gradient (stiffness parameter) of NCL for unsaturated soil and are both functions of suction
- p^c is a reference stress (see Figure 4.18a)

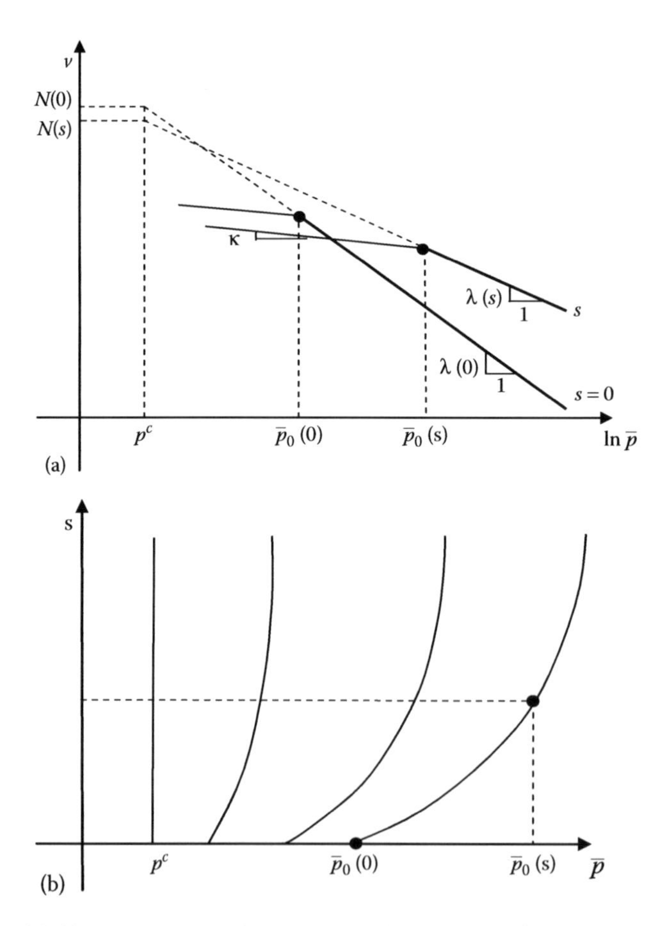

Figure 4.18 BBM model: (a) isotropic normal compression lines and (b) LC yield curve.

On unloading and reloading (at constant s) the soil is supposed to behave elastically:

$$dv = \kappa \frac{d\bar{p}}{\bar{p}} \tag{4.42}$$

where κ is the unloading stiffness parameter due to net stress loading.

Assume the suction unloading (wetting) occurs in the elastic domain and is given by a logarithmic expression similar to Equation 4.42:

$$dv = \kappa_s \frac{ds}{s + p_{at}} \tag{4.43}$$

where:

κ_s is the swelling stiffness parameter due to suction unloading

p_{at} is atmospheric pressure that is added to s to avoid infinite values as s approaches zero

The saturated preconsolidation stress is labeled $\bar{p}_0(0)$. An unsaturated sample at suction s will yield at a larger preconsolidation stress $\bar{p}_0(s)$. If both preconsolidation points belong to the same yield curve in a ($\bar{p}$, s) stress plane (Figure 4.18b), a relationship between the unsaturated yield stress, $\bar{p}_0(s)$ and the saturated value $\bar{p}_0(0)$ can be obtained by relating the specific volumes through virtual loading path and by assuming

$$N(s) - N(0) = \kappa_S \ln\left(\frac{s + p_{at}}{p_{at}}\right) \tag{4.44}$$

where $N(0)$ is the intercept of NCL in v-$\bar{p}$ plane for saturated soil ($s = 0$).

The yield curve in the ($\bar{p}$, s) plane can then be expressed as

$$\frac{\bar{p}_0(s)}{p^c} = \left(\frac{\bar{p}_0(0)}{p^c}\right)^{(\lambda(0)-\kappa)/(\lambda(s)-\kappa)} \tag{4.45}$$

where $\lambda(0)$ is the stiffness parameter of a saturated soil. This equation explains not only the apparent increase in preconsolidation stress associated with increasing suction, but also the collapse phenomena observed in wetting paths. For this reason, the yield curve is named LC yield curve. Figure 4.18b shows that the LC yield curve is vertical in the ($\bar{p}$, s) plane when $\bar{p}_0(0) = p^c$; and the yield curve then becomes increasingly inclined as the LC yield curve expands.

In the BBM model, the value of $\lambda(s)$ decreased with increasing suction and is assumed to vary exponentially with suction

$$\lambda(s) = \lambda(0)\left[(1 - r_{BBM})\exp(-\beta_{BBM}s) + r_{BBM}\right] \tag{4.46}$$

where r_{BBM} and β_{BBM} are soil constants. The value of r_{BBM} is less than 1.

To consider the effects of irreversible shrinkage when suction increases above a threshold value and increase of shear strength due to suction, the yield surface is further modified as shown in Figure 4.19. It is proposed that whenever the soil reaches a maximum previously attained value of the suction s_0, irreversible strains will begin to develop. The following yield condition is adopted:

$$s = s_0 \tag{4.47}$$

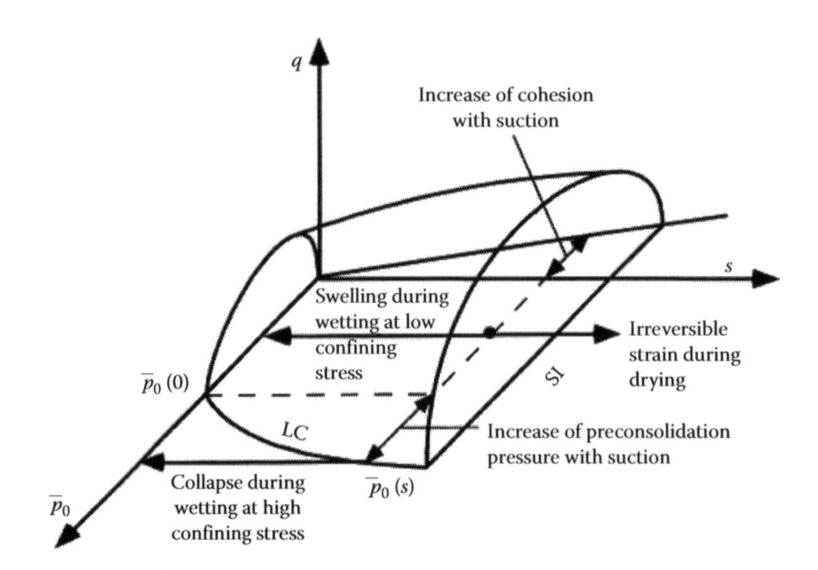

Figure 4.19 The yield locus of the BBM model in *p–q–s* plane. (Modified from *Engineering Geology*, 54, Alonso, E. E. et al., Modelling the mechanical behaviour of expansive clays, 173–183, Copyright 1999, with permission from Elsevier.)

where s_0 is the maximum past suction ever experienced by the soil and bounds the transition from the elastic state to the virgin range when suction is increased.

The effect of suction on shear strength is represented by an increase in cohesion, maintaining the slope of the critical state line (CSL), M, for saturated conditions. If the increase in cohesion follows a linear relationship with suction (Figure 4.19), the ellipses of yield will intersect the $\bar{p}$ axis at a point:

$$\bar{p} = -p_s = -k_{\text{BBM}} \cdot s \tag{4.48}$$

where:

k_{BBM} is a constant

p_s is the intercept of the ellipse of yield surface at the $\bar{p}$ axis at suction s

The equation for the ellipsoidal yield surface of BBM model is expressed as

$$F = q^2 - M^2\left(\bar{p} + p_s\right)\left(\bar{p}_0(s) - \bar{p}\right) = 0 \tag{4.49}$$

It assumes elastic behavior if the soil remains inside the yield surface and plastic strains are induced once the yield surface is reached. Several key features of the behavior of unsaturated soil can be modeled by the BBM model, for example, an increase of shear strength with suction and the irreversible volumetric compression on wetting. On reaching saturation, the model becomes a conventional critical state model.

An incremental stress–strain relation should be developed to be implemented in a numerical method to solve boundary value problems. For unsaturated soils, these incremental relations should be written according to the stress variables. For example, as in Sheng (2008a), the stress–strain relation based on independent stress variables can be written as

$$\begin{pmatrix} d\bar{\sigma} \\ ds \end{pmatrix} = \begin{pmatrix} \mathbf{D}^{ep} & \mathbf{W}^{ep} \\ \mathbf{R} & G \end{pmatrix} \begin{pmatrix} d\varepsilon \\ d\theta_w \end{pmatrix} \tag{4.50}$$

where:

$\bar{\sigma}$ is the net stress vector
ε is the strain vector
$\mathbf{D}^{ep}$ is the elastoplastic stress–strain matrix
$\mathbf{W}^{ep}$ is the elastoplastic suction–strain vector
$\mathbf{R}$ is a row vector
G is a scalar constant represent the change of suction due to the change of θ_w

4.3 SLOPE STABILITY ANALYSIS BASED ON COUPLED MODELING

4.3.1 Stress analysis

There are two approaches to consider pore-water pressure due to infiltration in stress analysis. In the first approach, simple assumed pore pressure distribution is incorporated into stress analysis. Hence, this approach is considered as a simplified approach. The second approach considers both equilibrium of soil structure and mass conservation of pore fluids and includes matric suction/pore pressure in constitutive models. Therefore, this approach can be considered as a fully coupled approach.

4.3.1.1 Simplified approach

Some researchers adopted simplified assumptions about pore-water pressure distribution due to rainfall. With the simplified assumption of pore-water pressure during rainfall and the effective stress concept, the elastoplastic constitutive models for saturated soils can be used without complicated coupled formulations of unsaturated soils. Ye (2004) and Ye et al. (2005) conducted 2D and 3D finite element (FE) analyses for a large-scale failure in a soft-rock slope due to heavy rain. The modified elastoplastic model with Matsuoka–Nakai failure criterion was used. To simplify the effect of rainfall infiltration, in Ye et al. (2005), the matric suction of unsaturated soil in the slope was not considered. The slope is divided into soil columns similar to the approach in LEM method of slices. A constant increment of pore-water pressure head is applied to each column in an incremental approach and the effective stress is reduced gradually during the whole process of rain (Figure 4.20). Figure 4.21 shows the change of displacement vectors of the slope with time in the 3D analysis. The slope moves upwards during the first 4000 s due to the floating force from the increase pore-water pressure, then it moves downwards along the shear band when the shear strain reached a certain value. The change of displacement vectors with time in the 2D analysis is shown in Figure 4.22 with a similar behavior. The maximum displacement was 13 cm in 2D calculation and 8 cm in 3D calculation (Figure 4.23). The reason of this discrepancy may be that, in the 2D analysis under plane-strain condition, strain along one direction is fixed, which may restrict the further expansion of the soil in 2D analysis. Cheuk et al. (2005) assessed the behavior of loose fill slopes with and without soil nails subjected to the effects of rainfall infiltration using the finite difference software FLAC2D. The strain-softening characteristic of loose fill material was simulated by a user-defined constitutive model (named as SP model). The initial pore-water pressure in the slope is assumed to be hydrostatic with a groundwater table located at a certain depth below

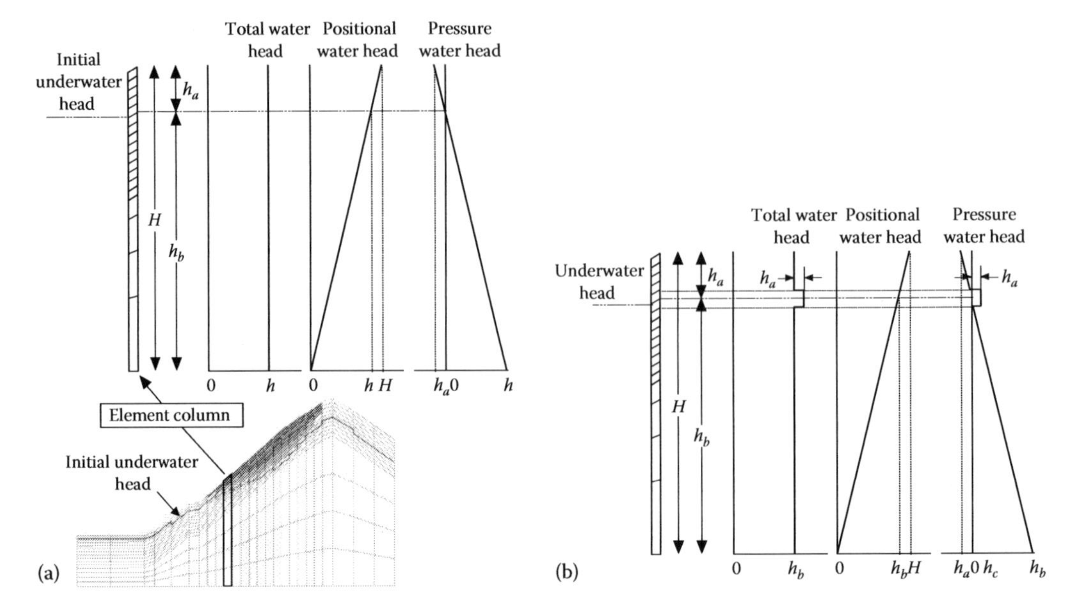

Figure 4.20 (a) Initial condition and (b) change in pore-water pressure due to heavy rain. (From Ye, G. L. et al., *Soils and Foundations*, 45, 1–15, 2005. With permission.)

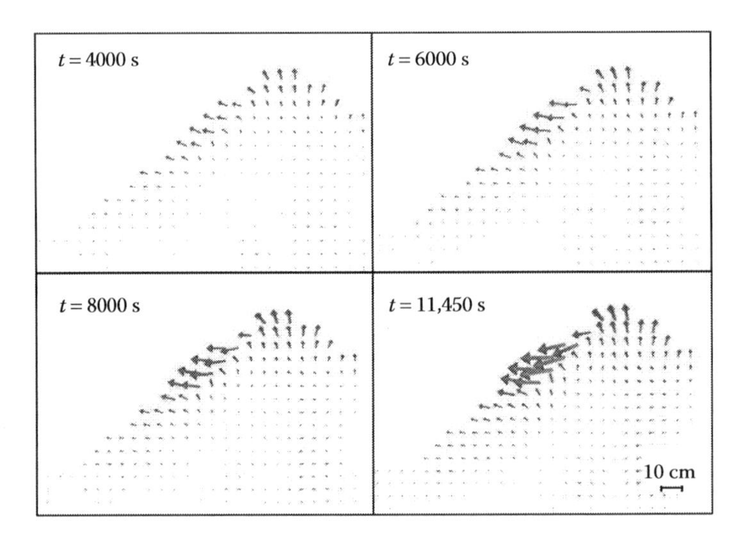

Figure 4.21 Development of displacement vector in the slope with time (central section plane of 3D analysis). (From Ye, G. L., *Numerical Study on the Mechanical Behavior of Progressive Failure of Slope by 2D and 3D FEM*. Ph.D. Thesis, Gifu University, Gifu, Japan, 2004. With permission.)

slope surface. The actual process of rainfall infiltration was not modeled in the numerical analysis. To consider the effects of stress changes in the soil caused by surface infiltration after heavy rainstorms, the suction profile in the top 3 m was replaced by a triangular pore-water pressure distribution with 0 at the slope surface and 10 kPa at the bottom of the 3 m wetted zone (Figure 4.24). For soil at depths below 3 m, the pore-water pressure (or suction) remained unchanged. This simulates the existence of a relatively impermeable layer at a depth of 3 m while a perched water table is allowed to build up.

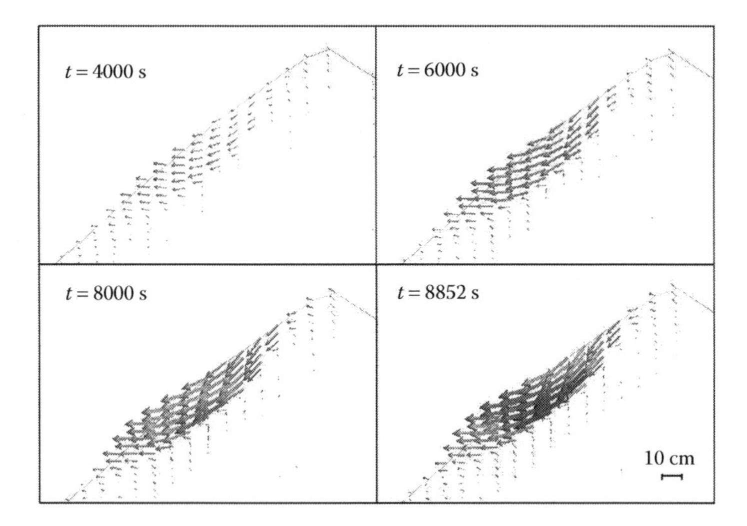

Figure 4.22 Development of displacement vector in the slope with time (2D analysis). (From Ye, G. L., *Numerical Study on the Mechanical Behavior of Progressive Failure of Slope by 2D and 3D FEM*. Ph.D. Thesis, Gifu University, Gifu, Japan, 2004. With permission.)

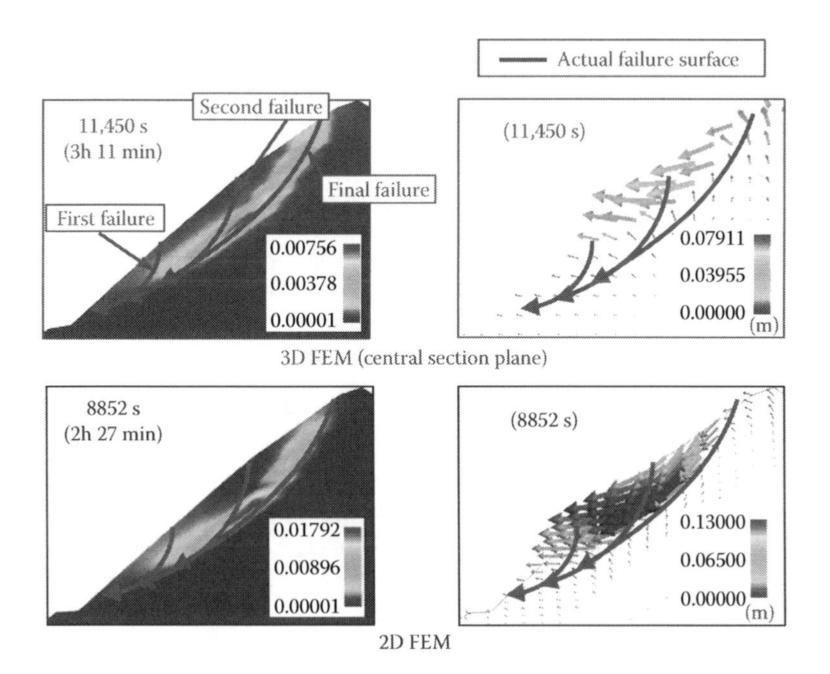

Figure 4.23 Comparison of the calculated shear zone and the observed failure zone. (From Ye, G. L., *Numerical Study on the Mechanical Behavior of Progressive Failure of Slope by 2D and 3D FEM*. Ph.D. Thesis, Gifu University, Gifu, Japan, 2004. With permission.)

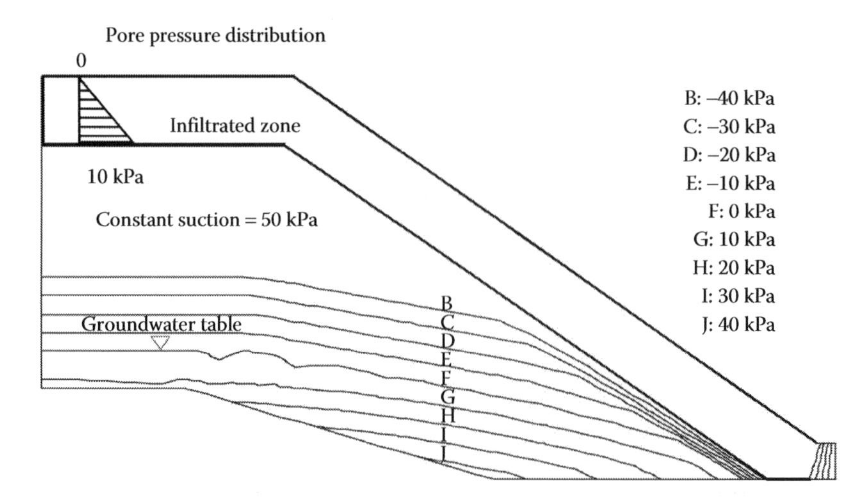

Figure 4.24 Pore pressure distribution to simulate the rainfall infiltration. (Modified from *Computers and Geotechnics*, 32, Cheuk, C.Y. et al., Numerical experiments of soil nails in loose fill slopes subjected to rainfall infiltration effects, 290–303, Copyright 2005, with permission from Elsevier.)

4.3.1.2 Fully coupled approach

The analytical solutions for the coupled hydro-mechanical modeling are very limited. Wu and Zhang (2009) presented an analytical solution to one-dimensional (1D) coupled water infiltration and deformation, which is derived by adopting the exponential functional forms for soil water characteristic curve and permeability function. The Fredlund and Rahardjo (1993) incremental elastic constitutive model for unsaturated soils is selected because the model is simple and can be implemented in the analytical solutions.

Cho and Lee (2001) examined the process of infiltration into a soil slope due to rainfall and the mechanical behavior of the slope using a 2D FE flow-deformation coupled analysis program. The stress–strain relationship was formulated under the framework of elastic modeling. Void ratio was related with suction and net mean stress using the state surface equation (Lloret and Alonso, 1985). The extended Mohr–Coulomb failure criterion was adopted for unsaturated soil shear strength. A hyperbolic model was assumed for the shear modulus to considering the stiffening effect with suction increase and reduction of shear stiffness with increase of stress ratio. Zhang et al. (2005) and Kim et al. (2012) adopted similar approach with incremental elastic and perfect plastic model in their studies.

Alonso and his colleagues (Lloret and Alonso, 1980; Olivella et al., 1996; Alonso et al., 2003) developed a series of coupled hydro-mechanical models with increasing complexity of constitutive relations to deal with problems of flow and mechanical interaction associated with changes in soil suction in saturated and unsaturated soil systems. The computer codes developed include NOSAT (Alonso et al., 1996) for coupled unsaturated flow-deformation analysis and CODE_BRIGHT (Olivella et al., 1996) for thermo-hydro-mechanical coupled analysis in unsaturated soils. Problems such as moisture transfer and deformation behavior of pavement under different climate conditions (Alonso et al., 2002), the deformation and stability of a slope in overconsolidated clays under rainfall (Alonso et al., 2003), and the deformation and seepage in an unsaturated expansive slope subjected to artificial rainfall infiltration (Zhan, 2003) have been solved using these computer codes.

Zhang et al. (2003) conducted a coupled FE analysis to investigate the progressive failure of a cut slope based on an elastoplastic model with strain hardening and softening.

Oka et al. (2010) conducted a multiphase (air–soil–water) coupled FE modeling for a river embankment considering unsaturated seepage flow based on the mixture theory. The saturated elasto-viscoplastic model proposed by Kimoto et al. (2007) was extended to unsaturated soil using the skeleton stress and suction as stress state variables. Garcia et al. (2011) conducted numerical analysis of 1D infiltration based on the multiphase coupled elasto-viscoplastic FE analysis formulation by Oka et al. (2010). Xiong et al. (2014) took the degree of saturation as a state variable and extended the finite-element–finite-difference (FE–FD) formulation in saturated condition proposed by Oka et al. (1994) to unsaturated condition in soil–water–air fully coupling scheme. They developed an FE program named SOFT based on a rational and simple constitutive model for unsaturated soil proposed by Zhang and Ikariya (2011) and simulated laboratory model tests of slope failure in unsaturated Shirasu soil.

Borja and White (2010), and Borja et al. (2012) developed a flow and deformation coupled numerical model for investigation a slope failure in a steep experimental catchment near Coos Bay, Oregon. The adopted constitutive model is an elastoplastic model developed by Borja (2004, 2006), and Borja & Koliji (2009). Smith (2003) investigated the behavior of unsaturated soil slopes under infiltration through a case study of a slope in Tung Chung, Hong Kong. The Imperial College Finite Element Program (ICFEP) (Potts and Zdravkovic, 1999) was used for the analysis. Chen et al. (2009) utilized the commercial FE software ABAQUS (Dassault Systèmes Simulia Corp., 2011). The Bishop's effective stress for unsaturated soil and the elastoplastic model with the Mohr–Coulomb yield criterion were adopted. Biot's consolidation theory is used to calculate the pore-water pressures in the soil. Similar approach was adopted by Cascini et al. (2010) using the code for coupled hydro-mechanical analysis GeHoMadrid (Mira McWilliams, 2001). Other research studies about coupled hydro-mechanical modeling of slope include Mori et al. (2011), Hu et al. (2011), and Sanavia (2009) among others.

Recently, advanced numerical techniques such as material point method (MPM) were used to solve fully coupled problems in unsaturated soil slopes. Bandara and Soga (2015) derived a fully coupled MPM formulation using a mixture-theory-based approach and applied the method to model progressive failure of river levees. Figure 4.25 shows the simulation of a levee failure with unsaturated zone above phreatic line. Initial shear failure generated very high excess pore-water pressures along the shear bands, and thus resulted in nearly zero mean effective stresses in the saturated region. More shear bands were developed with time and spread towards the levee, which lead to a rapid progressive failure. The unsaturated soil that lies above the saturated region undergoes discontinuous failure mechanisms, and the discrete soil blocks tend to flow above the saturated region that induced more instabilities. These results illustrated that the coupled MPM formulation can capture the failure, postfailure, and deposition stages when modeling large deformations and discontinuities.

4.3.2 Estimation of factor of safety for slope stability

Because the stress state and pore-water pressure of distributions can be obtained from the coupled hydro-mechanical modeling, a local safety factor at certain point in the slope can be obtained (Alonso et al., 2003). Alternatively, the enhanced limit method (also known as Kulhawy method; Kulhawy, 1969) that computes the global factor of safety, F_s, along a specified slip surface based on stress distributions from an FE analysis can be adopted (Cho and Lee, 2001; Zhang et al. 2005; Kim et al., 2012). The minimum factor of safety is obtained through a series of trial slip surfaces. The advantage of the enhanced limit method (ELM) is that it can fully utilize the information of the stress field obtained from numerical

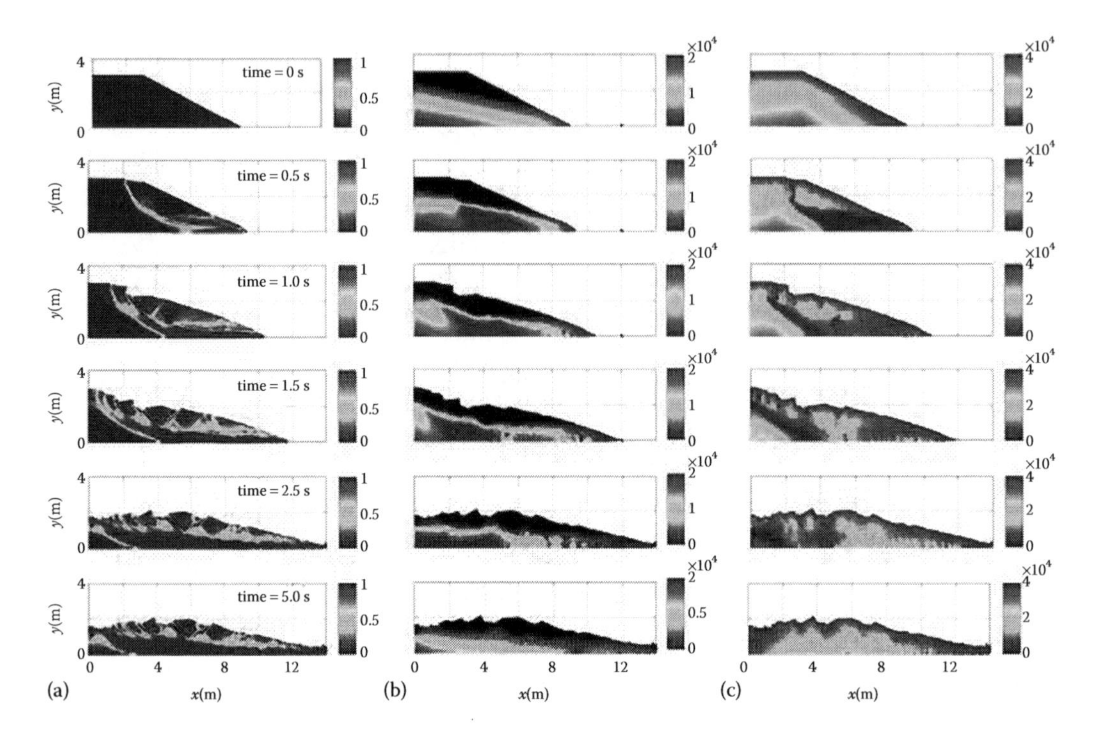

Figure 4.25 Levee failure with very large deformations by coupling of soil deformation and pore fluid flow using material point method: (a) deviatoric shear strain variation (b) pore-water pressure variation (Pa) and (c) vertical effective stress variation (Pa). (From *Computers and Geotechnics*, 63, Bandara, S., and Soga, K., Coupling of soil deformation and pore fluid flow using material point method, 199–214, Copyright 2015, with permission from Elsevier.)

modeling. In addition, the fundamental conditions associated with limit equilibrium methods of analysis are retained.

The basic procedure of the enhanced limit method is as follows:

1. Conduct a stress analysis for the soil slope using either FE or other numerical methods.
2. Assume a series of trial slip surfaces. Each trial slip surface is divided into many segments (Figure 4.26a). These segments correspond to the slices in the method of slices of LEM.
3. Compute the normal stress and shear stress at the center of the segment.

To compute the stresses at the center of a segment, a search must first be undertaken to locate the element within which the center of a segment is located. After the element encircling the center of the segment has been located, the stresses at the segment center can be obtained from the node point stresses based on the weighted area method (SoilVision System Ltd, 2010). Note that the stresses computed in an FE analysis generally correspond to the Gauss points and need to be transferred to the nodes of each FE.

Once the stress, σ_x, σ_y, and τ_{xy}, are known at the center of the base of each segment, the normal stress, σ_n, and mobilized shear stress, τ_n, can be calculated based on stress transformation (Das and Sobhan, 2013):

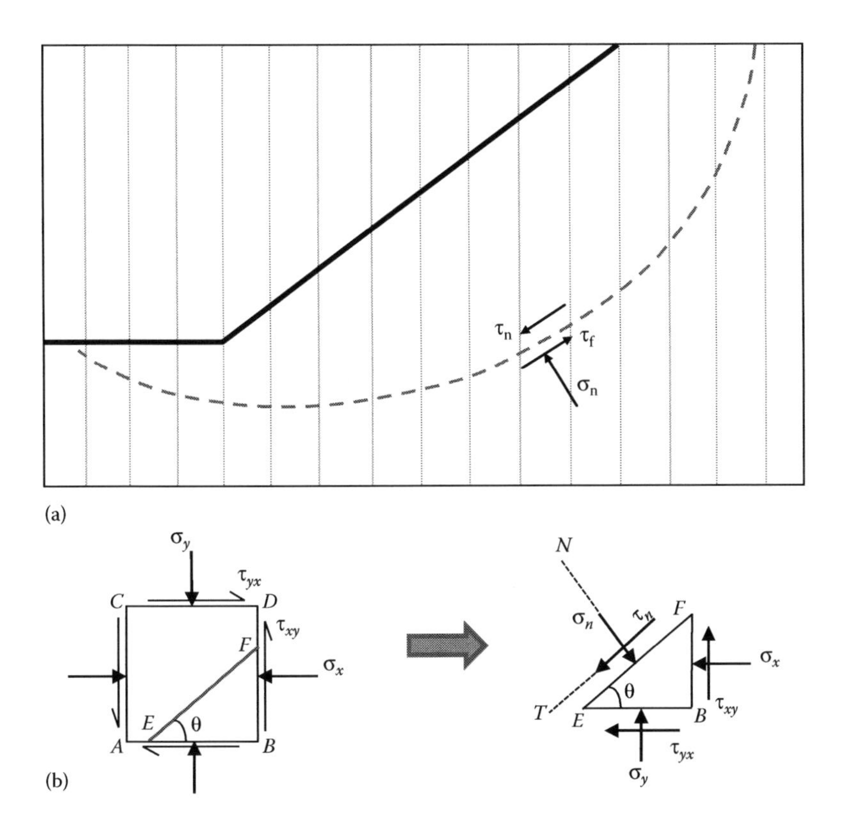

Figure 4.26 (a) Trial slip surface divided into segments in the enhanced limit method and (b) transformation of stresses.

$$\sigma_n = \frac{\sigma_x + \sigma_y}{2} + \frac{\sigma_x - \sigma_y}{2}\cos 2\theta + \tau_{xy}\sin 2\theta = \sigma_x \cos^2\theta + \sigma_y \sin^2\theta + 2\tau_{xy}\sin\theta\cos\theta$$

$$\tau_n = \frac{1}{2}(\sigma_y - \sigma_x)\sin 2\theta + \tau_{xy}\cos 2\theta \tag{4.51}$$

where:

σ_x is the normal stress in the x-direction at the center of a segment
σ_y is the normal stress in the y-direction at the center of a segment
τ_{xy} is the shear stress in the x plane and along the y-direction at the center of a segment
θ is the angle measured from the x-direction to the segment (Figure 4.26b)
σ_n is the normal stress perpendicular to the segment
τ_n is the mobilized shear stress parallel to the segment

4. The factor of safety F_s for each trial slip surface can be evaluated by (Kulhawy, 1969):

$$F_s = \frac{\sum_{i=1}^{n} \tau_{fi}\Delta L_i}{\sum_{i=1}^{n} \tau_{ni}\Delta L_i} \tag{4.52}$$

where:

n is the number of segments along the slip surface
τ_{ni} is the mobilized shear stress of the ith segment of the slip surface
τ_{fi} is the shear strength of the ith segment of the slip surface
ΔL_i is the length of the ith segment of the slip surface

The shear strength t_{fi} can be calculated based on the normal stress, the pore-water pressure from numerical modeling, and a shear strength model. The global safety factor is the minimum safety factor obtained from all the trial slip surfaces. Applying the same stability analysis approach at different times during a rainstorm, the stability of the slope during the whole rainfall process can be evaluated. The advantage of using the FE-based slope stability analysis for this study is that the stresses and pore-water pressures from the coupled numerical modeling can be utilized to evaluate the stability of a deformable unsaturated soil slope in a more meaningful way.

Let us use one benchmark example of a two-layer soil cut slope in Slope/W (Geo-slope International Ltd, 2012) to compares the result of LEM and the enhanced limit method. The slope is a cut slope at 2:1 (horizontal:vertical) with two soil layers. The upper soil layer is 5 m thick. The total height of the cut is 10 m. Bedrock exists 4 m below the base of the cut. The unit weight and shear strength parameters for the two layers of soils are shown in Figure 4.27. To obtain stress distribution for the enhanced limit method, the Young's modulus is assumed to be 20,000 kPa and the Poisson ratio is 0.33 in the FE stress analysis. The pore-water pressure distribution is hydrostatic based on the piezometric line. Figure 4.28 illustrates the results of Bishop's simplified method using Slope/W and the enhanced limit method using the code FESSA developed by Zhang et al. (2005). The contours of the safety factor and the critical slip surface are in good agreement. The minimum safety factors of the Bishop's method and the enhanced limit method are 1.508 and 1.452, respectively. The coordinates of the centers of rotation and the magnitudes of the radius for the circular critical slip surfaces by the Bishop's method, Morgenstern and Price method using Slope/W and ELM using FESSA are also listed in Figure 4.28b.

Alternatively, an FE method with the shear strength reduction technique (SSRFEM) (Matsui and San, 1992; Dawson et al., 1999; Griffiths and Lane, 1999) can be used to obtain the safety factor of the slope. The shear strength reduction factor just before the elastoplastic FE analysis cannot converge within a user-specified number of iterations is considered as the safety factor of the slope. The detailed procedures and applications of SSRFEM can be found in Griffiths and Lane (1999).

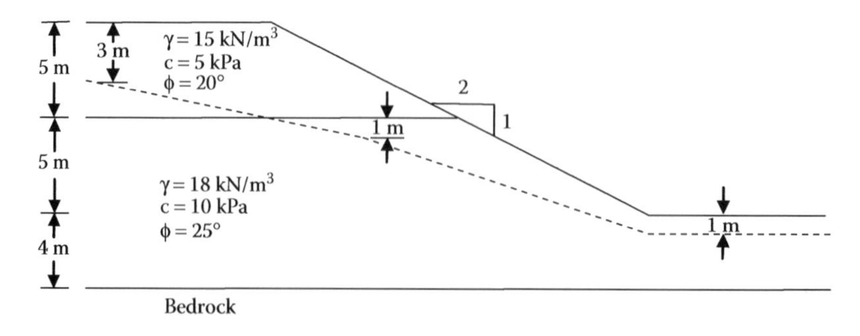

Figure 4.27 A benchmark example of two-layer soil slope (Example 3.3).

4.4 ILLUSTRATIVE EXAMPLES

4.4.1 A simple slope example based on elastoplastic model and effective stress concept

Perform an elastoplastic stress analysis for the simple geometry homogenous slope in Example 3.2 based on the effective stress concept of unsaturated soil. The commercial software ABAQUS is adopted for numerical modeling. In ABAQUS, the sign of stress follows the rules of continuity mechanics, that is, positive sign is used for tension, and negative sign represents compression for the stresses. For pore pressure, the positive sign means compression. The effective stress for unsaturated soil is defined as

$$\sigma' = \sigma + \left[\chi u_w + (1-\chi)u_a\right] \tag{4.53}$$

In ABAQUS, u_a is assumed to be zero and χ is assumed to be equal to degree of saturation.

To simulate the initial equilibrium condition before rainstorm, the pore-water pressure distribution and the stress distribution should be analyzed using a stress analysis with elastic model. In this example, the maximum initial suction is assumed to be 50 kPa. Therefore, a pore pressure distribution corresponding to the hydrostatic condition with a cutoff suction of 50 kPa should be obtained first and then the initial stress distribution corresponding to this pore pressure distribution is obtained using an elastic stress analysis. The initial distributions of stress and pore pressure are then imported as initial conditions in the GEOSTATIC step in the coupled hydro-mechanical analysis. Detailed procedures of assigning initial stress

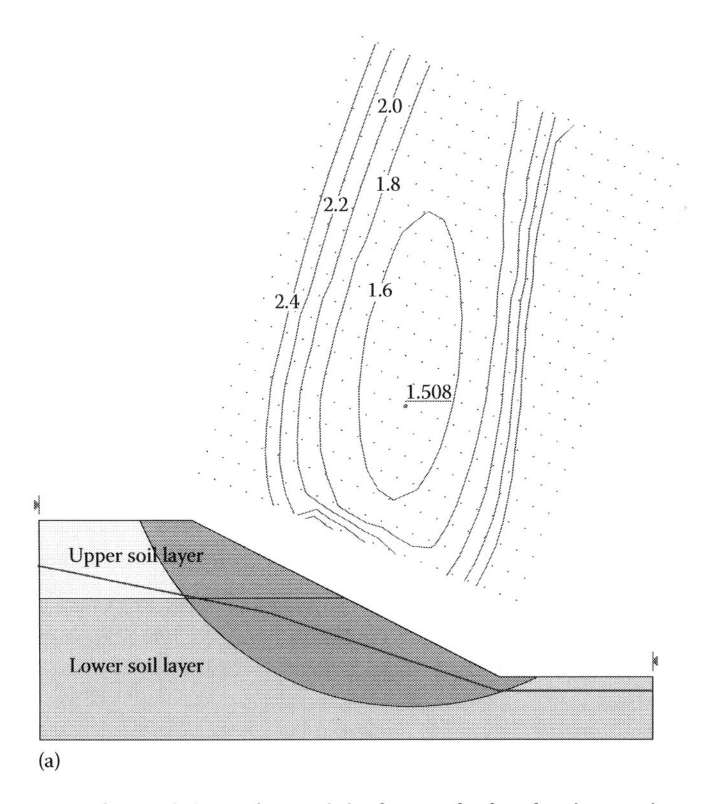

Figure 4.28 Comparison of critical slip surface and the factor of safety for the two-layer soil slope: (a) the Bishop method. *(Continued)*

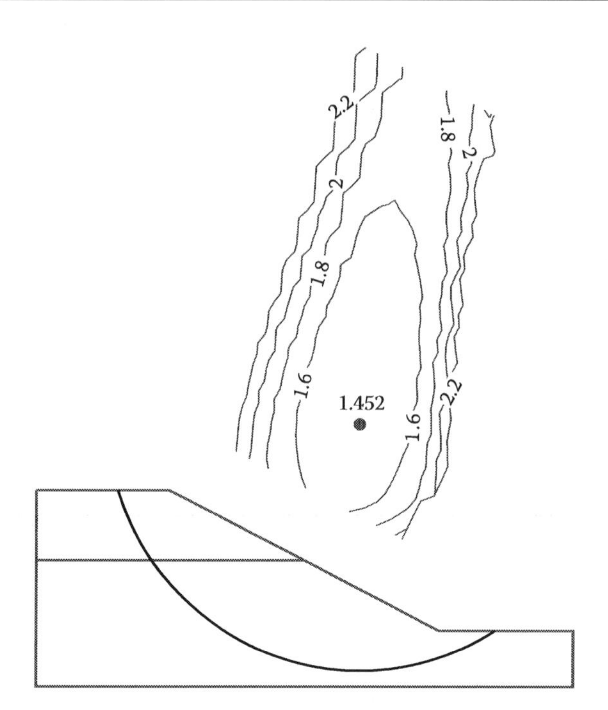

Methods	x	y	Radius	Factor of safety
Bishop	24.09	21.24	18.98	1.508
Morgenstern–Price	24.09	21.15	18.95	1.510
Enhanced limit method	24.12	20.12	19.00	1.452

(b)

Figure 4.28 (Continued) **Comparison of critical slip surface and the factor of safety for the two-layer soil slope: (b) the enhanced limit method.**

and saturation/pore pressure distributions through editing keywords can be referred to the manual of ABAQUS. A coupled flow and stress analysis is conducted in the SOILS step (Figure 4.29) with an elastoplastic model.

In this example, the elastic modulus of the soil is 10 MPa. The Poisson's ratio is 0.3. The unit weight of the soil is 18.0 kN/m^3. The cohesion is 10 kPa and the effective angle of internal friction is 35°. The Mohr–Coulomb model is used for the soil in the slope. The unsaturated hydraulic properties functions can be defined in the material behavior of pore fluid (Figure 4.30). The function between permeability and saturation represents unsaturated permeability function (Figure 4.31) and the one between pore pressure and saturation represents SWCC (Figure 4.32). In this example, the van Genuchten (1980) and Mualem (1976) (VGM) model is adopted for the unsaturated hydraulic function. The VGM model can be written in terms of effective degree of saturation S_e as follows:

$$S_e = (\theta_w - \theta_r) / (\theta_s - \theta_r) = \frac{1}{(1 + (a_{vgm} h_s)^{n_{vgm}})^{(1-n_{vgm})}} \tag{4.54}$$

$$k(S_e) = k_s \left\{ S_e^l \left[1 - \left(1 - S_e^{n_{vgm}/(n_{vgm}-1)}\right)^{1-1/n_{vgm}} \right]^2 \right\} \tag{4.55}$$

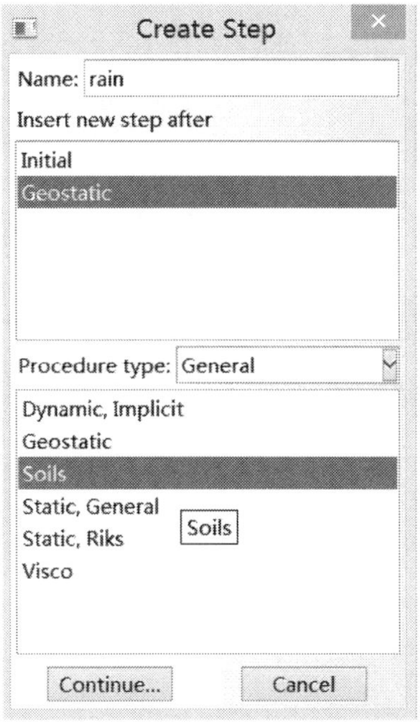

Figure 4.29 Create step for coupled hydro-mechanical modeling in ABAQUS.

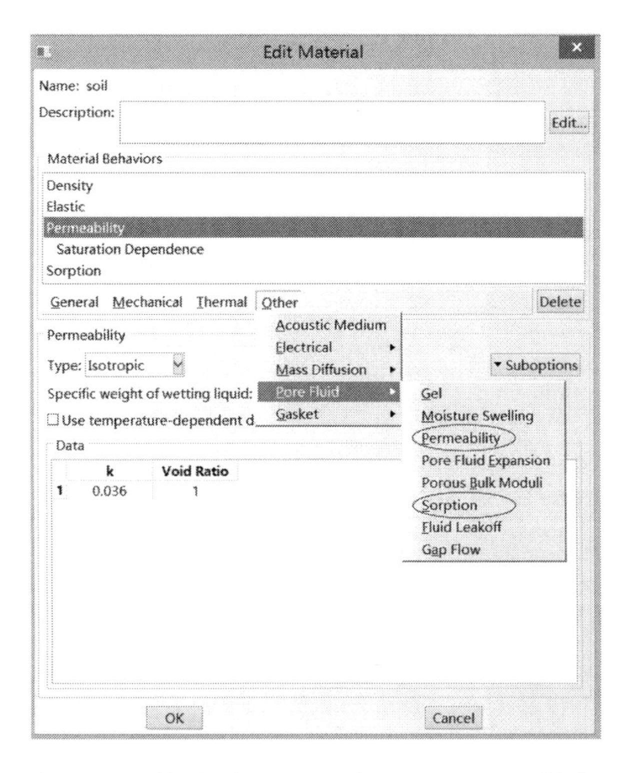

Figure 4.30 Definition of unsaturated hydraulic property functions in material behavior of ABAQUS.

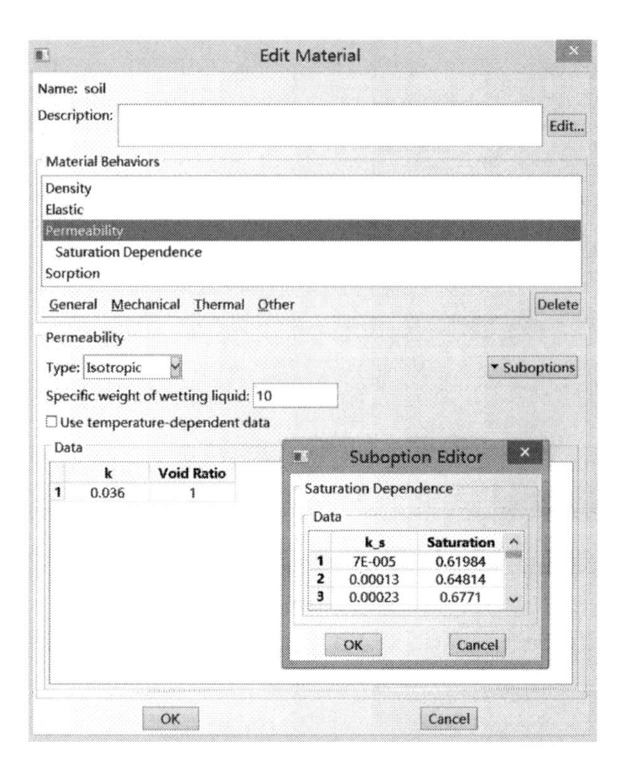

Figure 4.31 Definition of permeability–saturation function in ABAQUS.

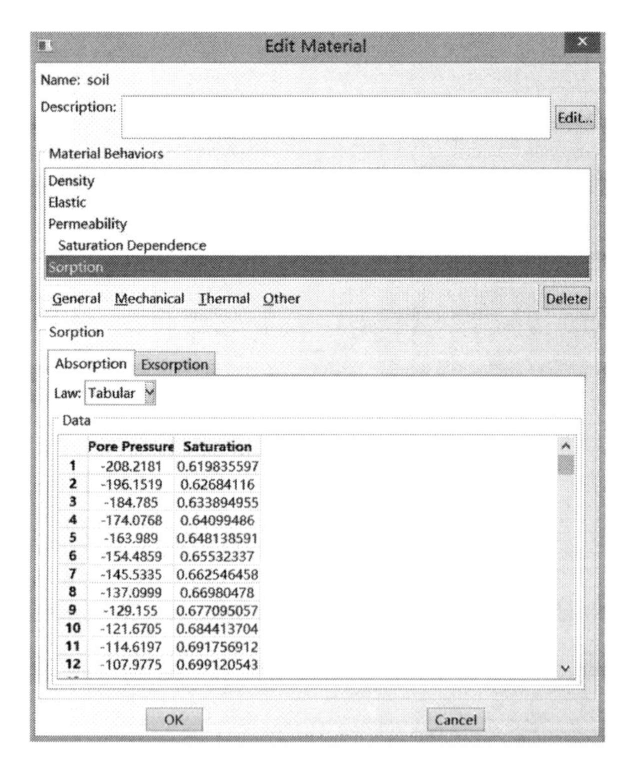

Figure 4.32 Definition of pore pressure–saturation function in ABAQUS.

where l is usually taken to be 0.5. The SWCC and unsaturated permeability function of the soil in this slope and the corresponding model parameters are presented in Figure 4.33. The element assigned for the finite mesh should be a Pore fluid/Stress type element such as CPE4P, which is a four-node plane strain quadrilateral, bilinear displacement, bilinear pore pressure element (Figure 4.34).

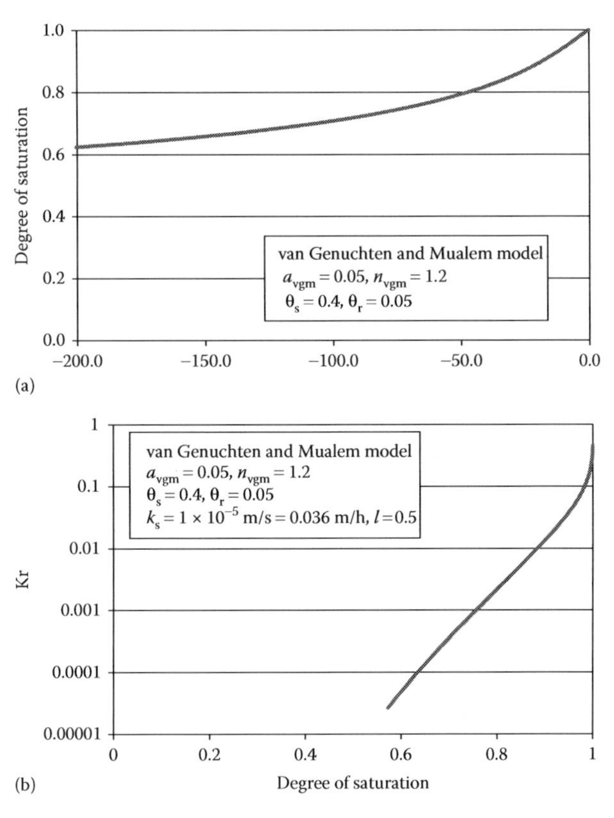

Figure 4.33 (a) Degree of saturation versus pore pressure and (b) relative permeability versus degree of saturation.

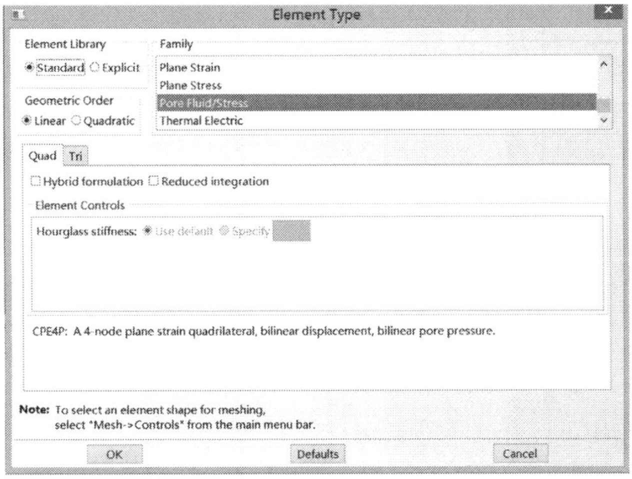

Element Type
Element Library
Standard Explicit
Family
Plane Strain
Plane Stress
Pore Fluid/Stress
Thermal Electric
Geometric Order
Linear Quadratic
Quad Tri
Hybrid formulation Reduced integration
Element Controls
Hourglass stiffness: Use default Specify
CPE4P: A 4-node plane strain quadrilateral, bilinear displacement, bilinear pore pressure.
Note: To select an element shape for meshing, select "Mesh->Controls" from the main menu bar.
OK Defaults Cancel

Figure 4.34 Definition of element type for coupled analysis in ABAQUS.

The rainfall lasts 24 h with the intensity shown in Figure 3.8. Figure 4.35 presents the pore pressure distribution in the soil slope after 24 h of rainfall. The displacement vectors after the rainfall is shown in Figure 4.36. As shown in the graph, the displacement vector is upwards because when the negative pore pressure is reduced due to infiltration, the effective stress is reduced. Hence, the soil expands due to reduction of effective stress. The contour plot of equivalent plastic strain shown in Figure 4.37 implies the plastic deformation mostly occurs at slope surface and near the slope toe where the soil is saturated.

The factor of safety of the slope can be obtained using the strength reduction method. A field variable can be defined in the Mohr–Coulomb model as shown in Figure 4.38.

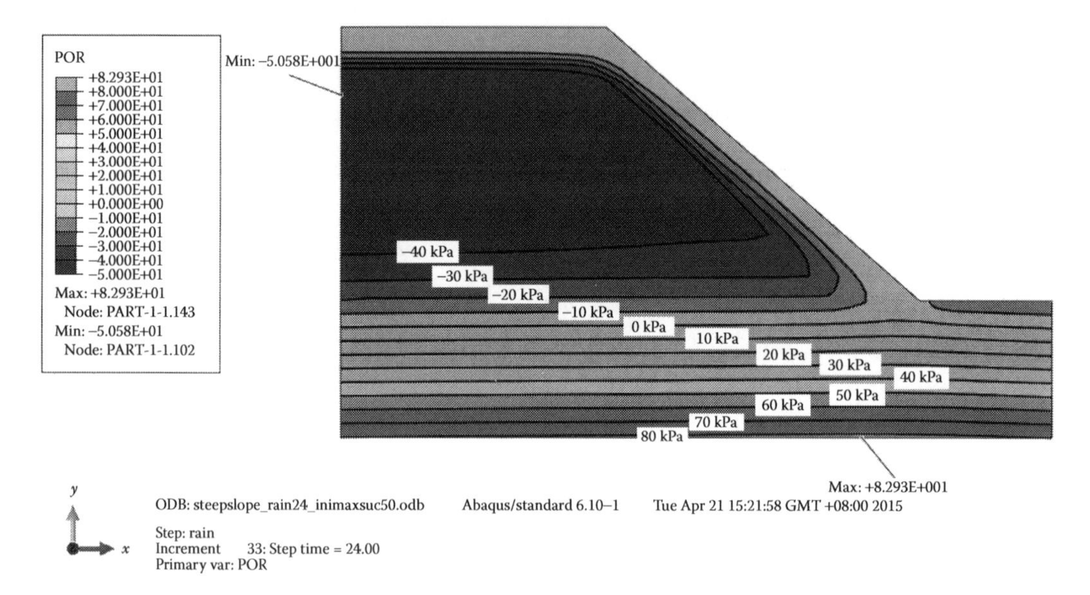

Figure 4.35 Pore-water pressure distribution after 24 h of rainfall.

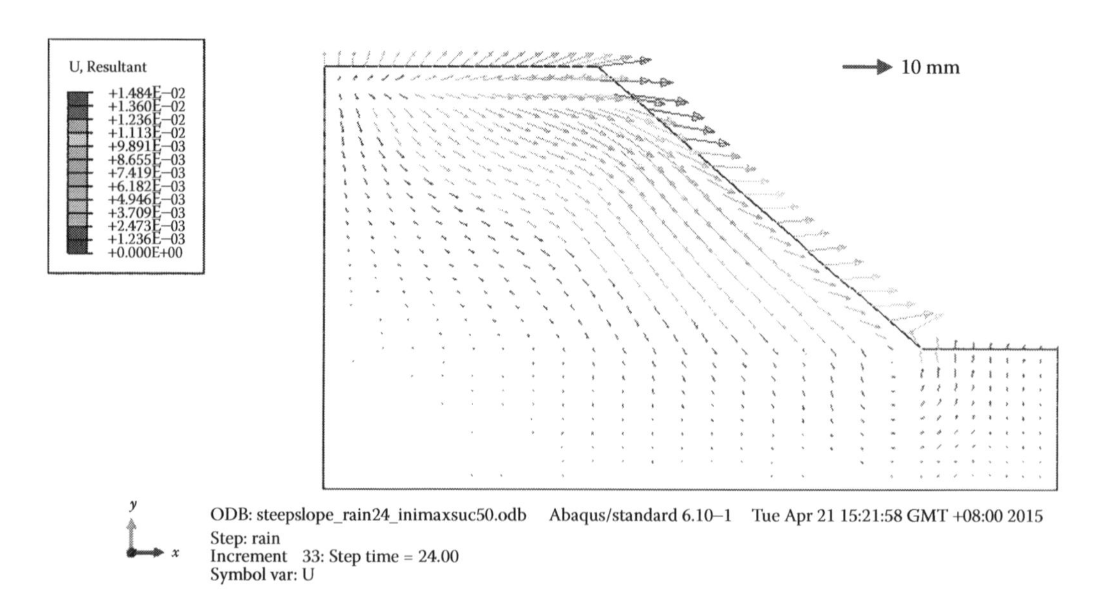

Figure 4.36 Displacement vectors after 24 h of rainfall.

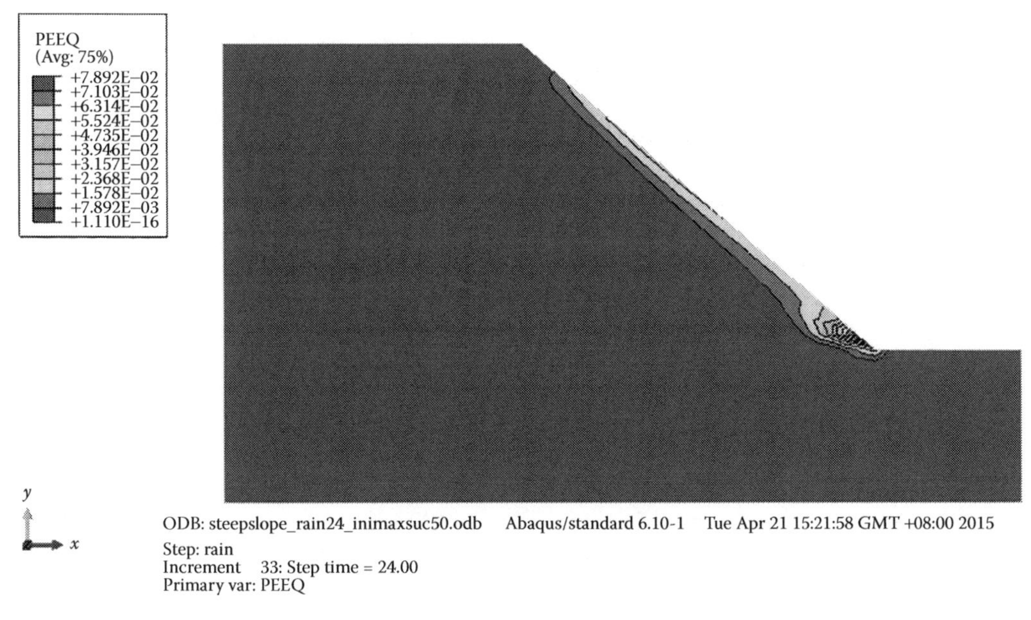

Figure 4.37 Equivalent plastic strain after 24 h of rainfall.

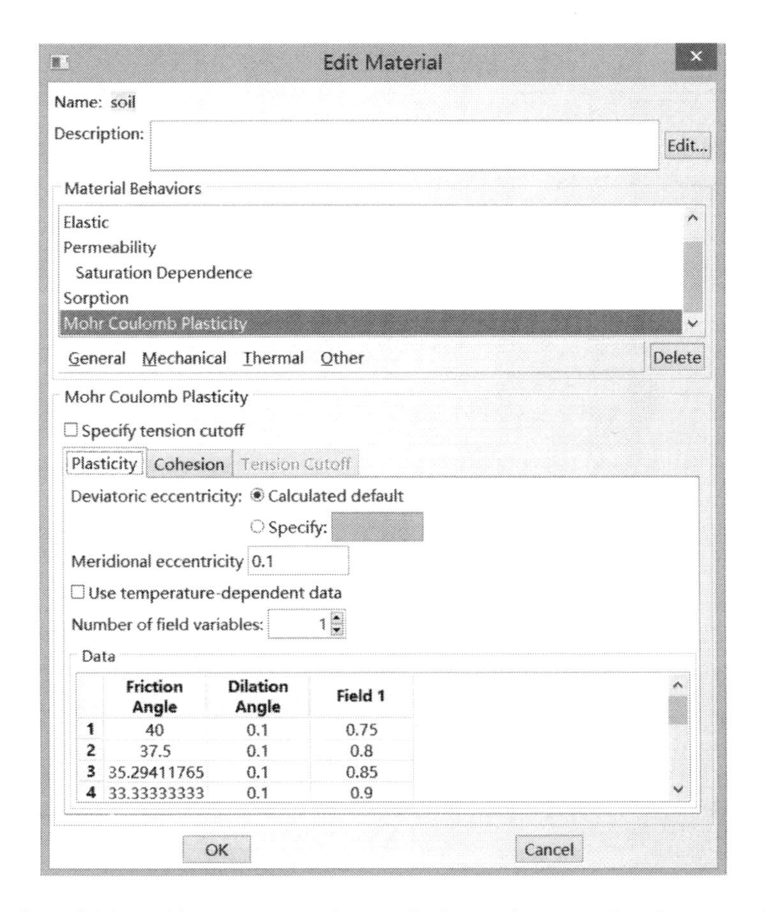

	Friction Angle	Dilation Angle	Field 1
1	40	0.1	0.75
2	37.5	0.1	0.8
3	35.29411765	0.1	0.85
4	33.33333333	0.1	0.9

Figure 4.38 Define a field variable to represent factor of safety in the strength reduction method.

The numerical analysis cannot reach convergence when the field variable is increased from 1.0 to 1.313. This means the factor of safety of the slope after 24 h of rainfall is 1.313. Figures 4.39 and 4.40 show the distribution of plastic strain and displacement vectors in the slope obtained using the strength reduction method. Compared with the results in Chapter 3 (Figure 3.11), the factor of safety and the critical slip surface is generally consistent with the result by LEM in Figure 3.11d.

Assume a constant rainstorm with the intensity of 0.012 m/h is applied on the slope surface for 60 h. The coupled numerical analysis cannot reach convergence when the rainfall infiltrates into the slope for 56.3 h. The equivalent plastic strain shown in Figure 4.41 shows a failure surface is formed after a long time of infiltration. The displacement vectors at failure

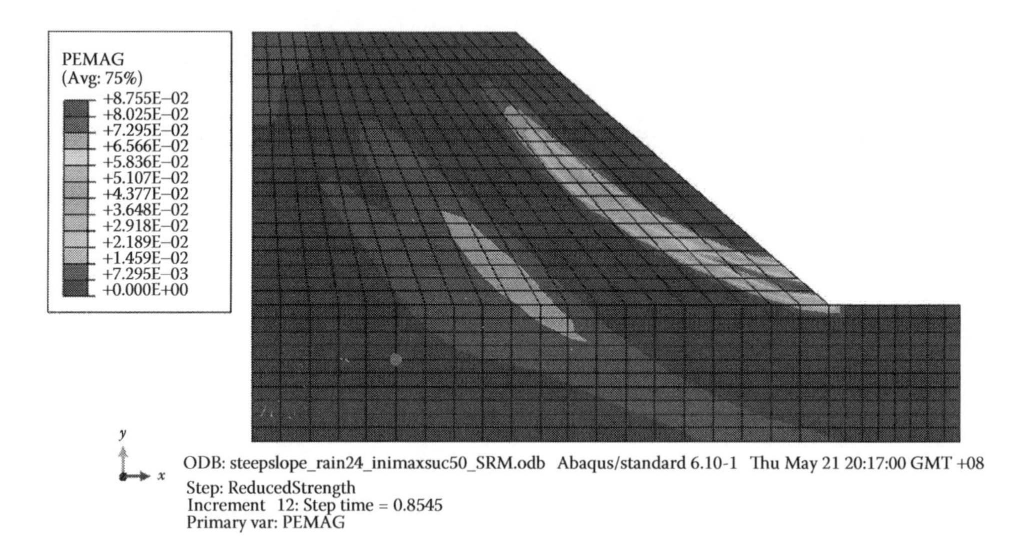

Figure 4.39 Plastic strain distribution using strength reduction method.

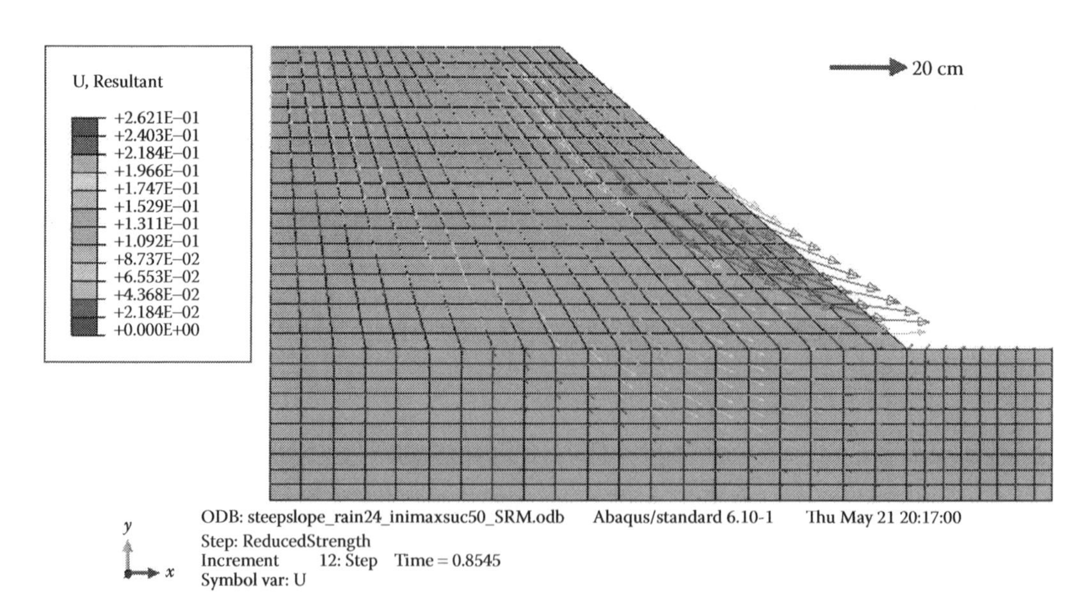

Figure 4.40 Displacement vectors obtained using strength reduction method.

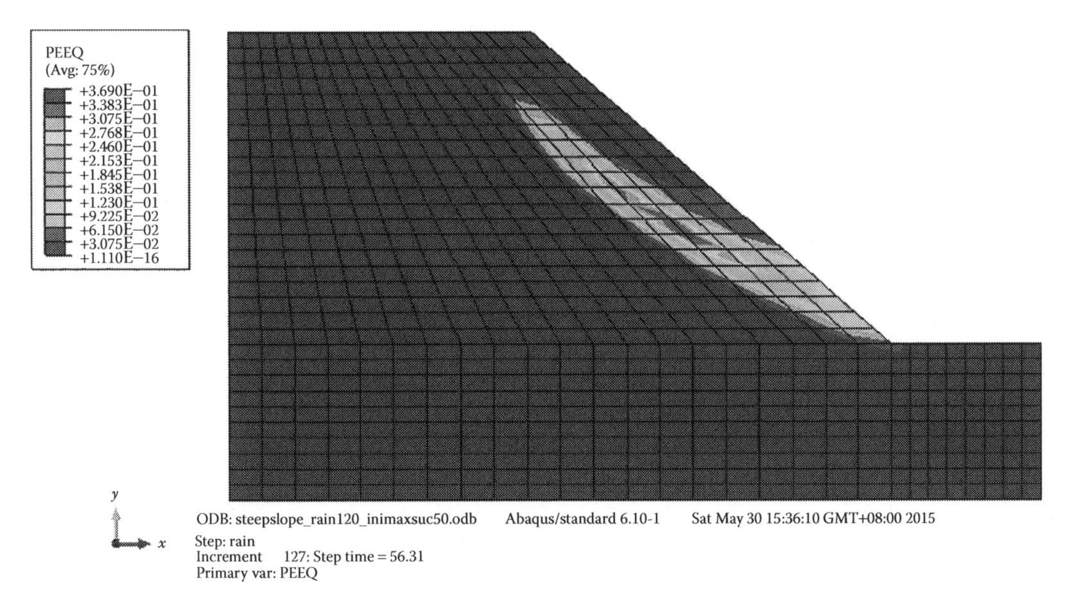

Figure 4.41 Equivalent plastic strain after 56.3 h of rainfall (constant $q = 0.012$ m/h).

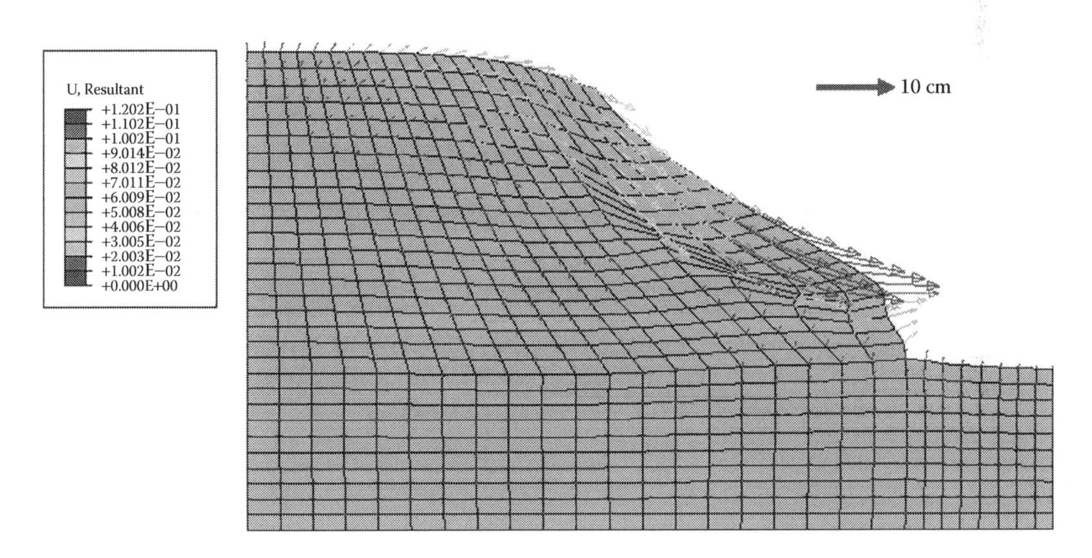

Figure 4.42 Displacement vectors after 56.3 h of rainfall (constant $q = 0.012$ m/h).

time is shown in Figure 4.42. As shown in the graph, the displacement vector turns from upwards to downwards because with further infiltration, the stress path of the soil moves gradually towards the yield surface of the constitutive model and the continuous slip surface is formed inside the slope.

4.4.2 Case study of 1976 Sau Mau Ping landslide

In the morning of 25 August 1976, following heavy rainfall associated with severe tropical storm Ellen, destructive landslides occurred at the Sau Mau Ping Estate, Hong Kong. In all, four landslides took place, all resulting from the failure of the side slopes of highway

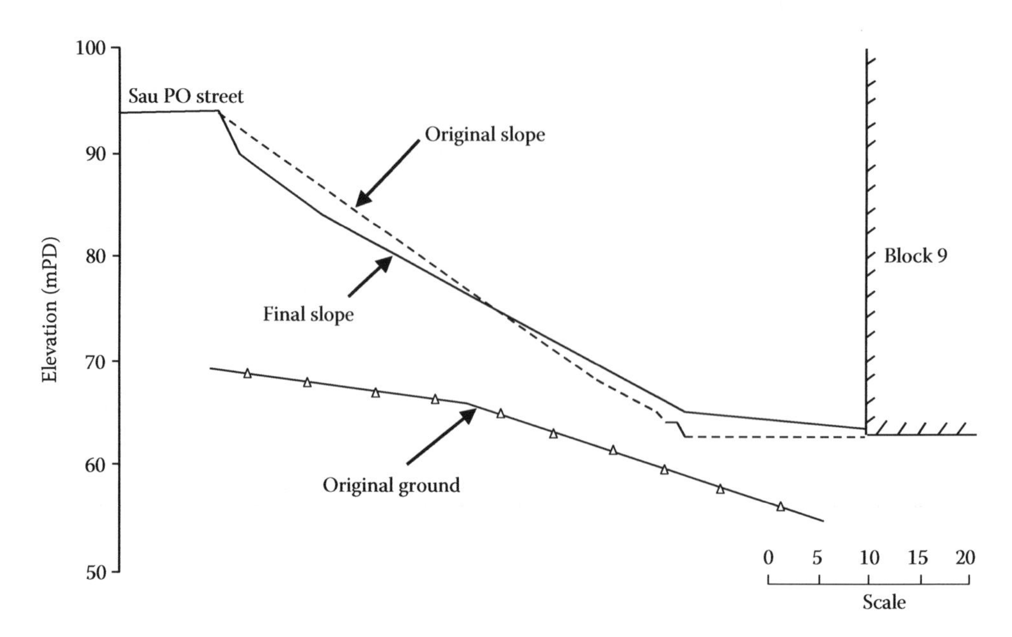

Figure 4.43 Typical section of the 1976 Sau Mau Ping landslide.

embankments formed of granitic fill material. The most fatal one occurred in a 30 m high, 33° embankment with a 2.5 m high toe wall. The debris of the landslide moved downwards and buried the ground floor of a residual building (Block 9) and killed 18 people. The volume of the failure was about 4000 m^3 and the failure zone was generally confined to the top 3 m of the slope. A typical cross section of the landslide is shown in Figure 4.43.

Investigations demonstrated that the failed slope at Sau Mau Ping was formed by end-tipping of fill in a loose condition. Subsequent investigation concluded that "these failures resulted from the development of a seepage condition within a wetted zone as water penetrated into the face of the slope. Consequent loss of strength in the fill resulted in downhill movement and an almost instantaneous conversion of the slope into a mud avalanche" (Hong Kong Government, 1977). In this section, a coupled hydro-mechanical analysis based on incremental elastic formulation by Fredlund and Rahardjo (1993) is conducted for the Sau Mau Ping landslide. The finite element partial differential equation solver, FlexPDE (PDE Solutions Inc., 2015), is used to develop a coupled numerical model.

4.4.2.1 FE model

The FE mesh for numerical modeling is shown in Figure 4.44. The height of the slope is 30 m and the slope inclination angle is 33°. The upper layer of soil is a fill of decomposed granite (DG). The average depth of the DG fill above the natural ground is around 6 m. Totally 504 triangular elements are used to generate the FE mesh. Denser mesh is assigned to the upper soil layer because the zone is influenced by rainfall infiltration more significantly.

Figure 4.44 illustrates the hydraulic boundaries conditions defined in the numerical modeling. The groundwater table is assumed to be at a considerable depth below the slope surface (from 79 mPD at the upstream vertical boundary AD to 54 mPD at the downstream vertical boundary CE). At cross section 2–2, the depth of the groundwater table is about 16 m. At the left and right boundaries, AD and CE, there is no flux above the groundwater table. Constant head boundary conditions are assumed for the parts below the groundwater table along AD and CE. The bottom boundary DE is assumed to be impermeable. Rainfall flux

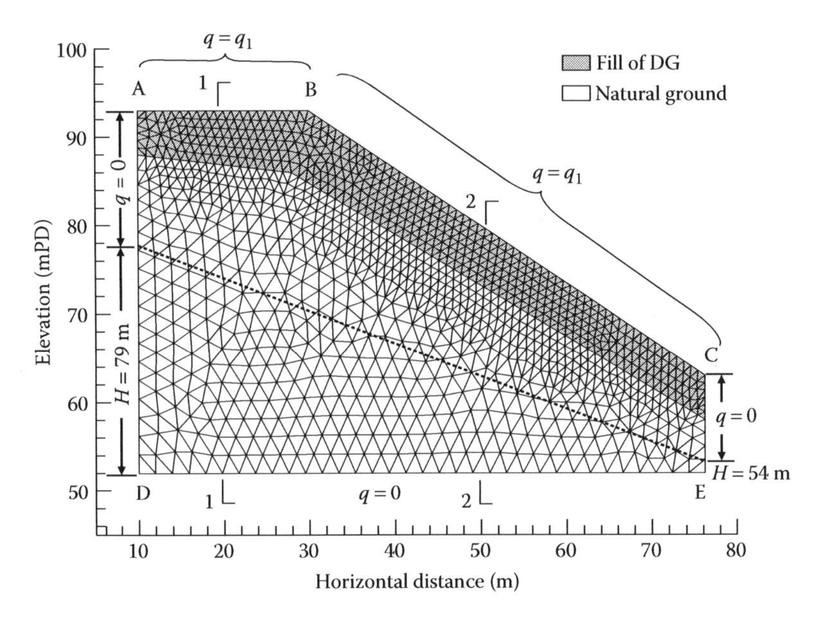

Figure 4.44 Finite element mesh and hydraulic boundary conditions in the numerical modeling of Sau Mau Ping slope.

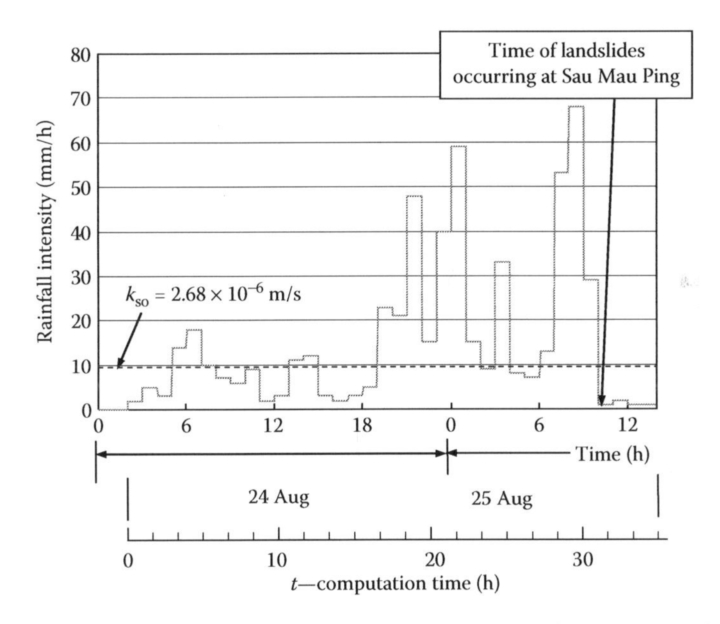

Figure 4.45 Rainstorm intensity during Aug 24–25 1976 recorded at Tate's Cairn and Hong Kong International Airport.

q_1 is applied along the slope surface from AB to BC. The rainfall intensity was based on the field measurements at Tate's Cairn and Hong Kong International Airport on the same day of landslide (Figure 4.45). The time when the rainstorm started (2:00 am 24 Aug 1976) is defined as the start time of modeling ($t = 0$ h). The time when the landslide occurred (10:00 am 25 Aug 1976) is the end time of the modeling ($t = 32$ h). Considering the surface runoff, when the rainfall intensity is less than the saturated permeability of the DG fill, the boundary condition along ABC is kept as a flux boundary. However, if the rainfall intensity is greater than the saturated permeability of the DG fill, a constant pressure boundary ($h = 0$) condition will be applied on the slope surface ABC. Along both upstream and downstream vertical boundaries AD and CE, deformations in the horizontal direction are restricted. The slope surface ABC is assumed to be free of motion. The bottom boundary DE is completely constrained in both vertical and horizontal directions.

The initial pore-water pressure condition is assumed to be hydrostatic with a maximum suction of 10 kPa. The initial void ratio of the slope is 0.89. The initial degree of saturation in the slope is about 78.6%, which is determined by the SWCC and the initial pore-water pressure. The initial unit weight of the soil is 17.4 kN/m^3, which is determined by the degree of saturation and volume mass relationship of the soil with a specific gravity of soil particle, $G_s = 2.66$. The initial stress distribution is established by switching on the self-weight of the soils. As the degree of saturation of the unsaturated soil increases due to rainfall infiltration, the self-weight of the soil also increases.

4.4.2.2 Soil properties

Figures 4.46 through 4.48 show in situ porosity (equal to saturated volumetric water content θ_s), saturated permeability and natural gravimetric water content measured at Sau Mau Ping site. The average values of the in situ porosity, saturated permeability, and natural water content are 0.47, 2.7×10^{-6} m/s, and 20.1%, respectively. If the trends with respect to depth are not considered, the coefficients of variation (COV) of porosity and natural water content are 9.8% and 25.5%, respectively. The variability of saturated permeability is more significant with a COV of 118%.

To consider the effect of deformation on soil hydraulic properties, the saturated permeability of DG fill is assumed to be a function of the porosity based on the Kozeny–Carman equation (Ahuja et al., 1989):

$$k_s = k_{so}\left(\theta_s/\theta_{so}\right)^4 \tag{4.56}$$

where:

θ_{so} is the initial porosity

k_{so} is the saturated permeability corresponding to the initial porosity

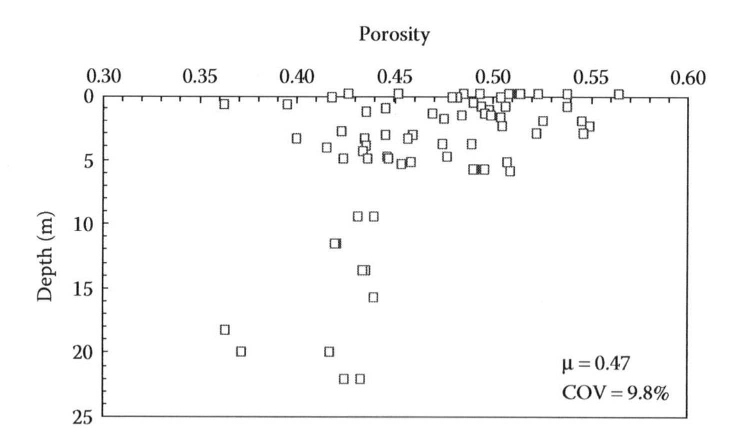

Figure 4.46 Measured in situ porosity from bore holes and trial pits at Sau Mau Ping site. (Plotted based on data from Hong Kong Government, *Report on the slope failures at Sau Mau Ping*, August 1976. Hong Kong Government Printer, Hong Kong, 1977.)

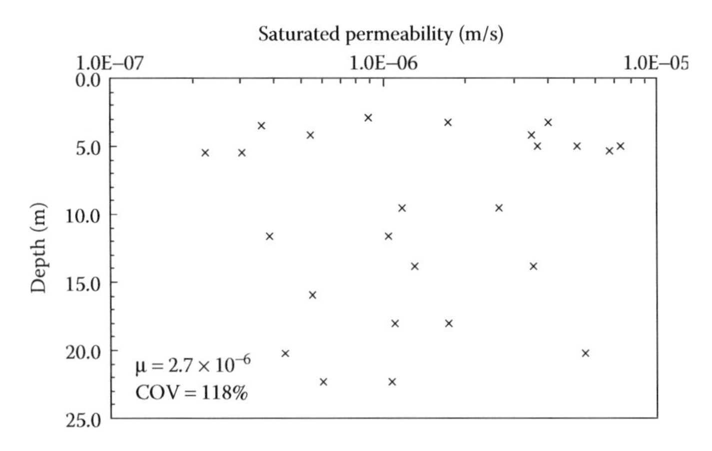

Figure 4.47 Measured in situ saturated permeability from bore holes using constant head method at Sau Mau Ping site. (Plotted based on data from Hong Kong Government, *Report on the slope failures at Sau Mau Ping*, August 1976. Hong Kong Government Printer, Hong Kong, 1977.)

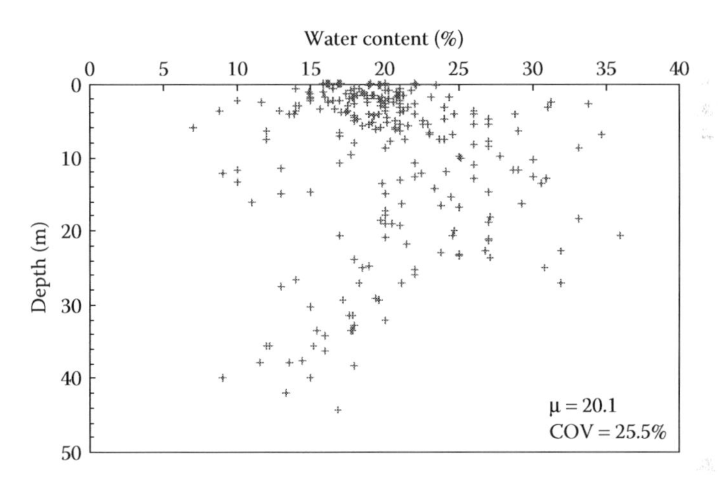

Figure 4.48 Measured natural water content from bore holes and trial pits at Sau Mau Ping site. (Plotted based on data from Hong Kong Government, *Report on the slope failures at Sau Mau Ping*, August 1976. Hong Kong Government Printer, Hong Kong, 1977.)

In this study, the initial porosity θ_{so} and k_{so} are taken as the average measured values at the Sau Mau Ping site, that is, $\theta_{so} = 0.47$ and $k_{so} = 2.7 \times 10^{-6}$ m/s.

There were no direct measurements for the SWCC and the permeability function of the DG fill at Sau Mau Ping. The SWCC of the soil is estimated based on the grain size distribution and the soil density (Fredlund et al., 2002). Then the estimated SWCC is fitted by the Fredlund and Xing (1994) model and the coefficient of permeability function can be estimated based on k_{so} and the fitted SWCC (Fredlund et al., 1994). Measured grain size distributions of the DG fill in Sau Mau Ping slope are presented in Figure 4.49. The average grain size distribution for DG fill is shown in Figure 4.50. Figure 4.51 shows the estimated SWCC and permeability function for numerical modeling. With changing of soil density due to deformation in the numerical modeling, the SWCC and permeability function will be modified based on the updated porosity and k_s.

There are no available experimental data of void ratio state surface for the soil at Sau Mau Ping site. Kam (1999) reported experimental data from 1D virgin wetting tests for

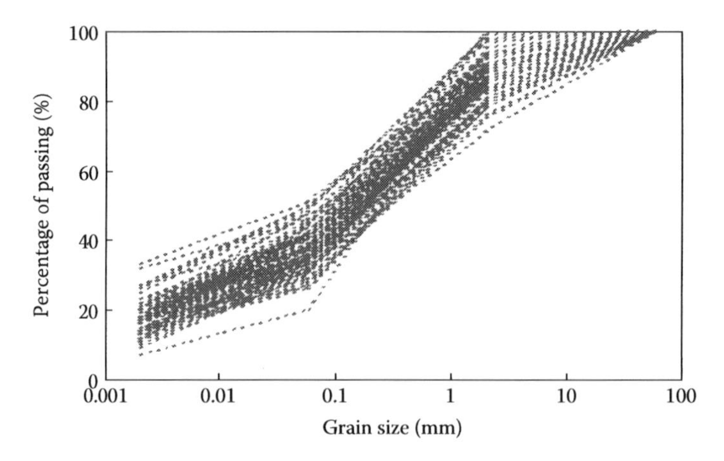

Figure 4.49 Measured grain size distributions of DG fill at Sau Mau Ping site. (Plotted based on data from Hong Kong Government, *Report on the slope failures at Sau Mau Ping*, August 1976. Hong Kong Government Printer, Hong Kong, 1977.)

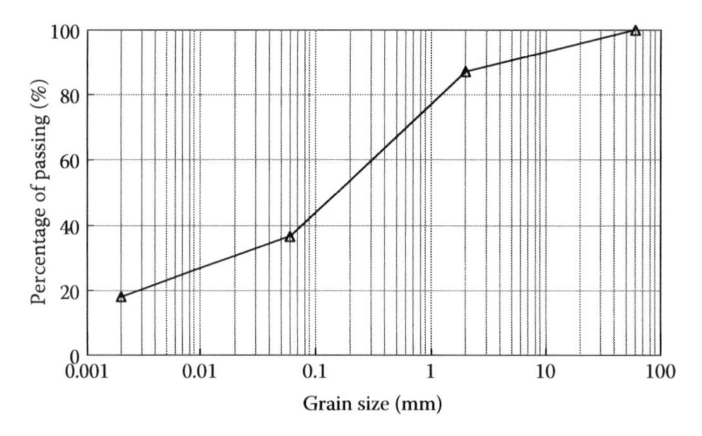

Figure 4.50 Average grain size distribution of DG fill at Sau Mau Ping site.

completely decomposed granite (CDG) in Cha Kwo Ling, Hong Kong. A void ratio state surface (Figure 4.52) can be established based on the experimental results by Kam (1999) and fitted using a model proposed by Lloret and Alonso (1985):

$$e = a_{la} + b_{la}\ln(\sigma_{mean} - u_a) + c_{la}(u_a - u_w) + d_{la}\ln(\sigma_{mean} - u_a)(u_a - u_w) \tag{4.57}$$

where a_{la}, b_{la}, c_{la}, and d_{la} are fitting coefficients. For Cha Kwo Ling CDG, the values of the fitting parameters are a_{la} = 1.2187, b_{la} = −0.04412, c_{la} = −0.00239 and d_{la} = 0.000747. In this study, this set of parameters will be used for DG fill in Sau Mau Ping slope as an approximation.

When CDG soil is fully saturated, it exhibits a strain-softening behavior under undrained condition (Ng and Chiu, 2003; Ng et al., 2004). In the $p' - q$ plane shown in Figure 4.53a, in which p' is the mean effective stress and q is the deviator stress, the stress path of a strain-softening soil under undrained shearing will reach a peak state first and then the critical state. The collapse surface is defined by a straight line that joins the peak stress points and passes the critical state stress point under undrained shearing (Sladen et al., 1985). The CSL is defined by a line that joints the critical stress state points and passes the origin.

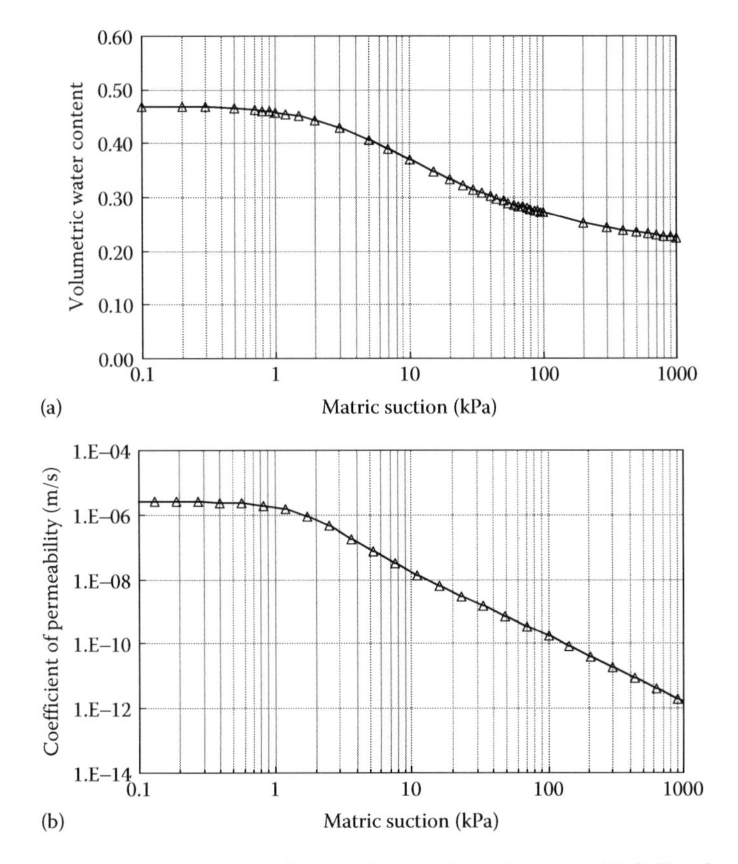

Figure 4.51 Estimated (a) SWCC and (b) coefficient of permeability function of DG fill in Sau Mau Ping slope ($\theta_s = \theta_{so} = 0.47$, $k_s = k_{so} = 2.7 \times 10^{-6}$ m/s).

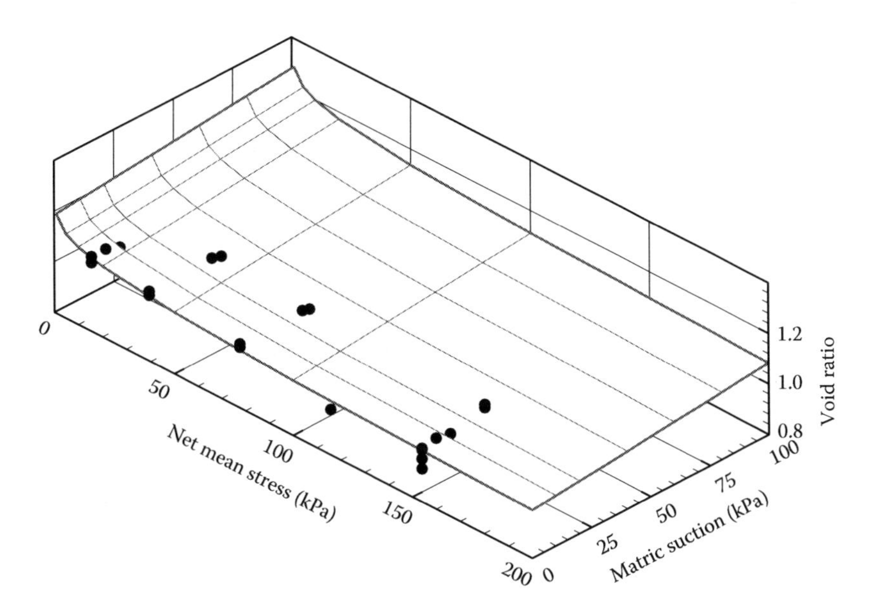

Figure 4.52 Void ratio state surface of DG fill.

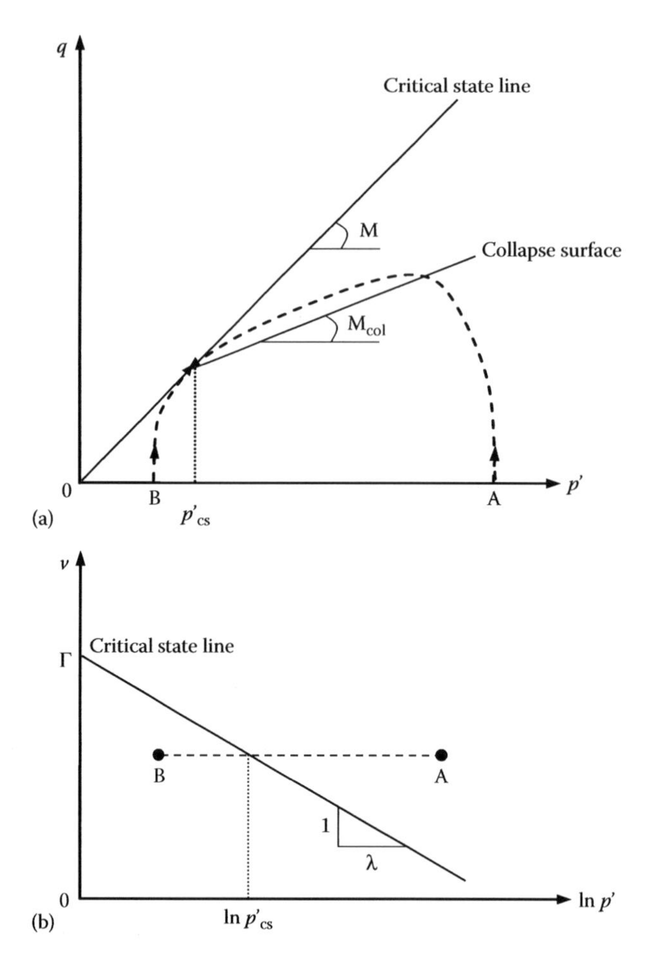

Figure 4.53 (a) Critical state line, collapse surface, and undrained shearing stress paths in the q–p' plane and (b) critical state line in the v–ln p' plane.

The slopes of the CSL and the collapse surface in the $p' - q$ plane are M and M_{col}, respectively. Figure 4.53b shows the CSL in the $v - \ln p'$ plane. Γ is the intercept of CSL in $v - \ln p'$ plane. λ is the slope of CSL in $v - \ln p'$ plane.

The true cohesion of DG fill can be assumed to be zero. The friction angle of the unsaturated soil is assumed to be the critical state friction angle. When the soil is saturated and the stress state is on the wet side of the CSL (i.e., above the CSL in $v - \ln p'$ plane, say point A in Figure 4.53), the shear strengths of the soils are determined by the collapse surface (Cheuk, 2001). When the CDG soil is saturated and the stress state is on the dry side of the CSL (i.e., below the CSL in $v - \ln p'$ plane, say point B in Figure 4.53), the shear strength of the soil is determined by the CSL.

Considering the contribution of soil suction, the shear strength model for DG fill can be expressed as follows:

$$\tau_f = \begin{cases} (\sigma_n - u_a)\tan\phi'_{cs} + (u_a - u_w)\tan\phi^b & \text{(unsaturated)} \\ c'(e) + (\sigma_n - u_w)\tan\phi'_{col} & \text{(saturated, } p' > p'_{cs}) \\ (\sigma_n - u_w)\tan\phi'_{cs} & \text{(saturated, } p' \le p'_{cs}) \end{cases} \tag{4.58}$$

where:

ϕ'_{cs} = the critical state friction angle where $\phi'_{cs} = \sin^{-1}[3M/(6+M)]$

ϕ'_{col} = the friction angle of the collapse surface where $\phi'_{col} = \sin^{-1}[3M_{col}/6+M_{col}]$

c′(e) is the intercept of the collapse surface line on the shear stress axis, which is function of void ratio

p'_{cs} is the mean effective stress at critical state

The function $c'(e)$ and p'_{cs} can be determined from the CSL and the collapse surface line:

$$p'_{cs} = \exp\left(\frac{\Gamma - 1 - e}{\lambda}\right) \tag{4.59}$$

$$c'(e) = (M - M_{col})/M_{col} \tan\phi'_{col} \exp\left(\frac{\Gamma - 1 - e}{\lambda}\right) \tag{4.60}$$

As the void ratio changes in the couple analysis, the effective cohesion and p'_{cs} also change.

The elastic modulus is assumed to be a hyperbolic function of the stress ratio:

$$E = \frac{3(1-2\nu)}{m_s^1}\left(1 - R_f \frac{q}{q_f}\right)^2 \tag{4.61}$$

where:

R_f is a constant coefficient

q is the acting deviator stress

q_f is the deviator stress at failure, which can be calculated from the shear strength Equation 4.58

In the above shear strength model for DG fill, it is assumed when the matric suction of a soil element is less than the AEV of the soil, the soil is saturated. Otherwise, the soil is unsaturated. The AEV of the soil is assumed to be equal to the parameter a_f in the Fredlund and Xing (1994) SWCC model.

For the natural ground, the SWCC and the permeability function are assumed to be the same as those of DG fill. The shear strength model for the natural ground is the extended Mohr–Coulomb shear strength model with $c' = 0$, $\phi' = 36.8°$ and $\phi^b = 15°$. Assuming the natural ground is dense and less deformable than the DG fill, the elastic modulus E is taken as -2×10^4 kPa and H_F is taken as 4×10^5 kPa. Table 4.3 presents the parameters of soil properties for the DG fill and the natural ground.

4.4.2.3 Results and discussion

Figure 4.54 illustrates the calculated results of the wetting fronts at two cross sections in the slope at different moments of the rainstorm. At cross section 1–1, the depth of the wetting front is about 2.5 m below the ground surface when the landslide occurred (t = 32 hours). At cross section 2–2, the depth of the wetting front is approximately 2.8 m below the slope surface at the moment of failure.

Laboratory infiltration tests on vertical soil columns of DG fill material showed that the penetration depth was between about 2 m (dry density 1.5 t/m^3, initial degree of saturation 50%) to about 6 m (dry density 1.4 t/m^3, initial degree of saturation 70%) (Knill et al., 1999).

Table 4.3 Summary of input soil parameters of Sau Mau Ping slope

Soil properties	*DG fill*	*Natural ground*
SWCC	$\theta_{so} = 0.47$, $a_f = 3.4$, $n_f = 1.3$, $m_f = 0.37$	Same as DG fill
Saturated permeability	$k_{so} = 2.68 \times 10^{-6}$ m/s	Same as DG fill
Shear strength	$c' = 0$, $\phi^b = 15°$, $\Gamma = 2.22$, $\lambda = 0.123$, $M = 1.5$ ($\phi'_{cs} = 36.8°$), $M_{col} = 0.984$ ($\phi'_{col} = 25°$)	$c' = 0$, $\phi^b = 15°$, $\phi' = 36.8°$
Void ratio	$\alpha_{la} = 1.2187$, $b_{la} = -0.04412$, $c_{la} = -0.00239$, and $d_{la} = 0.000747$	N.A.
Poisson's ratio	$\nu = 0.33$	$\nu = 0.33$
Elastic modulus E	Equation 4.61 with $R_f = 0.8$	$E = -2 \times 10^4$ kPa
Elastic modulus H	$H_F = \frac{3}{m_s^2}$	$H_F = 4 \times 10^5$ kPa

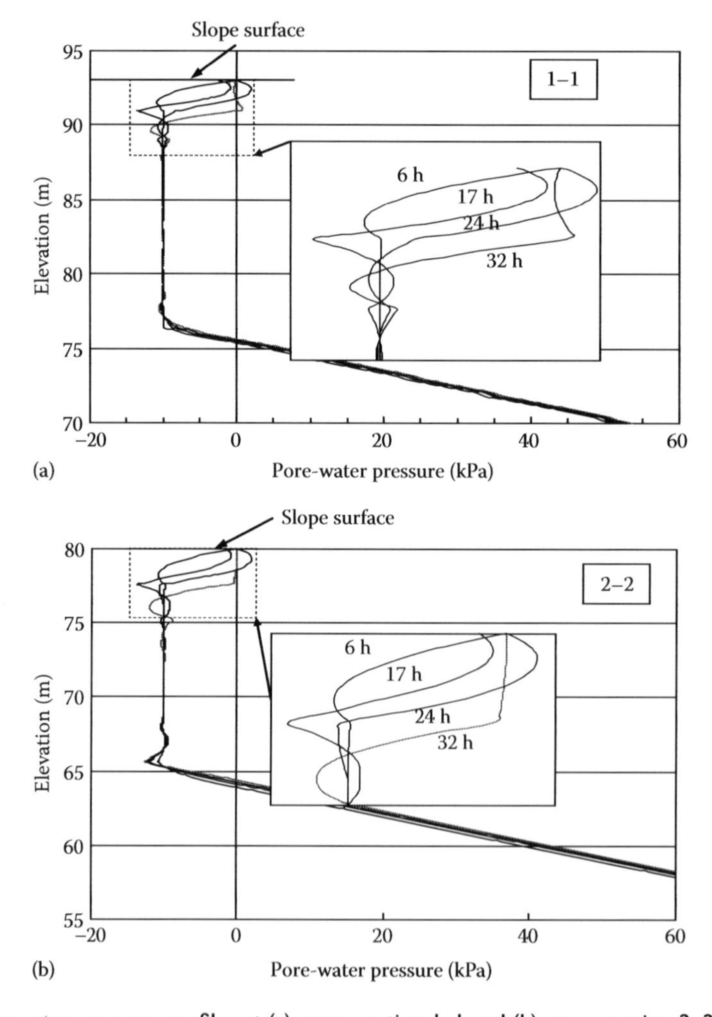

Figure 4.54 Pore-water pressure profiles at (a) cross section 1–1 and (b) cross section 2–2.

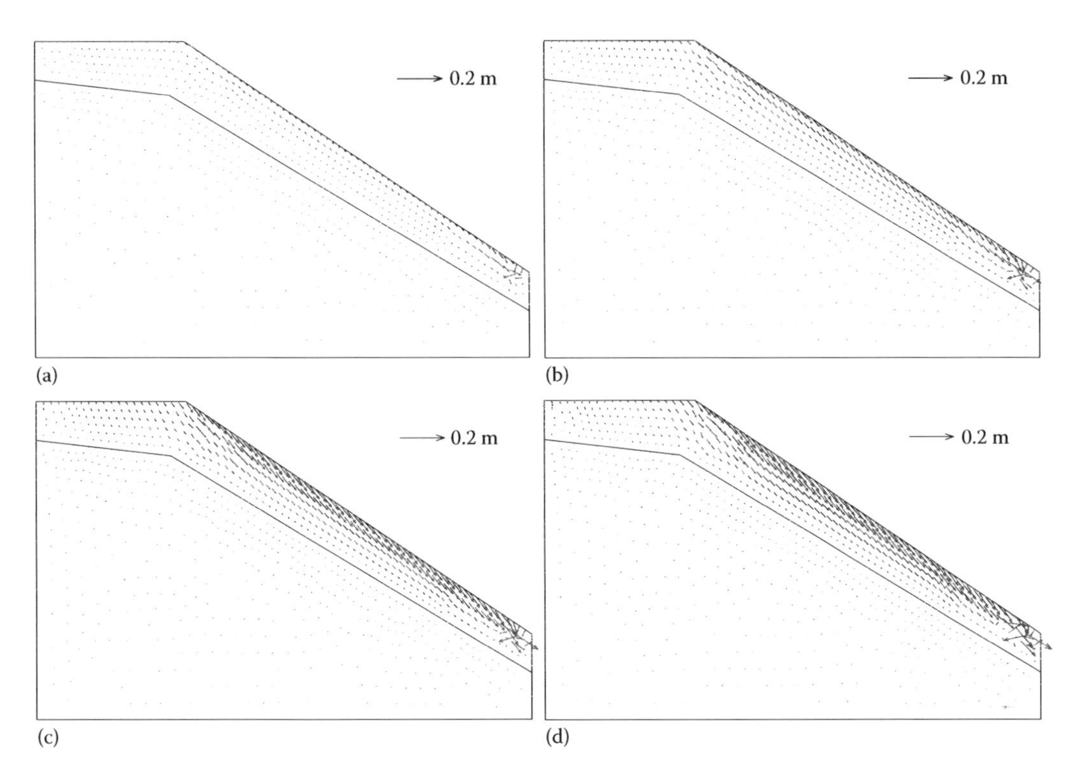

Figure 4.55 Displacement vectors in the slope at different moments of rainfall: (a) t = 6 h, (b) t = 17 h, (c) t = 24 h, and (d) t = 32 h.

Therefore, the calculated depths of the wetting front in the numerical modeling agree well with the laboratory test results.

Figure 4.55 illustrates the calculated displacement vectors at different times. At $t = 6$ h, the deformation mainly occurs at the slope surface. The maximum displacement is about 0.05 m, which occurs at the toe of the slope. As time passes, the deformation extends to soils at deeper depths and the displacement continues to increase. At the time when the landslide occurred, the deformation is mainly concentrated within 3 m depth below the slope surface. The maximum displacement at that time is about 0.17 m.

Figure 4.56 illustrates the variation of safety factor with time. The factors of safety are obtained using the enhanced limit method based on the results of the coupled analysis. Before the start of the rainstorm, the calculated global safety factor is 1.396. The safety factor starts to decrease after the rainstorm starts. After about 12 h of rainstorm, the safety factor of the slope is 1.181. Then the safety factor starts to increase although the rainstorm continues. The trend of increasing of safety factor stops at $t = 24$ h and the safety factor decreases again. At the moment when the landslide occurred, the calculated safety factor is 1.093.

Figure 4.57 shows the critical slip surfaces before the rain started and at the moment when the landslide occurred. The depth of the critical slip surface is about 2.3 m before the rain started, and is reduced to about 1.5 m at the moment when the landslide occurred. Because of the reduction of matric suction, the depth of the critical slip surface becomes shallower. According to the field investigation, the failure zone of Sau Mau Ping was generally confined to the top 3 m of the slope. The calculated results of the deformation and the critical slip surface show the failure of the slope is confined within the 3 m depth of the soil, which agrees with the field investigation.

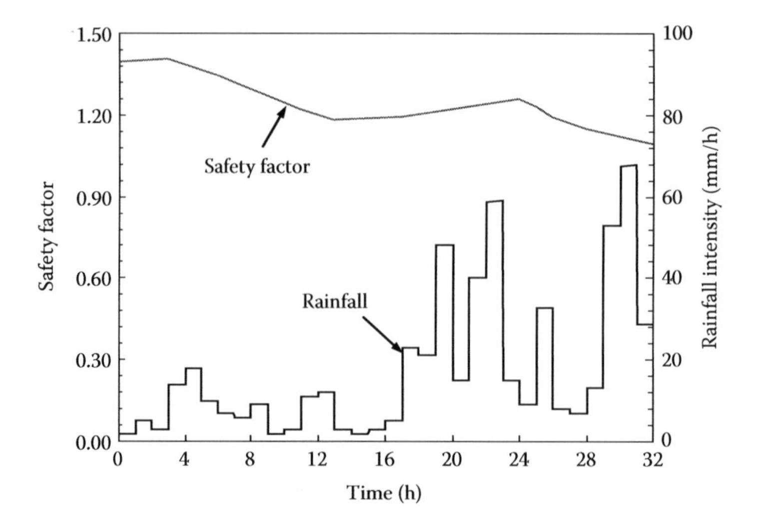

Figure 4.56 Safety factor versus time for Sau Mau Ping slope.

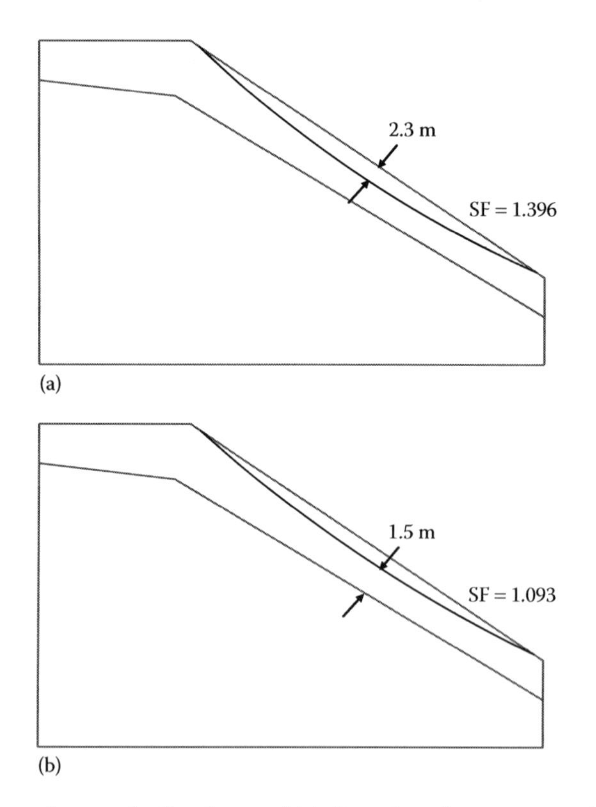

Figure 4.57 Critical slip surfaces and safety factors (a) before the rainstorm started ($t = 0$) and (b) at the moment when the landslide occurred ($t = 32$ h).

4.4.3 Case study of a field test of rainfall-induced landslide

4.4.3.1 Site description

A comprehensive field test was performed to study the behavior of a soil nailed loose CDG fill slope under artificial rainfall infiltration (Li, 2003; Li et al., 2008). The test site was located at Kadoorie Agricultural Research Centre, Hong Kong. The area of the test site was approximately 45 m by 35 m. The natural terrain was a moderately gentle sloping ground with an average angle of not more than 20° and composed of a surface layer of colluvium underlain by completely DG. No ground water was encountered within the depth of interest.

The experimental slope was designed at an angle of 33°, a height of 4.75 m (Figure 4.58). The slope was formed by dumping the fill on the base up to 3 m thick with hardly any compaction, simulating the old way with which loose fill slopes were constructed. The fill material is CDG, which was obtained from a construction site in Beacon Hill, Hong Kong. The maximum dry density of the soil was 19 kN/m^3 at the optimum moisture content of 12.7%. The initial measured average bulk density of the soil was 15.6 kN/m^3. The average relative compaction and moisture content measured at situ were 75% and 11%, respectively.

A 75 mm blinding layer with A252 steel mesh was designed for the fill slope base to isolate the fill material from the natural ground. A layer of asphalt was applied on the surface of the blinding layer as a further watertight measure. A layer of 150 mm no-fines concrete, a coarse free-draining material, was placed above the blinding layer forming a drainage layer. Permeable geotextile was applied between the no-fines concrete and the fill material dumped above to prevent the migration of fine-grained particles of the soil as water flowed across the drainage interface. A 300-diameter U channel covered with cast iron and steel mesh was provided at the toe of the slope to collect seepage water. A toe apron was constructed at the toe of the loose fill slope.

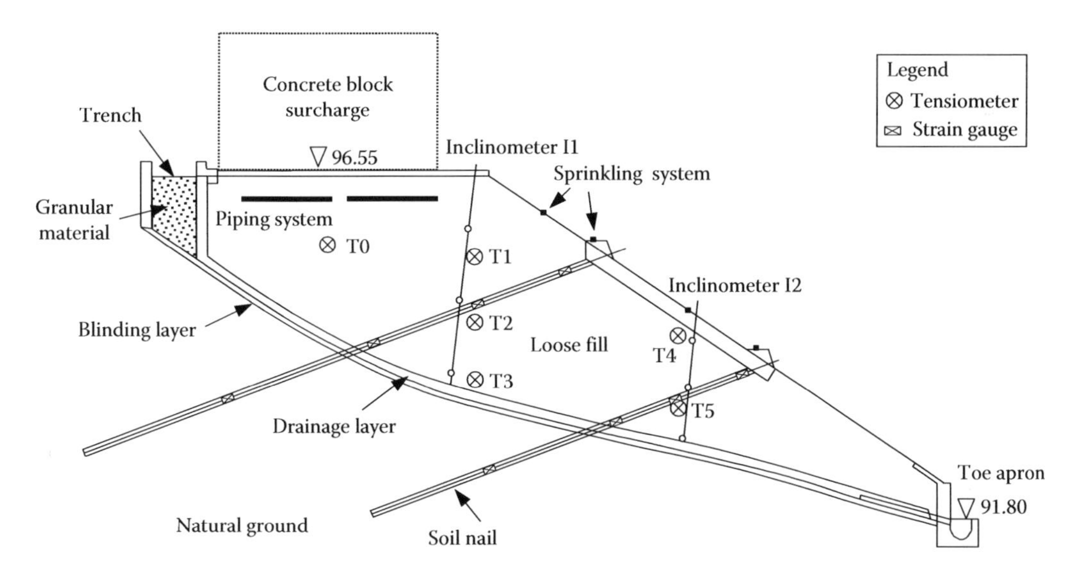

Figure 4.58 Cross section of the instrumented nailed slope at the Kadoorie Agricultural Research Centre.

Two rows of soil nails were installed in the loose fill slope. Boreholes of 100 mm in diameter were drilled at an inclination of 20° to the horizontal. The boreholes were spaced 1.5 m horizontally and 1.5 m vertically. The upper and lower rows of boreholes were 8 m and 6 m long, respectively. A 25 mm diameter ribbed steel bar was inserted into each hole. The holes were then filled with cement grout. Two types of nail head had been provided, namely independent head and grillage beams. No facing is constructed on the slope surface.

Two inclinometers were installed, one (I1) at the crest of the slope and the other (I2) in the middle of the two rows of nails, to measure the lateral displacements. Six tensiometers (T0–T5) were used to measure soil suctions at the specific locations. The forces in the nails were monitored using strain gauges, which were welded to the nails (Figure 4.57).

The fill slope was saturated using a recharge system comprised of a crest recharge trench, a buried piping system, and a surface sprinkling system, which could be operated independently. The crest recharge trench was directly connected with the drainage layer, through which water could be supplied to the soil from bottom. The piping system consisted of 10 sets of perforated bronze pipes laid 300 mm beneath the crest. Sprinklers were installed on the slope surface to simulate artificially the rainfall events.

4.4.3.2 *Experiment procedure*

Five layers of concrete blocks with the size of 1 m × 1 m × 0.6 m were placed on the central area of the crest as surcharge loading from 1 November 2002 to 20 November 2002. Starting from the morning of 20 November, an artificial rainfall with an intensity of 8.2 mm/h was created on the slope surface using the sprinklers for about four hours. During the daytime of the next day, another rainfall event was simulated. Starting from the morning of 22 November, the piping system embedded beneath the soil on the crest of the slope worked together with the sprinkling system. The surcharge concrete blocks eventually collapsed due to excess deformation at the crest of the slope at around 19:30 in the evening of 23 November. The artificial rainfall by the piping system and the sprinkling system had the same magnitude of intensity. Figure 4.59 shows the artificial rainfall applied on the slope by the sprinklers and the piping system.

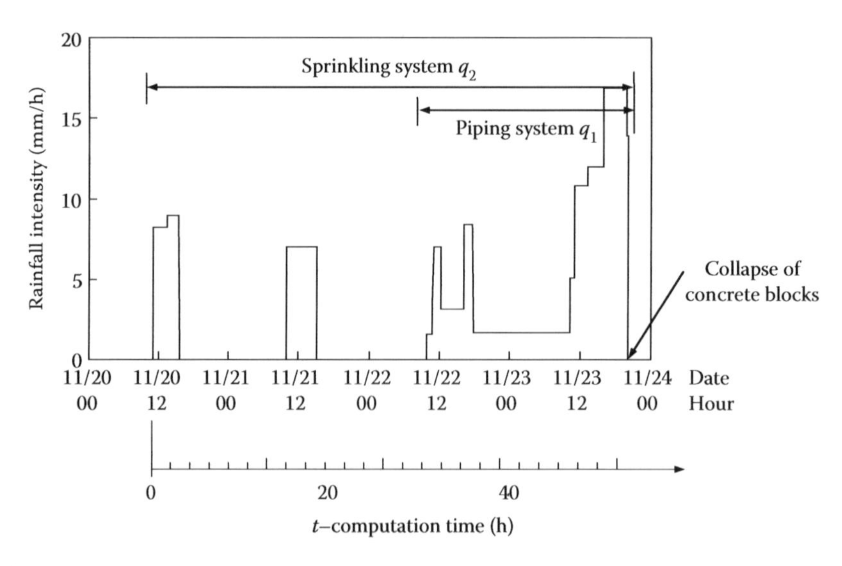

Figure 4.59 Artificial rainfall applied on the slope by the piping and sprinkler system.

4.4.3.3 Numerical model

Assuming that the soil nails may not influence water infiltration significantly, the FE meshes for seepage analysis and stress analysis are defined separately to reduce the computation load.

The FE mesh for stress analysis is shown in Figure 4.60. The upper layer of soil is CDG fill. The underlying soil layer is the natural ground. Second-order triangular elements are used in the FE mesh. Denser triangular elements with different properties are used to approximate the soil nails. The total number of elements in the FE mesh is 8322. On the vertical boundaries AD and CE, deformations in the horizontal direction are restricted. The slope surface ABC is assumed to be free of motion except that part subject to surcharge loading. The bottom boundary DE is completely constrained in both vertical and horizontal directions.

Figure 4.61 illustrates the FE mesh and the hydraulic boundary conditions defined in the seepage analysis. Soil nails are not defined in this mesh because the influence of soil nailing on the infiltration can be omitted. A blinding layer with very low permeability is defined between the fill material and the natural ground. The total amount of elements of the mesh is 540. Constant head boundary conditions are assumed for CI and FG. Constant pressure boundary conditions with $u_w = 0$ are assumed for GHI to simulate the effect of no-fines concrete. Along AB, AD, and IE and the bottom boundary DE, there is no flux. The flux rate by the sprinkling system and the piping system shown in Figure 4.59 are obtained by averaging the total amount of the water over the sloping surface (BC) and the area under the piping system (JK), respectively. The sprinklers were installed above the slope surface and the water was spread uniformly. Therefore, the flux rate q_2 can be applied along BC to simulate the recharge by the sprinkling system approximately. The piping system consisted of one main pipe and 10 branch pipes. For the soils right beneath the branch pipes, the actual flux rate can be greater than q_1. Therefore, cases with two different flux rates applied along a line JK, q_1 and $3q_1$, are investigated in this example.

The initial unit weight of the soil is 15.6 kN/m^3. As the degree of saturation of the soil increases, the self-weight of the soil also increases. The initial void ratio of the soils in the slope is 0.86. The measurements from the six tensiometers showed that the initial negative

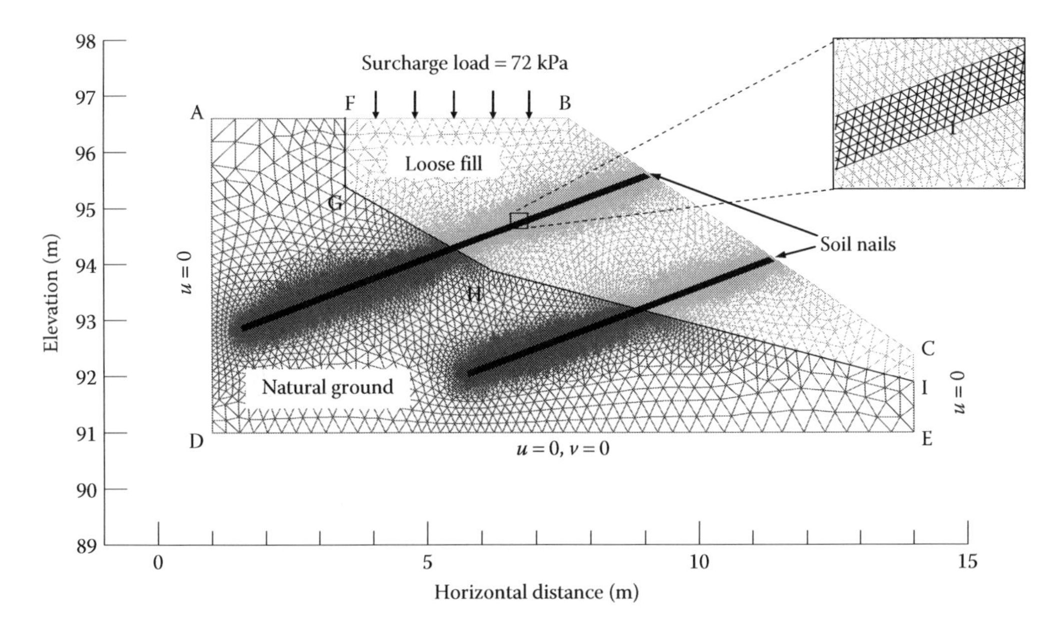

Figure 4.60 Finite element mesh and mechanical boundary conditions in stress analysis.

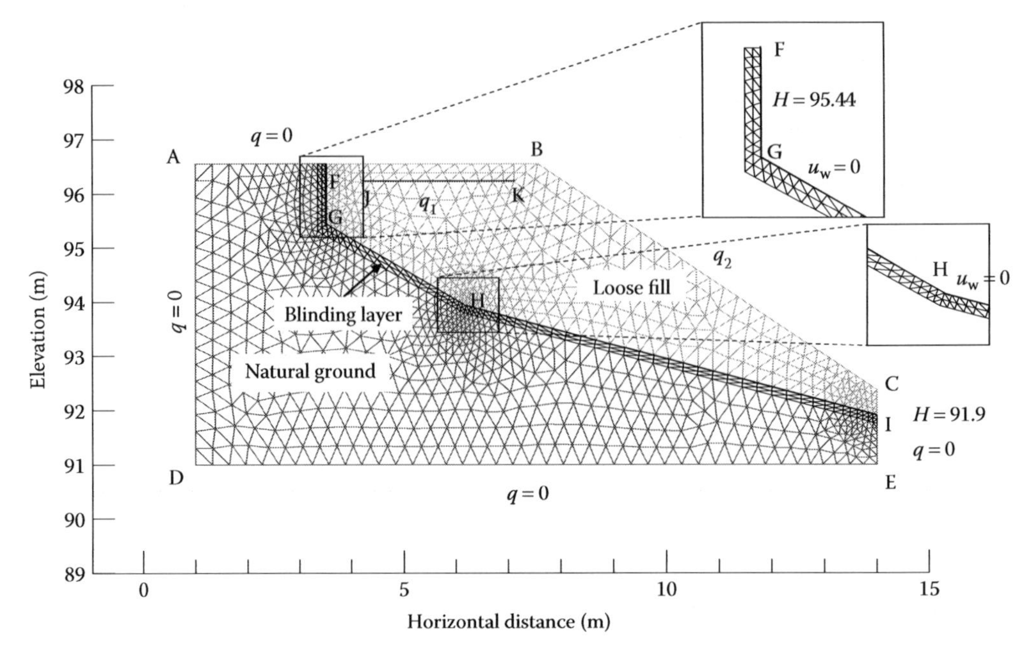

Figure 4.61 Finite element mesh and hydraulic boundary conditions in seepage analysis.

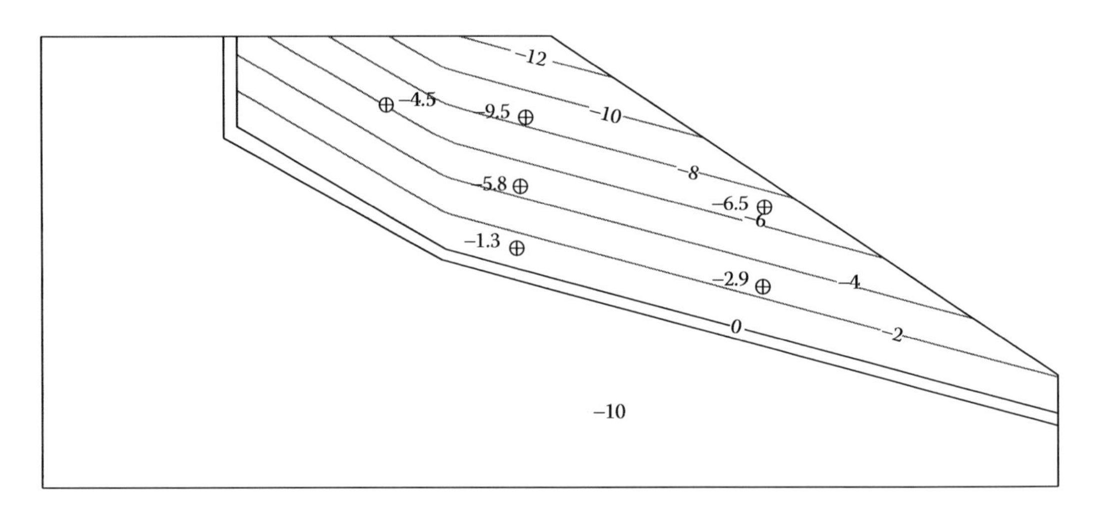

Figure 4.62 Initial pore-water pressure distribution (in kPa) assumed in the slope.

pore-water pressures were all less than 10 kPa. The initial pore-water pressure distribution of the numerical model (Figure 4.62) is assumed based on the field measurement.

4.4.3.4 Soil properties

Several research studies have involved with CDG soil from Beacon Hill, Hong Kong (Fung, 2001; Yeung, 2002; Li, 2003). Fung (2001) measured the SWCC of the CDG from Beacon Hill (Figure 4.63). Fitting the experimental SWCC data using the Fredlund and Xing (1994) SWCC model, the values of the fitting parameters are $q_s = 0.439$, $a_f = 0.212$, $n_f = 1.181$, and $m_f = 0.46$. Figure 4.64 shows the grain size distributions of CDG soils by Yeung (2002),

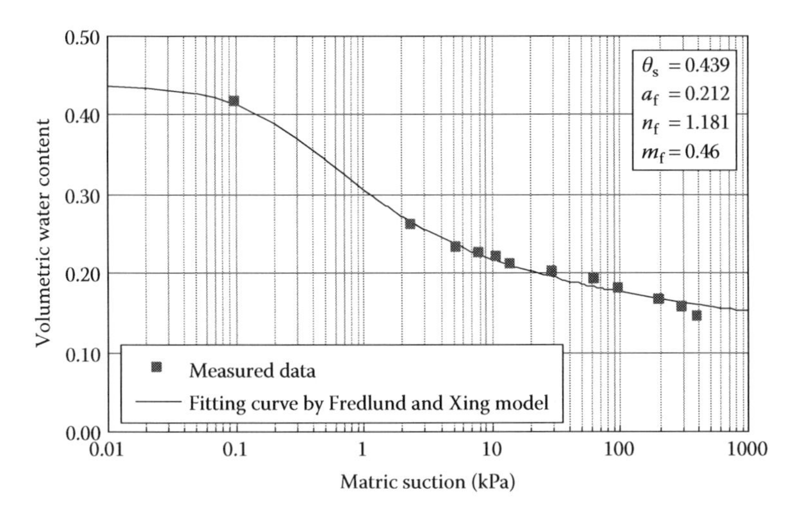

Figure 4.63 Experimental measurement of SWCC of CDG from Beacon Hill (plotted based on data from Fung, W. T., *Experimental Study and Centrifuge Modeling of Loose Fill Slope*. M. Phil Thesis, Hong Kong University of Science and Technology, Hong Kong, 2001) and the best-fitting curve.

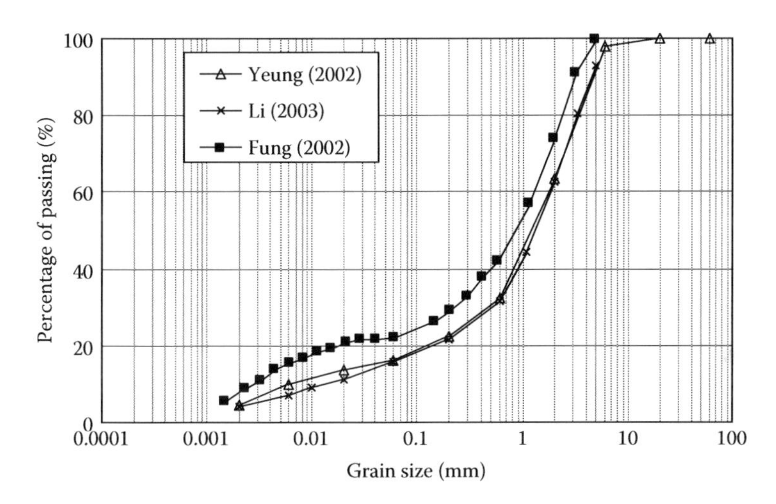

Figure 4.64 Measured grain size distributions of CDG from Beacon Hill.

Li (2003), and Fung (2001). Figure 4.65 shows the measured SWCC and estimated SWCCs based on grain size distributions. The estimated SWCCs (S2 and S3) have greater air-entry values than the lab measured SWCC (S1) by Fung (2001).

The saturated permeabilities of Beacon Hill CDG with different relative compactions were measured using permeameters (Yeung, 2002; Li, 2003; Pradhan, 2003; Pradhan et al., 2006). Figure 4.66 shows the experimental results of lab permeability tests. The variation of measured k_s can be two orders of magnitude. Field permeability tests carried out in 700 mm drill holes at the crest of the slope using a Guelph permeameter indicate that the saturated coefficient of permeability of the fill is 2×10^{-5} m/s in average (Li, 2003). The permeability functions estimated from the different SWCCs (S1 and S2) and different values of k_s are shown in Figure 4.67. Similar to the previous study of the Sau Mau Ping slope, with changing soil density due to deformation in the numerical modeling, the SWCC, and permeability function in each case will be modified accordingly.

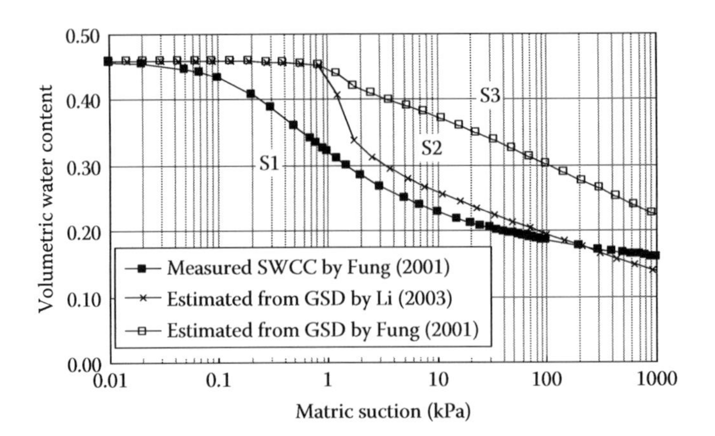

Figure 4.65 Predicted and estimated SWCCs of CDG soil in Kadoorie slope.

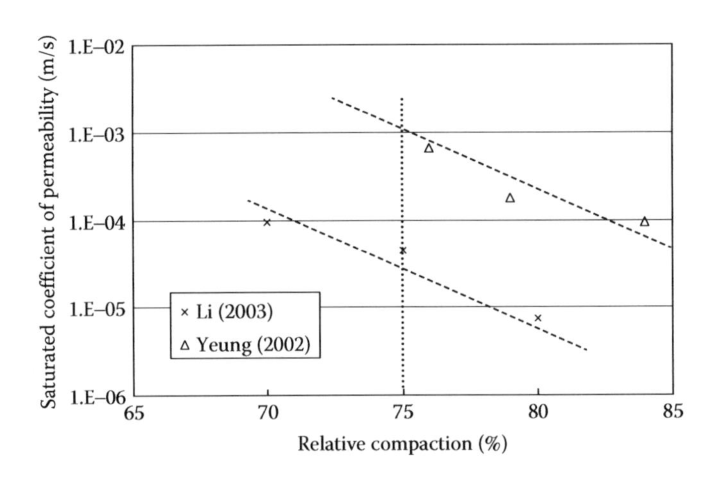

Figure 4.66 Saturated permeability k_s versus relative compaction of Beacon Hill CDG.

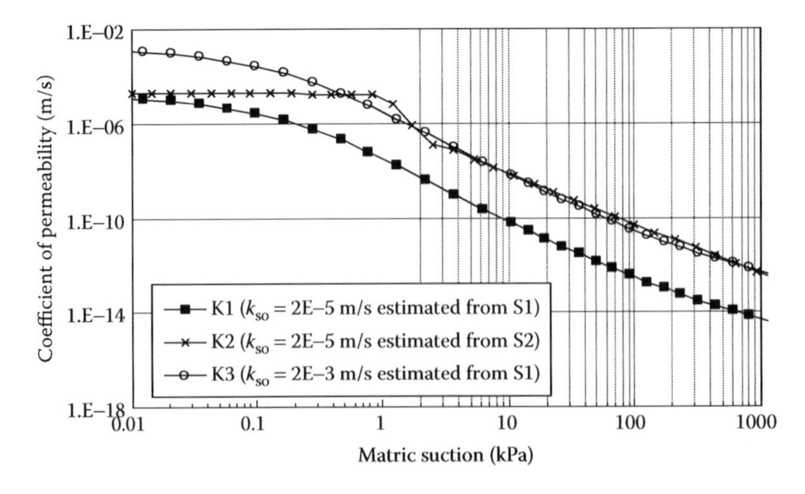

Figure 4.67 Permeability functions corresponding to different k_s and SWCC of CDG fill in Kadoorie slope.

For the fill material in Kadoorie slope, the same void ratio state surface (Equation 4.52) as that for the DG fill in Section 4.4.2 is used. The shear strength model is the same as that for the DG fill in the Sau Mau Ping slope except that parameters M, M_{col}, Γ, and λ are based on the experimental results on Beacon Hill CDG soil. Fung (2001) reported $M = 1.6$, $\Gamma = 1.173$, and $\lambda = 0.052$ for the CDG from Beacon Hill. The slope of collapse surface M_{col} is estimated to be 1.05.

For the soil in the natural ground, the SWCC and permeability function are assumed to be the same as those of the CDG fill. The shear strength model for the natural ground is the extended Mohr–Coulomb shear strength model with $c' = 0$, $\phi' = 39.2°$, and $\phi^b = 15°$. The elastic modulus E is taken as -2×10^4 kPa and H_F is taken as 4×10^5 kPa. Table 4.4 presents the parameters of soil properties for the CDG fill and the natural ground.

A layer of elements with extremely low permeability is defined in the seepage analysis to simulate the blinding layer. The permeability of the material is assumed to be 1×10^{-10} m/s. The SWCC of the material in the blinding layer is assumed to be the same as that of CDG fill.

The nails with steel bars and grout are modeled as a composite solid material in the stress analysis. The Young's modulus of the steel bar and the grout are 200 and 30 GPa, respectively. Considering that the 3D soil nail is approximately modeled as a plate under the plane-strain condition, the modulus of the composite material is taken as the average modulus of the steel bar, the grout and the surrounding soil, which is averaged based on their corresponding cross section areas. The Poisson ratio of the material is assumed to be 0.25. The effect of soil–nail interface is not considered.

Considering the variability in soil hydraulic properties and the uncertainty in the flux provided by the piping system, five cases (Table 4.5) with different combinations of SWCC, coefficient of permeability function, and input flux of piping system are investigated in this

Table 4.4 Summary of input soil parameters of Kadoorie field test

Soil properties	*DG fill*	*Natural ground*
SWCC	$\theta_{so} = 0.46, a_f = 0.212, n_f = 1.181, m_f = 0.46$ (S1) $\theta_{so} = 0.46, a_f = 1.092, n_f = 20, m_f = 0.123$ (S2)	Same as DG fill
Saturated permeability	$k_{so} = 2.0 \times 10^{-5}$ m/s (K1 and K2) $k_{so} = 2.0 \times 10^{-3}$ m/s (K3)	Same as DG fill
Shear strength	$c' = 0, \phi^b = 15°, \Gamma = 1.173, \lambda = 0.052, M = 1.6$ $(\phi'_{cs} = 39.2°), M_{col} = 1.05$ $(\phi'_{col} = 26.5°)$	$c' = 0, \phi^b = 15°, \phi' = 39.2°$
Void ratio	$a_{la} = 1.2187, b_{la} = -0.04412, c_{la} = -0.00239$, and $d_{la} = 0.000747$	N.A.
Poisson's ratio	$\nu = 0.40$	$\nu = 0.33$
Elastic modulus E	Equation 4.61 with $R_f = 0.5$	$E = -2 \times 10^4$ kPa
Elastic modulus H	$H_F = \frac{3}{m_s^2}$	$H_F = 4 \times 10^5$ kPa

Table 4.5 Study scheme for Kadoorie slope

Case	*SWCC of DG fill*	*Permeability function*	*Flux rate of piping system*
1	S2	K2	$3q_1$
2	S2	K2	q_1
3	S1	K1	$3q_1$
4	S1	K1	q_1
5	S1	K3	$3q_1$

example. A surcharge loading of 72 kPa is first applied along the crest of the slope. After that, fluxes are applied though BC and JK to simulate the artificial rainfall by the recharge system.

4.4.3.5 Results and discussion

4.4.3.5.1 Pore-water pressure responses

In this study, the measurements of pore-water pressures by the six tensiometers are compared with the calculated results for all five cases. The results of Cases 1 and 5 have the best agreement with the field measurements. In this section, only the results of Cases 1, 3, and 5 are presented. Figure 4.68a shows the measured and calculated suction at tensiometer T0, which is located 0.7 m below the piping system. After the rainfall of the piping system starts, the measured and calculated suctions are reduced very rapidly to around 0.5 and 1.5 kPa, respectively. The increase of calculated suction right before this rapid

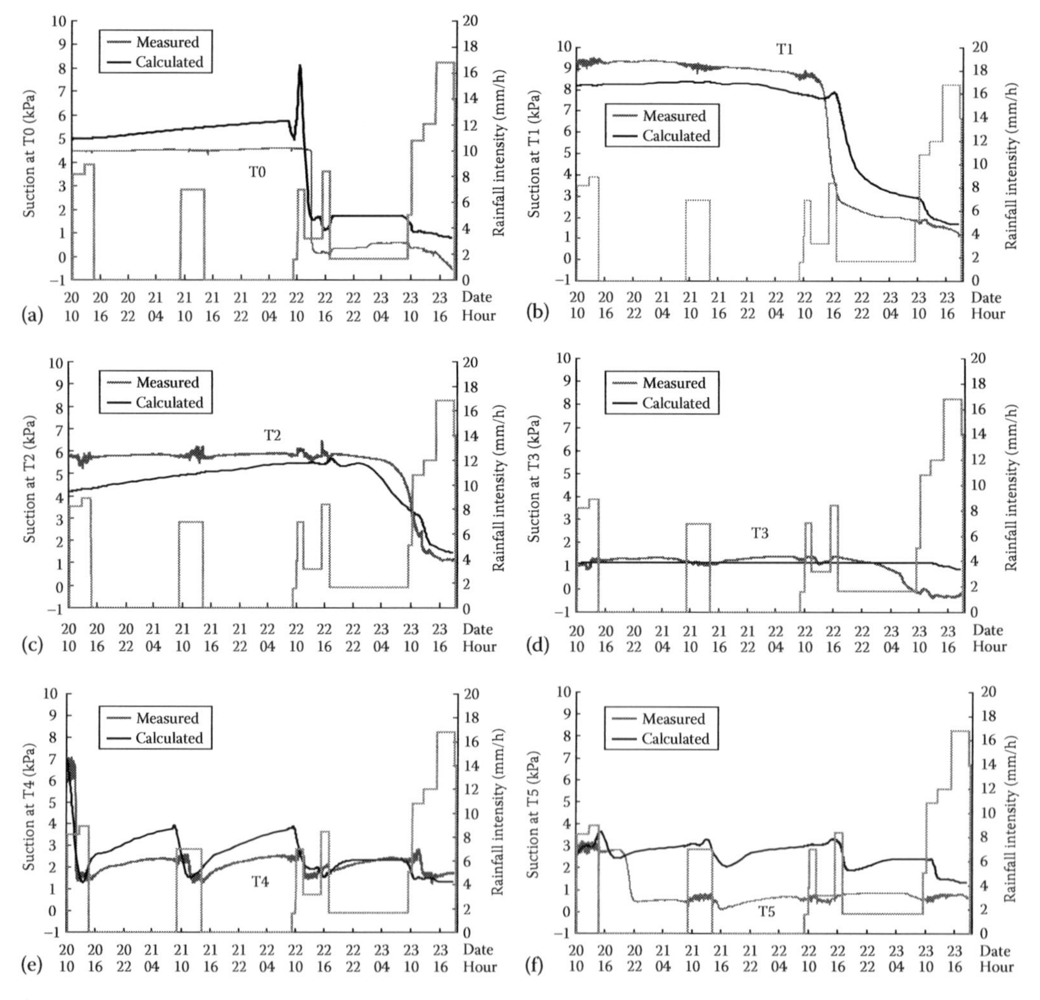

Figure 4.68 Comparison of measured and calculated suctions at tensiometers (Case 1): (a) T0, (b) T1, (c) T2, (d) T3, (e) T4, and (f) T5.

reduction is due to the numerical oscillation. Figure 4.68b shows the comparison of the suctions at tensiometer T1 for Case 1. The suction is not significantly changed after the two artificial rainfall events created by the sprinkling system. When the piping system starts, the suction starts to reduce. However, the response of the calculated suction value has about 3 h of delay compared with the measurement. Figure 4.68c illustrates the suction at tensiometer T2. This tensiometer is under deeper depth than T1, therefore the response of matric suction is slower than T1. Both the measurement and the calculated results show this delayed response. For tensiometer T3 (Figure 4.68d), the reduction of the calculated suction also lags behind the measurement. Figures 4.68e and f illustrate the comparison of the calculated suctions and the measurements at T4 and T5, respectively. The fluctuating features of the suction responses at both locations agree well with that of the measurements.

Figure 4.69 present the variations of suction for Case 3. The responses of the calculated suctions do not agree well with the measurements. Comparing the results of Case 3

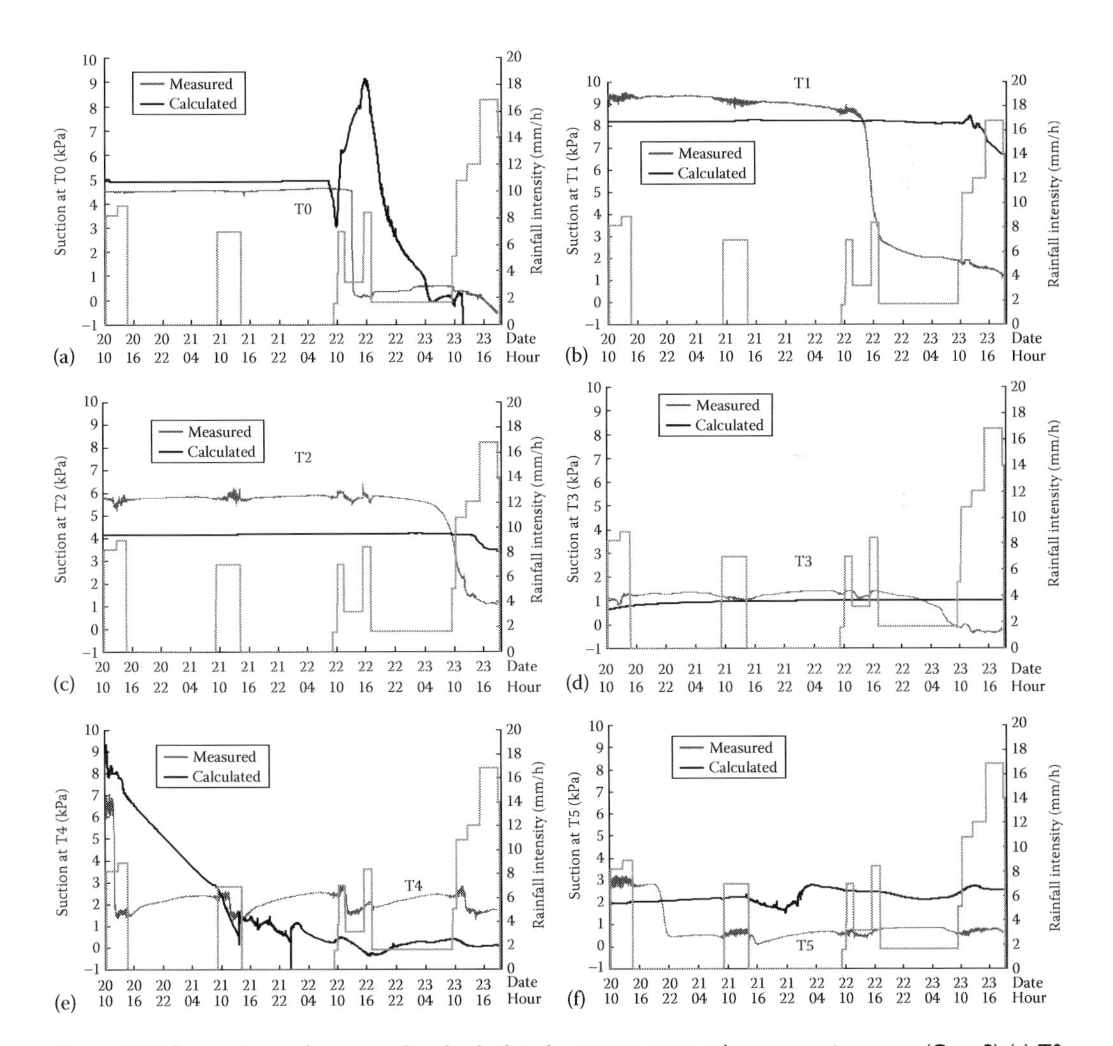

Figure 4.69 Comparison of measured and calculated matric suction values at tensiometers (Case 3) (a) T0, (b) T1, (c) T2, (d) T3, (e) T4, and (f) T5.

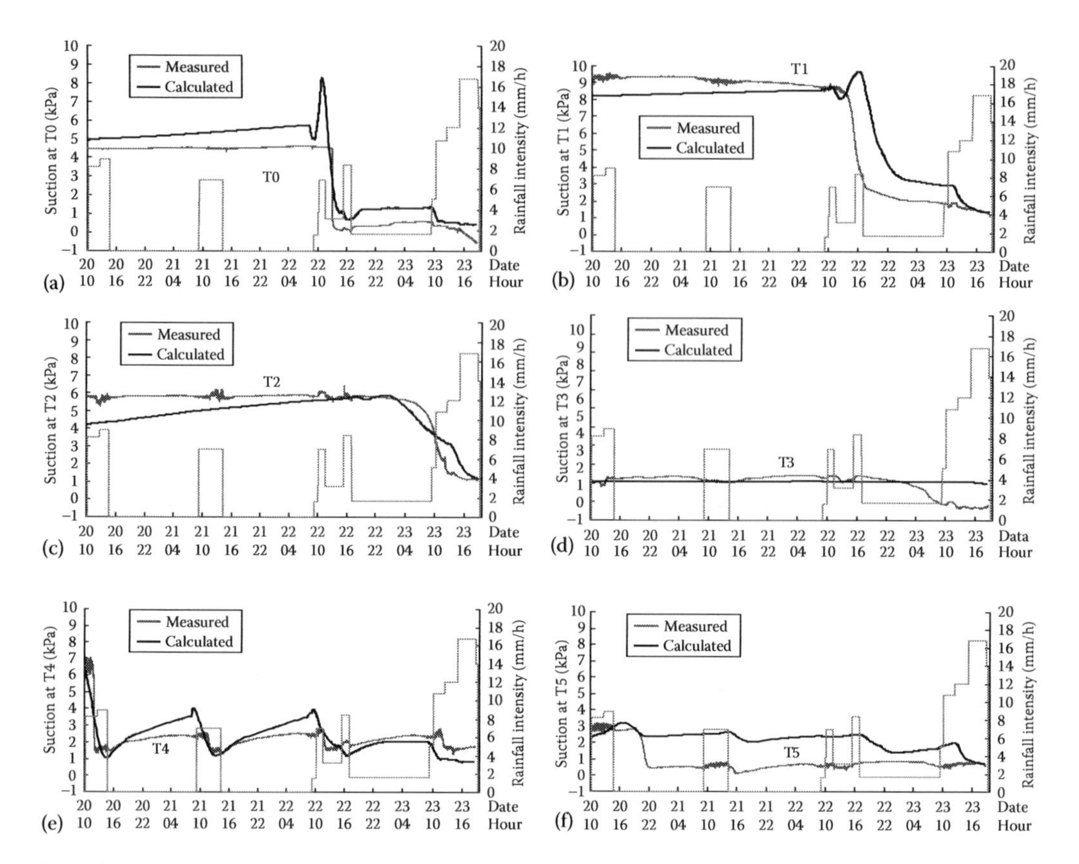

Figure 4.70 Comparison of measured and calculated matric suction values at tensiometers (Case 5): (a) T0, (b) T1, (c) T2, (d) T3, (e) T4, and (f) T5.

with those of Case 1, the response of the change of suction in a soil with a smaller AEV (SWCC is S1 in Case 3) is slower than that in a soil with a larger AEV (SWCC is S2 in Case 1).

Figure 4.70 illustrate the results of Case 5. Comparing the results of Case 5 with the results of Case 1, the responses of the suctions in the two cases are very similar. This is mainly because the permeability functions from the suction range of 0–10 kPa are quite close for these two cases (Figure 4.67). The saturated permeability of 2×10^{-3} m/s in Case 5 is assumed based on the trend line of permeability measurements (Figure 4.66). However it is an extremely high value for CDG. Because the permeability for in situ completely DGs is in the range of 10^{-5}–10^{-7} m/s (GEO, 1993).

4.4.3.5.2 Displacement

As the pore pressure results of Case 1 generally agree well with the measured, only the results of displacement for Case 1 are presented and compared with field measurements. Figure 4.71 illustrates the displacement vectors in the slope before the rainfall started. As the surcharge loading is applied on the crest of the slope, the deformation occurs mostly on the upper part of the slope. Figure 4.72 present the calculated lateral displacements and

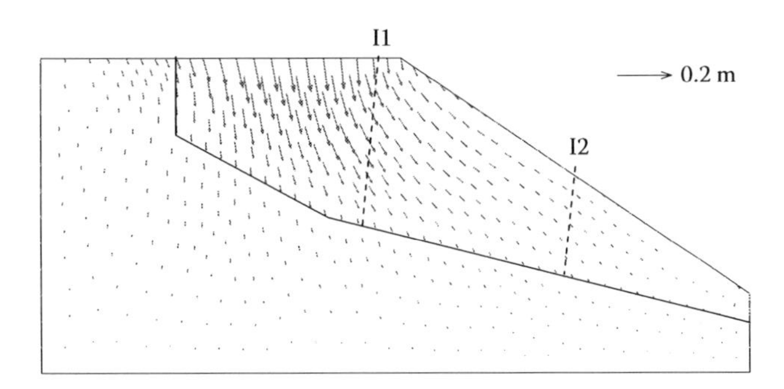

Figure 4.71 Calculated displacement vectors in Kadoorie slope due to surcharge loading before rain started.

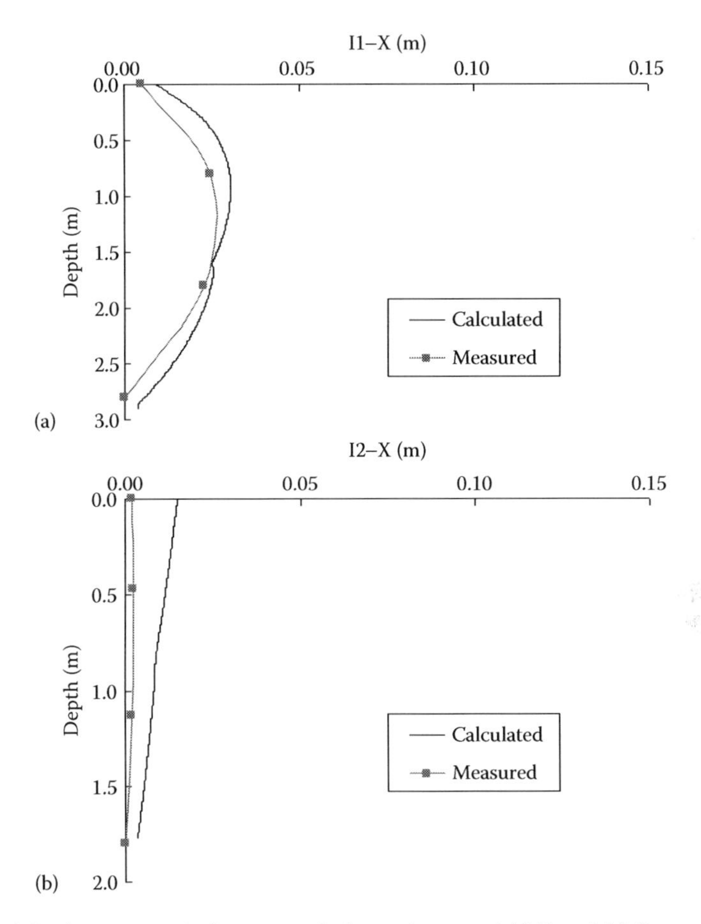

Figure 4.72 Lateral displacement at inclinometers before rain started (a) I1 and (b) I2.

the measurements by inclinometers I1 and I2 before the start of rainfall. The calculated displacement at I1 agrees well with the measurement at I1. The maximum calculated and measured horizontal displacements occur at about 0.8 m below the slope surface. If considering the reference point for the inclinometer is the lowest point where the displacement is assumed to be zero, the calculated and measured results agree even better. For

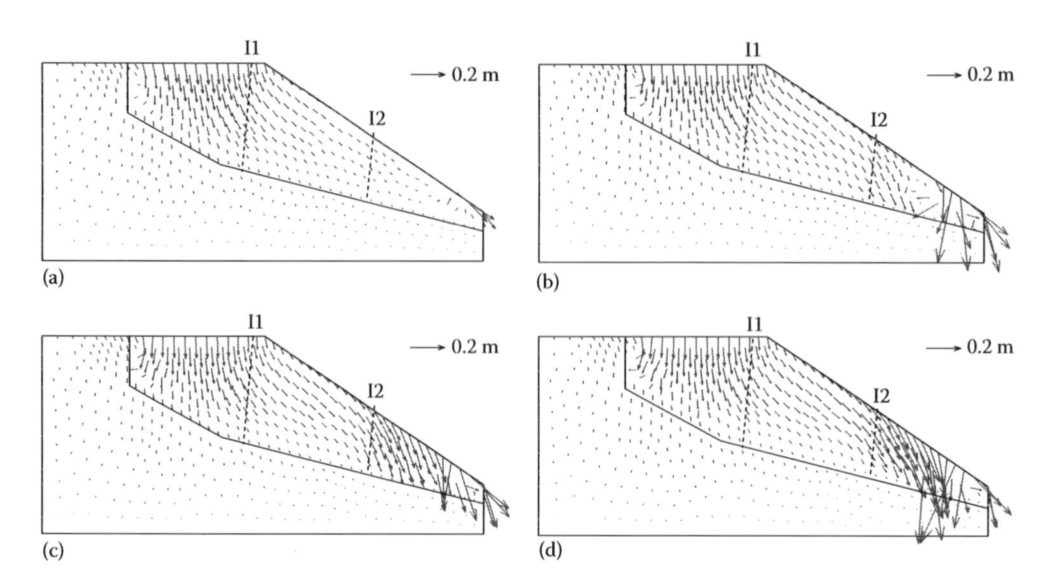

Figure 4.73 Displacement vectors in Kadoorie slope at different moments (Case I) (a) $t = 30$ h, (b) $b = 48$ h, (c) $t = 72$ h, and (d) $t = 81.5$ h.

inclinometer 2, the maximum horizontal displacements occur at the slope surface. The calculated result is greater than the measurement.

Figure 4.73 shows the deformation in the slope after the rainfall starts. After 30 h of rainfall applied along the sloping surface by the sprinkler system, the soils near the toe are saturated first and the maximum displacement is about 15 cm. After 48 h, the piping system starts to add water into soils and the rainfall intensity became larger. More soils in the slope are getting saturated and the shear strength and stiffness of these soils are reduced significantly. Therefore, the loading from surcharge on the slope crest is transferred to the other part of the slope.

Figure 4.74 illustrate the calculated and measured horizontal displacements when the concrete blocks fell down the slope. The calculated displacement is less than the measurement at I1. It is because the volume change of the soil under the piping system induces mostly vertical displacement in the numerical modeling as shown in Figure 4.73. This is primarily due to the assumptions of isotropic behavior and ignorance of plastic deformation in the model. The calculated horizontal displacement at I2 is greater than the measurement because shearing effect is more significant than the contractive volume change in the sloping part of the slope.

4.4.3.5.3 Slope stability

The change of the safety factor of the slope with respect to time is shown in Figure 4.75. The safety factor of the slope is 2.152 before the rain starts. As the artificial rainfall was created only intermittently, during the first 48 h, the safety factor fluctuates with time. On the third day, the safety factor continues to decrease to a value of 1.129. Figure 4.76 shows the calculated critical slip surfaces before the rainfall started and at the moment when the concrete blocks collapsed. The critical slip surface is shallower at the moment when the concrete blocks collapsed.

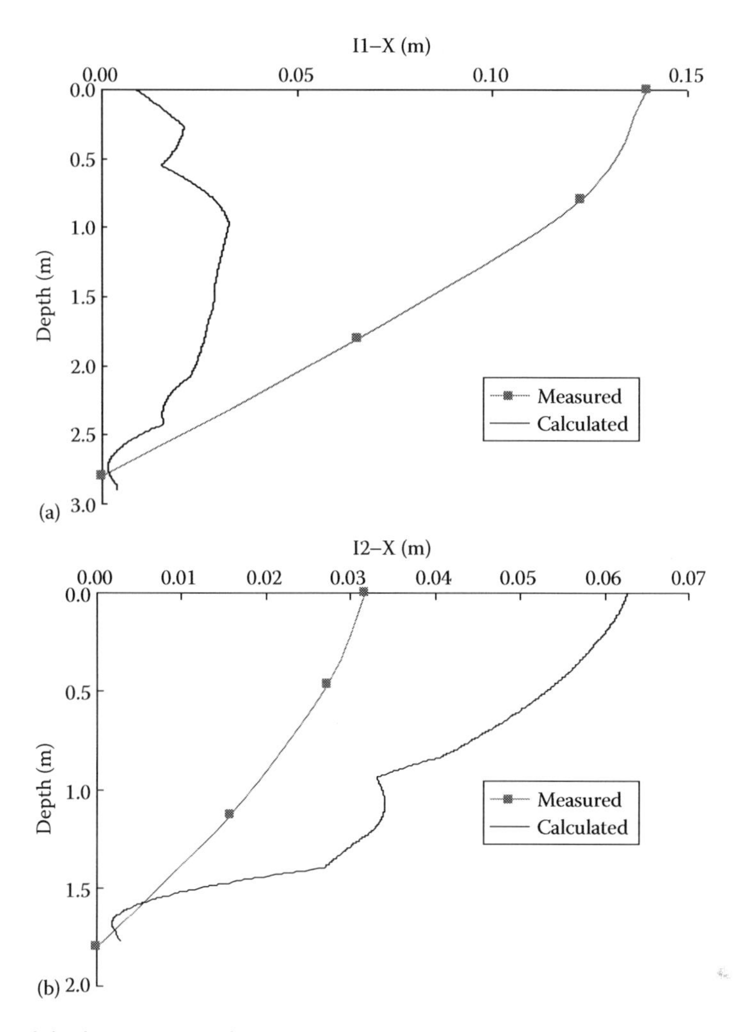

Figure 4.74 Lateral displacement at inclinometers at the moment when the slope collapsed (Case 1): (a) I1 and (b) I2.

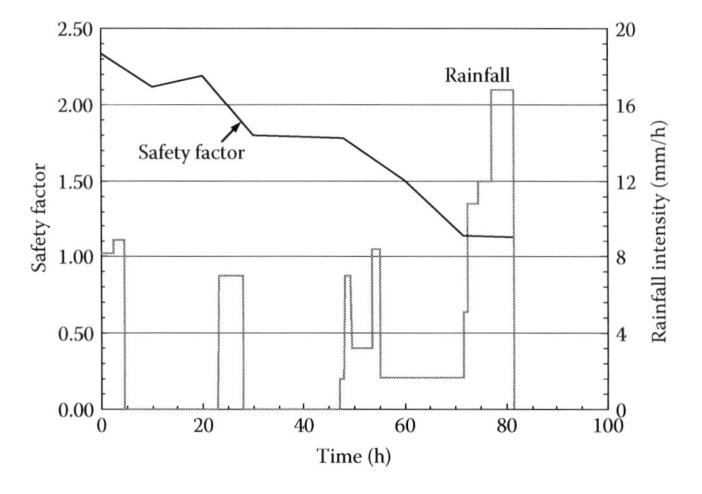

Figure 4.75 Safety factor versus time for Kadoorie field test (Case 1).

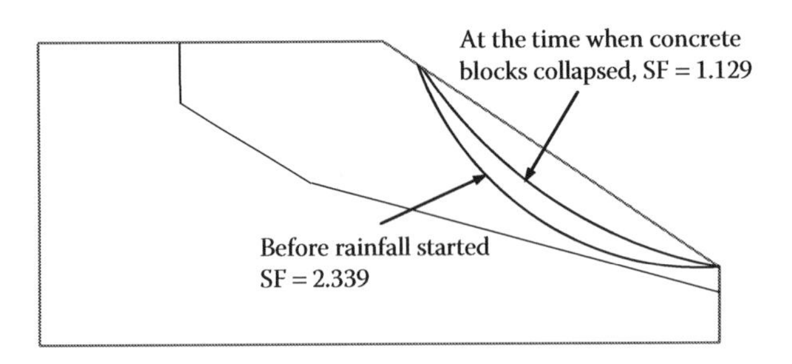

Figure 4.76 Critical slip surfaces and the safety factors of Kadoorie field test (Case 1).

REFERENCES

Ahuja, L. R., Cassel, D. K., Bruce, R. R., and Barnes, B. B. (1989). Evaluation of spatial distribution of hydraulic conductivity using effective porosity data. *Soil Science*, 148 (6), 404–411.

Alonso, E. E., Canete, A., and Olivella, S. (2002). Moisture transfer and deformation behaviour of pavements: effect of climate, materials and drainage. In: Jucá, J.F.T., de Campos, T.M.P., Marinho, F.A.M. (eds.), *Unsaturated Soils, Proceedings of the 3rd International Conference on Unsaturated Soils (UNSAT)*, Recife, Brazil, 10–13 March. Swets & Zeitlinger, Lisse, the Netherlands, Vol. 2, pp. 671–677.

Alonso, E. E., Gens, A., and Delahaye, C. H. (2003). Influence of rainfall on the deformation and stability of a slope in overconsolidated clays: a case study. *Hydrogeology Journal*, 11 (1), 174–192.

Alonso, E. E., Gens, A., and Hight, D. W. (1987). Special problem soils. General report. In: Hanrahan, E.T., Orr, T.L.L. and Widdis, T.F (eds.), *Proceedings of the 9th European Conference on Soil Mechanics and Foundation Engineering*, Dublin, Ireland, 31 August–3 September. Balkema, Rotterdam, the Netherlands, Vol. 3, pp. 1087–1146.

Alonso, E. E., Gens, A., and Josa, A. (1990). A constitutive model for partially saturated soils. *Géotechnique*, 40 (3), 405–430.

Alonso, E. E., Lloret, A., and Romero, E. (1996). Rainfall induced deformation of road embankments. In: Senneset, K. (ed.), *Landslides, Proceedings of the 7th International Symposium on Landslides*, Trondheim, Norway. Balkema, Rotterdam, Vol. 1, pp. 97–108.

Alonso, E. E., Vaunat, J., Gens, A. (1999). Modelling the mechanical behaviour of expansive clays. *Engineering Geology*, 54, 173–183.

Arairo, W., Prunier, F., Djéran-Maigre, I., and Darve, F. (2013). A new insight into modelling the behaviour of unsaturated soils. *Canadian Geotechnical Journal*, 37 (16), 2629–2654.

Bandara, S., and Soga, K. (2015). Coupling of soil deformation and pore fluid flow using material point method. *Computers and Geotechnics*, 63, 199–214.

Biot, M. A. (1941). General theory for three-dimensional consolidation. *Journal of Applied Physics*, 12 (2), 155–164.

Bishop, A. W. (1959). The principle of effective stress. *Teknisk Ukeblad*, 106 (39), 859–63.

Bishop, A. W., and Blight, G. E. (1963). Some aspects of effective stress in saturated and unsaturated soils. *Géotechnique*, 13 (3), 177–197.

Bolzon, G., Schrefler, B. A., and Zienkiewicz, O. C. (1996). Elastoplastic soil constitutive laws generalised to partially saturated states. *Géotechnique*, 46 (2), 279–289.

Borja, R. I. (2004). Cam-Clay plasticity. Part V: A mathematical framework for three-phase deformation and strain localization analyses of partially saturated porous media. *Computer Methods in Applied Mechanics and Engineering*, 193 (48), 5301–5338.

Borja, R. I. (2006). On the mechanical energy and effective stress in saturated and unsaturated porous continua. *International Journal of Solids and Structures*, 43 (6), 1764–1786.

Borja, R. I., and Koliji, A. (2009). On the effective stress in unsaturated porous continua with double porosity. *Journal of the Mechanics and Physics of Solids*, 57 (8), 1182–1193.

Borja, R. I., Liu, X. Y., and White, J. A. (2012). Multiphysics hillslope processes triggering landslides. *Acta Geotechnica*, 7 (4), 261–269.

Borja, R. I., and White, J. A. (2010). Continuum deformation and stability analyses of a steep hillside slope under rainfall infiltration. *Acta Geotechnica*, 5 (1), 1–14.

Cascini, L., Cuomo, S., Pastor, M., and Sorbin, G. (2010). Modelling of rainfall-induced shallow landslides of the flow-type. *Journal of Geotechnical and Geoenvironmental Engineering*, 136 (1), 85–98.

Chen, R. H., Chen, H. P., Chen, K. S., and Zhung, H. B. (2009). Simulation of a slope failure induced by rainfall infiltration. *Environmental Geology*, 58 (5), 943–952.

Cheuk, C. Y. (2001). *Investigations of Soil Nails in Loose Fill Slopes*. M. Phil. Thesis, Hong Kong University of Science and Technology, Hong Kong.

Cheuk, C. Y., Ng, C. W. W., and Sun, H. W. (2005). Numerical experiments of soil nails in loose fill slopes subjected to rainfall infiltration effects. *Computers and Geotechnics*, 32 (4), 290–303.

Childs, E. C., and Collis-George. N. (1950). The permeability of porous material. *Proceedings of the Royal Society* A, 201, 392–405.

Chiu, C. F., and Ng, C. W. W. (2003). A state-dependent elasto-plastic model for saturated and unsaturated soils. *Géotechnique*, 53 (9), 809–829.

Cho, S. E., and Lee, S. R. (2001). Instability of unsaturated soil slopes due to infiltration. *Computers and Geotechnics*, 28, 185–208.

Coleman, J. D. (1962). Stress/strain relations for partly saturated soils. *Géotechnique*, 12 (4), 348–350.

Cui, Y. J., and Delage, P. (1996). Yielding and plastic behaviour of an unsaturated compacted silt. *Géotechnique*, 46 (2), 291–311.

Das, B., and Sobhan, K. (2013). *Principles of Geotechnical Engineering*, 8th edn. Cengage Learning, Stamford, CT, USA.

Dassault Systèmes Simulia Corp. (2011). *ABAQUS User Documentation* Version 6.9. Dassault Systèmes Simulia Corp. Providence, RI, USA.

Dawson, E. M., Roth, W. H., and Drescher, A. (1999). Slope stability analysis by strength reduction. *Géotechnique*, 49 (6), 835–840.

Fredlund, D. G., and Morgenstern, N. R. 1977. Stress state variables for unsaturated soils, *Journal of Geotechnical Engineering Division ASCE*, 103 (GT5), 447–466.

Fredlund, D. G., and Rahardjo, H. (1993). *Soil Mechanics for Unsaturated Soils*. John Wiley & Sons, New York.

Fredlund, D. G., Rahardjo, H., and Fredlund, M. D. (2012). *Unsaturated Soil Mechanics in Engineering Practice*. John Wiley & Sons, New York.

Fredlund, D. G., and Xing, A. Q. (1994). Equations for the soil-water characteristic curve. *Canadian Geotechnical Journal*, 31 (4), 521–532.

Fredlund, D. G., Xing, A. Q., and Huang, S. Y. (1994). Predicting the permeability function for unsaturated soils using the soil-water characteristic curve. *Canadian Geotechnical Journal*, 31(4), 533–546.

Fredlund, M. D., Wilson, G. W., and Fredlund, D. G. (2002). Use of the grain-size distribution for estimation of the soil-water characteristic curve. *Canadian Geotechnical Journal*, 39 (5), 1103–1117.

Fung, W. T. (2001). *Experimental Study and Centrifuge Modeling of Loose Fill Slope*. M. Phil Thesis, Hong Kong University of Science and Technology, Hong Kong.

Gallipoli, D., Gens, A., Sharma, R., and Vaunat, J. (2003). An elastoplastic model for unsaturated soil incorporating the effects of suction and degree of saturation on mechanical behaviour. *Géotechnique*, 53 (1), 123–135.

Gamnitzer, P., and Hofstetter. G. (2015). A smoothed cap model for variably saturated soils and its robust numerical implementation. *International Journal for Numerical and Analytical Methods in Geomechanics*, 39(12), 1276–1303.

Garcia, E., Oka, F., and Kimoto, S. (2011). Numerical analysis of a one-dimensional infiltration problem in unsaturated soil by a seepage–deformation coupled method. *International Journal for Numerical and Analytical Methods in Geomechanics*, 35 (5), 544–568.

Gardner, W. R. (1958). Some steady state solutions of the unsaturated moisture flow equation with application to evaporation from a water table. *Soil Science*, 85 (4), 228–232.

Gens, A., and Alonso, E. E. (1992). A framework for the behaviour of unsaturated expansive clays. *Canadian Geotechnical Journal*, 29, 1013–1032.

Gens, A., Guimaraes, L., Sánchez, M., and Sheng, D. (2008). Developments in modelling the generalised behaviour of unsaturated soils. In: Toll et al., eds. *Unsaturated Soils: Advances in Geo-engineering*, Taylor & Francis Group, London, pp. 53–61.

GEO. (1993). *Guide to Retaining Wall Design (2nd Edition)*. Geotechnical Engineering Office, Civil Engineering Department, the Government of Hong Kong, Hong Kong.

Geo-slope International Ltd. (2012). *Slope/W for Slope Stability Analysis, User's Guide*, Geo-slope International Ltd, Calgary, Canada.

Griffiths, D. V., and Lane, P. A. (1999). Slope stability analysis by finite elements. *Géotechnique*, 49 (3), 387–403.

Hong Kong Government. (1977). *Report on the slope failures at Sau Mau Ping*, August 1976. Hong Kong Government Printer, Hong Kong.

Hu, R., Chen, Y. F., Liu, H. H., and Zhou, C. B. (2015). A coupled stress–strain and hydraulic hysteresis model for unsaturated soils: Thermodynamic analysis and model evaluation. *Computers and Geotechnics*, 63, 159–170.

Hu, R., Chen, Y. F., and Zhou, C. B. (2011). Modeling of coupled deformation, water flow and gas transport in soil slopes subjected to rain infiltration. *Science China Technological Sciences*, 54 (10), 2561–2575.

Kam, W. T. (1999). *A Study of One-dimensional Deformation of Hong Kong Soils Subjected Loading and Inundation*. Final Year Project. Hong Kong University of Science and Technology, Hong Kong.

Kim, J., Jeong, S., and Regueiro, R. A. (2012). Instability of partially saturated soil slopes due to alteration of rainfall pattern. *Engineering Geology*, 147–148, 28–36.

Kim, J. M. (1996). *A Fully Coupled Model for Saturated-unsaturated Fluid Flow in Deformable Porous and Fractured Media*. Ph.D. Thesis, Pennsylvania State University, University Park, Pennsylvania.

Kim, J. M. (2000). A full coupled finite element analysis of water-table fluctuation and land deformation in partially saturated soils due to surface loading. *International Journal for Numerical Methods in Engineering*, 49 (9), 1101–1119.

Kim, J. M. (2004). Fully coupled poroelastic governing equations for groundwater flow and solid skeleton deformation in variably saturated true anisotropic porous geologic media. *Geosciences Journal*, 8 (3), 291–300.

Kimoto, S., Oka, F., Fujiwaki, M., and Fujita, Y. (2007). Numerical analysis of deformation of methane hydrated contained soil due to the dissociation of gas hydrate. In: Exadaktylos, G., and Vardoulakis, I. G., eds. *Bifurcations, Instabilities, Degradation in Geomechanics*, Springer-Verlag, Berlin, pp. 361–380.

Knill, J. L., Lumb, P., Mackey, S., de Mello, V. F. B., Morgenstern, N. R., and Richards, B. G. (1999). *Report of the Independent Review Panel on Fill Slopes*. Geotechnical Engineering Office, Civil Engineering Department, Hong Kong.

Kohgo, Y., Nakano, M., and Miyazaki, T. (1993). Theoretical aspects of constitutive modeling for unsaturated soils. *Soils and Foundations*, 33 (4), 49–63.

Kohler, R., and Hofstetter, G. (2008). A cap model for partially saturated soils. *International Journal for Numerical and Analytical Methods in Geomechanics*, 32 (8), 981–1004.

Kulhawy, F. H. 1969. *Finite Element Analysis of the Behavior of Embankments*. Ph.D. Thesis, The University of California at Berkeley, California, USA.

Li, J. (2003). *Field Study of A Soil Nailed Loose Fill Slope*. Ph.D. Thesis, The University of Hong Kong, Hong Kong.

Li, J., Tham, L. G., Junaideen, S. M., Yue, Z. Q., and Lee, C. F. (2008). Loose fill slope stabilization with soil nails: Full-scale test. *Journal of Geotechnical and Geoenvironmental Engineering*, 134 (3), 277–288.

Li, X. S. (2007). Thermodynamics-based constitutive framework for unsaturated soils. 1: Theory. *Géotechnique*, 57 (5), 411–422.

Lloret, A., and Alonso, E. E. (1980). Consolidation of unsaturated soils including swelling and collapse behaviour. *Géotechnique*, 30 (4), 449–477.

Lloret, A., and Alonso, E. E. (1985). State surfaces for partially saturated soils. *Proceedings of the 11th International Conference on Soil Mechanics and Foundation Engineering*, Balkema, Rotterdam, the Netherlands, pp. 557–562.
Loret, B., and Khalili, N. (2002). An effective stress elastic–plastic model for unsaturated porous media. *Mechanics of Materials*, 34 (2), 97–116.
Mašın, D., and Khalili, N. (2008). A hypoplastic model for mechanical response of unsaturated soils. *International Journal for Numerical and Analytical Methods in Geomechanics*, 32 (15), 1903–1926.
Matsui, T., and San, K. C. (1992). Finite element slope stability analysis by shear strength reduction technique. *Soils and Foundations*, 32 (1), 59–70.
Matyas, E. L., and Radhakrishna, H. S. (1968). Volume change characteristics of partially saturated soils. *Géotechnique*, 18 (4), 432–448.
Mira McWilliams, P. (2001). *Analisis por Elementos Finitos de Problemas de Rotura en Geomateriales*. Ph.D. Thesis, Escuela Tecnica Superior de Ingenieros de Caminos, Canales y Puertos, Unviersidad Politecnica de Madrid. (in Spanish).
Mori, T., Uzuoka, R., Chiba, T., Kamiya, K., and Kazama, M. (2011). Numerical prediction of seepage and seismic behavior of unsaturated fill slope. *Soils and Foundations*, 51 (6), 1075–1090.
Mualem, Y. (1976). A new model for predicting the hydraulic conductivity of unsaturated porous media. *Water Resources Research*, 12 (3), 593–622.
Narasimhan, T. N., and Witherspoon, P. A. (1977). Numerical model for saturated—unsaturated flow in deformable porous media 1. Theory. *Water Resources Research*, 13 (3), 657–664.
Ng, C. W. W., and Chiu, A. C. F. (2003). Laboratory study of loose saturated and unsaturated decomposed granitic soil. *Journal of Geotechnical and Geoenvironmental Engineering*, 129 (6), 550–559.
Ng, C. W. W., Fung, W. T., Cheuk, C. Y., and Zhang, L. (2004). Influence of Stress Ratio and Stress Path on Behavior of Loose Decomposed Granite. *Journal of Geotechnical and Geoenvironmental Engineering*, 130 (1), 36–44.
Noorishad, J., Mehran, M., and Narasimhan, T. N. (1982). On the formulation of saturated-unsaturated fluid flow in deformable porous media. *Advances in Water Resources*, 5 (1), 61–62.
Oka, F., Kimoto, S., Takada, N., Gotoh, H., and Higo, Y. (2010). A seepage-deformation coupled analysis of an unsaturated river embankment using a multiphase elasto-viscoplastic theory. *Soils and Foundations*, 50 (4), 483–494.
Oka, F., Yashima, A., Shibata, T., Kato, M., and Uzuoka, R. (1994). FEM-FDM coupled liquefaction analysis of a porous soil using an elasto-plastic model. *Applied Scientific Research*, 52 (3), 209–245.
Olivella, S., Gens, A., Carrera, J., and Alonso, E. E. (1996). Numerical formulation for a simulator (CODE_BRIGHT) for the coupled analysis of saline media. *Engineering Computations*, 13 (7), 87–112.
Pastor, M., Zienkiewicz, O., and Chan, A. (1990). Generalized plasticity and the modelling of soil behaviour. *International Journal for Numerical and Analytical Methods in Geomechanics*, 14, 151–190.
PDE Solutions Inc., 2015. *FlexPDE 6 User Manual Version 6.37*. PDE Solutions Inc. Washington, USA.
Pereira, J. H. F. (1996). *Numerical Analysis of the Mechanical Behavior of Collapsing Earth Dams During First Reservoir Filling*. Ph.D. Thesis, University of Saskatchewan, Saskatchewan, Saskatoon, Canada.
Potts, D. M., and Zdravkovic, L. T. (1999). *Finite Element Analysis in Geotechnical Engineering: Theory*. Thomas Telford Limited, London.
Pradhan, B. (2003). *Study of Pullout Behaviour of Soil Nails in Completely Decomposed Granite Fill*. M.Phil. Thesis, The University of Hong Kong, Hong Kong.
Pradhan, B., Tham, L. G., Yue, Z. Q., Junaideen, S. M., and Lee, C. F. (2006). Soil–nail pullout interaction in loose fill materials. *International Journal of Geomechanics*, 6 (4), 238–247.
Russell, A. R., and Khalili, N. (2006). A unified bounding surface plasticity model for unsaturated soils. *International Journal for Numerical and Analytical Methods in Geomechanics*, 30 (3),181–212.

Sanavia, L. (2009). Numerical modelling of a slope stability test by means of porous media mechanics. *Engineering Computations*, 26 (3), 245–266.

Sánchez, M., Gens, A., Guimars, L., and Olivella, S. (2005). A double structure generalized plasticity model for expansive materials. *International Journal for Numerical and Analytical Methods in Geomechanics*, 29 (3), 751–787.

Sheng, D. (2011). Review of fundamental principles in modelling unsaturated soil behaviour. *Computers and Geotechnics*, 38,757–776.

Sheng, D., Fredlund, D. G., and Gens, A. (2008a). A new modelling approach for unsaturated soils using independent stress variables. *Canadian Geotechnical Journal*, 45 (4), 511–534.

Sheng, D., Gens, A., Fredlund, D. G., and Sloan, S. W. (2008b). Unsaturated soils: from constitutive modelling to numerical algorithms. *Computers and Geotechnics*, 35 (6), 810–824.

Sheng, D., Sloan, S. W., and Gens, A. (2004). A constitutive model for unsaturated soils: thermomechanical and computational aspects. *Computational Mechanics*, 33 (6), 453–65.

Sladen, J. A., D'Hollander, R. D. D., and Krahn, J. (1985). The liquefaction of sands, a collapse surface approach. *Canadian Geotechnical Journal*, 22 (4), 564–578.

Smith, P. G. C. (2003). *Numerical Analysis of Infiltration into Partially Saturated Soil Slopes*. Ph.D. Thesis, Imperial College of Science, Technology and Medicine, University of London, London, UK.

SoilVision System Ltd. (2010). *SVSlope User's Manual*. SoilVision System Ltd, Saskatoon, Canada.

Sun, D. A., Sheng, D., and Sloan, S. W. (2007). Elastoplastic modelling of hydraulic and stress-strain behaviour of unsaturated soil. *Mechanics of Materials*, 39 (3), 212–221.

Tamagnini, R. (2004). An extended cam-clay model for unsaturated soils with hydraulic hysteresis. *Géotechnique*, 54 (3), 223–228.

Tang, G. X., and Graham, J. (2002). A possible elastoplastic framework for unsaturated soils with high plasticity. *Canadian Geotechnical Journal*, 39 (4), 894–907.

Tarantino, A., and Tombolato, S. (2005). Coupling of hydraulic and mechanical behaviour in unsaturated compacted clay. *Géotechnique*, 55 (4), 307–317.

Thu, T. M., Rahardjo, H., and Leong, E. C. (2007). Elastoplastic model for unsaturated soil with incorporation of the soil-water characteristic curve. *Canadian Geotechnical Journal*, 44 (1), 67–77.

Toll, D. G. (1990). A framework for unsaturated soil behaviour. *Géotechnique*, 40 (1), 31–44.

Tripathy, S., Rao, K. S., and Fredlund, D. G. (2002). Water content-void ratio swell-shrink paths of compacted expansive soils. *Canadian Geotechnical Journal*, 39 (4), 938–959.

van Genuchten, M. T. (1980). A closed-form equation for predicting the hydraulic conductivity of unsaturated soils. *Soil Science Society of America Journal*, 44 (5), 892–898.

Vecchia, G. D., and Romero, E. (2013). A fully coupled elastic–plastic hydromechanical model for compacted soils accounting for clay activity. *International Journal for Numerical and Analytical Methods in Geomechanics*, 37 (5), 503–535.

Vu, H. Q. (2003). *Uncoupled and Coupled Solutions of Volume Change Problems in Expansive Soils*. Ph.D. Dissertation, University of Saskatchewan, Saskatoon, Canada.

Vu, H. Q., and Fredlund, D. G. (2004). The prediction of one-, two-, and three-dimensional heave in expansive soils. *Canadian Geotechnical Journal*, 41 (4), 713–737.

Wheeler, S. J., Sharma, R. S., and Buisson, M. S. R. (2003). Coupling of hydraulic hysteresis and stress–strain behaviour in unsaturated soils. *Géotechnique*, 53 (1), 41–54.

Wheeler, S. J., and Sivakumar, V. (1995). An elasto-plastic critical state framework for unsaturated soil. *Géotechnique*, 45 (1), 35–53.

Wu, L. Z., and Zhang, L. M. (2009). Analytical solution to 1D coupled water infiltration and deformation in unsaturated soils. *International Journal for Numerical and Analytical Methods in Geomechanics*, 33 (6), 773–790.

Xiong, Y., Bao, X., Ye, B., and Zhang, F. (2014). Soil–water–air fully coupling finite element analysis of slope failure in unsaturated ground. *Soils and Foundations*, 54 (3), 377–395.

Ye, G. L. (2004) *Numerical Study on the Mechanical Behavior of Progressive Failure of Slope by 2D and 3D FEM*. Ph.D. Thesis, Gifu University, Gifu, Japan.

Ye, G. L., Zhang, F., Yashima, A., Sumi, T., and Ikemura, T. (2005). Numerical analyses on progressive failure of slope due to heavy rain with 2D and 3D FEM. *Soils and Foundations*, 45 (2), 1–15.

Yeung, F. J. (2002). *Modelling of the Behaviour of Saprolitic Soil Slopes Under Rainfall*. M.Phil. Thesis, University of Cambridge, Cambridge, UK.

Zhan, L. T. (2003). *Field and Laboratory Study of An Unsaturated Expansive Soil Associated with Rain-induced Slope Instability*. Ph.D. Thesis, Hong Kong University of Science and Technology, Hong Kong.

Zhang, F., and Ikariya, T. (2011). A new model for unsaturated soil using skeleton stress and degree of saturation as state variables. *Soils and Foundations*, 51 (1), 67–81.

Zhang, F., Yashima, A., Osaki, H., Adachi, T., and Oka, F. (2003). Numerical simulation of progressive failure in cut slope of soft rock using a soil–water coupled finite element analysis. *Soils and Foundations*, 43 (5), 119–131.

Zhang, L. L., Zhang, L. M., and Tang, W. H. (2005). Rainfall-induced slope failure considering variability of soil properties. *Géotechnique*, 55 (2), 183–188.

Zhang, X., and Lytton, R. L. (2009). Modified state-surface approach to the study of unsaturated soil behavior. Part II: General formulation. *Canadian Geotechnical Journal*, 46 (5), 553–570.

Chapter 5

Stability of soil slope with cracks

5.1 INTRODUCTION

Cracks are widely present in slope soils. Cracks can be classified in several ways, namely, structural, desiccation, shearing, and freeze–thaw cracks. It is well recognized that cracks can induce slope instability (Baker, 1981; Lee et al., 1988; Chowdhury and Zhang, 1991; Yao et al., 2001; Wang et al., 2012; Michalowski, 2013; Utili, 2013). One of the main effects of cracks is that they provide preferential pathways for water flow, which significantly increases the hydraulic conductivity of the soil mass and, in turn, pore-water pressures in the soil mass (Zhang et al., 2012; Galeandro et al., 2013; Suryo, 2013). As a result, the shear strength of the soil decreases. Water infiltration into the cracked soil often starts from unsaturated conditions; thus, water retention ability and hydraulic conductivity under unsaturated conditions are important factors. The water retention ability and hydraulic conductivity of cracked soil vary with the soil suction under unsaturated conditions.

Soil hydraulic property functions, such as soil–water characteristic curve (SWCC) and permeability function, significantly affect the pore-water pressure distribution in soils. Although many methods have been developed to measure or to predict the hydraulic property functions of soils, few studies have been conducted on the hydraulic property functions of cracked soils (Reitsma and Kueper, 1994; Persoff and Pruess, 1995; National Research Council, 2001). There are two reasons for this: (1) cracks in soils, which significantly increase the permeability of soil, are randomly distributed and hard to describe and (2) the cracks will deform during a drying or wetting process. Both the presence of cracks and the change of cracks will significantly affect the hydraulic properties of the cracked soil.

In this chapter a method to predict the SWCC and permeability function for a cracked soil considering the crack volume change is proposed firstly. Then an example is presented to predict the hydraulic property functions based on the field measurement information. Finally, numerical analysis of flow in unsaturated cracked soil and stability of a cracked soil slope are conducted.

5.2 PREDICTION OF UNSATURATED HYDRAULIC FUNCTIONS OF CRACKED SOIL

In this section, a prediction method for estimating the SWCC and permeability function for a cracked soil is presented. The characteristics of pore-size distributions for cracked soil are introduced in Section 5.2.1. The water retention curves for cracks and the soil matrix are estimated based on the pore-size distributions of the crack system and the soil matrix system in Section 5.2.2. The SWCC for the nondeformable cracked soil is predicted by combining the water retention curve for the cracks and the soil matrix in Section 5.2.3. The permeability functions for the cracks, the soil matrix, and the cracked soil with a certain crack network are predicted in Section 5.2.4. Hysteresis models for the SWCC and permeability function are presented in Section 5.2.5. Finally, the crack volume changes are considered and the hysteretic SWCC and permeability function for the cracked soil considering crack volume changes are predicted in Section 5.2.6.

5.2.1 Pore-size distribution of cracked soil

The hydraulic properties of the cracked soil are closely related to the pore-size distribution of the cracked soil. Pore-size distribution is the distribution of pore volume with respect to the pore size. Alternatively, it can be defined by the related distribution of pore area with respect to the pore size. A cracked soil consists of two pore systems: cracks and soil pores. The two pore systems are characterized by different pore-size distributions.

The pore-size distribution for cracks can be obtained by calculating the percentage of crack volume. Assuming that the cracks do not vary along the depth, the crack area instead of crack volume can be used. The crack area for a crack is obtained by the product of the crack aperture and the crack length. As the aperture varies along the persistent direction, the average aperture is obtained by averaging aperture values at different locations of the crack. The area of all cracks divided by the total area of the soil is defined as the crack porosity. Figure 5.1 shows a schematic pore-size distribution of cracks. The pore size of a crack is represented by the crack aperture. When the cracks are rough with their apertures distributing in a wide range, the pore-size distribution is scattered.

5.2.2 Prediction of SWCC for desiccation crack networks

5.2.2.1 A general method to estimate SWCC based on the capillary law

Fredlund and Xing (1994) established a theoretical basis for SWCCs based on the pore-size distribution curve for the soil. The soil is regarded as a set of interconnected pores that are

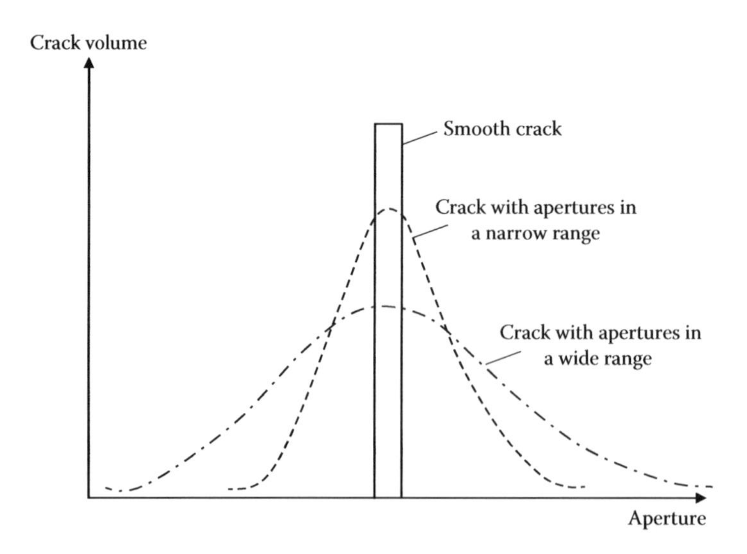

Figure 5.1 Crack pore-size distributions.

randomly distributed. The pores are characterized by pore radius r that is described by a pore-size distribution function $f_R(r)$ with the following identity

$$\int_{R_{min}}^{R_{max}} f_R(r)dr = \theta_s \tag{5.1}$$

where:

R_{max} and R_{min} are the maximum and minimum pore radii in the soil
θ_s is the volumetric water content when the soil is fully saturated (equal to the total porosity of the soil mass n), that is, $\theta_s = n$
$f_R(r)dr$ is the relative volume of pores of radius r to $r + dr$ in the total volume

When all the pores with radii less than or equal to R are filled with water, the volumetric water content $\theta(R)$ can be expressed as

$$\theta(R) = \int_{R_{min}}^{R} f_R(r)dr \tag{5.2}$$

Equation 5.2 implies that water tends to fill small pores first. This assumption can be justified using the capillary law

$$(u_a - u_w) = \frac{2T_s}{r}\cos\alpha_c \text{ or } r = \frac{C}{\psi} \tag{5.3}$$

where:

u_a and u_w are pore-air and pore-water pressures, respectively; $\psi = (u_a - u_w)$ is suction
T_s is the surface tension of water; α_c is the angle of contact between water and soil particle
$C = 2T_s \cos\alpha_c$

Based on this law, the suction is inversely proportional to the pore radius. Thus, smaller pores would be associated with higher suctions and would be filled earlier than larger pores.

According to Equation 5.3, two extreme suction conditions can be defined:

$$\psi_{max} = \frac{C}{R_{min}}$$
$$\psi_{min} = \frac{C}{R_{max}} \tag{5.4}$$

where ψ_{max} and ψ_{min} are the maximum and minimum suctions, respectively. In fact, ψ_{min} is the air-entry value of the soil (Fredlund and Xing, 1994). Substituting the capillary law (i.e., Equation 5.3) into Equation 5.2, the volumetric water content can be expressed as

$$\theta(\psi) = \int_{\psi_{max}}^{\psi} f_R\left(\frac{C}{s}\right)d\left(\frac{C}{s}\right) = \int_{\psi}^{\psi_{max}} f_R\left(\frac{C}{s}\right)\frac{C}{s^2}ds \tag{5.5}$$

where s is the variable (i.e., suction) of integration.

This equation can be written further into a general relationship between volumetric water content and suction (Fredlund and Xing, 1994)

$$\theta(\psi) = \theta_s \int_{\psi}^{\psi_{\max}} f_{\psi}(s)ds \tag{5.6}$$

Although Equation 5.6 is initially intended for soil, it should be valid for any porous medium containing pores that can be described by the capillary law. Therefore, Equation 5.6 may be extended to the water retention characteristics of a number of materials such as cracked soils and fractured rocks, where $f_{\psi}(s)$ is the pore-size distribution as a function of suction.

5.2.2.2 SWCC for soil matrix with known pore-size distribution

The pore-size distribution of soil matrix, $f_m(r)$ can be assumed to follow a modified lognormal distribution,

$$f_m(r) = \frac{\theta_{sm}}{\sqrt{2\pi}\zeta_m r} \exp\left[-\frac{1}{2}\left(\frac{\ln r - \lambda_m}{\zeta_m}\right)^2\right] \tag{5.7}$$

where:

λ_m and ζ_m are the mean and standard deviation of ln*(r)* of the soil matrix
θ_{sm} is the volumetric water content when the soil matrix is fully saturated

Note that the meaning of θ_{sm} in Equation 5.7 is different from θ_s in Equation 5.6. θ_s is equal to the porosity of the soil where the volume of cracks is included. However, θ_{sm} is the ratio of the pore volume inside the soil matrix (with the volume of cracks excluded) to the total volume of the soil mass. The volumetric water content and the pore-size distribution are assumed to be isotropic. Accordingly, the SWCC of the soil matrix can be written as, following Equation 5.2,

$$\begin{aligned}\theta(\psi) &= \int_{R_{\min}}^{R} \frac{\theta_{sm}}{\sqrt{2\pi}\zeta_m r} \exp\left[-\frac{1}{2}\left(\frac{\ln r - \lambda_m}{\zeta_m}\right)^2\right] dr \\ &= \int_{\left(\frac{\ln R_{\min} - \lambda_m}{\zeta_m}\right)}^{\left(\frac{\ln R - \lambda_m}{\zeta_m}\right)} \frac{\theta_{sm}}{\sqrt{2\pi}} \exp\left[-\frac{1}{2}y^2\right] dy \\ &= \theta_{sm}\left[\Phi\left(\frac{\ln(\frac{C}{\psi}) - \lambda_m}{\zeta_m}\right) - \Phi\left(\frac{\ln(\frac{C}{\psi_{\max}}) - \lambda_m}{\zeta_m}\right)\right]\end{aligned} \tag{5.8}$$

where Φ is the cumulative function of the standard normal distribution. When ψ_{max} approaches infinity, Equation 5.8 reduces to

$$\theta(\psi) = \theta_{sm}\Phi\left(\frac{\ln(\frac{C}{\psi}) - \lambda_m}{\zeta_m}\right) \tag{5.9}$$

5.2.2.3 Water retention curve for cracks

If a crack is not fully saturated, the water in the crack, which is in contact with both walls of the crack, will be held under tension. If the pore air is in connection with the atmosphere, on the basis of Equation 5.3, the height of water column h_c which the surface tension can hold in the crack is

$$h_c = \frac{2T_s}{\gamma_w x_c}\cos\alpha_c \tag{5.10}$$

where:

γ_w is the unit weight of water
x_c is the crack aperture

Where the gauge air pressure u_a is not zero, an isolated water phase can exist in the crack and the capillarity effect can be expressed as

$$(u_a - u_w) = \frac{2T_s}{x_c}\cos\alpha_c \tag{5.11}$$

This equation is similar to Equation 5.3. Therefore, the general equation (Equation 5.6) is also valid for soil cracks, with $f_\psi(s)$ can be substituted by the distribution of aperture of soil cracks.

A crack volumetric water content, θ_c, which is the ratio of the volume of water in the cracks to the total volume of the soil mass containing these cracks, is defined. In addition, the distribution of the crack aperture is also assumed to be lognormal

$$f_c(x_c) = \frac{\theta_{sc}}{\sqrt{2\pi}\zeta_c x_c}\exp\left[-\frac{1}{2}\left(\frac{\ln x_c - \lambda_c}{\zeta_c}\right)^2\right] \tag{5.12}$$

where:

λ_c and ζ_c are the mean and standard deviation of $\ln(x)$ of the crack aperture
θ_{sc} is the crack volumetric water content when the cracks are fully saturated

Given the distribution of the crack aperture, the water retention curve for the cracks can be expressed as

$$\theta(\psi) = \theta_{sc}\left[\Phi\left(\frac{\ln(\frac{C}{\psi}) - \lambda_c}{\zeta_c}\right) - \Phi\left(\frac{\ln(\frac{C}{\psi_{max}}) - \lambda_c}{\zeta_c}\right)\right] \tag{5.13}$$

When ψ_{max} approaches infinity, Equation 5.13 reduces to

$$\theta(\psi) = \theta_{sc}\Phi\left(\frac{\ln(\frac{C}{\psi}) - \lambda_c}{\zeta_c}\right) \tag{5.14}$$

The above equations are derived based on the capillary law. Equation 5.11 indicates that the soil suction decreases with increasing crack aperture. When the crack aperture is too large the capillary law cannot be applied. Therefore, there is a maximum aperture that the capillary law can be applied. Let us consider the vertical force equilibrium of the capillary water in the crack plane in Figure 5.2. The contact angle α_c is assumed to be zero. Assume the minimum capillary height, corresponding to the maximum aperture x_{cm}, is one half of the crack aperture. The vertical resultant of the surface tension is responsible for holding the weight of the water column,

$$(x_{cm} \cdot L_c \cdot \frac{x_{cm}}{2} - \frac{1}{2} \cdot \pi \cdot \frac{x_{cm}}{4}^2 \cdot L_c)\rho_w g = T_s \cdot 2L_c \tag{5.15}$$

where:

L_c is the length of the aperture
x_{cm} is the maximum crack aperture that the capillary law can be applied
ρ_w is the density of water

Substituting the values of T_s and ρ_w ($T_s = 0.0728$ N/m and $\rho_w = 1000$ kg/m^3) into Equation 5.15, the largest aperture to apply the capillary law is 11.76 mm.

5.2.3 Prediction of SWCC for nondeformable cracked soil

The suction inside a crack may not be equal to the suction inside the soil matrix near the crack at a nonequilibrium condition. The crack network system and the soil matrix system interact by exchanging water in response to pressure head differences. When under equilibrium conditions the pressures in the crack network and the soil matrix are assumed to be identical (e.g., Dykhuizen, 1987; Peters and Klavetter, 1988). Peters and Klavetter (1988) investigated this assumption and concluded that the assumption is reasonable for both saturated and unsaturated conditions when equilibrium in the soil is reached. Thus, the suction over the matrix system and the crack system can be assumed to be the same.

As mentioned in Section 5.2.1, the pores in the cracked soil, consisting of the matrix phase and the crack phase, should be the sum of the pores in the two material phases. Therefore, the pore-size distribution of the cracked soil can be expressed as

$$f_R(r) = f_m(r) + f_c(r) \tag{5.16}$$

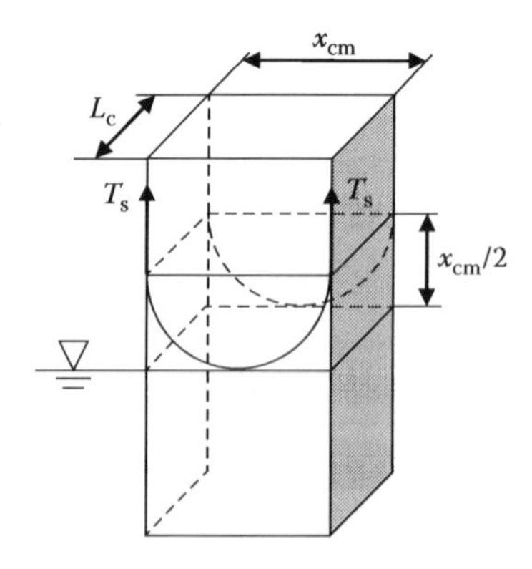

Figure 5.2 Physical model to determine the maximum crack aperture that the capillary law holds. (After Li, J. H. et al., *Proceedings of ICE-Geotechnical Engineering*, 164, 211–221, 2011. With permission.)

The distributions of the pore size of the soil matrix $f_m(r)$ and the aperture of the soil crack $f_c(r)$ have been defined in Equations 5.7 and 5.12, respectively. Using Equations 5.2 and 5.16 and considering that the water storages in the matrix blocks and in the crack network constitute the total water storage, the volumetric water content of the cracked soil is

$$\theta(\psi) = \int_{R_{min}}^{R} f_m(r)dr + \int_{X_{min}}^{X} f_c(x)dx$$

$$= \theta_{sm}\left[\Phi\left(\frac{\ln(\frac{C}{\psi}) - \lambda_m}{\zeta_m}\right) - \Phi\left(\frac{\ln(\frac{C}{\psi_{max}}) - \lambda_m}{\zeta_m}\right)\right] \tag{5.17}$$

$$+ \theta_{sc}\left[\Phi\left(\frac{\ln(\frac{C}{\psi}) - \lambda_c}{\zeta_c}\right) - \Phi\left(\frac{\ln(\frac{C}{\psi_{max}}) - \lambda_c}{\zeta_c}\right)\right]$$

When the maximum suction approaches infinity, this equation reduces to

$$\theta(\psi) = \theta_{sm}\Phi\left(\frac{\ln(\frac{C}{\psi}) - \lambda_m}{\zeta_m}\right) + \theta_{sc}\Phi\left(\frac{\ln(\frac{C}{\psi}) - \lambda_c}{\zeta_c}\right) \tag{5.18}$$

Equation 5.18 shows that the SWCC of the cracked soil is the sum of the SWCCs of the two material phases weighted by their volume porosities.

It can be shown that the volumetric water content of a fully saturated soil mass is

$$\theta_s = \theta_{sm} + \theta_{sc} \tag{5.19}$$

5.2.4 Prediction of tensorial permeability function for nondeformable cracked soil

5.2.4.1 *Saturated permeability of desiccation cracks*

The saturated permeability of desiccation cracks, k_{sc}, here refers to the saturated permeability of cracks when the volume change of cracks is ignored. The network of desiccation cracks can be generated according to digital pictures obtained from a field survey. The saturated permeability tensor of the crack network can then be calculated using numerical simulation (Li et al., 2009).

5.2.4.2 *Permeability function for desiccation cracks*

The coefficient of permeability for a crack network is not a constant. It depends on the saturated permeability and the volumetric water content or the suction. Therefore, the permeability function of desiccation cracks represents the relationship between the coefficient of permeability and crack suction. There are two approaches to obtain the permeability function of an unsaturated soil: (1) empirical equations (Richards, 1931; Brooks and Corey, 1964) and (2) statistical models (Childs and Collis-George, 1950; Burdine, 1953; Fredlund et al., 1994). These approaches can be applied to cracks if the cracks are viewed as a porous medium. The permeability function can be predicted when the saturated permeability of the cracks and the water retention curve are available.

Take the Brooks and Corey method (1964) as an example, the procedure to predict the permeability function for soils is shown in Figure 5.3 (Fredlund and Rahardjo, 1993). First, a residual degree of saturation (S_r), which is defined as the degree of saturation at which an increase in matric suction does not produce a significant change in the degree of saturation, is estimated. Then effective degree of saturation (S_e) is calculated based on the estimated residual degree of saturation as

$$S_e = \frac{S - S_r}{1 - S_r} \tag{5.20}$$

The effective degree of saturation is then plotted against matric suction. A horizontal and a sloping line can be drawn through the calculated effective degree of saturation points as shown in Figure 5.3. If the points at high value of matric suction are not lying on the straight line, the residual degree of saturation could be estimated again to force these points lying on the line. Then, a new plot of matric suction versus effective degree of saturation curve can be obtained.

Similarly the prediction of the permeability function for desiccation cracks can be conducted. The residual degree of saturation of cracks, S_{rc}, is used to calculate the effective degree of saturation of cracks, S_{ec},

$$S_{ec} = \frac{S_c - S_{rc}}{1 - S_{rc}} \tag{5.21}$$

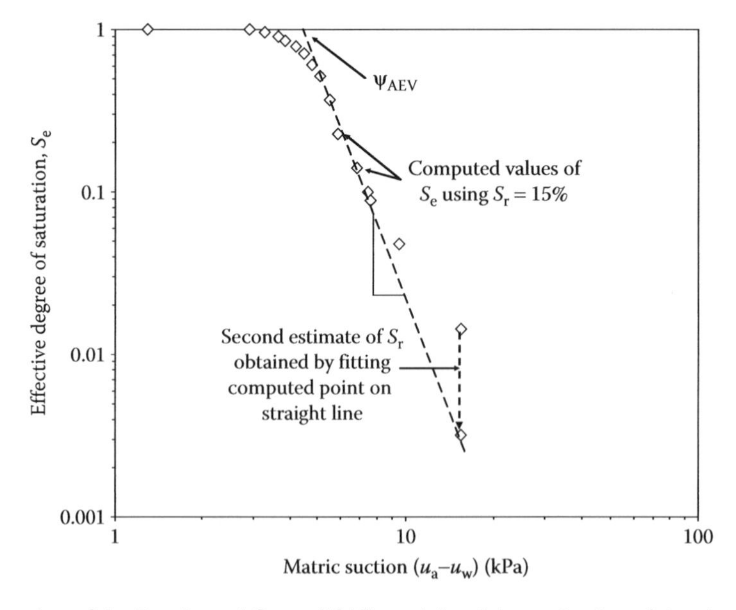

Figure 5.3 Illustration of the Brooks and Corey (1964) model and determination of the air-entry suction.

where:

S_{ec} is the effective degree of saturation of cracks
S_c is the degree of saturation of cracks

S_c is assumed to be isotropic. The water retention curve for cracks can then be estimated following the Brooks and Corey method (1964).

The air-entry value of the crack, ψ^c_{AEV}, is the matric suction value that must be exceeded before air enters into the cracks. It is a measure of the maximum crack aperture. The intersection point between the straight sloping line and the saturation ordinate defines the air-entry value of the cracks (Figure 5.3). The sloping line for the points having matric suctions greater than the air-entry value can be described by

$$S_{ec} = \left\{ \frac{\psi^c_{AEV}}{\psi} \right\}^{\lambda_{bc}} \quad \text{for } \psi > \psi^c_{AEV} \tag{5.22}$$

where λ_{bc} is a pore-size distribution index of the cracks, which is defined as the negative slope of the curve between the effective degree of saturation, S_{ec}, and the matric suction, ψ. Here, the superscript "c" in ψ^c_{AEV} denotes crack and the subscript "bc" in λ_{bc} denotes the Brooks and Corey (1964) model.

The coefficient of permeability for the cracks with respect to the water phase, k_c, can be predicted from the $S_{ec} - \psi$ curves as follows (Brooks and Corey 1964)

$$k_c = k_{sc} \text{ for } \psi \le \psi^c_{AEV} \tag{5.23}$$

$$k_c = k_{sc} \cdot (S_{ec})^{\delta_c} \text{ for } \psi > \psi^c_{AEV} \tag{5.24}$$

where k_c is the coefficient of permeability with respect to the water phase for cracks; k_{sc} is the saturated coefficient of permeability with respect to the water phase for cracks; δ_c is an empirical constant, which is related to the pore-size distribution index:

$$\delta_c = \frac{2 + 3\lambda_{bc}}{\lambda_{bc}} \tag{5.25}$$

5.2.4.3 Prediction of permeability function for cracked soil

Water flows through a cracked soil mainly along two paths: one path is across the soil matrix where the permeability of the soil matrix is the dominant factor; another path is along the interconnected cracks where the permeability of the crack network dominates. Hence, the permeability of the cracked soil can be expressed as the weighted superposition of the permeability along the two paths. The detailed derivation of the permeability function for the cracked soil can be referred to Li et al. (2009). The permeability function for the cracked soil is

$$k(\psi) = k_c(\psi) + (1 - w_c)k_m(\psi) \tag{5.26}$$

where:

k is the permeability function for the cracked soil
k_c is the permeability function for the crack network
k_m is the permeability function for the soil matrix
w_c is a volumetric weighting factor for the crack network, which is defined as the ratio between the total volume of the crack pore system and the total volume of the medium (i.e., including the crack network and the soil matrix)

In general, there is a complementary permeability tensor, $\mathbf{k}_m$, for the soil matrix in addition to the permeability tensor, $\mathbf{k}_c$, for the crack network. If the water retention curves are isotropic, the tensorial form of the permeability function for the cracked soil is (Li et al., 2009)

$$\mathbf{k}(\psi) = \mathbf{k}_c(\psi) + (1 - w_c)\mathbf{k}_m(\psi) \tag{5.27}$$

5.2.5 Hysteresis models for SWCC and permeability function

The nonuniform void space in a rough-walled crack can result in hysteresis in the SWCC and permeability function. For a given soil, there exists a boundary drying SWCC and a boundary wetting SWCC (Figure 5.4). These two curves serve as the bounds of all scanning curves. Besides these curves there exists an initial drying curve (Feng et al., 2002). An infinite number of scanning curves exist, resulting from different wetting and drying histories and different material states.

Pham et al. (2005) proposed a scaling method for estimating hysteretic SWCCs. The scaling method uses one of the following three curves, namely, the boundary drying curve, the boundary wetting curve, and the initial drying curve to estimate the other two curves (as shown in Figure 5.4). There are four important parameters associated with any SWCC; namely, the degree of saturation at zero soil suction (S_u) of the boundary curves, the air-entry value (ψ_{AEV}), the slope of the boundary curve (SL), and the residual degree of saturation (S_r) as illustrated in Figure 5.5. Once the information on these four parameters is available, it is possible to construct the entire SWCC. The boundary drying curve and the initial drying curve can be interchanged by changing the parameter that controls the water content at zero soil suction. Similarly, the boundary drying curve and the boundary wetting curve can be interchanged by changing the slope and the distance between the two curves as shown in Figure 5.5. The distance D_{SL} between the two boundary curves are calculated by a statistical analysis involving 34 soils (Pham et al., 2005). It is suggested that the distance is 20% log cycle for sand and 50% log cycle for silt loam and clay loam in the semilogarithmic suction coordinate system (i.e., ψ in log scale). The scaling method is used to predict hysteretic SWCCs in this section.

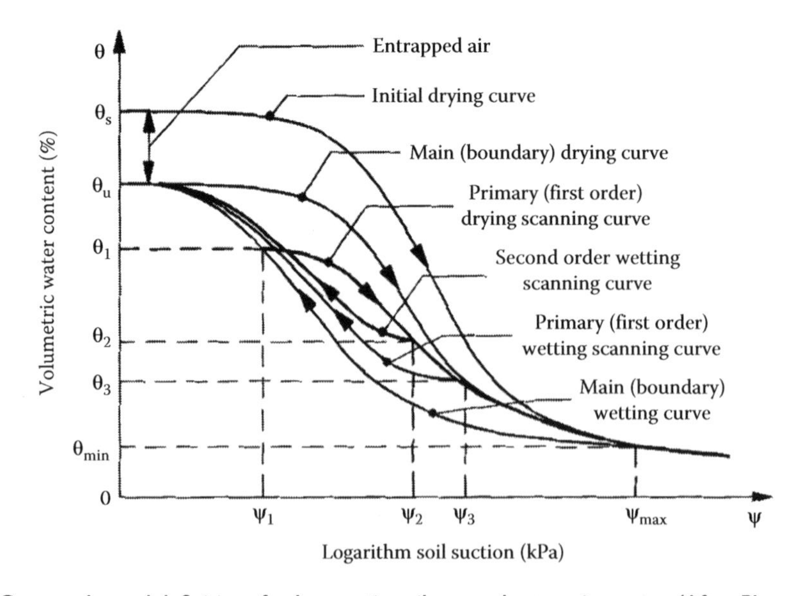

Figure 5.4 Commonly used definitions for hysteretic soil–water hysteresis curves. (After Pham, H. Q. et al., *Canadian Geotechnical Journal*, 42, 1548–1568, 2005. With permission.)

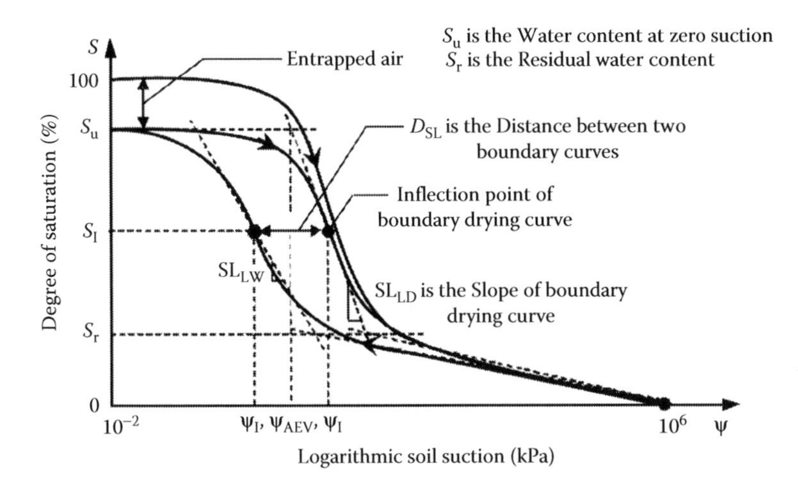

Figure 5.5 Schematic illustration of the slope and distance between two boundary hysteresis SWCCs. (After Pham, H. Q. et al., *Canadian Geotechnical Journal*, 42, 1548–1568, 2005. With permission.)

The wetting permeability function can be predicted using the wetting SWCC curve and the saturated permeability, which can be assumed to be the same as that in the drying process. The Fredlund et al. estimation (Fredlund et al., 1994), van Genuchten estimation (van Genuchten, 1980), or Brooks and Corey (Brooks and Corey, 1964) estimation can be used to predict the wetting permeability function.

Several researchers have attempt to obtain the hysteretic water retention curve and permeability function for cracks. However, a limited number of theoretical models are developed to estimate hysteretic water retention curves for cracks. Hu et al. (2001) conducted numerical simulations for the process of unsaturated seepage during drainage and imbibition in a rough-walled fracture with position dependent aperture. The water degree of saturation and unsaturated permeability under different capillary pressures are obtained during drainage and imbibition conditions (as shown in Figure 5.6). Zhou et al. (1998) obtained the drainage and imbibition processes for a rock joint and found that the unsaturated hydraulic conductivity of a single joint has certain similarities with that of porous media. Reitsma and Kueper (1994) measured the hysteretic relationship between capillary pressure and fluid saturation for a rough-walled rock fracture. The measured capillary pressure curves were found to be

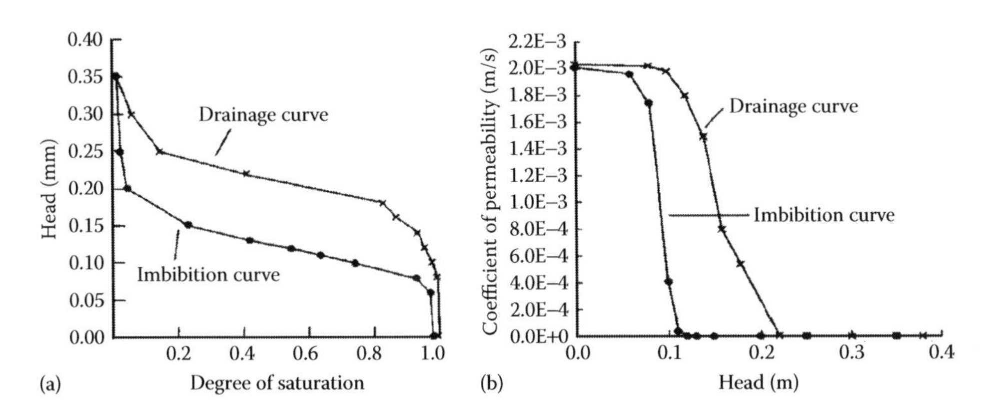

Figure 5.6 Numerical results of: (a) hysteresis water retention curve and (b) permeability function. (After Hu, Y. J. et al., *Chinese Journal of Geotechnical Engineering*, 23, 284–287 [in Chinese], 2001. With permission.)

well represented by a Brooks–Corey porous media capillary pressure function. Therefore, the hysteretic relationship in fractures is similar to that in soils. It is possible to use the models that were originally developed for soils to predict the hysteretic water retention curves and permeability function for the crack network.

A statistical analysis is conducted to obtain the distance between the drainage curve and the imbibition curve for the cracks based on the data from Zhou et al. (1998), Hu et al. (2001), and Reitsma and Kueper (1994). The distance between the drainage curve and the imbibition curve is in the range of 20%–40% log cycles.

5.2.6 Prediction of SWCC and permeability function considering crack volume change

5.2.6.1 Desiccation-induced crack volume change

Both laboratory tests (Mizuguchi et al., 2005; Kodikara and Choi, 2006; Peron et al., 2006) and field experiments (Konrad and Ayad, 1997a,b) were conducted to study desiccation-induced cracks. Li and Zhang (2010, 2011b) have conducted a field survey to study the desiccation-induced cracks of a silty clay focusing on the crack pattern and geometry change of cracks with water content. The test results showed that crack development follows three stages during a drying process. In stage 1, water content decreases slowly but no new cracks appear. In stage 2, new cracks appear and develop quickly as water content continues to decrease. In stage 3, water content decreases slowly and the cracks tend to be steady. In the steady state, the aperture and length of cracks kept unchanged and no new cracks appeared. In a wetting process, the aperture of cracks tends to be smaller when the water content increases. Cracks tend to close to a certain extend for highly plastic soils due to self-healing. On the other hand, cracks keep open for non- or low-plastic soils (Eigenbrod, 1996; Rayhani et al., 2008).

Obviously, the crack development is closely related with the water content and the soil suction. If the crack porosity is used as an indicator to represent the crack development, the crack development can be described by the relation between the crack porosity (n_c) and the soil suction (ψ). Figure 5.7 shows a simplified relation between crack porosity and soil suction. The initial crack porosity is n_b when the suction is small. When there are no cracks initially or cracks close after wetting, the initial crack porosity n_b can be zero. Cracks start to develop after suction is larger than ψ_b. Cracks develop and the crack porosity increases with the increase of suction. Finally, cracks tend to be steady when suction reaches ψ_s and the crack porosity at the steady state is n_s. Figure 5.7 shows a linear relationship between crack porosity and suction. The relationship can be simulated by other types of more realistic curves.

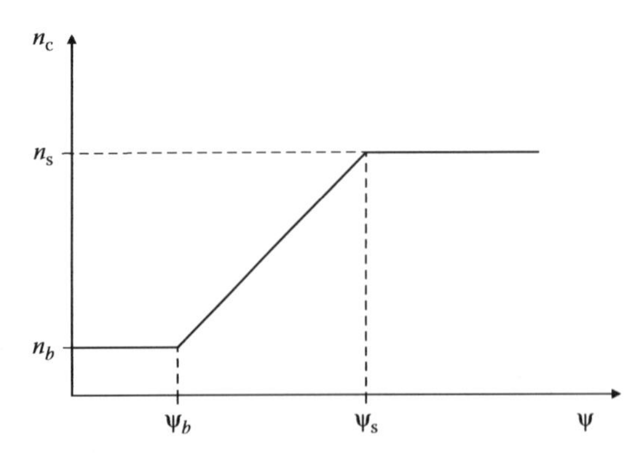

Figure 5.7 Crack development path. (After Li, J. H. et al., *Proceedings of ICE-Geotechnical Engineering*, 164, 211–221, 2011. With permission.)

5.2.6.2 *SWCC and permeability function for cracked soil considering crack volume change*

In Sections 5.2.3 and 5.2.4, the SWCC and permeability function for a cracked soil ignoring the crack volume change have been presented. The cracks at a certain moment are assumed to be constant during the drying-wetting process. The SWCCs at all moments can be combined to give a SWCC for the cracked soil considering crack volume change. Figure 5.8 gives a three-dimensional diagram to show the procedure to obtain the permeability function considering crack volume change. The procedure for getting the SWCC is similar to that for getting the permeability function considering crack volume changes. The procedure to obtain the permeability function (or water retention curve) is as follows:

1. At a certain time corresponding to a certain suction, obtaining the pore-size distributions for the cracks and the soil matrix.
2. Estimating the water retention curves for the cracks and the soil matrix from the pore-size distributions at the specified time.
3. Measuring or calculating the saturated permeability tensors for the cracks and the soil matrix at the specified time.
4. Predicting the permeability functions for the cracks and the soil matrix at the specified time based on the water retention curves and the saturated permeability tensors.
5. Combining the permeability functions (or water retention curves) for the cracks and the soil matrix to get the permeability function (or SWCCs) for the cracked soil at the suction. Figure 5.8 shows three permeability functions at different moments in the k–ψ plane. The permeability functions are for the cracked soils with nondeformable cracks at crack porosity n_{ci}. The saturated permeability of the cracked soil with crack porosity n_{ci} is k_{si}. The suction is ψ_i where the crack porosity is n_{ci}.

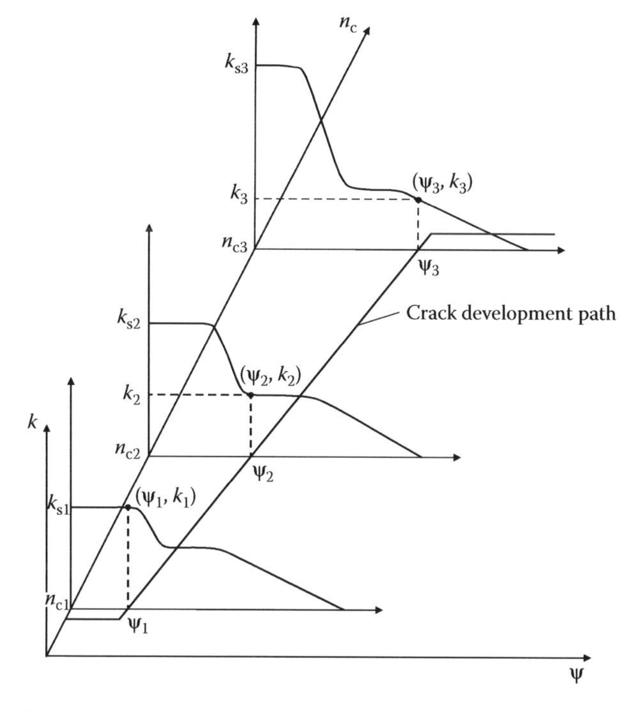

Figure 5.8 Relationship between permeability of a cracked soil and soil suction considering crack volume changes. (After Li, J. H. et al., *Proceedings of ICE-Geotechnical Engineering*, 164, 211–221, 2011. With permission.)

6. Obtain the crack development path that shows the relationship between the crack porosity and the suction. The crack development path intersects with the permeability functions (or water retention curves) at suctions ψ_i, which are the suctions when the cracks with crack porosity n_{ci} are presented. Then a pair of (ψ_i, k_i) (or (ψ_i, θ_i)) can be recognized on the corresponding permeability function (or water retention curve). An assumed crack development path is shown in the n_c–ψ plane in Figure 5.8.
7. Combining these pairs of (ψ_i, k_i) (or (ψ_i, θ_i)) in each permeability function curve (or water retention curve) will give the permeability function (or SWCC) for the cracked soil considering crack volume change.

5.3 EXAMPLE

5.3.1 Prediction of SWCC and permeability function for cracked soils considering crack volume change

A field test was conducted at the foot of the Huangshan Slope in Zhenjiang, downstream of Yangtze River, China. The purpose of the field test is to investigate the initiation, development of cracks and the hydraulic properties of the cracked soil. The soil in field is classified as silty clay with 34% of silt, 35% of clay, 23% of sand, and 8% of gravel. The crack development was continuously observed on the ground of the survey site. A digital imaging method was used to investigate the cracks. Images were taken of a plot of 180 mm by 210 mm periodically. Each image was taken with minimized distortion by fixing the camera on a tripod and using two scales placed around the plot as a reference.

Figure 5.9 shows the pictures of a cracked soil on four days during a drying process. The crack porosity values for the cracked soils on the four days were 0.41%, 0.95%, 2.61%, and 3.10%, respectively. The water contents in the four days were also measured. The corresponding matric suction can then be obtained using the SWCC of soils. When the soil was saturated no cracks appeared. When the suction of the soil reached about 47 kPa some cracks appeared and the crack porosity was 0.41% as shown in Figure 5.9a. Then cracks developed abruptly and reached a crack porosity of 2.61% at a suction of 8300 kPa as shown in Figure 5.9c. The cracks reached a steady state with a crack porosity of 3.10% at suction about 52,000 kPa as shown in Figure 5.9d. The crack development drying path for the cracked soil is shown in Figure 5.10. To obtain the hydraulic properties during a wetting process a crack development wetting path is needed. The aperture of cracks decreased with increasing water content (i.e., decreasing suction) and tended to close when saturated. Therefore, a wetting path was assumed as shown in Figure 5.10. The crack porosity on the wetting path is larger than that on the drying path (Shen and Deng, 2004). The measured crack pore-size distributions for the four crack networks are shown in Figure 5.11. The SWCCs for these crack networks can be predicted according to the capillary theory.

Take the cracked soil shown on November 12, 2007 (Figure 5.9d) as an example, the drying water retention curve for the crack network can be estimated by the corresponding pore-size distribution using Equation 5.14. The estimated drying water retention curve is shown in Figure 5.12. The wetting water retention curve for the crack network can be obtained by a scaling method introduced in Section 5.2.5 (Pham et al., 2005). The distance between the wetting curve and the drying curve is assumed to be 20% log cycle. The slope of the wetting water retention curve is assumed to be the same as that of the drying water retention curve. The estimated wetting water retention curve is also shown in Figure 5.12.

The saturated permeability tensors for the crack network can be calculated by simulating water flow through the crack network. The crack network for the cracked soil on November 12

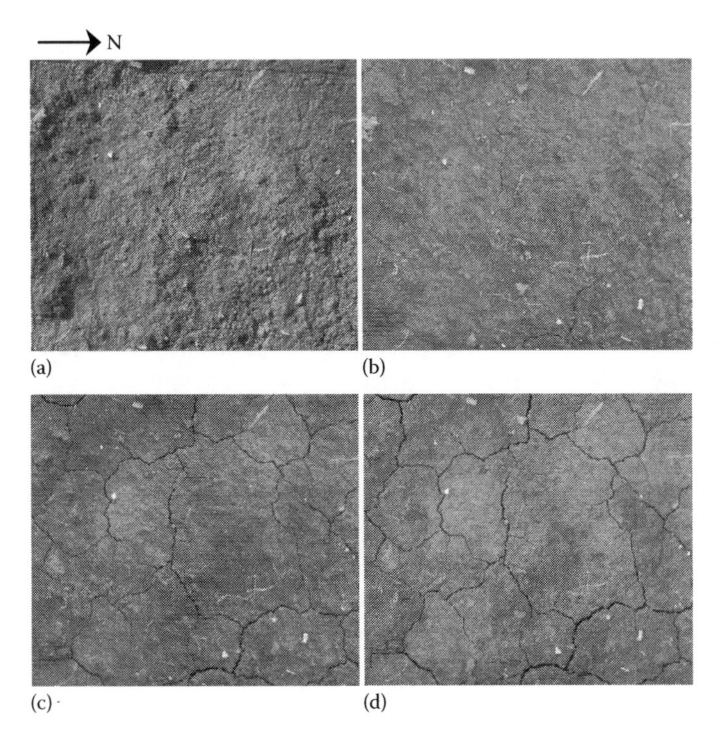

Figure 5.9 Cracked soils at four different moments in a drying process. (a) 02 Nov. 2007, (b) 04 Nov. 2007, (c) 06 Nov. 2007, and (d) 12 Nov. 2007. (After Li, J. H. et al., *Proceedings of ICE-Geotechnical Engineering*, 164, 211–221, 2011. With permission.)

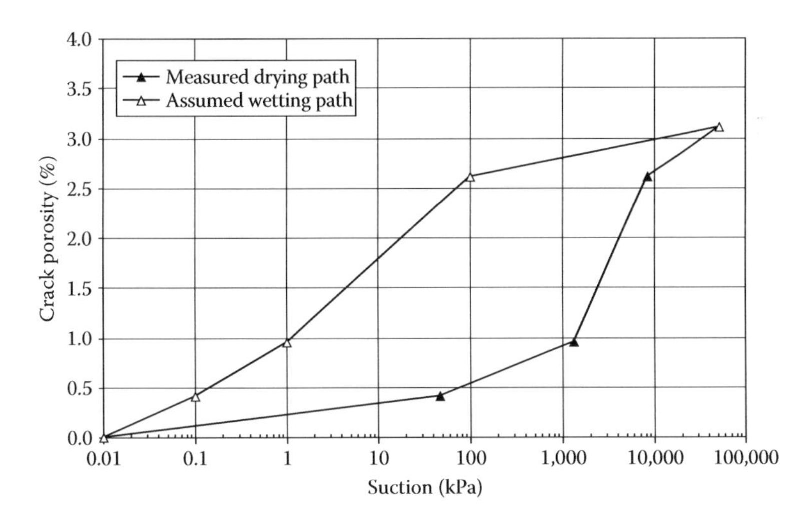

Figure 5.10 Crack development drying path and wetting path with suction change. (After Li, J. H. et al., *Proceedings of ICE-Geotechnical Engineering*, 164, 211–221, 2011. With permission.)

in the field survey was imported into the numerical model developed by Li et al. (2009). Water flow through the crack network was simulated and the permeability tensor for the saturated crack network was obtained. The detailed procedure is referred to Li et al. (2009). The saturated permeability values are 7.0×10^{-3} m/s along the north–south direction, 9.0×10^{-3} m/s along the west–east direction, and 1.35×10^{-2} m/s along the vertical direction for the crack network.

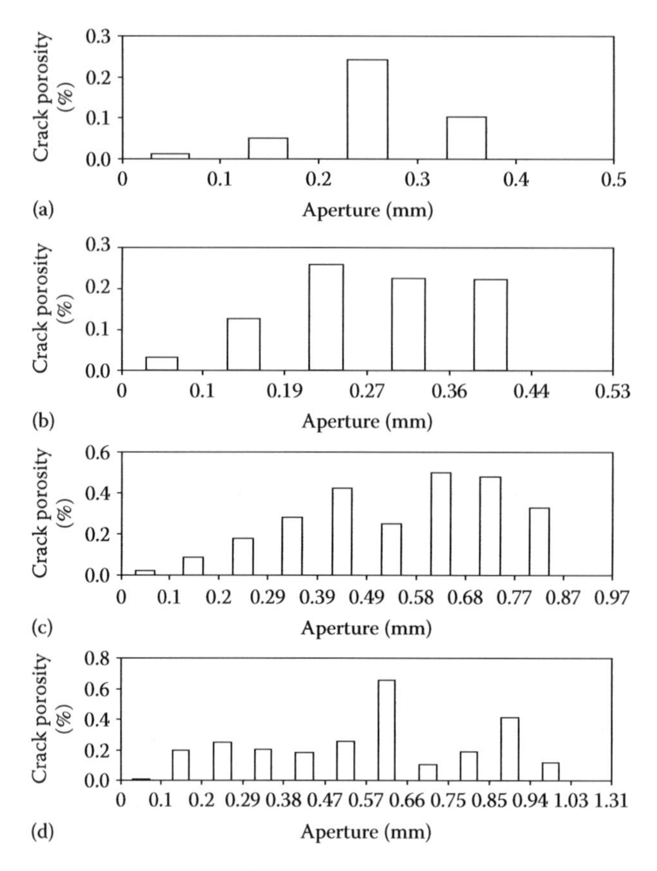

Figure 5.11 Crack pore-size distributions of the four snapshots in Figure 5.9: (a) 02 Nov. 2007, (b) 04 Nov. 2007, (c) 06 Nov. 2007, and (d) 12 Nov. 2007. (After Li, J. H. et al., *Proceedings of ICE-Geotechnical Engineering*, 164, 211–221, 2011. With permission.)

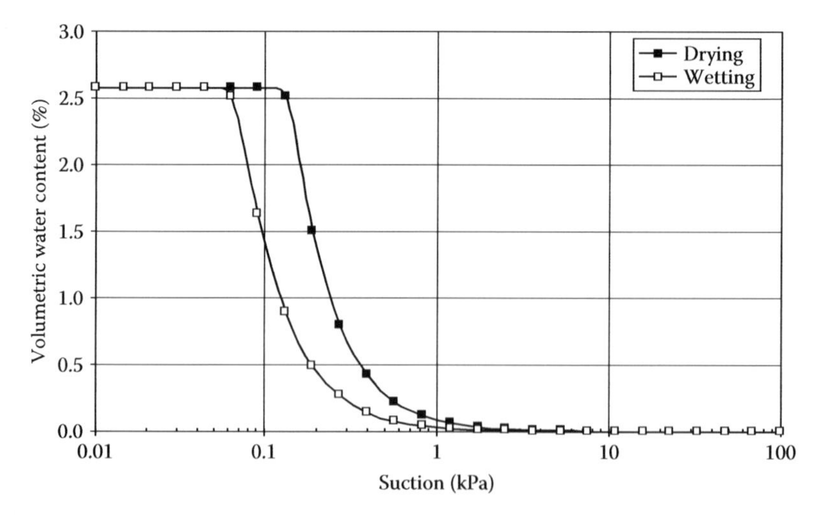

Figure 5.12 Hysteresis water retention curves for the crack network shown in Figure 5.9d. (After Li, J. H. et al., *Proceedings of ICE-Geotechnical Engineering*, 164, 211–221, 2011. With permission.)

In the horizontal direction, the major principal saturated permeability is 1.1×10^{-2} m/s along 34° and the minor one is 5×10^{-3} m/s in the direction perpendicular to the major principal direction. If take the permeability along north–south direction as k_{11}, the permeability along east–west direction as k_{22}, and that along vertical direction as k_{33}, the permeability tensor at saturated condition of the crack network is

$$k_{s,ij}^{c} = \begin{bmatrix} 7.0\times10^{-3} & 2.9\times10^{-3} & 0 \\ 2.9\times10^{-3} & 9.0\times10^{-3} & 0 \\ 0 & 0 & 1.35\times10^{-2} \end{bmatrix} \tag{5.28}$$

The tensorial form of the permeability function for the crack network can then be estimated by Brooks and Corey (1964) model as

$$\mathbf{k}_{c} = \mathbf{k}_{sc}\left(S_{ec}\right)^{\delta_c} \tag{5.29}$$

The drying and wetting permeability functions for the crack network can be obtained from the drying and wetting SWCCs, respectively. The hysteresis permeability function is shown in Figure 5.13.

The drying SWCC curve for the soil matrix has been measured in the laboratory and is shown in Figure 5.14. This measured drying SWCC is fitted using the Brooks and Corey (1964) model to give the drying SWCC over the entire suction range and is also shown in Figure 5.14. The drying SWCC is used to predict the wetting SWCC by the scaling method as shown in Figure 5.14. The distance between the drying curve and the wetting curve is assumed to be 50% log cycle (Pham et al., 2005). The slope is assumed to be the same as that of the drying curve.

The hysteresis permeability function for the soil matrix can be obtained using the hysteresis SWCC and the saturated permeability for the soil matrix. The saturated permeability values of the soil matrix were measured in the laboratory, which are 6.0×10^{-7} m/s along the north–south direction, 3.0×10^{-6} m/s along the west–east direction, and 3.5×10^{-6} m/s along the vertical direction. Then a tensorial form of permeability function for the soil matrix is

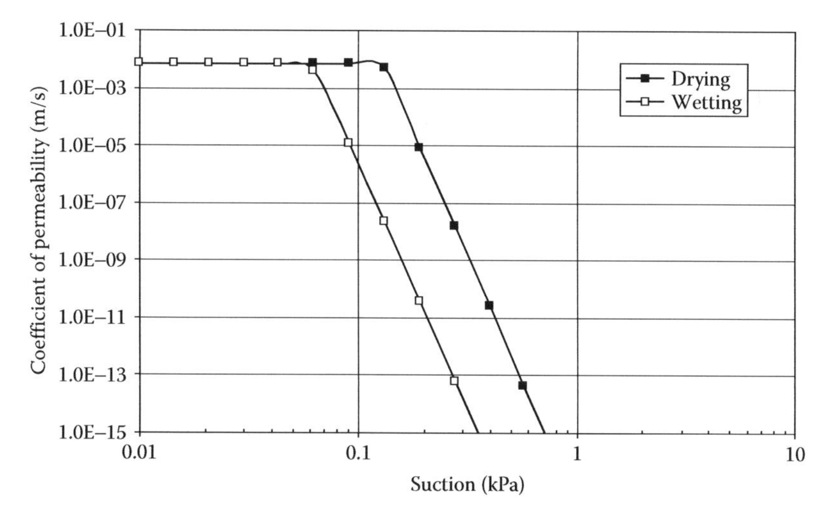

Figure 5.13 Estimated permeability functions for the crack network shown in Figure 5.9d. (After Li, J. H. et al., *Proceedings of ICE-Geotechnical Engineering*, 164, 211–221, 2011. With permission.)

$$\mathbf{k}_{\mathrm{m}} = \begin{bmatrix} 0.6\times10^{-6} & 0 & 0 \\ 0 & 3\times10^{-6} & 0 \\ 0 & 0 & 3.5\times10^{-6} \end{bmatrix} \left(S_{\mathrm{em}}\right)^{\delta_{\mathrm{m}}} \tag{5.30}$$

where δ_m is the empirical constant related with the pore-size distribution index for soil matrix.

The porosity of the soil matrix is 44.3%. The hysteresis permeability function along the north–south direction is shown in Figure 5.15.

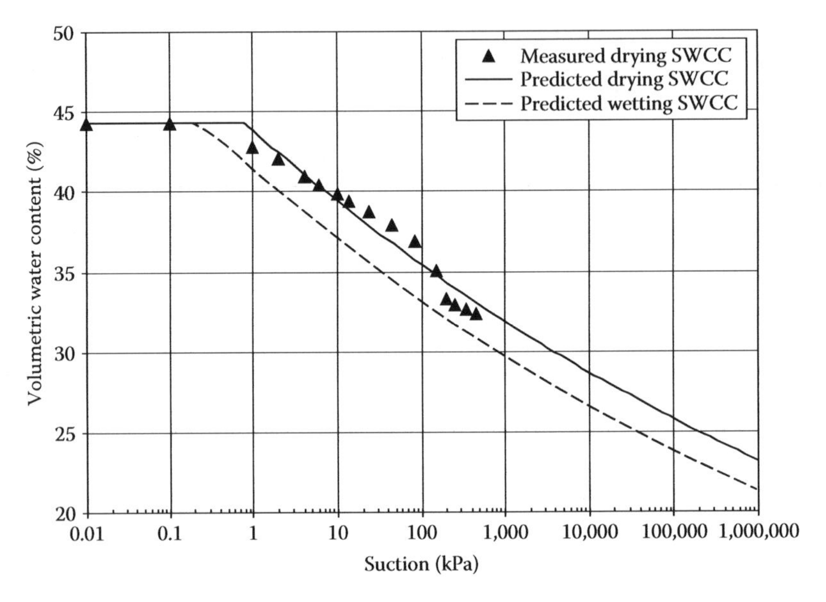

Figure 5.14 Measured drying SWCC and predicted hysteresis SWCCs for the soil matrix. (After Li, J. H. et al., *Proceedings of ICE-Geotechnical Engineering*, 164, 211–221, 2011. With permission.)

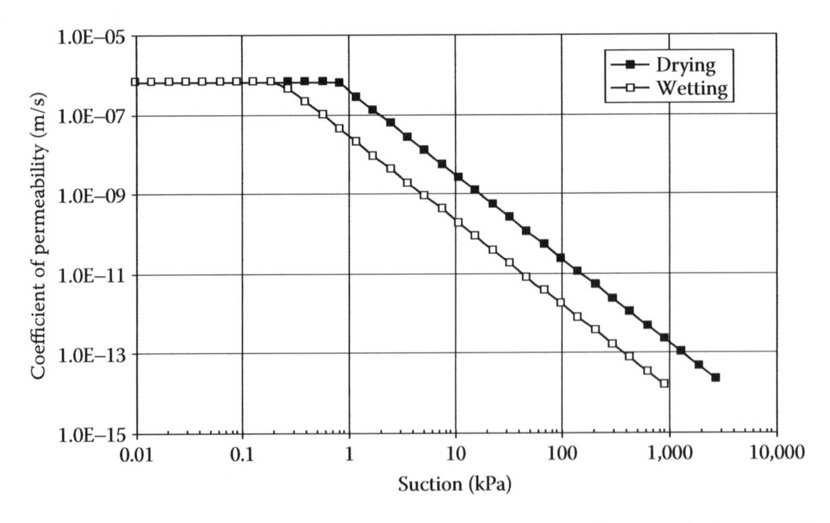

Figure 5.15 Estimated permeability functions along the north–south horizontal direction for the soil matrix. (After Li, J. H. et al., *Proceedings of ICE-Geotechnical Engineering*, 164, 211–221, 2011. With permission.)

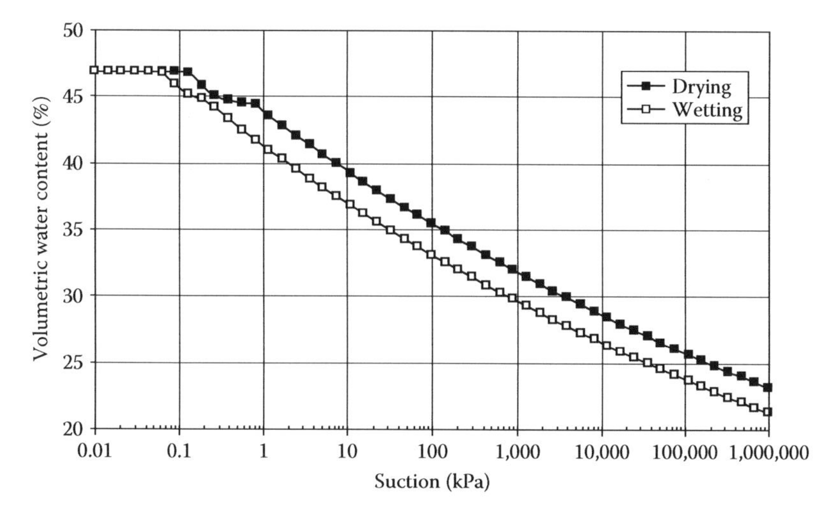

Figure 5.16 SWCCs for the cracked soil shown in Figure 5.9d. (After Li, J. H. et al., *Proceedings of ICE-Geotechnical Engineering*, 164, 211–221, 2011. With permission.)

The drying SWCC for the cracked soil can be obtained by combining the drying SWCCs for the crack network and for the soil matrix using Equation 5.17, as shown in Figure 5.16. Similarly, the wetting SWCCs for the crack network and the soil matrix can be combined to give the wetting SWCC for the cracked soil (also shown in Figure 5.16). The tensorial form of permeability function for the cracked soil can be obtained according to Equation 5.27. The volumetric weighting factor for the crack network, w_c, is about 0.03 according to the field survey. Hence, the tensorial form of permeability function is

$$\mathbf{k} = \begin{bmatrix} 7\times10^{-3} & 2.9\times10^{-3} & 0 \\ 2.9\times10^{-3} & 9\times10^{-3} & 0 \\ 0 & 0 & 1.35\times10^{-2} \end{bmatrix} \left(S_{ec}\right)^{\delta_c} + 0.97 \begin{bmatrix} 0.6\times10^{-6} & 0 & 0 \\ 0 & 3\times10^{-6} & 0 \\ 0 & 0 & 3.5\times10^{-6} \end{bmatrix} \left(S_{em}\right)^{\delta_m} \tag{5.31}$$

The hysteresis permeability function for the cracked soil can be obtained by combining the hysteresis permeability functions for the cracks and for the soil matrix using Equation 5.27. Figure 5.17 shows the permeability function along the north–south direction for the cracked soil. The SWCC for the cracked soil is only slightly different from the SWCC for the soil matrix at the low suction range. The small difference in SWCC results from the relatively small crack porosity (i.e., 3.10%) contrasting to the soil matrix porosity (i.e., 44.3%). However, the permeability for the cracked soil becomes a bimodal curve. The coefficients of permeability of the cracked soil at low suctions are much larger than that of the soil matrix. The significant difference in permeability results from the large permeability of cracks in the low suction range. The coefficients of permeability of the cracks are several orders of magnitude larger than that of the soil matrix in the low suction range. Therefore, the cracks will increase the coefficients of permeability greatly at low suctions. The SWCCs and permeability functions for the cracked soil for the other three moments can be obtained following the same procedure for the cracked soil on November 12, 2007 and are shown in Figures 5.18

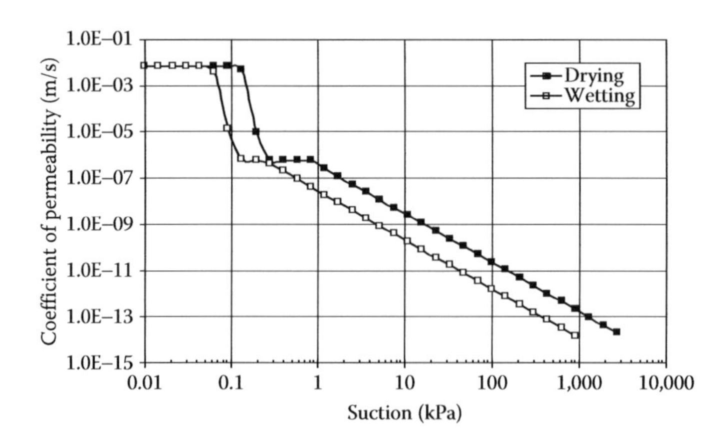

Figure 5.17 Permeability functions for the cracked soil along the north–south horizontal direction shown in Figure 5.9d. (After Li, J. H. et al., *Proceedings of ICE-Geotechnical Engineering*, 164, 211–221, 2011. With permission.)

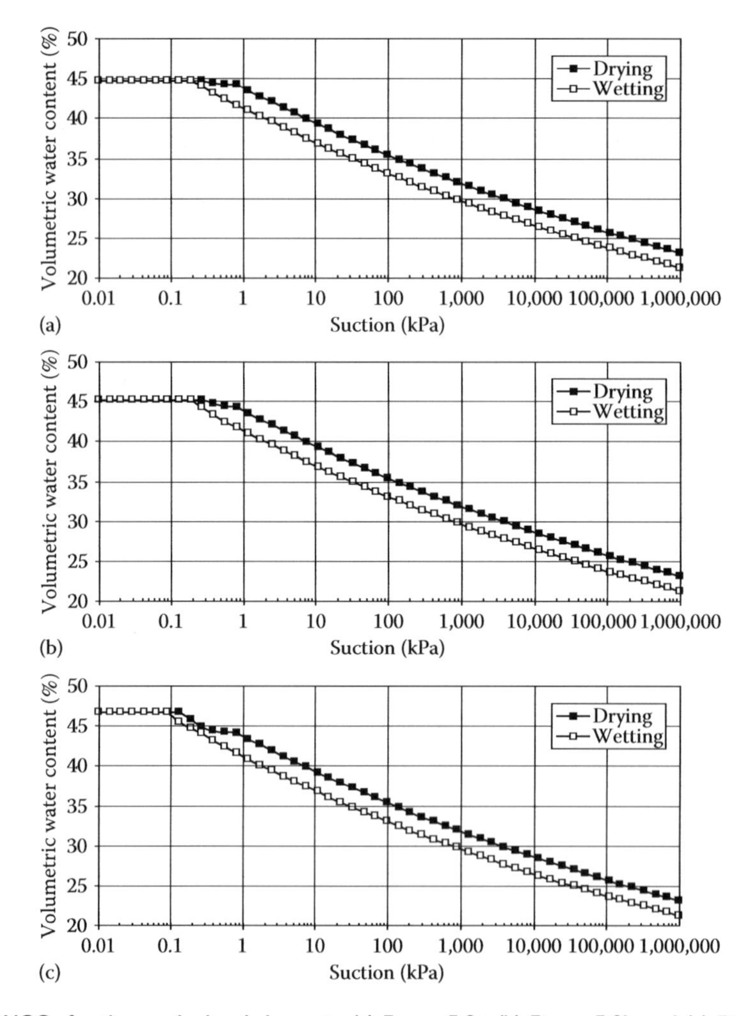

Figure 5.18 SWCCs for the cracked soil shown in: (a) Figure 5.9a, (b) Figure 5.9b, and (c) Figure 5.9c. (After Li, J. H. et al., *Proceedings of ICE-Geotechnical Engineering*, 164, 211–221, 2011. With permission.)

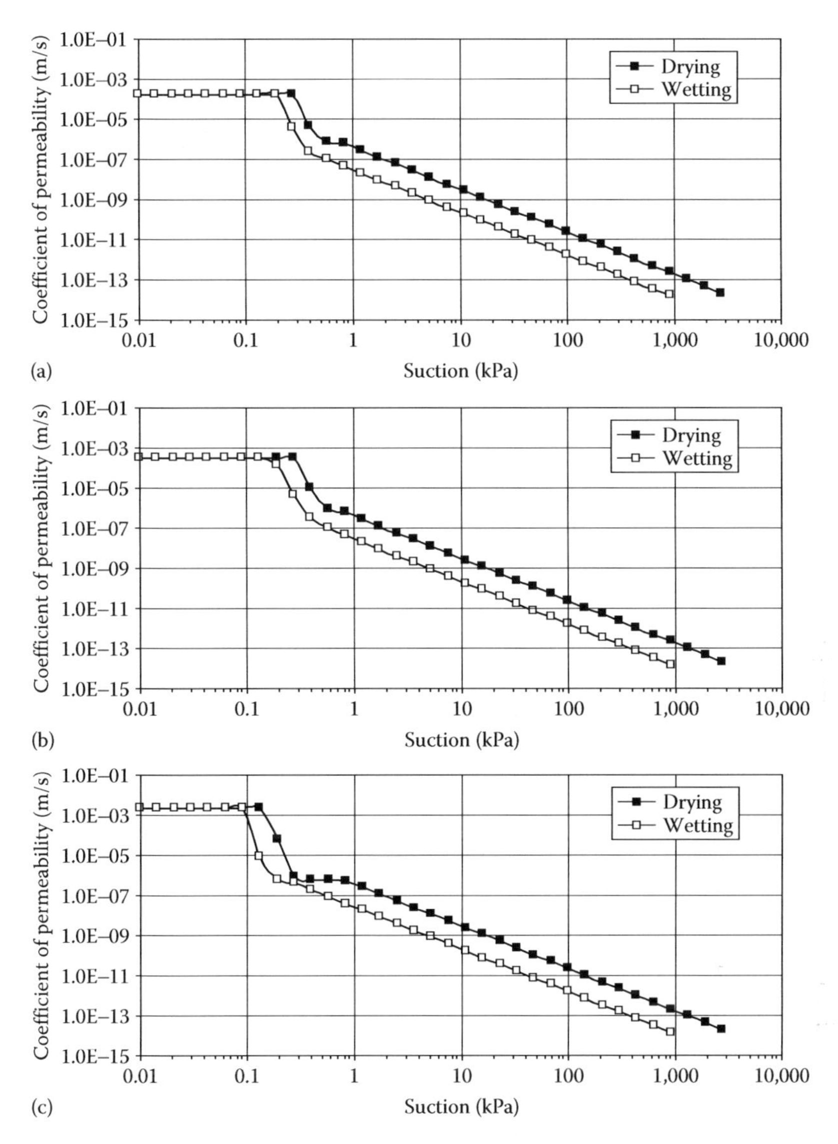

Figure 5.19 Permeability functions for the cracked soil shown in: (a) Figure 5.9a, (b) Figure 5.9b, and (c) Figure 5.9c. (After Li, J. H. et al., *Proceedings of ICE-Geotechnical Engineering*, 164, 211–221, 2011. With permission.)

and 5.19. As the cracks develop and grow during the drying process, the saturated permeability increases for the cracked soils.

Following the crack development drying path in Figure 5.10, the drying permeability function considering crack volume change can be obtained using the drying permeability functions of the nondeformable cracked soils at different snapshots following the method presented in Section 5.2.6. Similarly, the wetting permeability function can be obtained using the wetting permeability functions at different snapshots following the crack development wetting path in Figure 5.10. The hysteresis SWCCs and permeability functions considering crack volume change are shown in Figures 5.20 and 5.21, respectively. The hysteresis SWCCs considering crack volume change are similar to that for the soil matrix as shown in Figure 5.20. When the suction is low the permeability during drying is similar to that for the soil matrix. On the other hand, the permeability during wetting is much

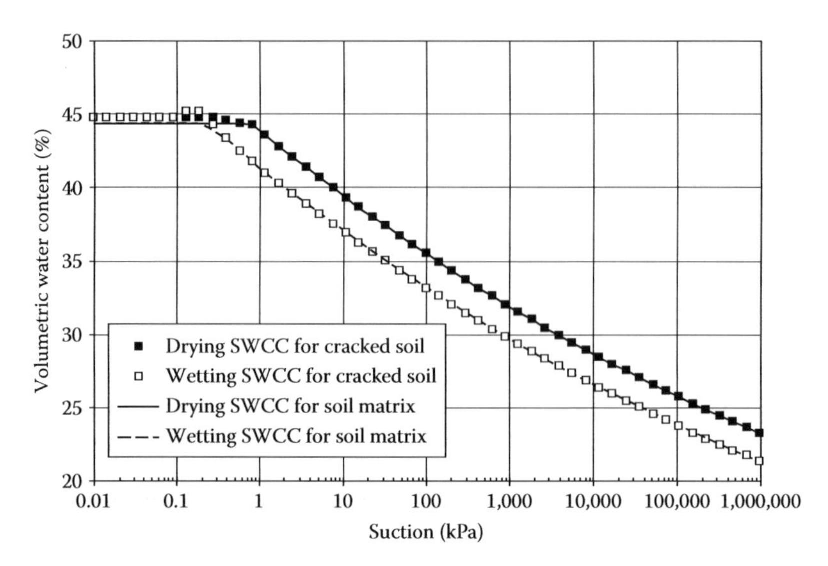

Figure 5.20 SWCCs for the cracked soil considering crack development. (After Li, J. H. et al., *Proceedings of ICE-Geotechnical Engineering*, 164, 211–221, 2011. With permission.)

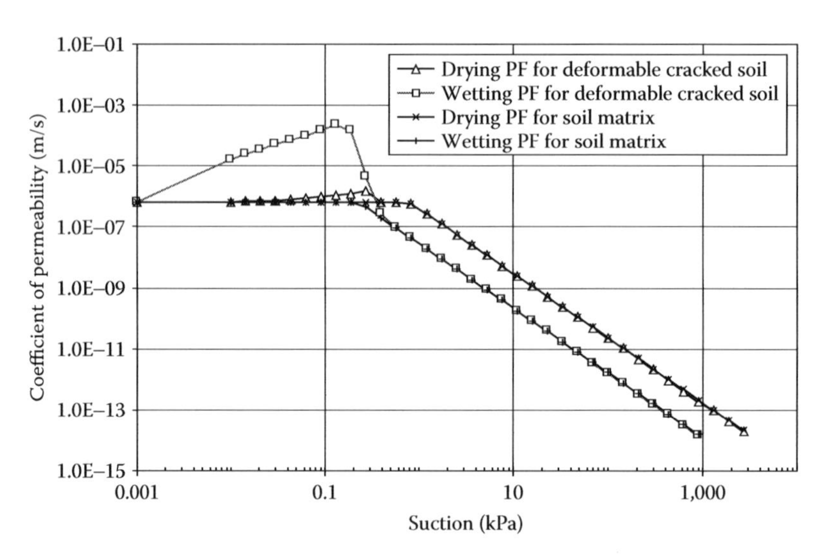

Figure 5.21 Permeability functions for the cracked soil considering crack development. (After Li, J. H. et al., *Proceedings of ICE-Geotechnical Engineering*, 164, 211–221, 2011. With permission.)

higher than that for the soil matrix. The difference between the two cases is caused by different crack development paths (Figure 5.10). On the crack development drying path, cracks open at a relatively larger suction (47 kPa), which indicates no cracks contribute to the permeability until the suction reach 47 kPa. On the crack development wetting path, some cracks are still open at low suctions (0.1 kPa), which means the cracks can contribute to the permeability in the low suction range. Because the permeability of the crack networks in low suction range (e.g., <1 kPa) is several orders of magnitude larger than that of the soil matrix, the permeability of the cracked soil is dominated by that of the crack networks in the low suction range in the wetting process. When the suction is high, the permeability is similar to that for the soil matrix in both drying process and wetting process. When the soil suction is large, the crack aperture becomes large. However, the large apertures are incapable of conducting water at large suctions. The small pores in the soil matrix are the primary paths for water flow at large suctions. Therefore, the permeability at high suctions is dominated by the permeability of the soil matrix. In this example, the film flow and intermittent flow are not considered.

5.3.2 Parametric study and discussions

Obviously, the crack development path influences the permeability functions significantly. The following part discusses the permeability functions following different crack development paths.

In addition to the crack development path shown in Figure 5.10, there are other possibilities of crack development. Figure 5.22 shows five crack development paths and Figures 5.23

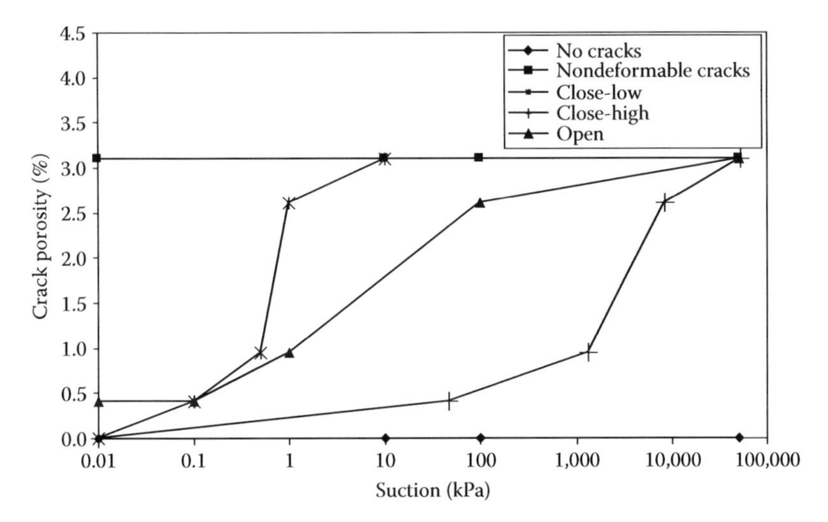

Figure 5.22 Assumed crack development paths. (After Li, J. H. et al., *Proceedings of ICE-Geotechnical Engineering*, 164, 211–221, 2011. With permission.)

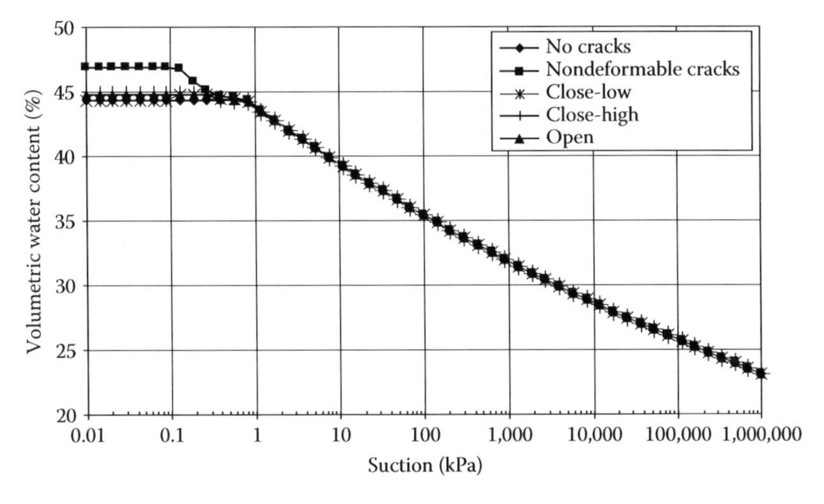

Figure 5.23 SWCCs of the soils corresponding to different crack development paths. (After Li, J. H. et al., *Proceedings of ICE-Geotechnical Engineering*, 164, 211–221, 2011. With permission.)

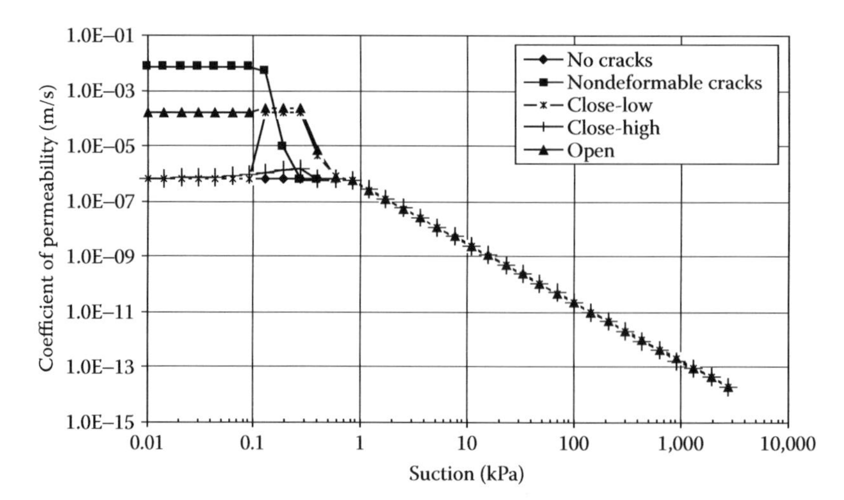

Figure 5.24 Permeability functions corresponding to different crack development paths. (After Li, J. H. et al., *Proceedings of ICE-Geotechnical Engineering*, 164, 211–221, 2011. With permission.)

and 5.24 present the SWCCs and permeability functions for the five cases, respectively. The following observations can be made:

1. If no cracks appear during the drying-wetting process (denoted as "No cracks"), the permeability function is obviously the same as that of the soil matrix. It is the lower boundary of the permeability functions for the cracked soil.
2. The second case is that the cracks do not deform during the drying-wetting cycles and the crack porosity is always the steady crack porosity (denoted as "Nondeformable cracks"). The permeability function in this case is a bi-modal function. The permeability at low suctions is dominated by the crack network but the permeability at high suctions is controlled by the soil matrix. The permeability function for this case is the same as that of the cracked soil at the steady state without considering crack volume change. The air-entry value is relatively small in this case because the crack aperture is larger than the following cases.
3. The third case is that the crack aperture becomes smaller upon wetting (denoted as "Open"). However, there are still open cracks when the soil is almost fully saturated. The saturated permeability is the weighted sum of the saturated permeability values of the initial crack network and the soil matrix. Because the saturated permeability of the crack networks is several orders of magnitude larger than that of the soil matrix, the saturated permeability of the cracked soil is dominated by that of the crack networks. When the soil suction is larger, the crack aperture becomes larger. However, the larger apertures are incapable of conducting water at larger suctions. The permeability at high suctions is dominated by the permeability of the soil matrix.
4. The fourth case is that the cracks close completely when saturated and open at low suctions (denoted as "Close-low"). The saturated permeability is dominated by the soil matrix as no cracks appear initially. At a certain low suction, cracks appear and water can flow through the crack network. Then the permeability at low suctions may be dominated by that of the crack networks. When suction is high, water cannot flow through the cracks because of the larger apertures of cracks. The permeability at high suctions is also controlled by the soil matrix. It is similar to the crack development wetting path mentioned earlier.

5. The fifth case is that the cracks close when saturated and open at high suctions (denoted as "Close-high"). Because there are no cracks at low suctions, the permeability is dominated by the soil matrix at low suctions. When the suction is high the cracks appear. However, water cannot flow through cracks at high suctions. The permeability of the soil matrix is still controlled by the soil matrix at high suction. Therefore, the permeability function of the cracked soil in this case is similar to that of the soil matrix. It is similar to the crack development drying path mentioned earlier.

Although composite hydraulic functions of crack soil can account for a significant increase in the soil permeability near saturation, they do not by themselves lead to preferential flow. When used in a single-domain model based on the Richards equation, simulations will still produce uniform wetting front with some accelerated advance of the wetting front for surface ponding conditions because of the higher saturated permeability.

5.4 STABILITY OF SOIL SLOPE WITH CRACKS

5.4.1 Methodology

Cracks in a slope will affect the pore-water pressure distribution and the slope stability. This section investigates the stability of slopes with cracked soils and without cracked soils using Seep/W and Slope/W (Geo-slope International Ltd., 2012a,b). The cracked soils investigated in the above field study in Section 5.3.1 are used in the following numerical simulation. More details on the field study and the cracked soil can be found in Li and Zhang (2010), Li et al. (2011), and Li and Zhang (2011a,b). The average aperture of the cracks is 0.42 mm, to which the capillary law can be applicable. The representative elementary area size of the cracked soil on the horizontal surface is about 165 mm as investigated by Li and Zhang (2010). The representative elementary area size in different directions is assumed to be the same as that on the horizontal surface. Hence, the representative elementary volume of the cracked soil is about $165 \times 165 \times 165$ mm^3.

The slope considered is 20 m high with an inclination angle of 42°. Figure 5.25 shows the geometry and finite element mesh of the slope. The initial water table is also shown in Figure 5.25. The total head below the initial water table is 7.3 m along the left edge and

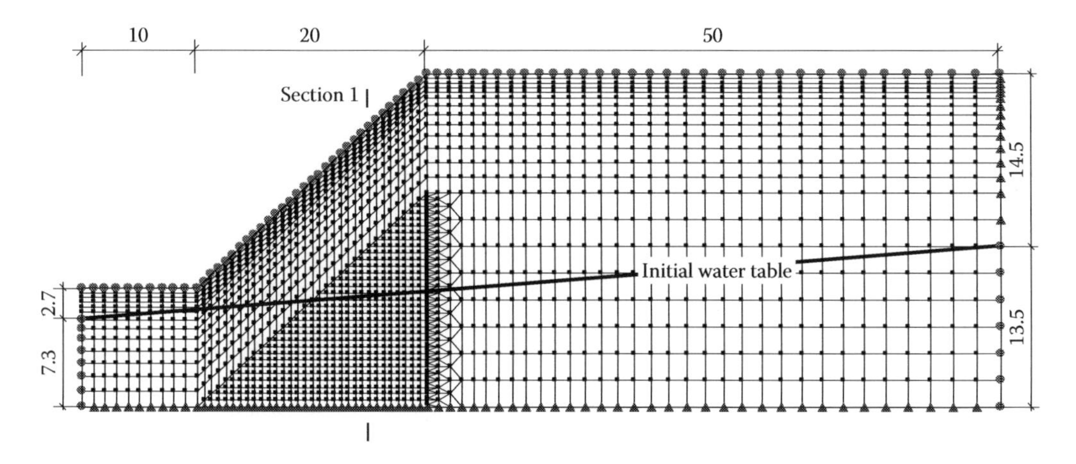

Figure 5.25 Geometry and finite element mesh of the slope (unit: meter).

13.5 m along the right edge. It is assumed that the bedrock below the slope is impermeable. The average annual rainfall is 1088 mm at the test site. Assuming 10% of rainfall infiltrates into the ground (considering runoff and evaporation), the precipitation rate will be 3.45×10^{-9} m/s. The initial conditions of the slope are obtained by applying this precipitation rate on the slope surface until the steady state (shown in Figure 5.26a). Then, rainfall is modeled by a zero pressure head on the surface of the slope to study the pore-water pressure distributions in the slope and the stability of the slope.

Generally the cracks only develop at the shallow depths of soils. Hence, in the simulation, a cracked soil is used on the top 2 m of the slope and a uniform soil matrix without cracks is assumed in the deeper depths. As a reference, a homogeneous soil slope containing only the uniform soil matrix is also simulated. The SWCC and permeability function for the soil

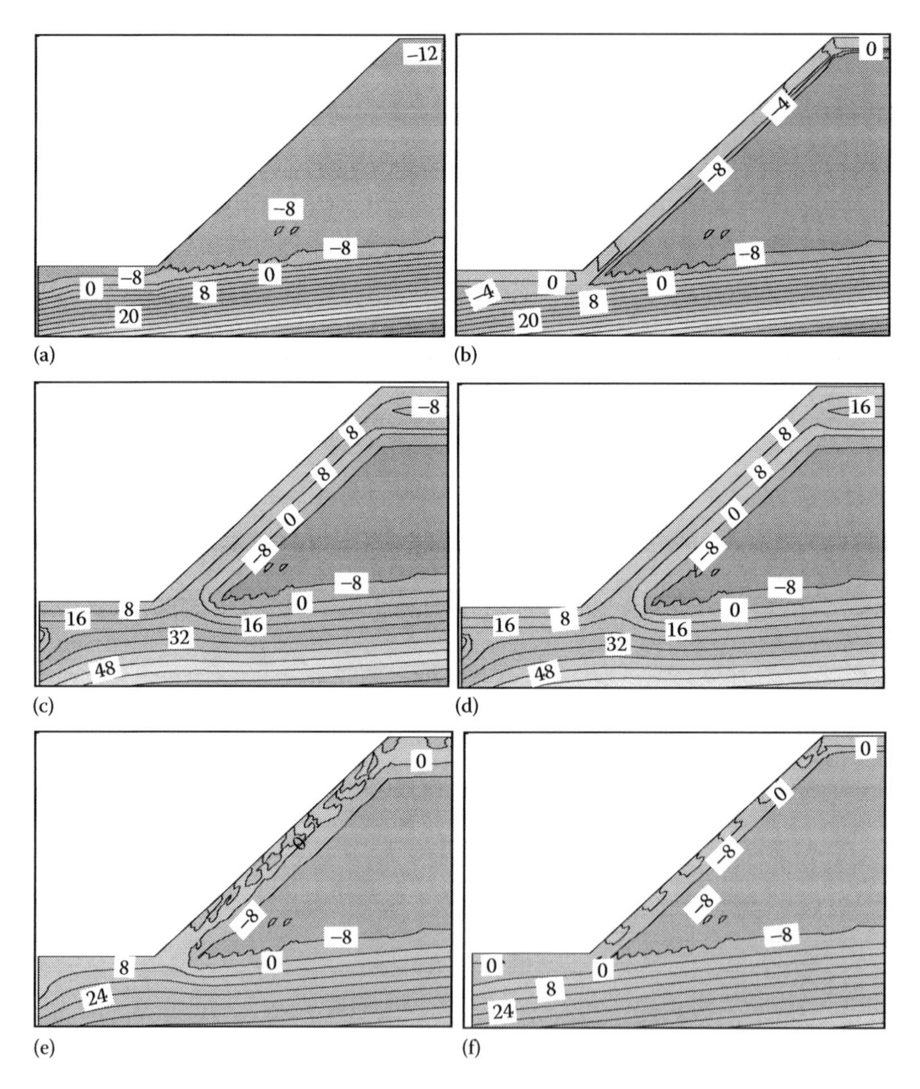

Figure 5.26 Pore-water pressure distributions (in kPa): (a) at initial condition, (b) in the slope with homogeneous soil after 24 hours of rainfall, (c) in the slope with cracked soil of the first case after 24 hours of rainfall, (d) in the slope with cracked soil of the second case after 24 hours of rainfall, (e) in the slope with cracked soil of the third case after 24 hours of rainfall, and (f) in the slope with cracked soil of the fourth case after 24 hours of rainfall.

matrix are shown in Figures 5.14 and 5.15. The cracked soil in the slope is modeled considering four scenarios:

1. The cracks do not deform during drying-wetting cycles and the crack porosity is always steady. This case corresponds to the second case (i.e., nondeformable cracks) in the parametric study presented in Section 5.3.2.
2. The crack aperture becomes smaller upon wetting. However, there are still open cracks when the soil is fully saturated. This case corresponds to the third case (i.e., Open) in the parametric study.
3. The cracks close completely when saturated but is still open at low suctions. This case corresponds to the fourth case (i.e., Close-low) in the parametric study.
4. The cracks close when saturated and exist only at high suctions. This case corresponds to the fifth case (i.e., Close-high) in the parametric study.

The scale of the slope is 28 m in height and 80 m in width. The smallest element dimension is 0.281 m, which is larger than the representative elementary size of the cracked soil (i.e., 0.165 m). Therefore, the cracked soil can be approximated as a homogeneous porous medium and an equivalent continuum method can be used to conduct the seepage analysis. For simplicity, the two types of soil are assumed to be homogeneous and isotropic. The permeability functions in the horizontal north–south direction are used in the analysis. The SWCCs and permeability functions for the four cases of cracked soil are shown in Figures 5.23 and 5.24. In addition, the adverse effect of cracks on the shear strength parameters of the soil is not considered in this chapter. Therefore, the effective cohesion, c', and friction angle, ϕ', are the same for the cracked soil and the soil matrix. The effective cohesion is taken as 11.7 kPa and the friction angle is taken as 24° in the simulation. The friction angle indicating the rate of increase in shear strength relative to the matric suction, ϕ^b, is taken as 12°.

5.4.2 Results

Figure 5.26a shows the pore-water pressure distributions at the initial condition for the slopes. Figure 5.26b shows the pore-water distributions in the homogeneous soil slope with no cracks after 24 hours of rainfall. Figure 5.26c–f show the pore-water distributions in the slopes of the four cracked soil scenarios presented in the previous section. There is a thin layer (about 0.8 m) of saturated zone in the homogeneous soil slope (Figure 5.26b). In contrast, when the cracks are nondeformable, the saturated zone in the slope with this cracked soil is much thicker (about 4 m) as shown in Figure 5.26c. This is deeper than the cracked zone (i.e., 2 m). The difference is attributed to the presence of cracks, which increase the coefficient of permeability of soil. The perched groundwater caused by a layer of cracked soil on the slope surface was also observed by Ng et al. (2003). When the crack aperture decreases upon wetting but remain open under saturated condition, the pore-water pressure distribution for the slope is shown in Figure 5.26d, which is similar to that in Figure 5.26c. When the cracks open at low suctions the saturated zone below the ground surface in the cracked soil slope is about 2 m (Figure 5.26e). When the cracks open only at high suctions, the pore-water pressure distribution in the slope is similar to that in the homogeneous soil slope (Figure 5.26f). It indicates that, when the cracks open only at high suctions, the effect of cracked soil on the water infiltration is limited.

The pore-water pressure profiles along cross section 1 (shown in Figure 5.25) in the slopes with different kinds of soil are investigated. Figure 5.27 shows the pore-water pressure

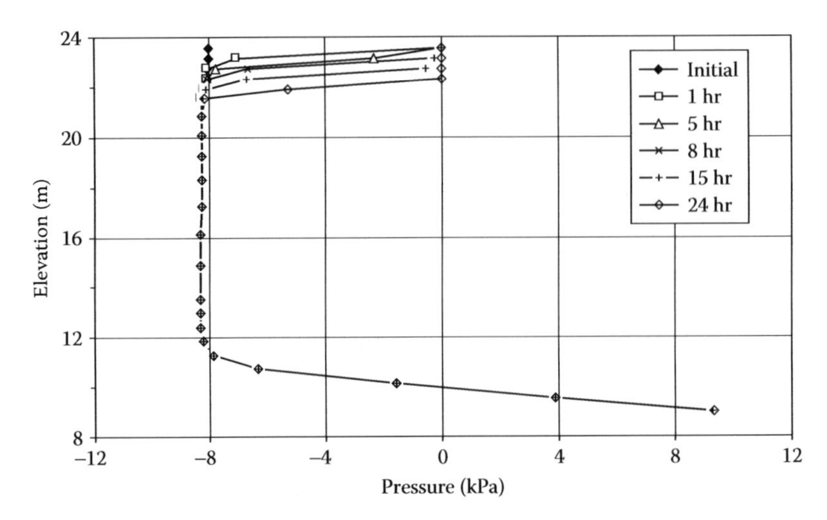

Figure 5.27 Pore-water pressure profiles along section I in the slope with homogeneous soil.

profile in the slope with the homogeneous soil. The pore-water pressure is initially negative and increases gradually in the top layer of the slope. Figure 5.28 shows the pore-water pressure profiles in the cracked soil slope with the nondeformable cracks. In the first five hours the pore-water pressures increase gradually and are still negative. After 8 hours of rainfall, there is a saturated zone on the top 2 m of the slope. Afterward, the saturated zone becomes deeper and is about 4 m thick after 24 hours of rainfall. The pore-water pressures change abruptly from 5 hours to 8 hours. Then the change in pore-water pressure is much slower from 8 hours to 24 hours. This phenomenon is also observed in a field infiltration test conducted at a hillside near Three Gorges Dam by Zhang et al. (2000). In their test, the pressures change from negative to positive after 3 hours to 11 hours of infiltration. Then the pore-water pressure changes are limited after 11 hours of infiltration. The reason for

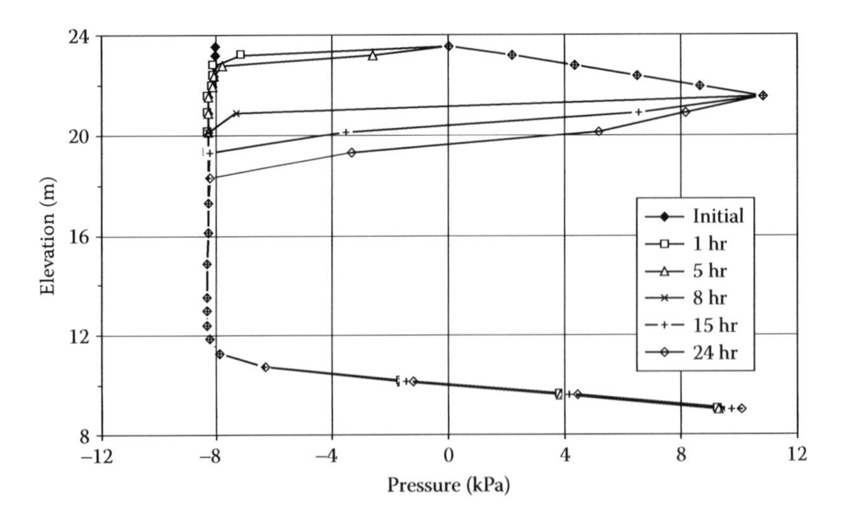

Figure 5.28 Pore-water pressure profiles along section I in the cracked soil slope with nondeformable cracks.

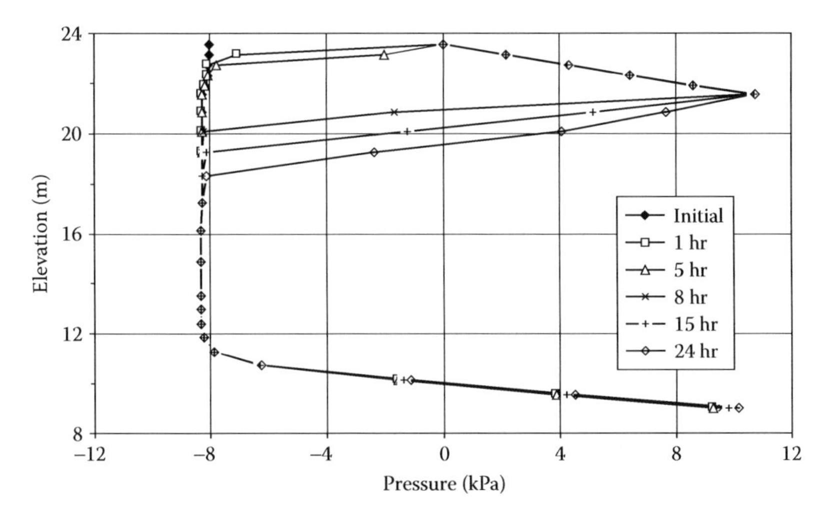

Figure 5.29 Pore-water pressure profiles along section I in the cracked soil slope with open cracks.

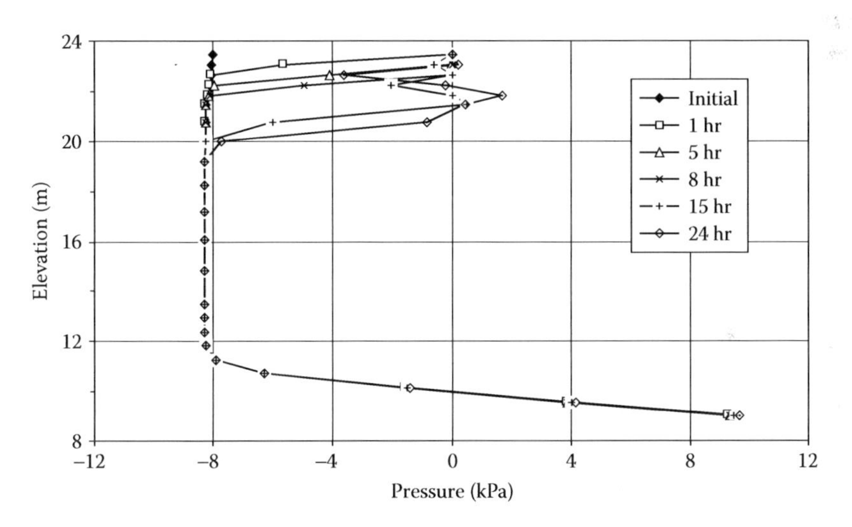

Figure 5.30 Pore-water pressure profiles along section I in the cracked soil slope with cracks open at low suctions ("Close-low" case).

this phenomenon is also the presence of rock joints. Figure 5.29 shows the same trend as Figure 5.28, which indicates that the pore-water pressure will change much faster as long as there are cracks under saturated conditions. Figure 5.30 shows the pore-water pressure profiles in the cracked soil slope when the cracks open at low suctions and close under saturated conditions. Two meters of saturated zone appears after 24 hours of rainfall, which is slower than that for the slope with cracks still open under the saturated conditions. When the cracks close upon saturation and only open under high suctions ("close-high" case), the pore-water distribution in the cracked soil slope is similar to that in the homogeneous soil slope (as shown in Figure 5.31).

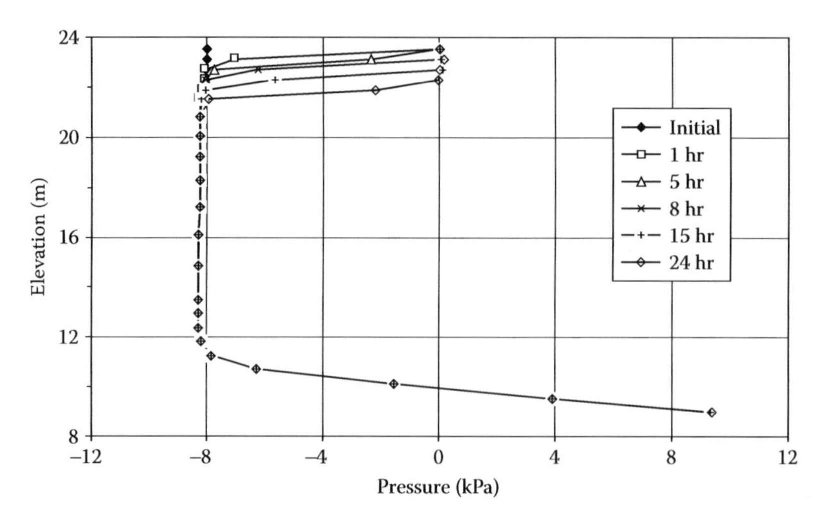

Figure 5.31 Pore-water pressure profiles along section I in the cracked soil slope with cracks open at high suctions ("Close-high" case).

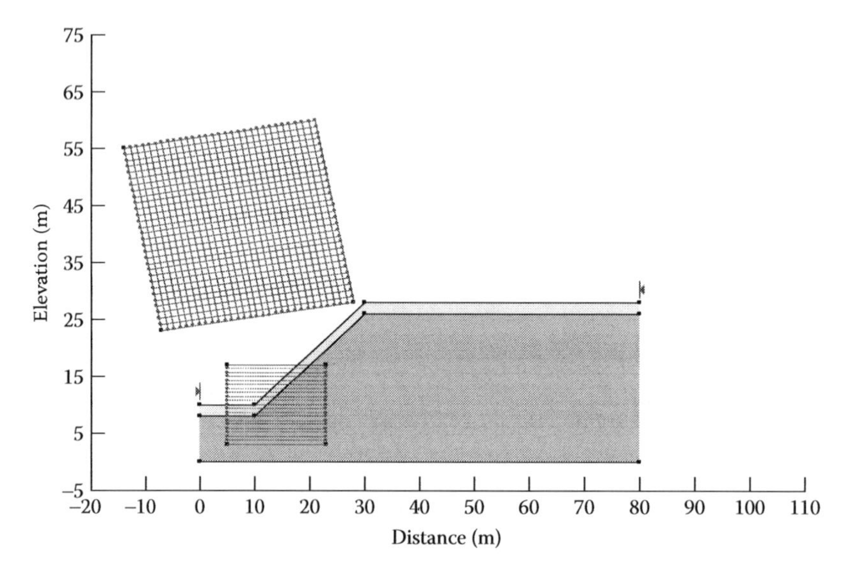

Figure 5.32 Centers and radii for searching critical slip surface.

The stability of the slopes is analyzed using Slope/W. The critical slip surface is searched by setting a grid of centers of rotation and a radius zone for the possible slip surfaces (Figure 5.32). The factor of safety, F_s, for the homogeneous soil slope and cracked soil slopes are summarized in Figure 5.33.

The factor of safety is 1.018 at the initial conditions. The factor of safety after 24 hours of rainfall decreases a little to 1.011 in the slope with the homogeneous soil only. The factor of safety decreases to 0.954 when there is a layer of cracked soil on the slope surface and the cracks are still open under saturated conditions. The factor of safety is 0.999 when the cracks are present at low suctions but close under saturated conditions. The change in the

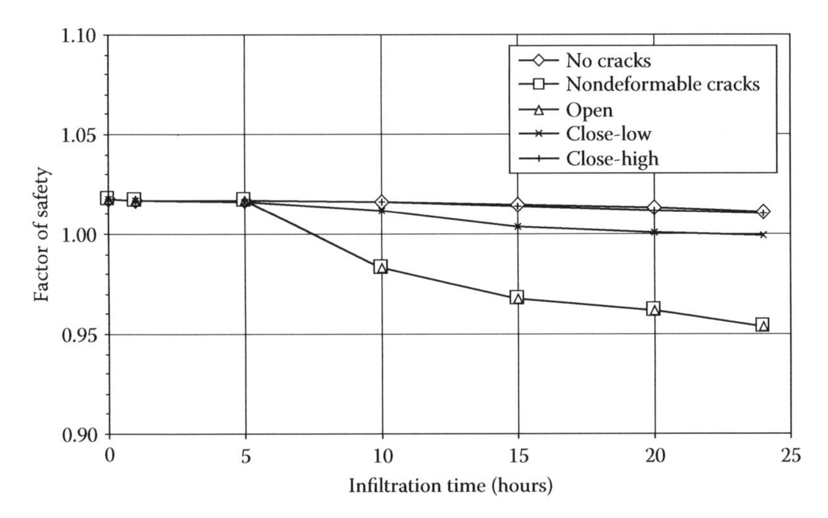

Figure 5.33 Changes of factor of safety of slopes with infiltration time.

factor of safety of the slope with the cracks open only at high suction values is similar to that of the homogeneous soil slope. The results indicate that, when the cracks are still open under saturated conditions, the factor of safety will decrease significantly after rainfall. The decrease in the factor of safety is due to the higher coefficient of permeability of the cracked soils. When the cracks are open only under high suction values the cracks have little effect on the stability of the slope.

REFERENCES

Baker, R. (1981). Tensile strength, tension cracks, and stability of slopes. *Soils and Foundations*, 21(2), 1–17.

Brooks, R. H., and Corey, A. T. (1964). Hydraulic properties of porous media, Hydrology Papers No. 3. Colorado State University, Fortcollins, CO.

Burdine, N. T. (1953). Relative permeability calculations from pore size distribution data. *Journal of Petroleum Technology*, 5(03), 71–78.

Childs, E. C., and Collis-George, N. (1950). The permeability of porous materials. *Proceedings of the Royal Society of London A*, 201, 392–405.

Chowdhury, R. N., and Zhang, S. (1991). Tension cracks and slope failure. *Slope Stability Engineering, Proceedings of the International Conference*, Isle of Wight, pp. 27–32.

Dykhuizen, R. C. (1987). Transport of solutes through unsaturated fractured media. *Water Research*, 21, 1531–1539.

Eigenbrod, K. D. (1996). Effects of cyclic freezing and thawing on volume changes and permeabilities of soft fine-grained soils. *Canadian Geotechnical Journal*, 33(4), 529–537.

Feng, M., Fredlund, D. G., and Shuai, F. S. (2002). A laboratory study of the hysteresis of a thermal conductivity soil suction sensor. *Geotechnical Testing Journal*, 25(3), GTJ20029678-253.

Fredlund, D. G., and Rahardjo, H. (1993). *Soil Mechanics for Unsaturated Soils*. Wiley, New York, NY.

Fredlund, D. G., and Xing, A. (1994). Equations for the soil-water characteristic curve. *Canadian Geotechnical Journal*, 31, 521–532.

Fredlund, D. G., Xing, A., and Huang, S. Y. (1994). Predicting the permeability function for unsaturated soils using the soil-water characteristic curve. *Canadian Geotechnical Journal*, 31, 533–546.

Galeandro, A., Simunek, J., and Simeone, V. (2013). Analysis of rainfall infiltration effects on the stability of pyroclastic soil veneer affected by vertical drying shrinkage fractures. *Bulletin of Engineering Geology and the Environment*, 72, 447–455.

Geo-slope International Ltd (2012a). *Seep/W for Finite Element Seepage Analysis, User's Guide.* Geo-slope International Ltd., Calgary, Canada.

Geo-slope International Ltd (2012b). *Slope/W for Slope Stability Analysis, User's Guide.* Geo-slope International Ltd., Calgary, Canada.

Hu, Y. J., Qian, R., and Su, B.Y. (2001). A numerical simulation method to determine unsaturated hydraulic parameters of fracture. *Chinese Journal of Geotechnical Engineering,* 23(3), 284–287 (in Chinese).

Kodikara, J. K., and Choi, X. (2006). A simplified analytical model for desiccation cracking of clay layers in laboratory tests. *Geotechnical Special Publication* No. 147, ASCE, Reston, VA, pp. 2558–2569.

Konrad, J. M., and Ayad, R. (1997a). An idealized framework for the analysis of cohesive soils undergoing desiccation. *Canadian Geotechnical Journal,* 34, 477–488.

Konrad, J. M., and Ayad, R. (1997b). Desiccation of a sensitive clay: field experimental observations. *Canadian Geotechnical Journal,* 34, 929–942.

Lee, F. H., Lo, K. W., and Lee, S. L. (1988). Tension crack development in soils. *Journal of Geotechnical Engineering,* 114(8), 915–929.

Li, J. H., and Zhang, L. M. (2010). Geometric parameters and REV of a crack network in soil. *Computers and Geotechnics,* 37, 466–475.

Li, J. H., and Zhang, L. M. (2011a). Connectivity of a network of random discontinuities. *Computers and Geotechnics,* 38, 217–226.

Li, J. H., and Zhang, L. M. (2011b). Study of desiccation crack initiation and development at ground surface. *Engineering Geology,* 123, 347–358.

Li, J. H., Zhang, L. M., and Kwong, C. P. (2011). Field permeability at shallow depth in a compacted fill. *Proceedings of ICE-Geotechnical Engineering,* 164(GE3), 211–221.

Li, J. H., Zhang, L. M., Wang, Y., and Fredlund, D. G. (2009). Permeability tensor and representative elementary volume of saturated cracked soil. *Canadian Geotechnical Journal,* 4, 928–942.

Michalowski, R. L. (2013). Stability assessment of slopes with cracks using limit analysis. *Canadian Geotechnical Journal,* 50(10), 1011–1021.

Mizuguchi, T., Nishimoto, A., Kitsunezaki, S., Yamazaki, Y., and Aoki, I. (2005). Directional crack propagation of granular water systems. *Physical Review E,* 71(5), 056122.

National Research Council. (2001). *Conceptual Models of Flow and Transport in The Fractured Vadose Zone.* National Academy Press. Washington, DC.

Ng, C. W. W., Zhan, L. T., Bao, C. G., Fredlund, D. G., and Gong, B. W. (2003). Performance of an unsaturated expansive soil slope subjected to artificial rainfall infiltration. *Géotechnique,* 53(2), 143–157.

Peron, H., Laloui, L., Hueckel, T., and Hu, L. (2006). Experimental study of desiccation of soil. *Geotechnical Special Publication,* 147, 1073–1084.

Persoff, P., and Pruess, K. (1995). Two-phase flow visualization and relative permeability measurement in natural rough-walled rock fractures. *Water Resources Research,* 31(5), 1175–1186.

Peters, R. R., and Klavetter, E. A. (1988). A continuum model for water movement in an unsaturated fractured rock mass. *Water Resources Research,* 24(3), 416–430.

Pham, H. Q., Fredlund, D. G., and Barbour, S. L. (2005). A study of hysteresis models for soil-water characteristic curves. *Canadian Geotechnical Journal,* 42, 1548–1568.

Rayhani, M. H. T., Yanful, E. K., and Fakher, A. (2008). Physical modeling of desiccation cracking in plastic soils. *Engineering Geology,* 97, 25–31.

Reitsma, S., and Kueper, B. H. (1994). Laboratory measurement of capillary pressure-saturation relationships in a rock fracture. *Water Resources Research,* 30(4), 865–878.

Richards, L. A. (1931). Capillary conduction of liquids through porous medium. *Journal of Applied Physics,* 1, 318–333.

Shen, Z. J., and Deng, G. (2004). Numerical simulation of crack evolution in clay during drying and wetting. *Rock and Soil Mechanics,* 25(s2), 1–6, 12 (in Chinese).

Suryo, E. A. (2013). *Real-Time Prediction of Rainfall Induced Instability of Residual Soil Slopes Associated with Deep Crack.* PhD Thesis, Queensland University of Technology.

Utili, S. (2013). Investigation by limit analysis on the stability of slopes with cracks. *Géotechnique*, 63(2), 140–154.

van Genuchten, M. T. (1980). A closed-form equation for predicting the hydraulic conductivity of unsaturated soils. *Soil Science Society of America Journal*, 44, 892–898.

Wang, Z. F., Li, J. H., and Zhang, L. M. (2012). Influence of cracks on the stability of a cracked soil slope. *5th Asia-Pacific Conference on Unsaturated Soils*, Thailand, Vol. 2, pp. 594–600.

Yao, H. L., Zheng, S. H., and Chen, S. Y. (2001). Analysis of the slope stability of expansive soils considering cracks and infiltration of rainwater. *Chinese Journal of Geotechnical Engineering*, 23(5), 606–609 (in Chinese)

Zhang, G., Wang, R., Qian, J., Zhang, J., and Qian, J. (2012). Effect study of cracks on behavior of soil slope under rainfall conditions. *Soils and Foundations*, 52(4), 634–643.

Zhang, J., Jiao, J. J., and Yang, J. (2000). In situ rainfall infiltration studies at a hillside in Hubei Province, China. *Engineering Geology*, 57(1–2), 31–38.

Zhou, C. B., Ye, Z. T., and Xiong, W. L. (1998). A study on unsaturated hydraulic conductivity of rock joints. *Chinese Journal of Hydraulic Engineering*, 3, 22–26 (in Chinese).

Chapter 6

Stability analysis of colluvium slopes

6.1 INTRODUCTION

Colluvium landslides refer to the landslides that occurred in the quaternary system or in the relative loose cumulative stratum before the quaternary period. They are a major concern in the construction and maintenance of railways, highways, waterways, reservoirs, and many other infrastructure projects. They are generally characterized by loose composition, wide gradation due to the specific composition of the sliding mass.

The colluvium landslides were frequently triggered by rainstorms. For example, in late June, 1998, a series of rainstorms dropped 16.5 cm of rain in a 72-hour period over southeastern Ohio, USA, triggered more than 60 shallow landslides along a 64-km stretch of Interstate 77 between Buffalo and Marietta, OH. Field observations indicated that the slides occurred in colluvial soils that mantle the road cuts along the interstate highway. Most of the slides were shallow and translational in nature, with the failure surfaces located along the contact between the colluviums and the unweathered bedrock (Shakoor and Smithmyer, 2005).

To analyze the stability of colluvium slopes comprehensively, the shear strength, hydraulic properties, and the properties of possible existing weak layer (Brand, 1981; Mair, 1993; He and Wang, 2006) need to be determined. Among these properties, the hydraulic properties of colluvial soils are required in the seepage analysis, which is helpful for the investigation of the triggering mechanisms and is the basis for evaluating the evolution in time of the stability of structures involving colluviums. The experimental study demonstrates that the hydraulic properties of colluvial soils are highly related to their coarse contents (Zhang and Li, 2010). A question to ask is how stable are slopes composed of colluvial soils with different gravel proportions under a rainfall event. In this chapter, the infiltration process and the stability of colluvium slopes are investigated by numerical simulation. A hypothetical case is considered in the numerical simulation and applied to the four colluvial soils. The hydraulic properties measured in the laboratory are used in the seepage analysis. The shear strength parameters of colluvium are majorly referred to Irfan and Tang (1993). The infiltration process during and the one after a rainstorm are simulated first. Then the stability of different slopes is evaluated based on the seepage analysis results.

6.2 HYDRAULIC PROPERTIES OF COLLUVIAL SOILS

Several researches have been conducted on tills from Canada, which are also typical colluvial soils and considered as excellent construction materials. Loiselle and Hurtubise (1976) studied tills from the Manicouagan and Outardes rivers area in Canada, and found saturated permeability values varying from 2×10^{-10} to 6×10^{-8} m/s; Pare et al. (1982) tested another till from Canada, and obtained saturated permeability values of about 5×10^{-6} m/s. The testing conditions, which are shown to play a major role in the measured saturated permeability, were poorly documented in both cases. Leroueil et al. (2002) tested six tills from northern Quebec. In those tests, only the fraction of the tills finer than 5 mm was tested for practical reasons. The fine particle content (smaller than 0.080 mm) was between 22% and 43%, and the clay-size particle content (smaller than 0.002 mm) varied from 2.5% to 12.6%. The uniformity coefficient, C_u, varied from 15 to 232. The compaction curves showed optimum water contents between 7.5% and 8.8% and optimum dry densities between 2005 and 2155 kg/m^3, which corresponds to void ratios ranging from 0.36 to 0.27. The degree of saturation at the optimum water content varied between 67% and 82%. These materials were nonplastic or low plastic. Leroueil et al. (2002) summarized the permeability of tills as follows:

1. Permeability of till must not be considered as a soil characteristic but rather as a parameter that varies widely depending on the compaction conditions.
2. For low-plasticity or nonplastic tills from northern Quebec, the soil fabric is aggregated when the soil is compacted at water contents lower than the optimum water content. This phenomenon results in a macroporosity that gives higher permeability than that obtained with a homogeneous fabric, when the soil is compacted at water contents higher than the optimum water content.
3. The permeability obtained with homogeneous fabric (i.e., soil compacted at water content higher than the optimum water content) varies with the clay fraction, typically decreasing by one order of magnitude per 2.5% increase of the clay fraction.
4. The permeability obtained with aggregated fabric (i.e., soil compacted at water contents lower than the optimum water content) increases with an increase in the clay fraction.

Feuerharmel et al. (2006) measured the soil–water characteristic curves (SWCCs) of two undisturbed colluvial soils, which were considered to be colluvial soils that showed clear bimodal grain size distributions (GSDs). The SWCCs measured by Feuerharmel et al. (2006) are found to have bimodal features. Burger and Shackelford (2001) measured other artificial colluvial soils, which were mixed from two types of processed diatomaceous earth. Because of the special structure of pellets, which are full of intraparticle pores, the mixed colluvial soils used by Burger and Shackelford (2001) are considered to have typical dual-porosity structure. The measured SWCCs for the artificial gap-graded diatomaceous mixtures show clear bimodal feature. These bimodal SWCCs are considered to be indicators of dual-porosity structure (i.e., bimodal pore-size distribution).

Both experimental research and modeling research are limited for colluvial soils. To evaluate the stability of structures involving colluvial soils, such as the colluvium landslides that are widely distributed in Hong Kong and TGRZ, it is important and in urgent need to study the hydraulic proprieties of colluvial soils.

To study the hydraulic properties of colluvial soils, four soils are artificially mixed from natural completely decomposed granite at a construction site at Beacon Hill, Hong Kong (Li, 2009). These soils are classified as silty gravels (denoted as SM with gravel), clayey gravels (denoted as SC with gravel), inorganic silts (denoted as sandy ML), and inorganic clays (denoted as CL with sand), as illustrated in Figure 6.1. The four soils have different gravel (particles with diameter in range of 4.75–75 mm according to ASTM D2487) proportions, that is, 44%, 29%, 15%, and 0%, respectively. The detailed parameters are shown in Table 6.1. The liquid limit, plastic limit, and shrinkage limit of CL with sand are 47%, 32%, and 8%, respectively. The specific gravity of these soils is 2.64.

The SWCCs for four colluvial soils were measured in the laboratory (Li, 2009) and fitted to Fredlund and Xing model (Fredlund and Xing, 1994), as illustrated in Figure 6.2. The saturated coefficients of permeability of four soils were measured using traditional constant head and falling head tests. The unsaturated coefficients of permeability of four soils were measured using the wetting front advancing method (Li et al., 2009). The measured coefficients of permeability of four soils are fitted to unimodal and bimodal permeability function model (Dane and Klute, 1977) and illustrated in Figure 6.3.

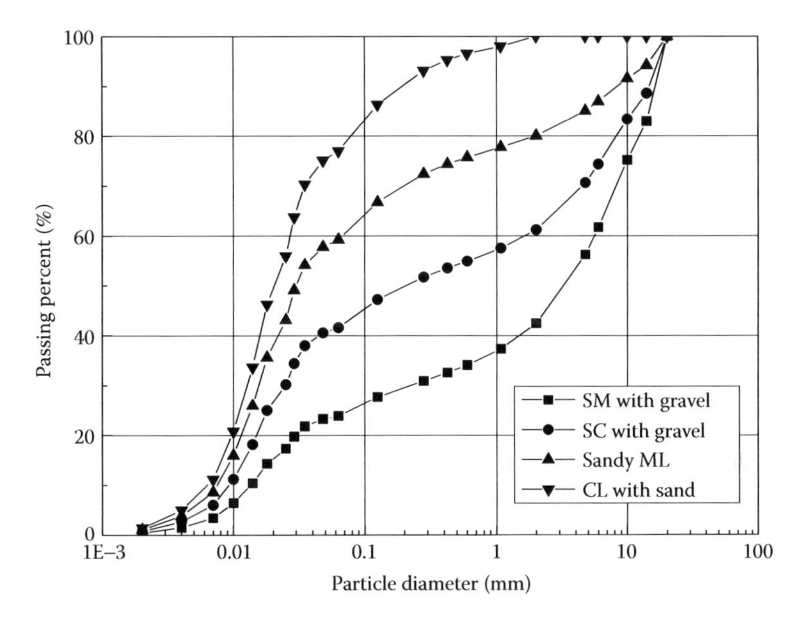

Figure 6.1 GSDs of the four colluvial soils used in laboratory tests.

Table 6.1 Soil properties of four colluvial soils

Group symbols (ASTM D2487)	*Maximum dry density (kg/m³)*	*Optimum water content (%)*	*Proportions*		
			Fines (%)	*Sand (%)*	*Gravel (%)*
SM with gravel	1970	11	24	32	44
SC with gravel	1900	13	42	29	29
Sandy ML	1710	18	60	25	15
CL with sand	1550	21	80	20	0

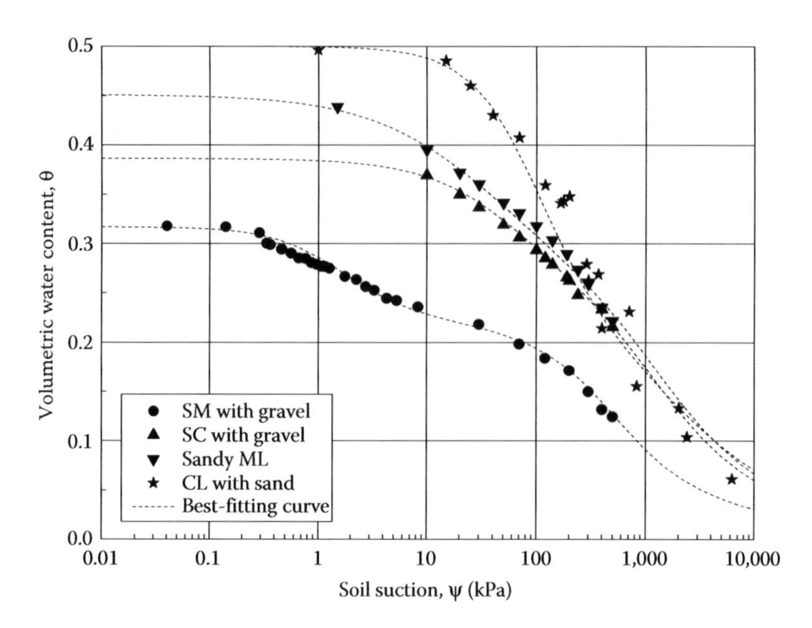

Figure 6.2 Measured and fitted SWCCs for four soils used in seepage analysis.

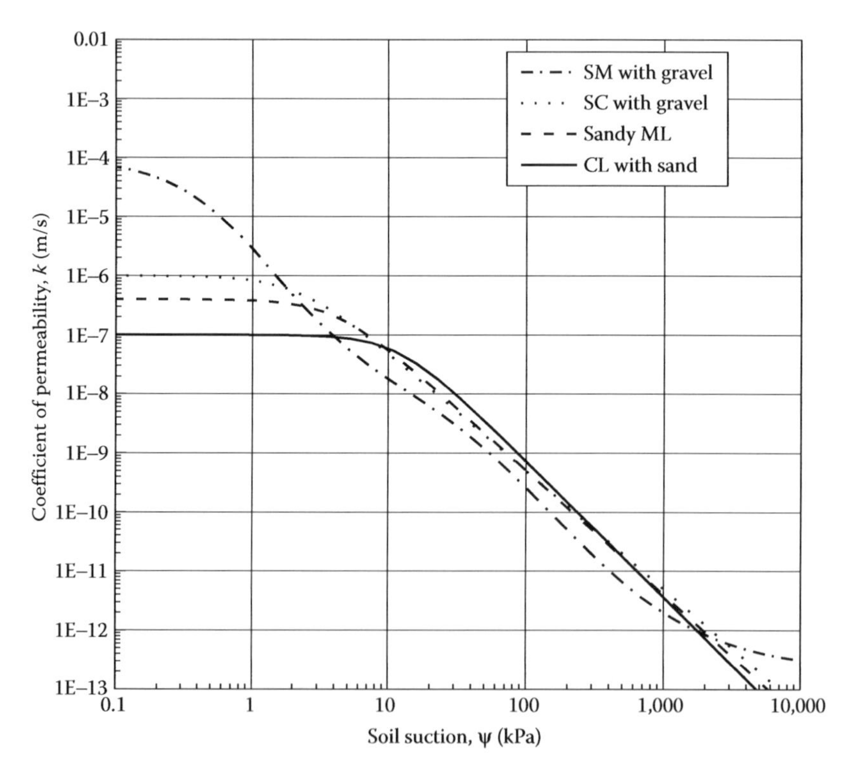

Figure 6.3 Fitted permeability functions for four soils used in seepage analysis.

6.3 SHEAR STRENGTH OF COLLUVIAL SOILS

Irfan and Tang (1993) studied the effect of coarse fractions on shear strength of colluvial soils, which are widely graded and the mixtures of granite aggregate and fine colluvium grains, and proposed a model to characterize the shear strength of colluvial soils, as

$$\tau_f = \begin{cases} c' + \sigma_n' \tan\phi' & \text{for } P_g < P_c \\ c' + \sigma_n' \tan(\Delta\phi + \phi') & \text{for } P_u > P_g > P_c \end{cases} \tag{6.1}$$

where P_g is the proportion of the granular materials, P_c is the critical proportion of the granular materials at which the shear strength of the mixtures starts to increase, P_u is the upper-bound critical proportion of granular material, at which the shear strength is similar to the granular material, $\Delta\phi$ is the increase in friction angle resulting from an increase in granular material proportion and is equal to

$$\Delta\phi = (P_g - P_c)\alpha_g \tag{6.2}$$

where α_g is a material constant that depends on the shape, size, and density of the granular material.

The colluvial soils used in Li (2009) have the similar GSDs and same mineral component with that mixture used by Irfan and Tang (1993). As recommended by Irfan and Tang (1993), the parameters, P_c and a_g, are chosen as 35% and 3.5° per 10% granular material proportion, respectively. Because Irfan and Tang (1993) did not conduct experiments on soils with coarse fractions above 52% and did not find the exact value of P_u, P_u is taken as 72% as recommended by Vallejo and Mawby (2000). The fines of CL with sand used in Li (2009) are finer than that used by Irfan and Tang (1993) which is a silty sand. Therefore, the CL with sand is considered to have a smaller friction angle comparing to the silty sand used by Irfan and Tang (1993). The effective friction angle of the CL with sand is chosen as $\phi' = 25°$, which is the same as that of the clayey silty sand used by Holtz and Gibbs (1956), but is smaller than that silty sand of the used by Irfan and Tang (1993), that is, 37.1°. The effective cohesion of the CL with sand is chosen as $c' = 5$ kPa, a recommended value for colluvial soils (Irfan and Tang, 1993). Using $c' = 5$ kPa, $\phi' = 25°$, $P_c = 35\%$, $\alpha = 3.5°$ per 10%, $P_u = 72\%$, the shear strength parameters of the four soils are estimated by Equations 6.1 and 6.2, as listed in Table 6.2. The angle indicating the increase rate in shear strength related to soil suction ϕ^b is taken as two thirds of the effective angles of internal friction.

Table 6.2 Soil parameters used for stability analysis

Group symbols (ASTM D2487)	*Gravel proportion (%)*	*Dry density (g/cm³)*	*Effective cohesion c′ (kPa)*	*Effective angle of internal friction φ′ (°)*	*Angle related to matric suction φb (°)*
SM with gravel	44	1.81	5	28.1	18.7
SC with gravel	29	1.62	5	25.0	16.7
Sandy ML	15	1.45	5	25.0	16.7
CL with sand	0	1.32	5	25.0	16.7

6.4 AN EXAMPLE

6.4.1 Methodology

Evans and King (1998) suggested that natural terrain deposits with a slope angle between 25° and 50° have a high degree of susceptibility to shallow failure. The slope simulated in this study is 16 m high at an inclination of 38.6° (inclination 1H: 0.8V). The slope geometry is shown in Figure 6.4. To simplify the problem, it is assumed that the slope is composed of a homogenous and isotropic colluvial soil. The left and right edges are located at a distance of 30 m from the slope crest and the toe, respectively, to avoid any influence of the boundary conditions on the seepage process within the slope area.

A commercially available software, Seep/W (Geo-slope International Ltd., 2012a), is used to simulate the infiltration process. A finite element mesh is designed to analyze the infiltration of water in the slope under rainstorms as shown in Figure 6.5. In the simulation, the required SWCC and permeability function of the four colluvial soils are inputted manually. The hysteresis of the hydraulic property functions is not considered in the simulation.

For better comparison of the hydraulic behavior of the slopes composed of the four soils, all the simulations start from a hydrostatic state corresponding to the same initial water table (as shown in Figure 6.4). It is assumed that the base soil has very low permeability

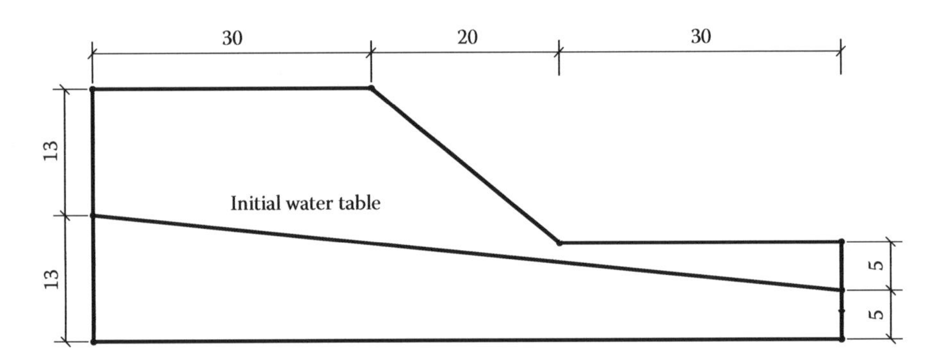

Figure 6.4 Slope geometry and initial water table (unit: meter).

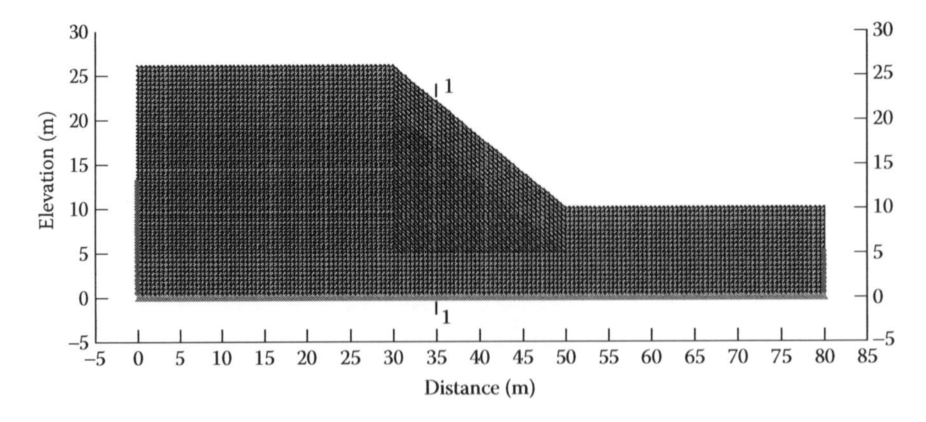

Figure 6.5 Finite element mesh and boundary conditions.

and therefore a zero nodal flow boundary is applied along the base. Along the left and right edges, constant water head boundaries are applied to define the initial groundwater level and to produce the initial pore-water pressure profile. The water heads are 13 m at the nodes on the left edge and 5 m at the nodes on the right edge.

Maximum rolling 24-hour rainfall (MR24HR) has been adopted by the GEO in establishing relations with the failure frequency of man-made slopes (Yu, 2004). The current slope warning criteria are based on the use of MR24HR (Pun et al., 1999; Chan et al., 2003). A rigorous assessment of the relationship between rainfall and slope failure in the period from 1985 to 2000 was carried out by Evans (1997) and Ko (2003). Their assessment shows that the failure frequency of natural slopes increases exponentially with the normalized MR24HR, which is defined as the ratio of the MR24HR and the location-specific mean annual rainfall. Evans and Yu (2001) calculated the MR24HR based on the data from 100 years of observations in Hong Kong, and found that the MR24HR of 10-year return period is 402 mm; the MR24HR of 50-year return period is 548 mm; the MR24HR of 100-year return period is 609 mm. The Geotechnical Manual for Slopes (GEO, 1984) recommends that slope design and stability assessments be carried out for the groundwater conditions that would result from a rainfall event with a 10-year return period. In this example, the MR24HR of 10-year return period in Hong Kong is taken for numerical simulation. The rainfall event is uniformly distributed as 16.75 mm/h (i.e., 4.7×10^{-6} m/s) along the slope surface and lasts for 24 hours. After 24 hours, the rainfall is stopped. To allow the slope to drain and to study the recovery of negative pore water pressures, the seepage simulation lasts for 111 days. No evaporation is taken into consideration during the no-rain period afterwards.

6.4.2 Seepage analysis results

The four colluvial soils are classified as soils with high fines fractions (SC with gravel, sandy ML, CL with sand) and soils with high coarse fractions (SM with gravel). As the same initial water table and initial steady state, the initial pore water pressure contours are the same for these five soils and are shown in Figure 6.6.

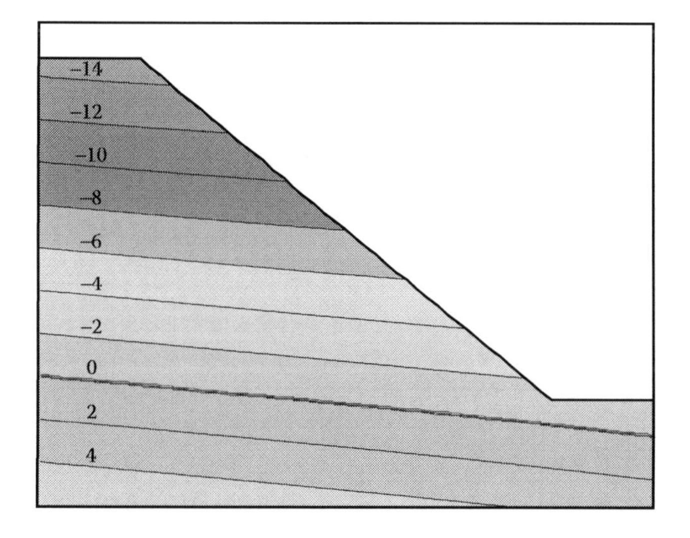

Figure 6.6 Initial pore-water pressure head (in meter) contours.

6.4.2.1 Slopes composed of soils with high fines fractions

During the rainfall, limited dissipation of pressure head is found in the SC with gravel slope, the sandy ML slope, and the CL with sand slope (as shown in Figures 6.7 through 6.9). Although the ground surface is saturated during the rainfall, pore water pressures change only in a limited zone near the ground surface during the rainfall. Because the rainfall intensity (4.7×10^{-6} m/s) is much larger than the saturated permeability of SC with gravel, sandy ML, and CL with sand, most water turn to runoff from the ground surface (i.e., 78% for SC with gravel, 91% for sandy ML, and 98% for CL with sand). Therefore, the dissipation rate

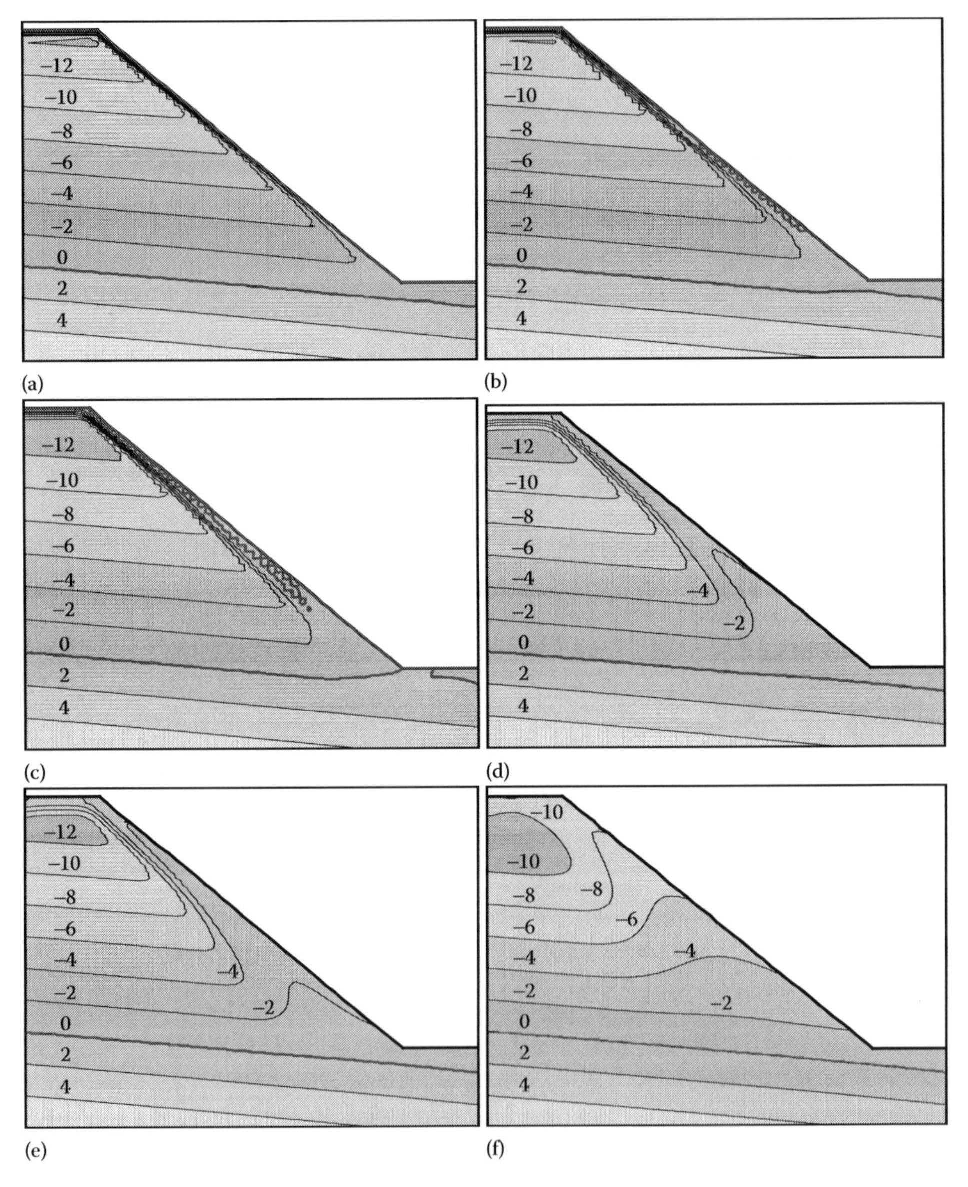

Figure 6.7 Evolution of pressure head contours in SC with gravel slope (unit: m): (a) Time = 0.1 day (during rainfall), (b) Time = 0.5 day (during rainfall), (c) Time = 1 day (end of rainfall), (d) Time = 6 days (after rainfall), (e) Time = 11 days (after rainfall), and (f) Time = 111 days (after rainfall).

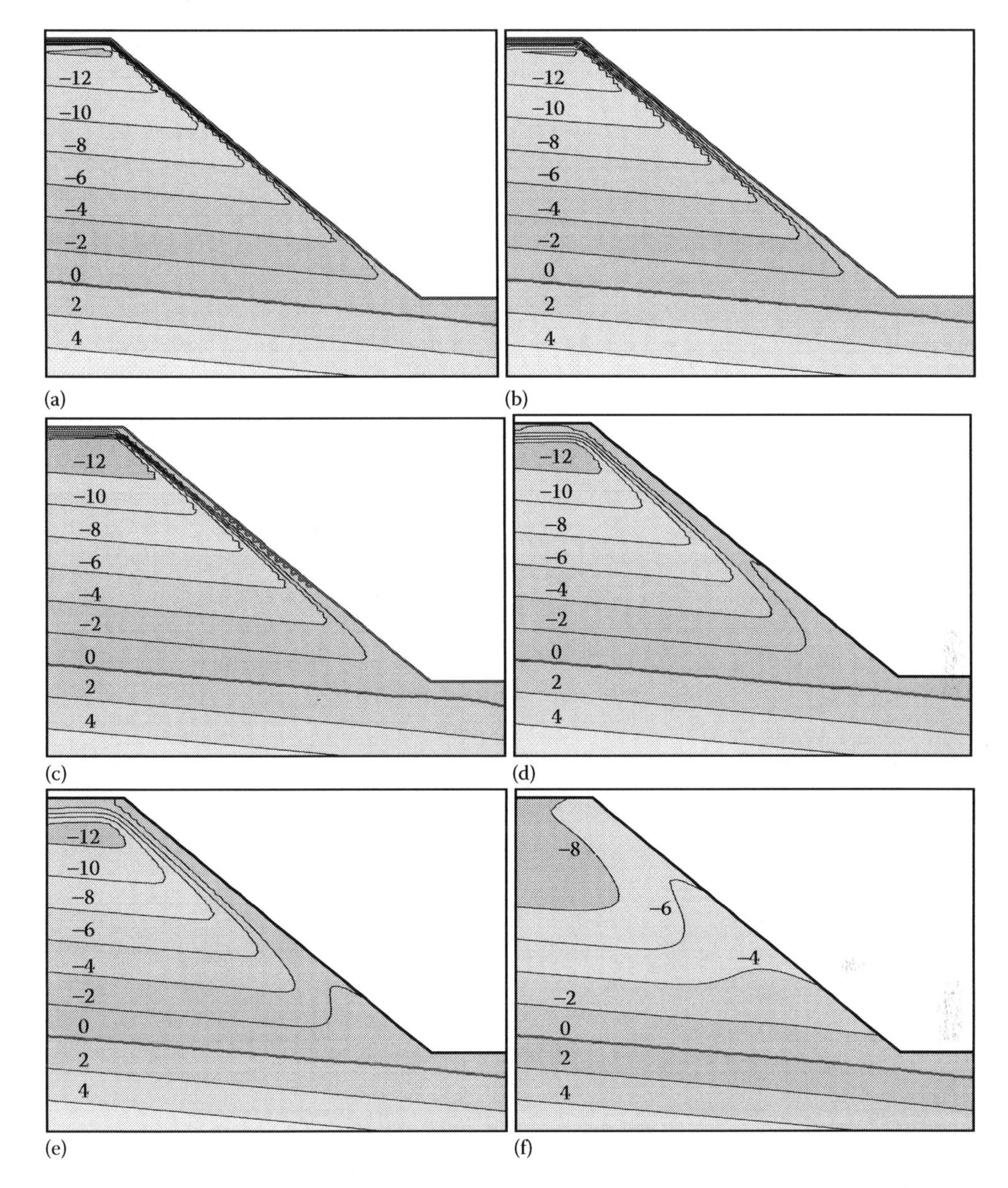

Figure 6.8 Evolution of pressure head contours in sandy ML slope (unit: m): (a) Time = 0.1 day (during rainfall), (b) Time = 0.5 day (during rainfall), (c) Time = 1 day (end of rainfall), (d) Time = 6 days (after rainfall), (e) Time = 11 days (after rainfall), and (f) Time = 111 days (after rainfall).

of suctions in these three slopes are mainly controlled by the soil saturated permeability and time of rainfall duration.

In Figure 6.7c, the water table rises at the toe of SC with gravel during the rainfall. This demonstrates that the water infiltrated into the soil during the rainfall event (i.e., 1 day) and saturated the soil at slope toe. Because the saturated permeability of SC with gravel (i.e., 1×10^{-6} m/s) is high, most water infiltrate into the soil and saturated the soil at slope toe. On the contrary, no rising of ground water table is found in the sandy ML slope and in the CL with sand slope (Figures 6.8 and 6.9). The saturated permeability values of sandy ML (3×10^{-7} m/s) and CL with sand (1×10^{-7} m/s) are much lower than that of SC with gravel.

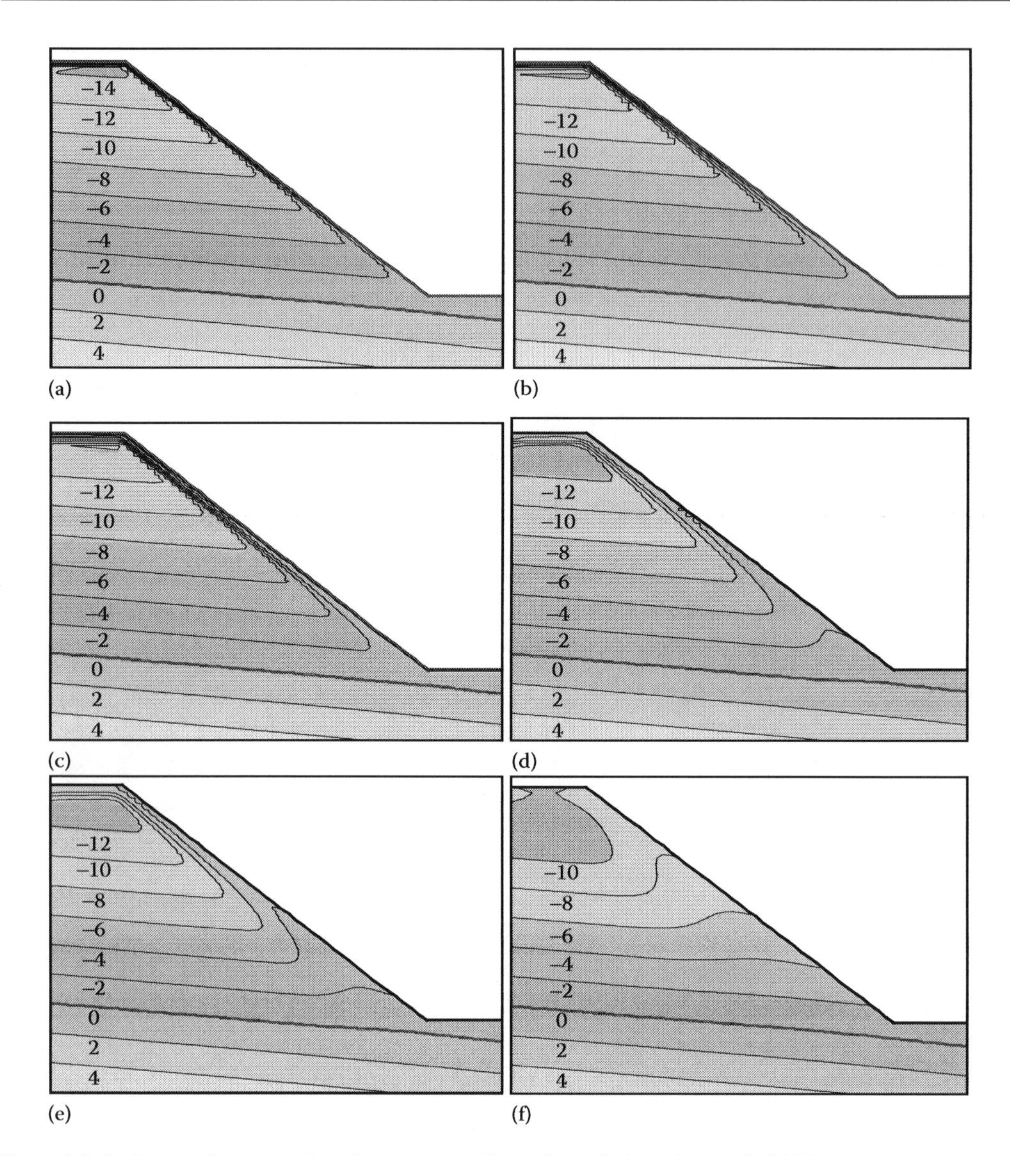

Figure 6.9 Evolution of pressure head contours in CL with sand slope (unit: m): (a) Time = 0.1 day (during rainfall), (b) Time = 0.5 day (during rainfall), (c) Time = 1 day (end of rainfall), (d) Time = 6 days (after rainfall), (e) Time = 11 days (after rainfall), and (f) Time = 111 days (after rainfall).

Therefore, the water infiltrated into soil is much fewer and most water runs away along the ground surface. After rainfall, the dissipation of pressure head in the SC with gravel slope, sandy ML slope and CL with sand slope continues for months due to the low unsaturated permeability of these three soils.

The pore-water profiles at different times along cross section 1 (referring to Figure 6.5) are shown in Figures 6.10 through 6.12. The evolution of pressure head profiles in these three slopes, the SC with gravel slope, the sandy ML slope, and the CL with sand slope, is similar during and after the rainfall. During the rainfall, the negative pore water pressures in soils within 1 m depth decrease gradually. After rainfall, the soils within 1 m depth show gradually increase of negative pore water pressure, whereas, the negative pore water pressures in soils between 1 m depth and 3 m depth continue to decrease.

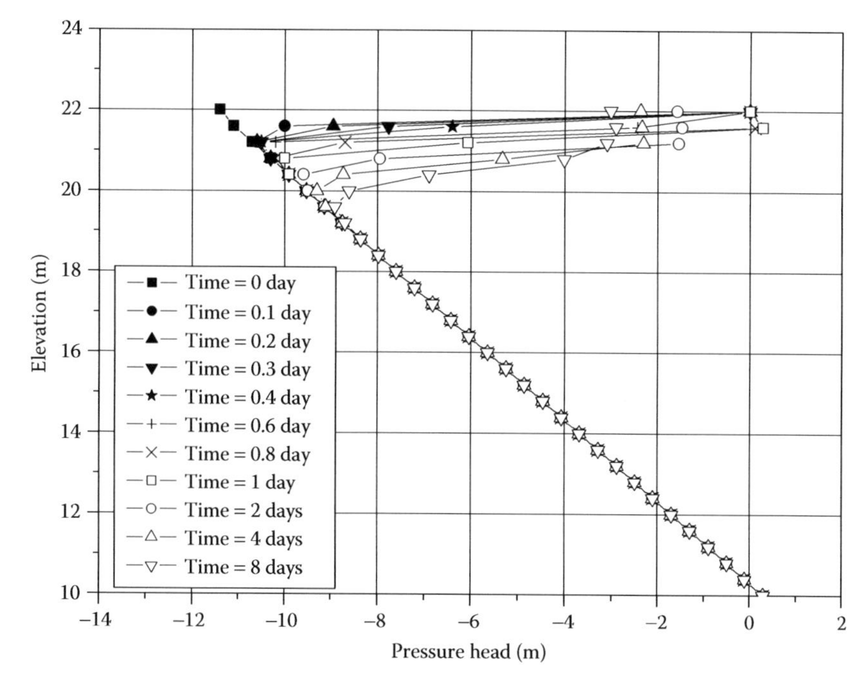

Figure 6.10 Pressure head profiles along cross section I in SC with gravel slope.

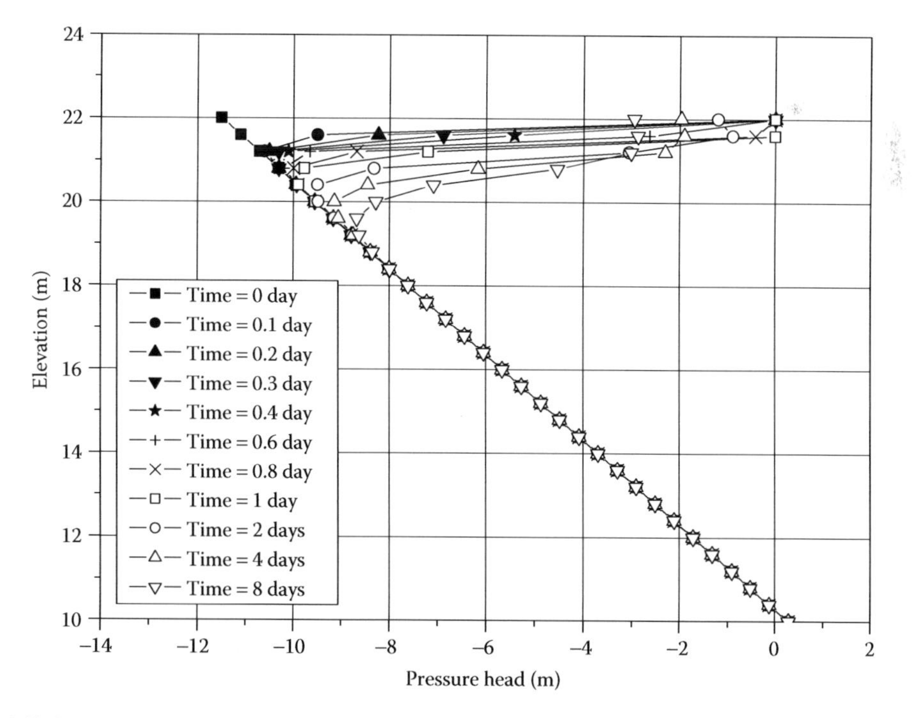

Figure 6.11 Pressure head profiles along cross section I in sandy ML slope.

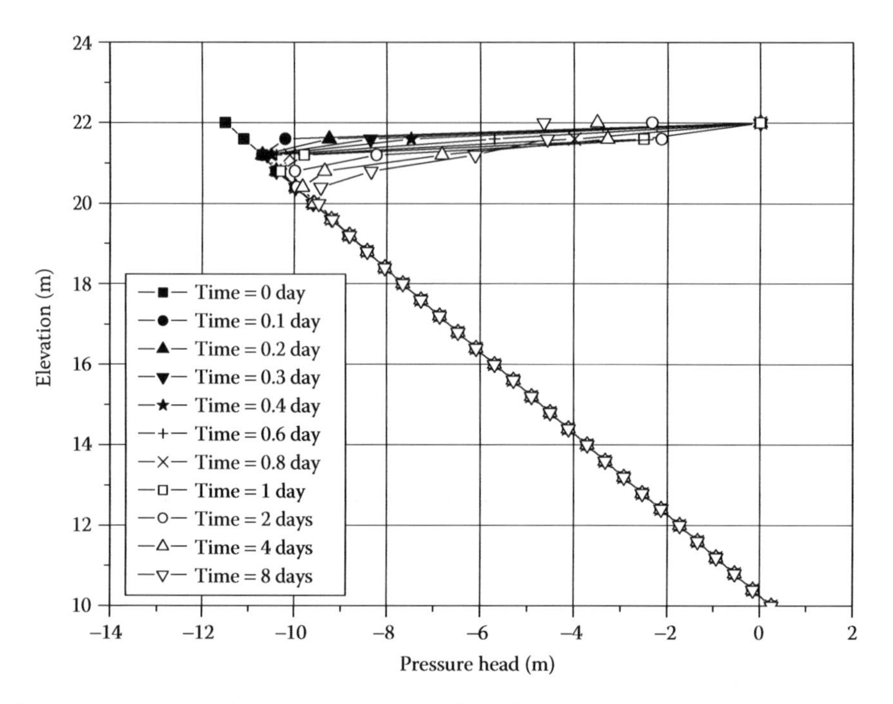

Figure 6.12 Pressure head profiles along cross section I in CL with sand slope.

Table 6.3 Risk of failure for colluvium slopes

Slope soil	*Saturated permeability*	*Unsaturated permeability in slope zone*[a]	*Controlling parameter of inflow*	*Influenced depth after 1-day rainfall*[b]	*Rising of groundwater table*	*Potential slip mode*
SM with gravel	High	High to low	Rainfall intensity	Whole slope	Large	Shallow seated
SC with gravel	Medium[c]	Medium to low	Saturated permeability	>3 m	Some	Nil
Sandy ML	Low[d]	Medium to low	Saturated permeability	~3 m	Limited	Nil
CL with sand	Low	Medium to low	Saturated permeability	~2.5 m	Limited	Nil

[a] Corresponding to suctions in 0~100 kPa.
[b] An average value.
[c] The order of permeability is similar to the average 24-hour rainfall intensity (therefore, about 10^{-6} m/s).
[d] The order of permeability is 10 times smaller than the average 24-hour rainfall intensity (therefore, about 10^{-7} m/s).

For colluvial soils with high fines fraction, limited pressure head changes occur during the rainfall. The inflow is mainly controlled by the saturated permeability and the rainfall duration. The pore water pressure changes after rainfall are mainly controlled by the unsaturated permeability of the soils in the slope. The hydraulic behavior of slopes is summarized in Table 6.3.

6.4.2.2 Slopes composed of colluvial soils with high coarse fractions

As shown in Figure 6.13, the dissipating rate of suctions is striking in the SM with gravel slope. A saturated zone forms below the ground surface at the beginning of the rainfall and expands continuously during the rainfall. A thin layer below the ground surface becomes quasi-saturated (pressure head smaller than 2 m) due to the rainfall infiltration (see Figure 6.13b). Then, the quasi-saturated zone expands rapidly. After 0.3 day (Figure 6.13c), the saturated zone merges with the ground water table.

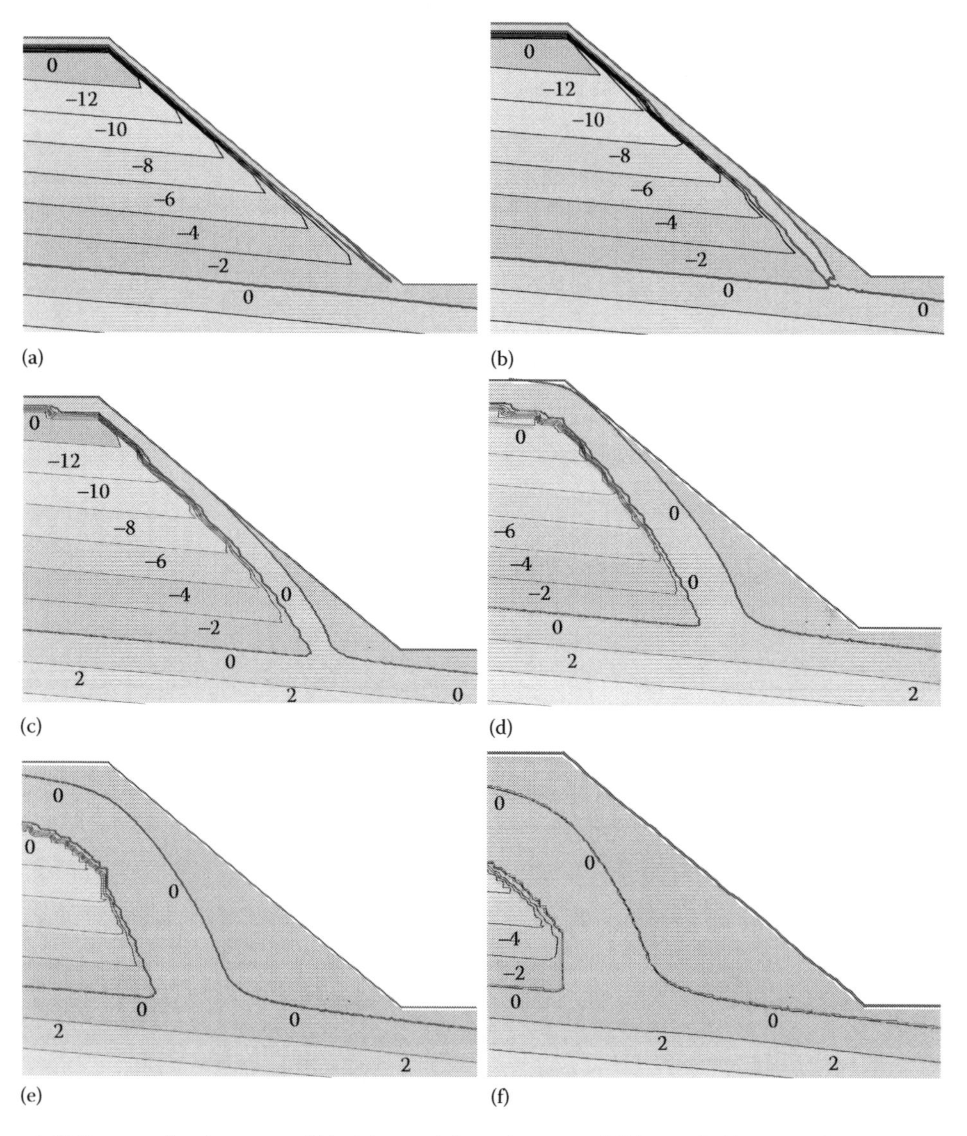

Figure 6.13 Pressure head contours SM with gravel slope during rainfall (unit: m): (a) Time = 0.1 day, (b) Time = 0.2 day, (c) Time = 0.3 day, (d) Time = 0.5 day, (e) Time = 0.8 day, and (f) Time = 1 day.

After rainfall, the saturated zone continues to move down (Figure 6.14). After 6 days, the saturated zone moves to the position of the initial water table (i.e., a straight line in Figure 6.14b). Drainage from the quasi-saturated zone continues for a long period and shows a recovery of suctions to the hydrostatic steady state. After 21 days (Figure 6.14d), the soil at the top surface has been recovered to pressure head values above 2 m.

The profiles of pore-water pressure head at different times along cross section 1 in the SM with gravel slope (referring to Figure 6.4) are shown in Figure 6.15. After about 0.6 day rainfalls (Figure 6.15), the negative pressure head above the initial water table dissipates and remains at values close to 0 during the rainfall. After rainfall, the pressure head above the initial water

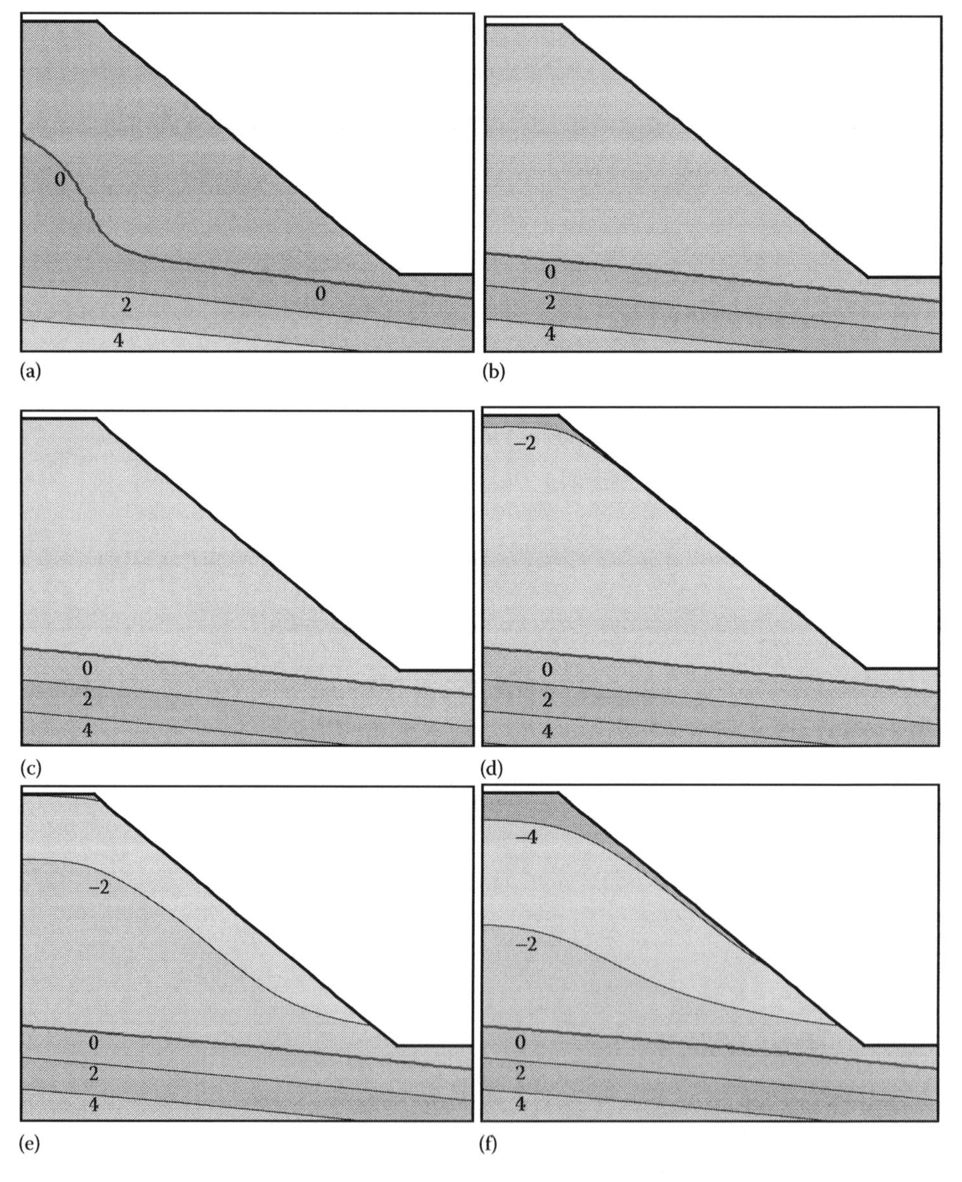

Figure 6.14 Pressure head contours in SM with gravel slope after rainfall (unit: m): (a) Time = 2 days, (b) Time = 6 days, (c) Time = 11 days, (d) Time = 21 days, (e) Time = 61 days, and (f) Time = 111 days.

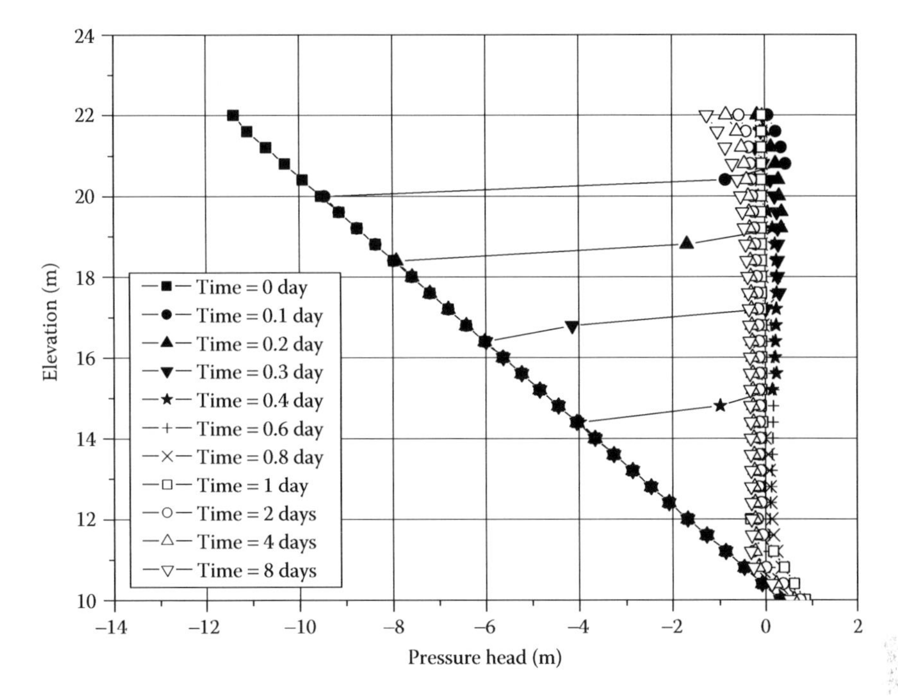

Figure 6.15 Pressure head profiles along section I in SM with gravel slope.

table recovers to the hydrostatic state gradually. For the SM with gravel slope, the striking wetting speed during rainfall and the remarkable drainage speed after rainfall are mainly contributed from the high unsaturated permeability of SM with gravel.

For SM with gravel, the inflow during rainfall is controlled by the rainfall intensity. The advancing speed of these saturated/quasi-saturated zones is substantially influenced by especially the unsaturated permeability of the soil in that zone (i.e., soils in a suction range of 0–100 kPa).

After rainfall, the saturated/quasi-saturated zones move down continuously (Figures 6.13 and 6.14a) until merging with the initial ground water table (Figure 6.14b). Then, the quasi-saturated zones drain slowly and recover gradually to the hydrostatic state (Figure 6.14c-f). This process is substantially controlled by the unsaturated permeability. The permeability of the soils with high fine fractions (SC with gravel, sandy ML, and CL with sand) in a negative pressure head range of 0.3–10 m are much higher than these of SM with gravel (i.e., about 5 times for SM with gravel, see Figure 8.3), the recovery process of negative pressure heads in the slopes composed of the soil with high fine fractions is much faster than that in the slope composed of SM with gravel. For soils with permeability in the order of 10^{-7} m/s, this process would continue for weeks. For soils with permeability in the order of 10^{-8} m/s, this process would continue for months. For soils with permeability in the order of 10^{-9} m/s, this process would continue for years. The hydraulic behavior of slopes composed of soils with high coarse fractions upon rainfall is also summarized in Table 6.3.

6.4.3 Stability analysis results

The stability of these slopes is analyzed using Slope/W (Geo-Slope International Ltd., 2012b). The settings for center of slip surface and radius of slip surface are shown in Figure 6.16. The potential slip surfaces during and after the rainfall as shown in Figures 6.17 and 6.18, respectively.

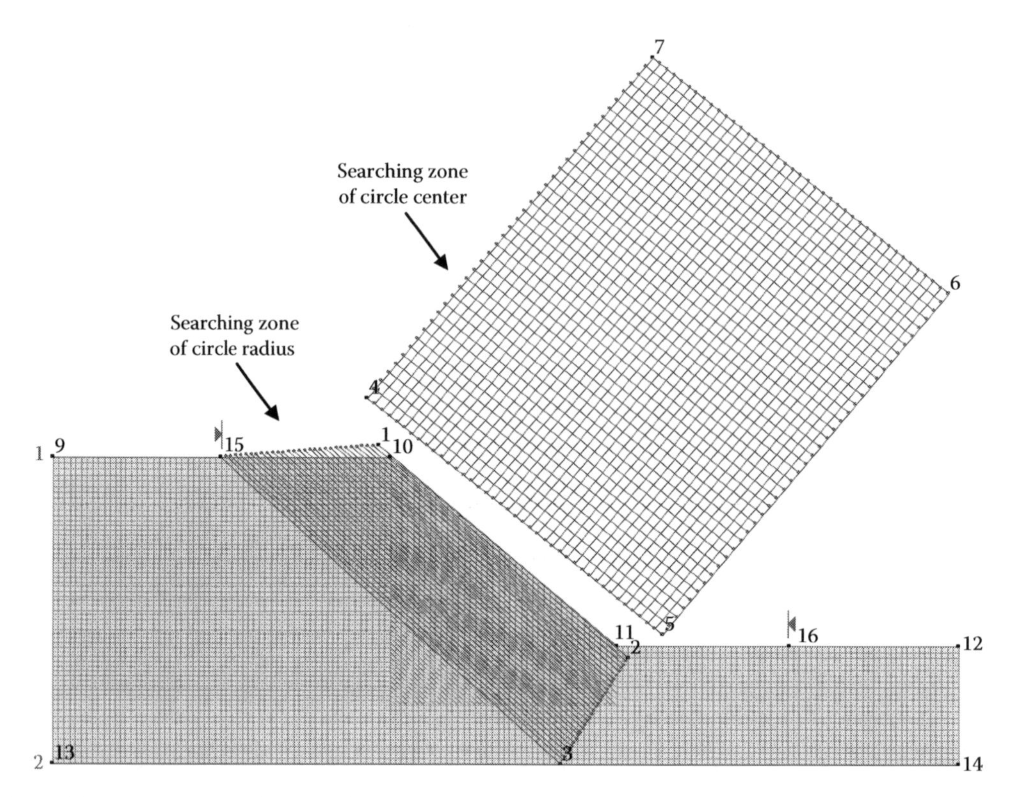

Figure 6.16 Searching critical slip surface.

No shallow failure or deep-seated failure is found for the slopes composed of soil with high fine fractions (Figure 6.17). Because the pore water pressure changes in these slopes are limited during and after the rainfall, the changes of factor of safety are also limited. It is anticipated, however, the factor of safety of these slopes may decrease significantly in the case of a sustained rainfall event.

The critical slip patterns in the SM with gravel slope during rainfall are shown in Figure 6.18. The factor of safety of the critical slip surface decreases rapidly from 1.440 to 0.943 after 0.5 day of rainfall. The critical slip surface is shallow-seated. The shallow failure zone expands quickly when the rainfall continues and finally develops into a global failure.

The response of these colluvium slopes with different coarse fractions during rainfall infiltration is quite different and can be summarized as:

1. The slopes composed of colluviums with high coarse fractions are very sensitive to rainfall. A saturated/quasi-saturated zone is produced at the ground surface at the beginning of rainfall and the zone expands rapidly during the rainfall. The inflow in such slopes is mainly controlled by the rainfall intensity. Both the saturated permeability and the unsaturated permeability of soils in the slopes (normally corresponding to a suction range of 0–100 kPa) have important influences on the advancing speed of the saturated/quasi-saturated zone.

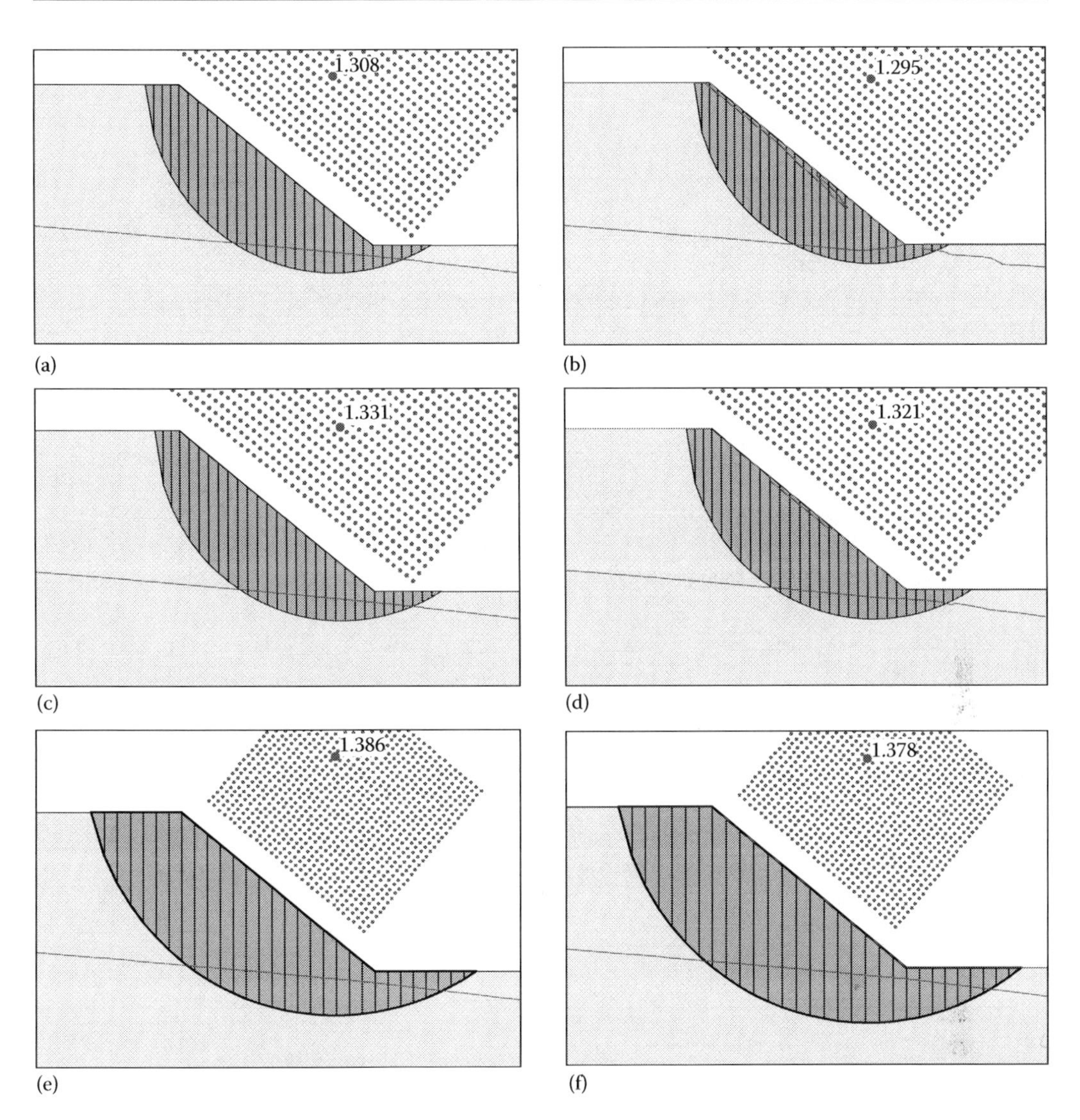

Figure 6.17 Critical slip patterns of slopes composed of soils with high fines fractions: (a) SC with gravel, time = 0 day (Fs = 1.308); (b) SC with gravel, time = 1 day (F_s = 1.295); (c) Sandy ML, time = 0 day (F_s = 1.331); (d) Sandy ML, time = 1 day (F_s = 1.321); (e) CL with sand, time = 0 day (F_s = 1.386); and (f) CL with sand, time = 1 day (F_s = 1.378).

2. The negative pore-water pressures decrease in the saturated/quasi-saturated zone, which may induce a considerable decrease of soil shear strength and hence a considerable decrease of factor of safety. Shallow-seated failure may occur in the saturated/quasi-saturated zone and expands continuously with the rainfall time.
3. The colluvium slopes with low coarse fractions are not sensitive to an intense rainfall event with short duration. Only limited dissipation of negative pore-water pressure occurs in a thin layer near the ground surface. The inflow in such slopes is mainly controlled by the saturated permeability of soils and the rainfall duration.

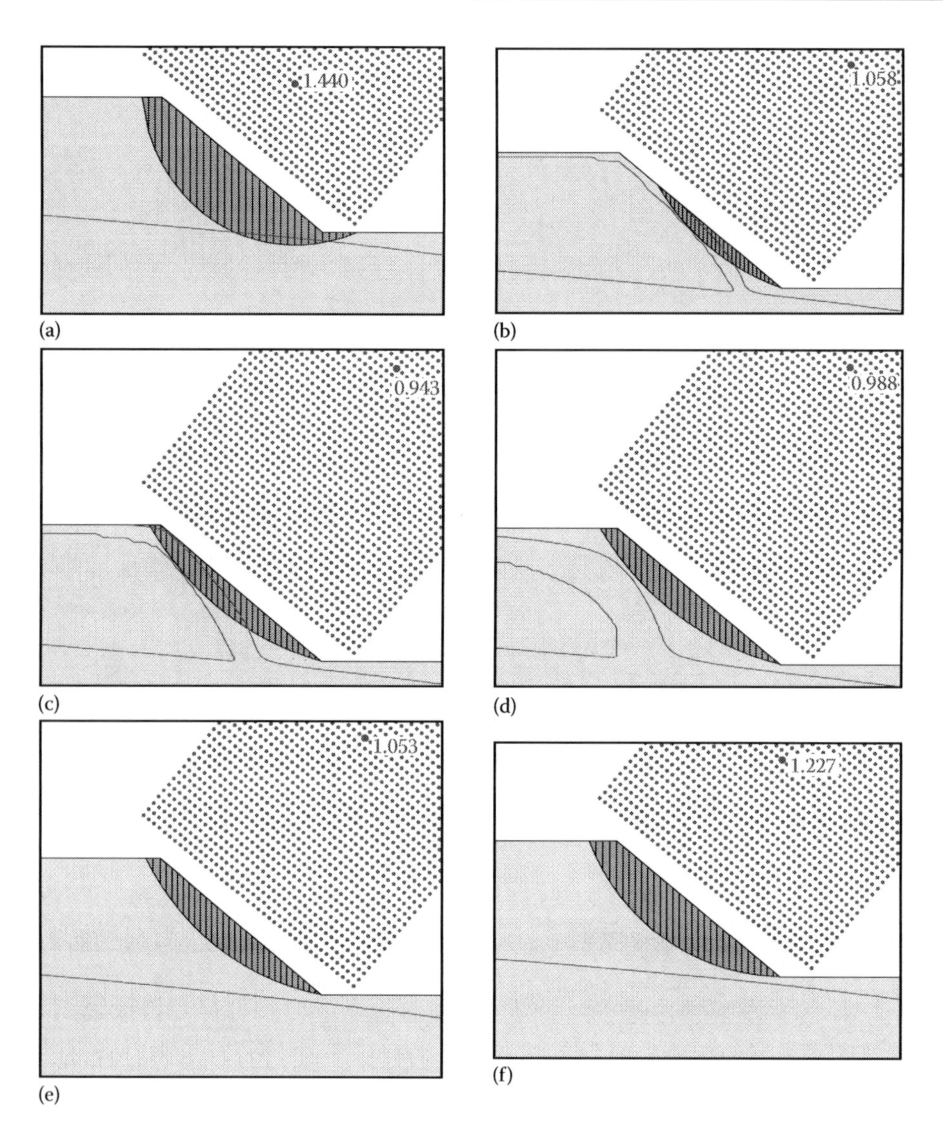

Figure 6.18 Critical slip patterns of SM with gravel slope: (a) Time = 0 day (F_s = 1.440), (b) Time = 0.3 day (F_s = 1.058), (c) Time = 0.5 day (F_s = 0.943), (d) Time = 1 day (F_s = 0.988), (e) Time = 11 days (F_s = 1.053), and (f) Time = 111 days (F_s = 1.227).

REFERENCES

Brand, E. W. (1981). Some thoughts on rain-induced slope failure. *Proceedings of the 10th International Conference on Soil Mechanics and Foundation Engineering*, Vol. 3. London, pp. 374–376.

Burger, C. A., and Shackelfor, C. D. (2001). Evaluating dual porosity of pelletized diatomaceous earth using bimodal soil-water characteristic curve functions, *Canadian Geotechnical Journal*, 38(1), 53–66.

Chan, W. L., Ho, K. K. S., and Sun, H.W. (1998). Computerised database of slopes in Hong Kong. *Proceedings of the Annual Seminar on Slope Engineering in Hong Kong*, Hong Kong. A.A. Balkema Publishers, Rotterdam, the Netherlands, pp. 213–220.

Dane, J. H., and Klute, A. (1977). Salt effects on the hydraulic properties of a swelling soil. *Soil Science Society of America Journal*, 41(6), 1043–1049.

Evans, N. C. (1997). *Preliminary Assessment of the Influence of Rainfall on Natural Terrain Slope Initiation* (Discussion Note No. DN 1/97). Geotechnical Engineering Office, Hong Kong.

Evans, N. C., and King, J. P. (1998). *The Natural Terrain Slope Study—Debris Avalanche Susceptibility* (Technical Note No. TN1/98). Geotechnical Engineering Office, Hong Kong, 96p.

Evans, N. C., and Yu, Y. F. (2001). *Regional Variation in Extreme Rainfall Values* (GEO Report No.115). Geotechnical Engineering Office, Hong Kong.

Feuerharmel, C., Gehling, W. Y. Y., and Bica A. V. D. (2006). The use of filter-paper and suction-plate methods for determining the soil-water characteristic curve of undisturbed colluvial soils. *Geotechnical Testing Journal*, 29(5), 419–425.

Fredlund, D. G., and Xing, A. Q. (1994). Equations for the soil-water characteristic curve. *Canadian Geotechnical Journal*, 31, 521–532.

Geo-slope International Ltd. (2012a). *Seep/W for Finite Element Seepage Analysis, User's Guide*. Geo-slope Ltd., Calgary, Alberta, Canada.

Geo-slope International Ltd. (2012b). *Slope/W for Slope Stability Analysis, User's Guide*. Geo-slope Ltd., Calgary, Alberta, Canada.

Geotechnical Engineering Office (GEO). (1984). *Geotechnical Manual for Slopes*, 2nd Edition, pp. 1–302.

He, K. Q., and Wang, S. J. (2006). Double-parameter threshold and its formation mechanism of the colluvial landslide: Xintan landslide, China. *Environmental Geology*, 49(5), 696–707.

Holtz, W. G., and Gibbs, H. J. (1956). Triaxial shear tests on pervious gravelly soils. *Journal of the Soil Mechanics and Foundations Division*, ASCE, 82, 1–22.

Irfan, T. Y., and Tang, K. Y. (1993). *Effect of the Coarse Fractions on the Shear Strength of Colluvium* (GEO Report No. 23). Geotechnical Engineering Office, Hong Kong.

Ko, F. W. Y. (2003). *Correlation Between Rainfall and Natural Terrain Slope Occurrence in Hong Kong* (GEO Report No. 168). Geotechnical Engineering Office, Hong Kong.

Leroueil, S., LeBihan, J., Sebaihi, S., and Alicescu, V. (2002). Hydraulic conductivity of compacted tills from northern Quebec. *Canadian Geotechnical Journal*, 39(5), 1039–1049.

Li, X. (2009). *Dual-porosity Structure and Bimodal Hydraulic Property Functions of Coarse Granular Soils*. Ph.D Thesis, the Hong Kong University of Science and Technology, Hong Kong.

Li, X., Zhang, L. M., and Fredlund, D. G. (2009). Wetting front advancing column test for measuring unsaturated hydraulic conductivity. *Canadian Geotechnical Journal*, 46(12), 1431–1445.

Loiselle, A. A., and Hurtubise, J. E. (1976). Properties and behaviour of till as construction material. In: Legget, R. F. (Ed.). *Glacial Till an Interdisciplinary Study*. Royal Society of Canada, Ottawa, pp. 346–363.

Mair, R. J. (1993). Developments in geotechnical engineering research. *Proceedings of the Institution of Civil Engineers*, 93, 30–37.

Pare, J. J., Ares, R., Cabot, L., and Garzon, M. (1982). Large scale permeability and filter tests at LG-3. *Proceedings of the 14th Congress on Large Dams*, Rio de Janeiro, 55(7), 103–121.

Pun, W. K., Wong, A. C. W., and Pang, P. L. R. (1999). *Review of Landslip Warning Criteria* 1998/1999 (Special Project Report SPR 4/99). Geotechnical Control Office, Hong Kong. http://www.cedd.gov.hk/eng/publications/geo_reports/index.htm.

Shakoor, A., and Smithmyer, A. J. (2005). An analysis of storm-induced landslides in colluvial soils overlying mudrock sequences, southeastern Ohio, USA. *Engineering Geology*, 78(3), 257–274.

Vallejo, L. E., and Mawby, R. (2000). Porosity influence on the shear strength of granular material-clay mixtures. *Engineering Geology*, 58(2), 125–136.

Zhang, L. M., and Li, X. (2010). Micro-porosity structure of coarse granular soils. *Journal of Geotechnical and Geoenvironmental Engineering*, 136(10), 1425–1436.

Part II

Probabilistic assessment

Chapter 7

Reliability analysis of slope under rainfall

7.1 INTRODUCTION

The factor of safety of a geotechnical structure is often estimated by the ratio of resistance and load. For the slope stability problem, the safety factor of a slope can be calculated by the ratio of shear resistances to shear stresses along a given slip surface using a limit equilibrium method. However, the resistance and load are subjected to uncertainties, making the factor of safety of a slope also uncertain. To apply the same nominal value of safety factor to the conditions that involves widely varying degrees of uncertainties may not be logical. Probabilistic approach and reliability analysis methods provide a rational way to account for the uncertainties from different sources and to estimate the probability of satisfactory performance in a systematic way. Extensive studies have been conducted on the probabilistic assessment of slope stability. The mean first-order reliability method (MFORM) is mainly used in early studies on slope reliability (e.g., Wu and Kraft, 1970; Tang et al., 1976; Vanmarcke, 1977; Li and Lumb, 1987; Liang et al., 1999; Christian et al., 1994). The advanced first-order reliability method (AFORM) is used to study both the location of the most critical slip surface and the reliability index of the most critical slip surfaces (e.g., Hassan and Wolff, 1999; Bhattacharya et al., 2003; Xue and Gavin, 2007). Sampling methods have been used for reliability analysis for a given slip surface (e.g., El-Ramly et al., 2002), or system reliability of slopes considering multiple slip surfaces (e.g., Ching et al., 2009; Zhang et al., 2011, 2013a), and reliability analysis of slopes considering the spatial variability of soil properties (e.g., Griffiths et al., 2009; Santoso et al., 2011; Li et al., 2015). The traditional response surface method (RSM), which can replace the computationally intensive deterministic models with an approximated less expensive surrogate model, has been developed to achieve computational efficiency (Li et al., 2011, 2015; Zhang et al., 2013b). In most of previous reliability studies of slope stability, limit equilibrium methods are generally used as a deterministic model. Complex failure mechanisms due to soil behavior such as progressive failure (Chowdhury et al., 1987; Gilbert and Tang, 1989; Liu et al., 2001) are usually considered by numerical models.

For reliability of rainfall-induced slope failures, the first-order reliability methods are most frequently adopted (Hu, 2000; Sivakumar Babu and Murthy, 2005; Penalba et al., 2009; Tan et al., 2013). Tung and Chan (2003), Zhang et al. (2005), Park et al. (2013), and Zhang et al. (2014) adopted the sampling methods.

In this chapter, the fundamental concept of reliability is introduced in Section 7.2. In Section 7.3, the reliability methods that have been commonly adopted in geotechnical engineering are briefly introduced together with example problems. The uncertainty of soil properties especially soil hydraulic properties are presented in Section 7.4. In Section 7.5, the effects of the uncertainty and correlation of soil hydraulic properties are illustrated. Reliability analysis of the 1976 Sau Mau Ping landslide is presented in Section 7.6 as an illustrative example of reliability analysis of rainfall-induced slope failure. Finally, the quantitative risk assessment of landslide is briefly introduced in Section 7.7.

7.2 FUNDAMENTAL CONCEPT OF RELIABILITY

The first step to evaluate the reliability of a system such as a slope is to decide the relevant uncertain input parameters in the geotechnical model, called the basic random variables X_i, and a relationship of these random variables with the performance of the system Z:

$$Z = g(\mathbf{X}) = g(X_1, X_2, \ldots, X_n) \tag{7.1}$$

where:

$\mathbf{X}$ is the vector of the random variables X_i
$g(\mathbf{X})$ is the performance function

If the failure of the structure is defined as $Z < 0$ and the safety of the structure is defined as $Z > 0$, then the limit state function is $Z = g(\mathbf{X}) = 0$, representing the boundary between the safe and unsafe regions. A performance function can be either an explicit or an implicit function of basic random variables. It is usually a nonlinear function for geotechnical structures. The probability of failure is the probability that the system cannot perform its intended function, presented by the integral:

$$p_f = P(Z < 0) = \int\int_{\mathbf{X} \in F} \cdots \int f_{\mathbf{X}}(x_1, x_2, \ldots, x_n) dx_1 dx_2 \cdots dx_n \tag{7.2}$$

where $f_{\mathbf{X}}(x_1, x_2, \ldots, x_n)$ is the joint probability density function (PDF) for $\mathbf{X}$. The integration is performed over the failure region F (i.e., $Z < 0$) as shown in Figure 7.1.

Reliability is defined as the probabilistic measure of the assurance of the performance (Ang and Tang, 1984). It is common to express reliability in a form of reliability index β, which is defined as the ratio of the mean to the standard deviation of the performance function:

$$\beta = \frac{\mu_Z}{\sigma_Z} \tag{7.3}$$

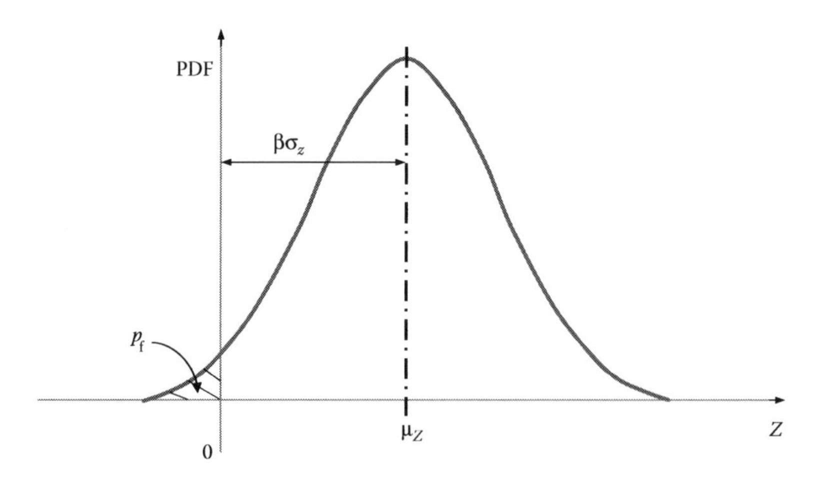

Figure 7.1 Definitions of reliability index and probability of failure (performance function is $Z = 0$).

where:

μ_Z is the mean of Z
σ_Z is the standard deviation of Z

If Z follows a normal distribution, the probability of failure is simply

$$p_f = P(Z < 0) = \Phi\left(\frac{0 - \mu_Z}{\sigma_Z}\right) = \Phi(-\beta) = 1 - \Phi(\beta) \quad (7.4)$$

where $\Phi(\cdot)$ is the cumulative distribution function (CDF) of the standard normal distribution.

It should be noted that Equations 7.2 and 7.3 are the definitions of the probability of failure and the reliability index. However, Equation 7.4 is only valid when the distribution of Z is normal. In the following example, we will illustrate the relationship between the probability of failure and reliability index for different distributions.

Example 7.1

Assume the performance function of a geotechnical structure is

$$Z = g(R, Q) = R - Q \quad (7.5)$$

where R is the resistance and Q is the load.

The mean and variance of Z can be obtained as

$$\mu_Z = \mu_R - \mu_Q \quad (7.6)$$

$$\sigma_Z^2 = \sigma_R^2 + \sigma_Q^2 - 2\rho_{RQ}\sigma_R\sigma_Q \quad (7.7)$$

where:

μ_R, μ_Q are mean values of R and Q, respectively
σ_R, σ_Q are standard deviations of R and Q, respectively
ρ is the correlation coefficient between R and Q

On the definition of reliability index in Equation 7.3, we can obtain

$$\beta = \frac{\mu_Z}{\sigma_Z} = \frac{\mu_R - \mu_Q}{\sqrt{\sigma_R^2 + \sigma_Q^2 - 2\rho_{RQ}\sigma_R\sigma_Q}} \quad (7.8)$$

Assuming R and Q are normal random variables, Z hence also follows a normal distribution. The probability of failure can be estimated as

$$p_f = P(Z < 0) = 1 - \Phi\left(\frac{\mu_R - \mu_Q}{\sqrt{\sigma_R^2 + \sigma_Q^2 - 2\rho_{RQ}\sigma_R\sigma_Q}}\right) \quad (7.9)$$

The performance function can also be defined in a form of factor of safety. The factor of safety is defined as

$$F_s = \frac{R}{Q} \quad (7.10)$$

Failure occurs when $F_s = 1$, and the performance function is defined by

$$Z = g(R, Q) = F_s - 1 \quad (7.11)$$

Then the reliability index is

$$\beta = \frac{\mu_Z}{\sigma_Z} = \frac{\mu_{F_s} - 1}{\sigma_{F_s}} \tag{7.12}$$

Assume that R and Q both follow lognormal distributions. Therefore, F_s also follows a log-normal distribution. The probability of failure is

$$p_f = P(Z < 0) = P(F_s < 1) = P(\ln F_s < 0) = \Phi\left(\frac{-\mu_{\ln F_s}}{\sigma_{\ln F_s}}\right) \tag{7.13}$$

where:

$\mu_{\ln F_s}$ is the mean of $\ln F_s$
$\sigma_{\ln F_s}$ is the standard deviation of $\ln F_s$

For a lognormal distribution F_s, the mean and standard deviation of $\ln F_s$ can be estimated as

$$\mu_{\ln F_s} = \ln \mu_{F_s} - \frac{1}{2}\ln(1 + \delta_{F_s}^2) \tag{7.14}$$

$$\sigma_{\ln F_s} = \sqrt{\ln(1 + \delta_{F_s}^2)} \tag{7.15}$$

where:

μ_{F_s} and σ_{F_s} are mean and standard deviation of F_s
δ_{F_s} is the coefficient of variation (COV) of F_s, $\delta_{F_s} = \sigma_{F_s}/\mu_{F_s}$

The mean of F_s can also be written as follows according to Equation 7.12

$$\mu_F = \frac{1}{(1 - \beta\delta_F)} \tag{7.16}$$

The probability of failure can then be expressed as a function of δ_{F_s} and the reliability index β (Baecher and Christian, 2003)

$$\begin{aligned} p_f &= \Phi\left(-\frac{\ln \mu_{F_s} - \frac{1}{2}\ln(1 + \delta_{F_s}^2)}{\sqrt{\ln(1 + \delta_{F_s}^2)}}\right) \\ &= \Phi\left(\frac{\ln\left(\sqrt{1 + \delta_{F_s}^2} / \mu_{F_s}\right)}{\sqrt{\ln(1 + \delta_{F_s}^2)}}\right) \\ &= \Phi\left[\frac{\ln\left(\sqrt{1 + \delta_{F_s}^2}\,(1 - \beta\delta_{F_s})\right)}{\sqrt{\ln(1 + \delta_{F_s}^2)}}\right] \end{aligned} \tag{7.17}$$

Figure 7.2 shows the reliability index and probability of failure for different distributions. As shown in the graph, for values of β less than 2, there is little difference between the results. The assumption of normal distribution is conservative for most of the range of reliability index. Based on the central limit theorem, the distribution of a sum of a large number of random variables converges to a normal distribution. Therefore, it is reasonable to assume a normal distribution for Z and use Equation 7.4 for estimation of probability of failure although this assumption will probably overestimate the probability of failure

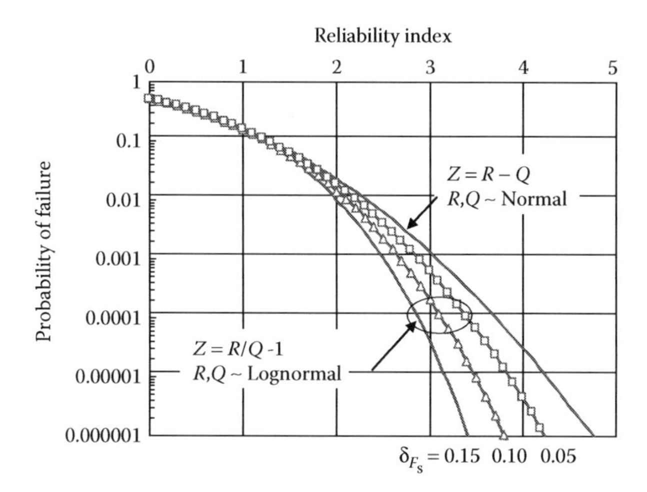

Figure 7.2 Relationship between reliability index and probability of failure for normal and lognormal distributions.

(Baecher and Christian, 2003). However, when β is much greater than 2, the discrepancy between the estimated probability of failure for different distributions can be significant.

7.3 RELIABILITY METHODS AND APPLICATIONS ON SLOPE STABILITY

In geotechnical engineering, the joint distribution of random variables $f_X(x_1, x_2, \ldots, x_n)$ is difficult to be obtained and the evaluation of the multiple integral (Equation 7.2) is extremely complicated. Therefore, approximation methods are proposed. Various reliability analysis methods were developed corresponding to the types and complexity of performance functions. In following sections, reliability methods that are commonly adopted in slope reliability analysis, that is, first-order reliability method (FORM), RSM, and sampling methods, will be introduced.

7.3.1 First-order reliability method

7.3.1.1 Mean first-order second-moment method

The mean first-order second-moment (MFOSM) method uses the first-order terms of a Taylor series expansion of the performance function to estimate the expected value and variance of the performance function. It is called a second-moment method because the variance is a second moment and is the highest order statistics used in this method. As the performance function is expanded at the mean values of the random variables, the method is named as the mean value first-order second-moment method.

A Taylor series expansion of the performance function about the mean values gives

$$Z = g(\mu_X) + \sum_{i=1}^{n} \left[\frac{\partial g}{\partial X_i} \right]_{\mu_X} (X_i - \mu_{X_i}) + \frac{1}{2} \sum_{i=1}^{n} \sum_{j=1}^{n} \left[\frac{\partial g}{\partial X_i} \right]_{\mu_X} \left[\frac{\partial g}{\partial X_j} \right]_{\mu_X} (X_i - \mu_{X_i})(X_j - \mu_{X_j}) + \cdots \tag{7.18}$$

where:

μ_X is the mean value vector of **X**

μ_{X_i} is the mean of each X_i

Truncating the series at the first-order terms, the first-order approximated mean and variance of Z are obtained as

$$\mu_Z \approx g(\mu_X) = g(\mu_{X_1}, \mu_{X_2}, \ldots, \mu_{X_n}) \tag{7.19}$$

$$\mathrm{var}(Z) \approx \sum_{i=1}^{n}\sum_{j=1}^{n}\left[\frac{\partial g}{\partial X_i}\right]_{\mu_X}\left[\frac{\partial g}{\partial X_j}\right]_{\mu_X} C(X_i, X_j) \tag{7.20}$$

where:

var[.] is the variance operator

C[.] represents the covariance operator

$C(X_i, X_j)$ is the covariance of X_i and X_j

If the random variables are uncorrelated, the variance of Z is

$$\mathrm{var}(Z) = \sigma_Z^2 = \sum_{i=1}^{n}\left[\left(\frac{\partial g}{\partial X_i}\right)_{\mu_X}\right]^2 \sigma_{X_i}^2 \tag{7.21}$$

where σ_{X_i} is the standard deviation of X_i.

The reliability index can be estimated as

$$\beta = \frac{\mu_Z}{\sigma_Z} = \frac{g(\mu_X)}{\sqrt{\sum_{i=1}^{n}\left[\left(\frac{\partial g}{\partial X_i}\right)_{\mu_X}\right]^2 \sigma_{X_i}^2}} \tag{7.22}$$

The MFOSM only utilizes the mean and variance of random variables but cannot consider the distribution of random variables. As will be shown in the following example, when the performance function is nonlinear, the estimated reliability index may subject to large errors. In addition, if the performance function is formulated in different forms, the reliability index obtained may also be different.

Example 7.2 A sliding slope failure with a circular slip surface

A soil slope is composed of two layers of soils (Figure 7.3). The slip surface is circular. The center of rotation is O and the radius of the circle is 10 m. W is the self-weight of the sliding mass. F_1 and F_2 are sliding resistances provided by soil layer no. 1 and 2, respectively. T is the surcharge load applied near the crest of the slope. Assume that W, T, F_1, and F_2 are independent normal random variables with mean and coefficient of variation as listed in Table 7.1. Calculate the probability of failure of the soil slope.

(1) Linear performance function

Based on the equilibrium of moment, the performance function of slope stability can be expressed as

$$g(\mathbf{X}) = 10F_1 + 10F_2 - 3W - 5T$$

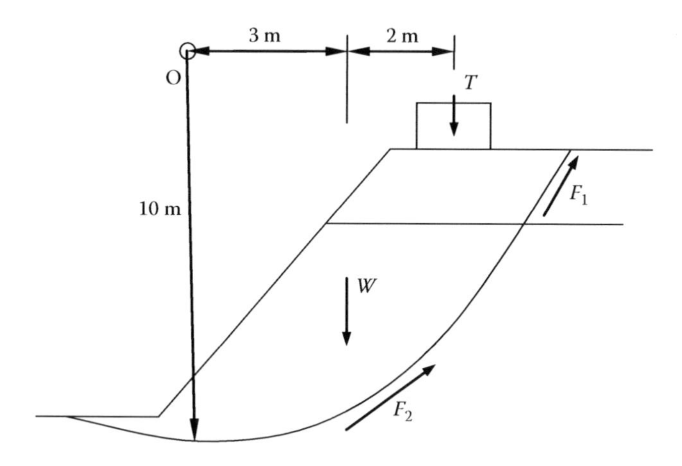

Figure 7.3 A soil slope failure with a circular slip surface.

Table 7.1 The statistics of random variables (Example 7.2)

Random variables	*Mean value (kN)*	*Coefficient of variation*
F_1	100	0.30
F_2	200	0.20
W	500	0.15
T	10	0.10

The derivatives of performance function at mean value are

$$\left(\frac{\partial g}{\partial F_1}\right)_{\mu X} = 10, \quad \left(\frac{\partial g}{\partial F_2}\right)_{\mu X} = 10, \quad \left(\frac{\partial g}{\partial W}\right)_{\mu X} = -3, \quad \left(\frac{\partial g}{\partial T}\right)_{\mu X} = -5$$

The standard deviations are

$$\sigma_{F_1} = 30, \quad \sigma_{F_2} = 40, \quad \sigma_W = 75, \quad \sigma_T = 1$$

According to Equation 7.19,

$$g(\mu_X) = 10\mu_{F_1} + 10\mu_{F_2} - 3\mu_W - 5\mu_T = 1450$$

Based on Equation 7.22, the reliability index of the slope stability is

$$\beta = \frac{g(\mu_X)}{\sqrt{\left[\left(\frac{\partial g}{\partial F_1}\right)_{\mu X}\right]^2 \sigma_{F_1}^2 + \left[\left(\frac{\partial g}{\partial F_2}\right)_{\mu X}\right]^2 \sigma_{F_2}^2 + \left[\left(\frac{\partial g}{\partial W}\right)_{\mu X}\right]^2 \sigma_W^2 + \left[\left(\frac{\partial g}{\partial T}\right)_{\mu X}\right]^2 \sigma_T^2}} = 2.644$$

The probability of failure is

$$p_f = \Phi(-\beta) = 0.0041$$

The MATLAB® code of the example is as follows:

```
% Example 7.2 % linear performance function
clear; clc;
mux = [100;200;500;10];
covx = [0.3;0.2;0.15;0.10];
sigmax = mux.*covx;
mug = 10*mux(1)+10*mux(2)-3*mux(3)-5*mux(4);
dgdx = [10;10;-3;-5];
beta = mug/norm(dgdx.*sigmax);
pf = normcdf(-beta,0,1)
```

(2) Nonlinear performance function

Define the factor of safety of sliding failure for the slope as

$$F_s = \frac{10F_1 + 10F_2}{3W + 5T}$$

Then, the performance function is

$$g(\mathbf{X}) = \frac{10F_1 + 10F_2}{3W + 5T} - 1$$

The derivatives at the mean value are

$$\left(\frac{\partial g}{\partial F_1}\right)_{\mu_X} = \left(\frac{10}{3W + 5T}\right)_{\mu_X} = 0.0065$$

$$\left(\frac{\partial g}{\partial F_2}\right)_{\mu_X} = \left(\frac{10}{3W + 5T}\right)_{\mu_X} = 0.0065$$

$$\left(\frac{\partial g}{\partial W}\right)_{\mu_X} = \left(-\frac{(10F_1 + 10F_2)\times 3}{(3W + 5T)^2}\right)_{\mu_X} = -0.0037$$

$$\left(\frac{\partial g}{\partial T}\right)_{\mu_X} = \left(-\frac{(10F_1 + 10F_2)\times 5}{(3W + 5T)^2}\right)_{\mu_X} = -0.0062$$

The reliability index is

$$\beta = \frac{g(\mu_X)}{\sqrt{\left[\left(\frac{\partial g}{\partial F_1}\right)_{\mu_X}\right]^2 \sigma_{F_1}^2 + \left[\left(\frac{\partial g}{\partial F_2}\right)_{\mu_X}\right]^2 \sigma_{F_2}^2 + \left[\left(\frac{\partial g}{\partial W}\right)_{\mu_X}\right]^2 \sigma_W^2 + \left[\left(\frac{\partial g}{\partial T}\right)_{\mu_X}\right]^2 \sigma_T^2}} = 2.187$$

which corresponds to the probability of failure

$$p_f = \Phi(-\beta) = 0.0144$$

The MATLAB code of the example with the nonlinear performance function is as follows:

```
% Example 7.2 %
% nonlinear performance function %
clear; clc;
mux = [100;200;500;10];
covx = [0.3;0.2;0.15;0.10];
sigmax = mux.*covx;
mug = (10*mux(1)+10*mux(2))/(3*mux(3)+5*mux(4))-1;
dgdx = [10/(3*mux(3)+5*mux(4));10/(3*mux(3)+5*mux(4));
-3*(10*mux(1)+10*mux(2))/(3*mux(3)+5*mux(4))^2;-5*(10*mux(1)+10*mux(2))/
(3*mux(3)+5*mux(4))^2];
beta = mug/norm(dgdx.*sigmax)
pf = normcdf(-beta,0,1)
```

As shown in Example 7.2, the reliability of a slope can be different when the performance function is formulated in different forms.

7.3.1.2 Advanced first-order second-moment method

The MFOSM method has the following basic shortcomings: (1) if the performance function is nonlinear and the linearization takes place at the mean values, errors may be introduced by neglecting higher order terms; (2) the method fails to be invariant to different equivalent formations of the same problem; (3) MFOSM cannot include information of distributions even if such information are available.

The second shortcoming can be avoided by using a procedure usually attributed to Hasofer and Lind (1974). Instead of expanding Taylor's series about the mean value point, the linearization point is taken at some point on the failure surface. On the failure surface, the performance function and its derivatives are independent of the formulations of the problem. In the Hasofer–Lind procedure, the basic random variables are transformed to standardized (dimensionless/reduced) variables with zero mean and unit variance by the following equation if Xi follows the normal distribution

$$Y_i = (X_i - \mu_{x_i}) / \sigma_{X_i} \tag{7.23}$$

To illustrate the geometric implication of reliability index in the standardized variable space, we use the linear performance function (7.5) as an example.

The standardized variables of the random variables R and Q are

$$Y_1 = \frac{R - \mu_R}{\sigma_R} \tag{7.24}$$

$$Y_2 = \frac{Q - \mu_Q}{\sigma_Q} \tag{7.25}$$

The performance function can then be expressed as

$$Z = R - Q = \sigma_R Y_1 - \sigma_Q Y_2 + \mu_R - \mu_Q \tag{7.26}$$

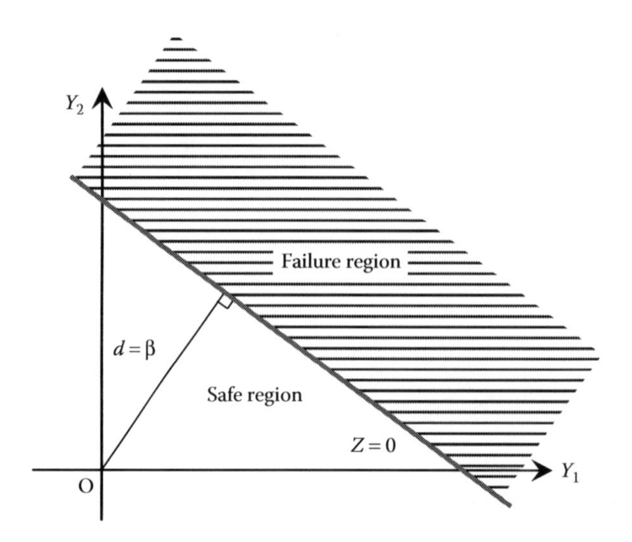

Figure 7.4 Reliability index in the standardized random variable space.

Figure 7.4 is a plot of the limit state function ($Z = 0$) on the standardized space. The origin is the point at which both R and Q equal their mean values (i.e., mean point). The distance d between the origin and the limit state surface/line is

$$d = \frac{\mu_R - \mu_Q}{\sqrt{\sigma_R^2 + \sigma_Q^2}} \tag{7.27}$$

According to Equation 7.8, if R and Q are uncorrelated, the reliability index is

$$\beta = \frac{\mu_Z}{\sigma_Z} = \frac{\mu_R - \mu_Q}{\sqrt{\sigma_R^2 + \sigma_Q^2}} \tag{7.28}$$

This result suggests that the reliability index can be interpreted geometrically as the distance between the origin and the limit state surface in the standardized space.

For a general form of a limit state function,

$$Z = g(\mathbf{X}) = 0 \tag{7.29}$$

For a design point $\boldsymbol{x}^* = (x_1^*, x_2^*, \ldots, x_n^*)^T$ at the limit state surface,

$$g(\boldsymbol{x}^*) = 0 \tag{7.30}$$

The Taylor series that is expanded at point $\boldsymbol{x}^*$ and truncated to the first order can be expressed as

$$Z \approx g(\boldsymbol{x}^*) + \sum_{i=1}^{n} \left(\frac{\partial g}{\partial X_i} \right)_{x_i^*} (X_i - \boldsymbol{x}^*) \tag{7.31}$$

For uncorrelated normal random variables, the mean and standard deviation of Z are

$$\mu_Z \approx g(\boldsymbol{x}^*) + \sum_{i=1}^{n} \left(\frac{\partial g}{\partial X_i} \right)_{x^*} (\mu_{x_i} - x_i^*) \tag{7.32}$$

$$\sigma_Z = \sqrt{\sum_{i=1}^{n}\left[\left(\frac{\partial g}{\partial X_i}\right)_{x^*}\right]^2 \sigma_{X_i}^2} \tag{7.33}$$

Substituting Equations 7.32 and 7.33 into Equation 7.3, the reliability index β is

$$\beta = \frac{\mu_Z}{\sigma_Z} = \frac{g(x_i^*) + \sum_{i=1}^{n}\left(\frac{\partial g}{\partial X_i}\right)_{x^*}(\mu_{x_i} - x_i^*)}{\sqrt{\sum_{i=1}^{n}\left[\left(\frac{\partial g}{\partial X_i}\right)_{x^*}\right]^2 \sigma_{X_i}^2}} \tag{7.34}$$

Write the limit state function (7.31) in terms of standardized variable Y_i and divide both sides of the equation using Equation 7.33,

$$\frac{g(x^*) + \sum_{i=1}^{n}\left(\frac{\partial g}{\partial X_i}\right)_{x^*}(\mu_{x_i} - x_i^*)}{\sqrt{\sum_{i=1}^{n}\left[\left(\frac{\partial g}{\partial X_i}\right)_{x^*}\right]^2 \sigma_{X_i}^2}} + \frac{\sum_{i=1}^{n}\left(\frac{\partial g}{\partial X_i}\right)_{x^*}(\sigma_{X_i} \cdot Y_i)}{\sqrt{\sum_{i=1}^{n}\left[\left(\frac{\partial g}{\partial X_i}\right)_{x^*}\right]^2 \sigma_{X_i}^2}} = 0 \tag{7.35}$$

Substituting Equation 7.34 into the above equation, we can obtain

$$-\beta - \frac{\sum_{i=1}^{n}\left(\frac{\partial g}{\partial X_i}\right)_{x^*}\sigma_{X_i}}{\sqrt{\sum_{i=1}^{n}\left[\left(\frac{\partial g}{\partial X_i}\right)_{x^*}\right]^2 \sigma_{X_i}^2}} Y_i = 0 \tag{7.36}$$

Define sensitivity coefficients of X_i as

$$\alpha_{X_i} = -\frac{\left(\frac{\partial g}{\partial X_i}\right)_{x^*}\sigma_{X_i}}{\sqrt{\sum_{i=1}^{n}\left[\left(\frac{\partial g}{\partial X_i}\right)_{x^*}\right]^2 \sigma_{X_i}^2}} \tag{7.37}$$

Then, the limit state function (7.36) can be expressed as

$$\sum_{i=1}^{n}\alpha_{X_i}Y_i - \beta = 0 \tag{7.38}$$

As shown in Equations 7.27 and 7.28, the distance from the origin to the limit state surface in the standardized space is β. The cosines of the line that is normal to the limit state surface is $\cos\theta_{Y_i} = \alpha_{X_i}$. Therefore, the design point $\boldsymbol{y}^*$ that is the closest point on the limit state surface to the origin in the standardized space (Figure 7.5) is

$$y_i^* = \beta\cos\theta_{Y_i} = \beta\alpha_{X_i}, \quad i = 1,2,\ldots,n \tag{7.39}$$

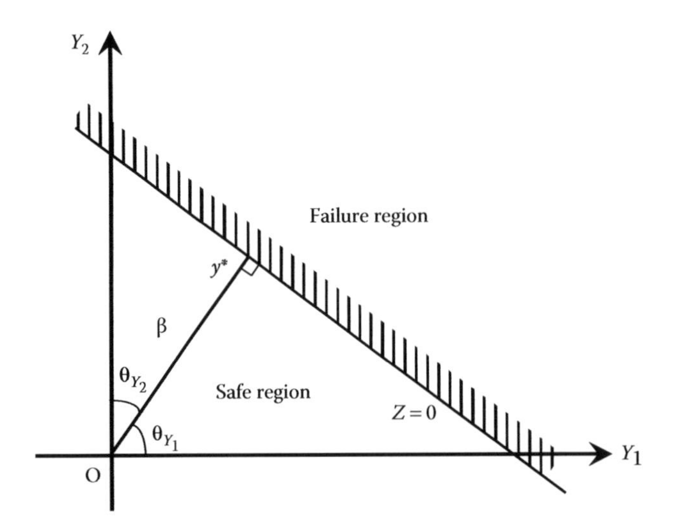

Figure 7.5 Design point of AFOSM and reliability index in the standardized random variable space.

Therefore, in the original space of **X**, the corresponding design point $\boldsymbol{x}^* = (x_1^*, x_2^*, \ldots, x_n^*)^T$ is

$$x_i^* = \mu_{X_i} + \beta\alpha_{X_i}\sigma_{X_i} \quad i = 1,2,\ldots,n \tag{7.40}$$

Ang and Tang (1984) suggested the basic algorithm of AFOSM to solve the Hasofer–Lind reliability index in an iterative way as follows:

1. Assume an initial value of $\boldsymbol{x}^*$, usually a starting point can be $\boldsymbol{x}^* = \mu_X$.
2. Compute α_{X_i} using Equation 7.37.
3. Compute β using Equation 7.34.
4. With the calculated value of β and α_{X_i}, calculate the new $\boldsymbol{x}^*$ using Equation 7.40.
5. Repeat steps 2–4 until the process converges. The convergence criterion can be the distance of $\boldsymbol{x}^*$ between two steps is less than a given allowable error ε.

The above process is the procedure of AFOSM for uncorrelated normal variables. If the variables are correlated or non-normal, adjustments are necessary. To handle the correlated variables, the Colesky decomposition approach or the eigenvalue and eigenvector approach can be used. For non-normal variables, transformation methods such as the Rosenblatt transformation can be adopted. For detailed information on these approaches, the readers can refer to Ang and Tang (1984), Baecher and Christian (2003), Tung et al. (2006), and Phoon (2008).

Example 7.3

Calculate the reliability index β and design point for Example 7.2 with the nonlinear performance function of the sliding failure using AFOSM.

The iterative results of AFOSM are listed in Table 7.2. The reliability index converges to $\beta = 2.644$. The corresponding coordinates of the design point is $F_1 = 56.6$ kN, $F_2 = 122.8$ kN, $W = 581.4$ kN, and $T = 10$ kN. The reliability index is slightly larger than that obtained in Example 7.2 using the MFOSM.

Table 7.2 Iterative results using AFOSM (Example 7.3)

Iteration no.	*Variables*	x_i^*	$\left(\frac{\partial g}{\partial X_i}\right)_{x*}$	α_{X_i}	β	*New* x_i^*
1	F_1	100	0.0065	–0.452	2.187	70.32
	F_2	200	0.0065	–0.603		147.24
	W	500	–0.0037	0.657		607.70
	T	10	–0.0062	0.015		10.03
2	F_1	70.32	0.0053	–0.532	2.662	57.53
	F_2	147.24	0.0053	–0.709		124.50
	W	607.70	–0.0019	0.463		592.49
	T	10.03	–0.0031	0.010		10.03
3	F_1	57.53	0.0055	–0.548	2.644	56.57
	F_2	124.50	0.0055	–0.730		122.79
	W	592.49	–0.0016	0.409		581.11
	T	10.03	–0.0027	0.009		10.02
4	F_1	56.57	0.0056	–0.547	2.644	56.59
	F_2	122.79	0.0056	–0.729		122.83
	W	581.11	–0.0017	0.410		581.39
	T	10.02	–0.0028	0.009		10.02
5	F_1	56.59	0.0056	–0.547	2.644	56.59
	F_2	122.83	0.0056	–0.730		122.83
	W	581.39	–0.0017	0.410		581.39
	T	10.02	–0.0028	0.009		10.02

The MATLAB code of this example is as follows.

```
% Example 7.3 %
% AFOSM %
clear; clc;
mux = [100;200;500;10];
covx = [0.3;0.2;0.15;0.10];
sigmax = mux.*covx;
x = mux;
normX = eps; % eps is the floating point relative accuracy
while abs(norm(x)-normX)/normX > 1e-6
    normX =  norm(x);
    g = (10*x(1)+10*x(2))/(3*x(3)+5*x(4))-1;
dgdx = [10/(3*x(3)+5*x(4));10/(3*x(3)+5*x(4));-3*(10*x(1)+10*x(2))/
(3*x(3)+5*x(4))^2;-5*(10*x(1)+10*x(2))/(3*x(3)+5*x(4))^2];
    gs = dgdx.*sigmax;
    alphax = -gs/norm(gs);
    beta = (g+dot(dgdx,(mux-x)))/norm(dgdx.*sigmax)
    x = mux+beta*sigmax.*alphax
end
pf = normcdf(-beta,0,1)
```

7.3.1.3 Spreadsheet method

Low and Tang (1997a, 2004) proposed a very efficient procedure that takes advantage of the optimization tool available in spreadsheet software packages. According to the definition of reliability index by Hasofer and Lind (1974), the forms of the performance function will not influence the estimation of reliability because the geometric shape of limit state surface and the distance from the origin in standardized space remain constant (Ditlevsen, 1981). The matrix formulation of Hasofer–Lind reliability index β can be expressed as follows when X is multivariate normal (Ditlevsen, 1981)

$$\beta = \min_{\mathbf{x} \in F} \sqrt{(\mathbf{X} - \boldsymbol{\mu}_\mathbf{X})^T \mathbf{C}_\mathbf{X}^{-1} (\mathbf{X} - \boldsymbol{\mu}_\mathbf{X})} \tag{7.41}$$

where:

$\mathbf{X}$ is the random variable vector
$\boldsymbol{\mu}_\mathbf{X}$ is the mean value vector
$\mathbf{C}_\mathbf{X}$ is the covariance matrix of $\mathbf{X}$
F is the failure region

In the original random variable space, the equation of a tilted one-standard-deviation (1–σ) dispersion ellipsoid is

$$(\mathbf{X} - \boldsymbol{\mu}_\mathbf{X})^T \mathbf{C}_\mathbf{X}^{-1} (\mathbf{X} - \boldsymbol{\mu}_\mathbf{X}) = 1 \tag{7.42}$$

Figure 7.6 shows an ellipse in a two-dimensional random variable space for different values of correlation coefficient ρ. The center of the ellipse is in the mean of the random variables. With the change of correlation coefficient ρ, the ellipse rotates and changes its aspect ratio.

Multiply β^2 on the right-hand side of Equation 7.42:

$$(\mathbf{X} - \boldsymbol{\mu}_\mathbf{X})^T \mathbf{C}_\mathbf{X}^{-1} (\mathbf{X} - \boldsymbol{\mu}_\mathbf{X}) = \beta^2 \tag{7.43}$$

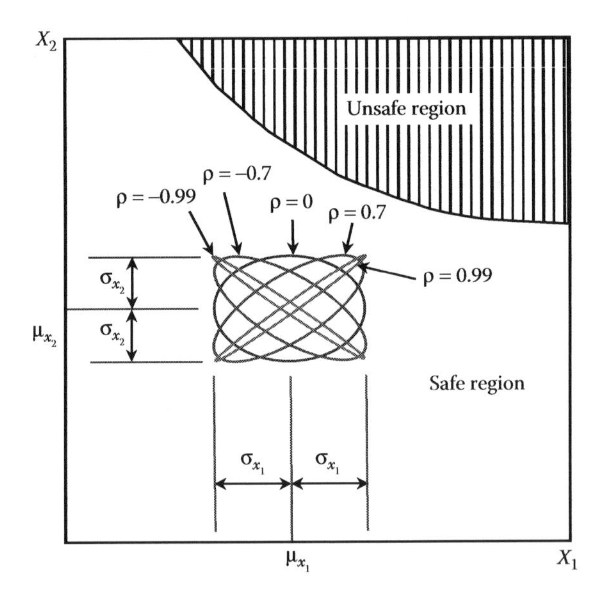

Figure 7.6 1–σ ellipse rotates as correlation coefficient ρ changes. (Modified from Low, B. K., and Tang, W. H., Efficient reliability evaluation using spreadsheet, *Journal of Engineering Mechanics*, 123(7), 749–752, 1997a. With permission from ASCE.)

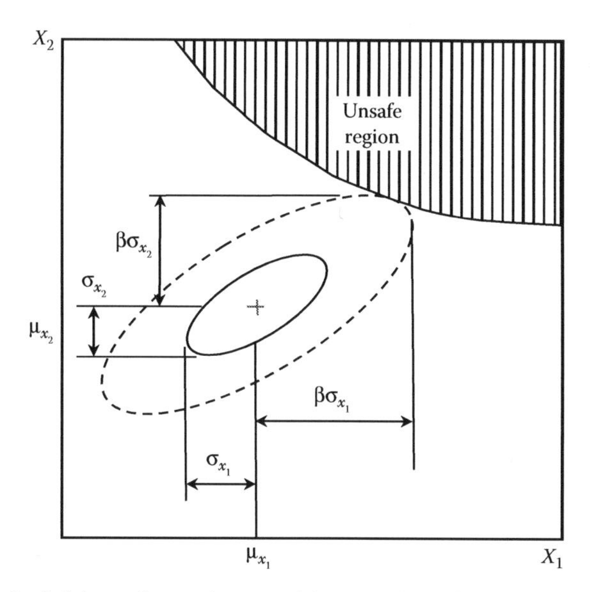

Figure 7.7 Illustration of reliability index in the spreadsheet method. (Modified from Low, B. K., and Tang, W. H., *Journal of Engineering Mechanics,* 123(7), 749–752, 1997a. With permission from ASCE.)

In Figure 7.7, the ellipse that is tangent to the failure surface is β times the (1 – σ) dispersion ellipsoid. This provides an intuitive meaning of the reliability index in the original space of random variable. In the spreadsheet method, conventional spreadsheet software packages such as Microsoft-Excel with built-in optimization tools are used to solve the optimization problem. The spreadsheet method does not require transformed space and the optimization problem is solved by the add-in optimization tool Solver in Excel, which is the most important advantage of the method. Low and Tang (1997b, c) and Low et al. (1998) showed the usefulness of the spreadsheet method in automatically searching critical slip surface and obtaining the reliability index in slope reliability problems.

Example 7.4

Assume that the random variables in Example 7.3 are correlated with $\rho_{F_1F_2} = 0.6$, $\rho_{WT} = 0.4$. Calculate reliability index using the spreadsheet method.

The random variable vector is $\mathbf{X} = (F_1, F_2, W, T)^T$.

The correlation matrix is

$$\rho_X = \begin{bmatrix} 1 & 0.6 & & \\ 0.6 & 1 & & \\ & & 1 & 0.4 \\ & & 0.4 & 1 \end{bmatrix}$$

The covariance matrix is

$$C_X = \begin{bmatrix} 900 & 720 & & \\ 720 & 1600 & & \\ & & 5625 & 30 \\ & & 30 & 1 \end{bmatrix}$$

The diagonal terms of the covariance matrix are variance of the random variables. The off-diagonal terms represent $\rho_{ij} X_i X_j$.

With Microsoft Excel, the following procedure can be used to compute the reliability index of the slope.

Step 1: Input vectors of the design point, the mean value, the standard deviation, and the covariance matrix of the random variables in a Excel sheet as shown in Figure 7.8.

Step 2: Calculate the row vector $(\mathbf{X}-\mu_{\mathbf{X}})^T$ and column vectors $(\mathbf{X}-\mu_{\mathbf{X}})$ (see Figure 7.9).

Step 3: The inverse of the covariance matrix is obtained using the Excel build-in function MINVERSE. Select the covariance matrix, that is, 4 × 4 cells and name it as "array" in the Name Box. Select a 4 × 4 blank cells for the inverse matrix and type "=MINVERSE(array)" in the equation. Press "Enter" while holding down "Ctrl" and "Shift." The formula bar will display "={MINVERSE (array)}." The result is shown in Figure 7.10.

Random variable	X value	Mean	Standard deviation
F_1	100	100	30
F_2	200	200	40
W	500	500	75
T	10	10	1

Covariance matrix			
900	720	0	0
720	1600	0	0
0	0	5625	30
0	0	30	1

Figure 7.8 Input statistics of the random variable and the initial value of design point.

Random variable	X value	Mean	Standard deviation
F_1	100	100	30
F_2	200	200	40
W	500	500	75
T	10	10	1

Covariance matrix			
900	720	0	0
720	1600	0	0
0	0	5625	30
0	0	30	1

$(X-\mu_x)^T$			
0	0	0	0

$(X-\mu_x)$
0
0
0
0

Figure 7.9 Calculate $(\mathbf{X}-\mu_{\mathbf{X}})^T$ and $(\mathbf{X}-\mu_{\mathbf{X}})$.

Name Box

R9 fx

Random variable	X value	Mean	Standard deviation
F_1	100	100	30
F_2	200	200	40
W	500	500	75
T	10	10	1

Covariance matrix			
900	720	0	0
720	1600	0	0
0	0	5625	30
0	0	30	1

$(X-\mu_x)^T$			
0	0	0	0

Inverse covariance matrix			
0.001736	−0.00078	0	0
−0.00078	0.000977	0	0
0	0	0.000212	−0.00635
0	0	−0.00635	1.1190476

$(X-\mu_x)$
0
0
0
0

Figure 7.10 Calculate the inverse matrix of covariance matrix.

Step 4: Use the built-in function MMULT for matrix multiplication to obtain $C_X^{-1}(X-\mu_X)$ and $(X-\mu_X)^T C_X^{-1}(X-\mu_X)$ (Figure 7.11). Press "Enter" while holding down "Ctrl" and "Shift" for matrix calculation.

Step 5: The equations of the reliability index β (Equation 7.41) and the performance function $g(\mathbf{X})$ are entered as shown in Figure 7.12.

Step 6: Click menu [Tool]|[Adds-Ins], select [Solver Add-in], and press OK to invoke Solver Add-in Calculator.

Step 7: Set minimizing β as the objective of optimization in Solver. The variable cells are set as the design point. The constraint is $g(\mathbf{X}) = 0$. Different optimization methods can be adopted in Solver.

The result of this example is shown in Figure 7.13. The reliability index β is 2.172. If the random variables are independent, as in Example 7.3, modify the covariance matrix in the spreadsheet. The reliability index β obtained using the spreadsheet method is 2.644

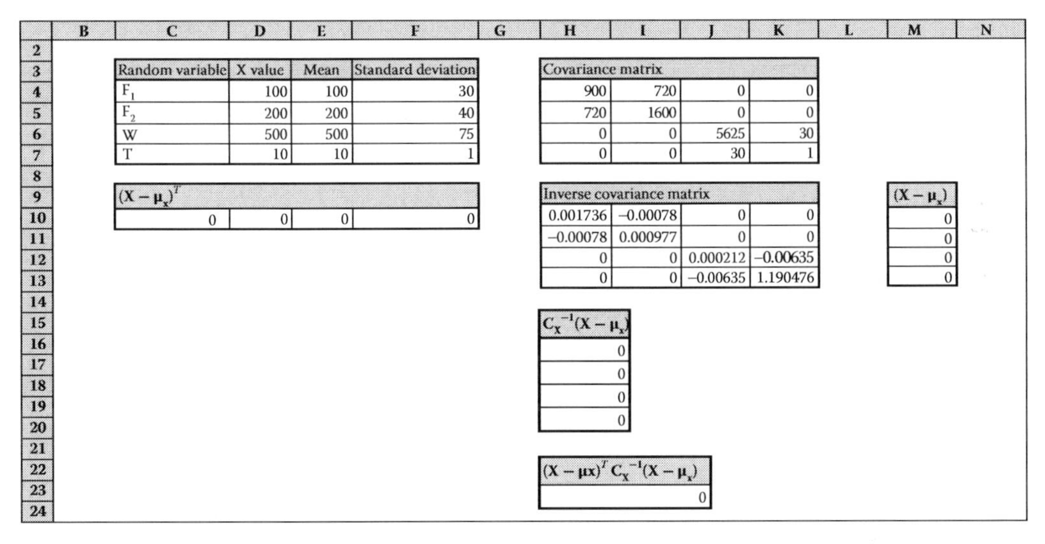

Figure 7.11 Solve $\mathbf{C}^{-1}(\mathbf{x}-\mathbf{m})$ and $(\mathbf{x}-\mathbf{m})^T\mathbf{C}^{-1}(\mathbf{x}-\mathbf{m})$.

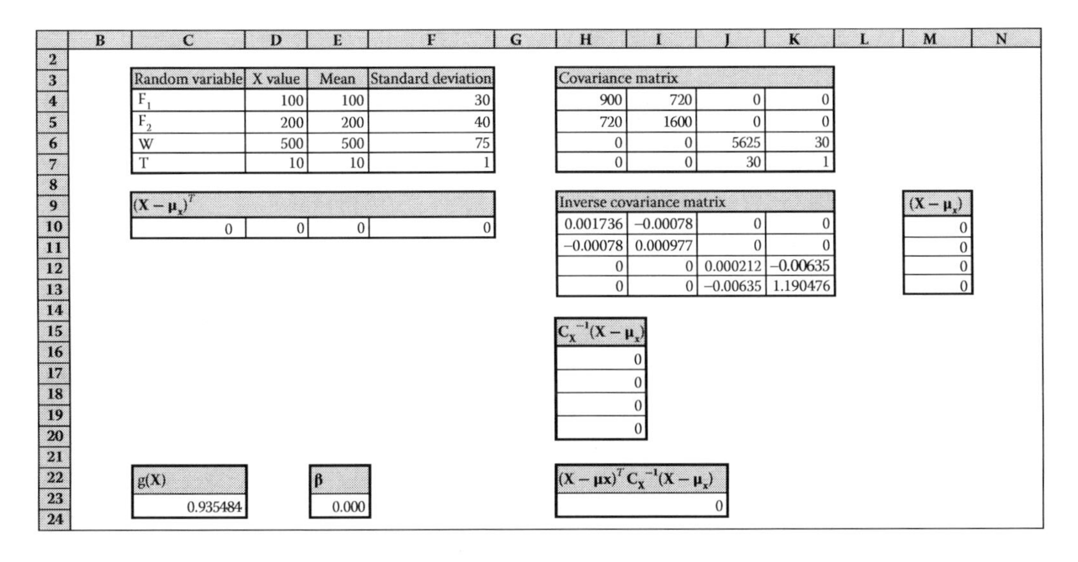

Figure 7.12 Calculate β and the performance function g(X).

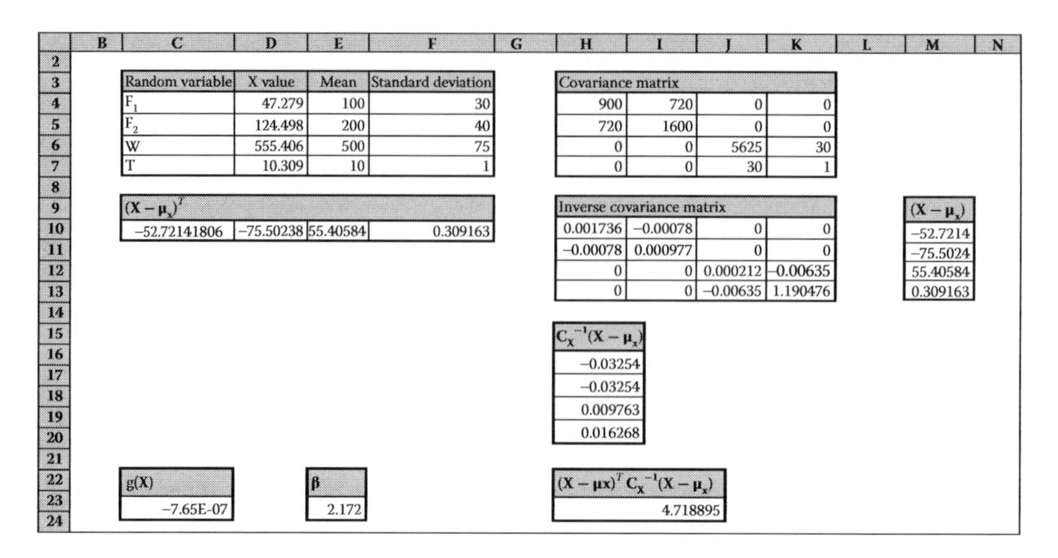

Random variable	X value	Mean	Standard deviation
F_1	47.279	100	30
F_2	124.498	200	40
W	555.406	500	75
T	10.309	10	1

Covariance matrix			
900	720	0	0
720	1600	0	0
0	0	5625	30
0	0	30	1

$(X-\mu_x)^T$			
−52.72141806	−75.50238	55.40584	0.309163

Inverse covariance matrix			
0.001736	−0.00078	0	0
−0.00078	0.000977	0	0
0	0	0.000212	−0.00635
0	0	−0.00635	1.190476

$(X-\mu_x)$
−52.7214
−75.5024
55.40584
0.309163

$C_X^{-1}(X-\mu_x)$
−0.03254
−0.03254
0.009763
0.016268

g(X)	β	$(X-\mu x)^T C_X^{-1}(X-\mu_x)$
−7.65E-07	2.172	4.718895

Figure 7.13 Reliability analysis with Excel for correlated normal variables.

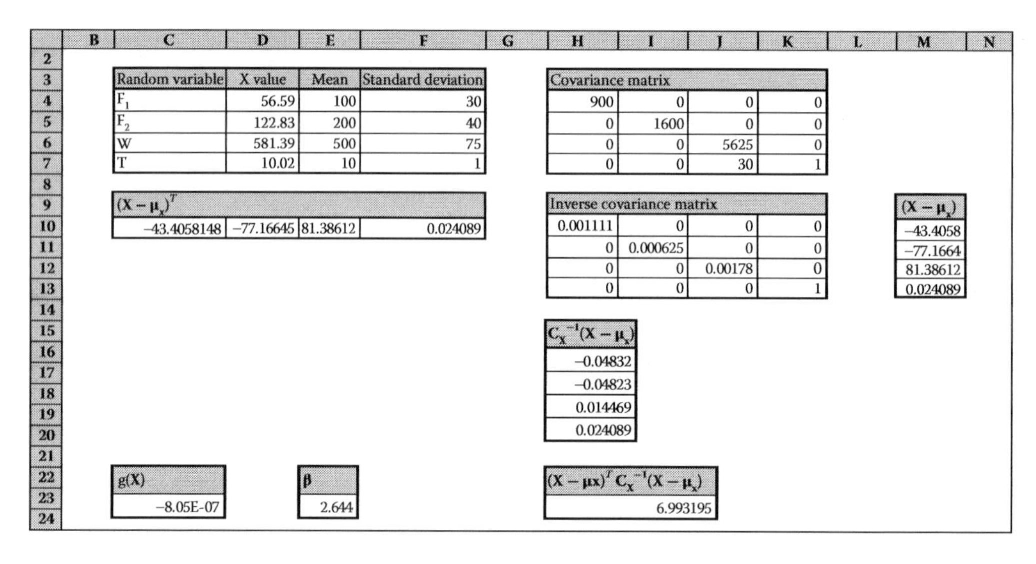

Random variable	X value	Mean	Standard deviation
F_1	56.59	100	30
F_2	122.83	200	40
W	581.39	500	75
T	10.02	10	1

Covariance matrix			
900	0	0	0
0	1600	0	0
0	0	5625	0
0	0	30	1

$(X-\mu_x)^T$			
−43.4058148	−77.16645	81.38612	0.024089

Inverse covariance matrix			
0.001111	0	0	0
0	0.000625	0	0
0	0	0.00178	0
0	0	0	1

$(X-\mu_x)$
−43.4058
−77.1664
81.38612
0.024089

$C_X^{-1}(X-\mu_x)$
−0.04832
−0.04823
0.014469
0.024089

g(X)	β	$(X-\mu x)^T C_X^{-1}(X-\mu_x)$
−8.05E-07	2.644	6.993195

Figure 7.14 Reliability analysis with Excel for independent normal variables.

as shown in Figure 7.14. The calculated result obtained by the spreadsheet method is the same as that obtained using Ang and Tang's (1984) iteration procedure.

7.3.2 Sampling methods

7.3.2.1 Monte Carlo simulation

Monte Carlo simulation was developed in the 1940s by scientists in the Los Alamos National Laboratory for nuclear weapon projects. The method is named after the city Monte Carlo in Monaco and its many casinos. Today Monte Carlo simulation methods are used in a wide array of applications, including physics, finance, and engineering.

When applied to reliability analysis, the Monte Carlo simulation is a method used to obtain the probability distribution of the output variable given the probability distributions of a set of input random variables. A key task in the Monte Carlo simulation is the generation of the appropriate values of the random variables (i.e., generation of random numbers) in accordance with the PDF of the input random variables. Fortunately, the tools for random number generation are now easily available in many commercial software packages such as MATLAB or @Risk, which greatly facilitate the application of the Monte Carlo simulation in practice.

To introduce the application of the Monte Carlo simulation for the evaluation of failure probability, note first that probability of the failure can be mathematically defined as follows:

$$p_f = \int \cdots \iint I(\mathbf{x}) f_X(\mathbf{x}) d\mathbf{x} \tag{7.44}$$

where $f_X(\mathbf{x})$ is the PDF of $\mathbf{X}$ and $I(\mathbf{x})$ is the indicator function with

$$I(\mathbf{x}) = \begin{cases} 1 & \mathbf{x} \text{ is in the failure region} \\ 0 & \mathbf{x} \text{ is not in the failure region} \end{cases} \tag{7.45}$$

Equation 7.44 shows that the failure probability is indeed the mean value of the indicator function $I(\mathbf{x})$, which can thus be estimated as follows:

$$\hat{p}_f \approx \frac{1}{N} \sum_{i=1}^{N} I(\mathbf{x}^i) \tag{7.46}$$

where:

$\hat{p}_f$ is the estimation of the probability of failure p_f
$\mathbf{x}^i$ denote the ith sample of $\mathbf{x}$
N is the number of random samples

The COV of the statistical error associated with the above estimator of failure probability can be evaluated as follows (e.g., Ang and Tang, 1984):

$$\text{cov}(\hat{p}_f) \approx \sqrt{\frac{(1-\hat{p}_f)}{N \cdot \hat{p}_f}} \tag{7.47}$$

Equation 7.47 shows that the COV of the estimator decreases with the sample number, that is, the statistical error reduces if more samples are adopted.

Example 7.5 Consider the slope in Example 7.4 with correlated random variables.

In this example, $X = \{F_1, F_2, W, T\}$. The mean and covariance matrix of $\mathbf{X}$ are

$$\mu_x = \{100, 200, 500, 10\} \text{ and } C_X = \begin{bmatrix} 900 & 720 & 0 & 0 \\ 720 & 1600 & 0 & 0 \\ 0 & 0 & 5625 & 30 \\ 0 & 0 & 30 & 1 \end{bmatrix}, \text{ respectively.}$$

The MATLAB code for calculating the failure probability of the slope is shown as follows, in which the MATLAB routine *mvnrnd.m* is used for sample generation. The failure probability of the slope is 1.44% based on 10,000 samples, which is consistent with the result in Example 7.4 with β is 2.172. Based on Equation 7.47, the COV associated with the estimated failure probability is 0.082.

```
% Example 7.5 %
% Matlab code for estimating slope failure probability using Monte Carlo
  simulation%
function MCS
nsamples=10000;
xm=[100 200 500 10];
xsd=[30 40 75 1]';
xr=[1 0.6 0 0;
    0.6 1 0 0;
    0 0 1 0.4;
    0 0 0.4 1];
Cx=(xsd*transpose(xsd)).*xr

for i=1:nsamples
x=mvnrnd(xm,Cx,1);
y=gfun(x);
if y<0
    I(i,1)=1;
else
    I(i,1)=0;
end
end
pf_m=mean(I)
pf_cov=sqrt((1-pf_m)/(nsamples*pf_m))

function y=gfun(x)
y=(10*x(1)+10*x(2))/(3*x(3)+5*x(4))-1;
```

7.3.2.2 Importance sampling

While versatile, Monte Carlo simulation often requires repeated evaluation of the performance function, which could be computationally intensive when the performance function is complex or when the failure probability is small. The importance sampling can sometimes be used to enhance the efficiency of the Monte Carlo simulation. In the conventional Monte Carlo simulation, samples are drawn from the x of **X**, that is, $f_{\mathbf{X}}(\mathbf{x})$ directly. In the impotence sampling, samples are drawn from a sampling function $s(\mathbf{x})$ instead of $f_{\mathbf{X}}(\mathbf{x})$. In such a case, Equation 7.44 can be written as follows:

$$p_f = \int \cdots \iint \left[I(\mathbf{x})w(\mathbf{x})\right]s(\mathbf{x})d\mathbf{x} \tag{7.48}$$

where $w(\mathbf{x})$ is a weighting function defined as below

$$w(\mathbf{x}) = \frac{f_X(\mathbf{x})}{s(\mathbf{x})} \tag{7.49}$$

The estimation for the failure probability and the statistical error associated with such an estimator can be written as follows (e.g., Ang and Tang, 1984):

$$\hat{p}_f \approx \frac{1}{N}\sum_{i=1}^{N} I(\mathbf{x}^i)w(\mathbf{x}^i) \tag{7.50}$$

$$\text{var}(\hat{p}_f) \approx \frac{1}{N}\left\{\frac{1}{N}\sum_{i=1}^{N}\left[I(\mathbf{x}^i)w(\mathbf{x}^i)\right]^2 - \hat{p}_f^2\right\} \tag{7.51}$$

To apply the importance of sampling, the key is to determine an appropriate sampling function. It can be proved that the sampling function is most efficient when $s(\mathbf{x})$ is proportional to $I(\mathbf{x})w(\mathbf{x})$ (e.g., Ang and Tang, 1984). In practice, however, it is difficult to find such a sampling function with closed-form. For failure probability estimation, one can first identify the design point $\mathbf{x}^*$ using the first-order reliability method, and then center the sampling function at the design point. For instance, if $f_X(\mathbf{x})$ is multivariate normal with mean of μ_x and covariance matrix of C_x, $s(\mathbf{x})$ can be a multivariate normal distribution with mean of $\mathbf{x}^*$ and a covariance matrix of C_x (Harbitz, 1986). The sampling function constructed in such a way has significant coverage in the failure domain of the sampling function and hence is efficient for failure probability estimation.

Example 7.6

In this example, we will solve the problem in Example 7.5 with importance sampling. Based on AFORM, the design point in this example is $\mathbf{x}^* = \{47.28\ 124.5\ 555.42\ 10.31\}^T$ as shown in Figure 7.13. Following the suggestion made by Harbitz (1986), $s(\mathbf{x})$ in this example can be a normal distribution with a mean of $\mathbf{x}^*$ and a covariance matrix of C_x. The MATLAB code for implementing the importance sampling is shown below.

```
% Example 7.6 %
% Matlab code for estimating slope failure probability using importance
  sampling %
function IS
nsamples=10000;
xD=[47.28 124.5 555.42 10.31]'
xm=[100 200 500 10]';
xsd=[30 40 75 1]';
xr=[1 0.6 0 0;
    0.6 1 0 0;
    0 0 1 0.4;
    0 0 0.4 1];
```

(*Continued*)

```
Cx=(xsd*transpose(xsd)).*xr;


for i=1:nsamples
   x=mvnrnd(xD',Cx,1);
    y=gfun(x);
    w(i,1)=mvnpdf(x,xm',Cx)/mvnpdf(x,xD',Cx);
    if y<0
        I(i,1)=1;
    else
        I(i,1)=0;
    end
end
I2=I.*w;
pf_m=mean(I2)
pf_var=(sum(I2.^2)/nsamples-pf_m^2)/nsamples
pf_std=sqrt(pf_var);
pf_cov=pf_std/pf_m

function y=gfun(x)
y=(10*x(1)+10*x(2))/(3*x(3)+5*x(4))-1;
```

With 10,000 samples the failure probability of the slope is found to be 1.54% with a COV of 0.015. We can see that when the number of the samples is the same, the COV associated with the estimated failure probability by the importance sampling is significantly smaller than the conventional Monte Carlo simulation, indicating that the importance sampling is more efficient.

7.3.2.3 Latin hypercube sampling

Latin hypercube sampling (LHS) is a stratified random sampling technique. The basic idea of the LHS technique is to select random samples for each input random variable over its range in a stratified manner such that the overall uncertainty of a model could be reasonably described by finite samples (Stein, 1987; Mckay, 1988; Pebesma and Heuvelink, 1999).

A simple example to illustrate the basic concept of LHS is shown in Figure 7.15. Suppose that it is desired to generate three random values for a random variable X. The CDF of X is $F(X)$. Select one value from the low third of the distribution, one from the middle third, and one from the upper third. Let u_1, u_2, and u_3 be three randomly generated values from uniform distribution in the interval (0, 1/3). The three selected values of X are

$$x_i = F^{-1}\left(\frac{i-1}{3} + u_i\right), \quad i = 1, 2, 3$$

Then random permutation technique can be used to order the three values.

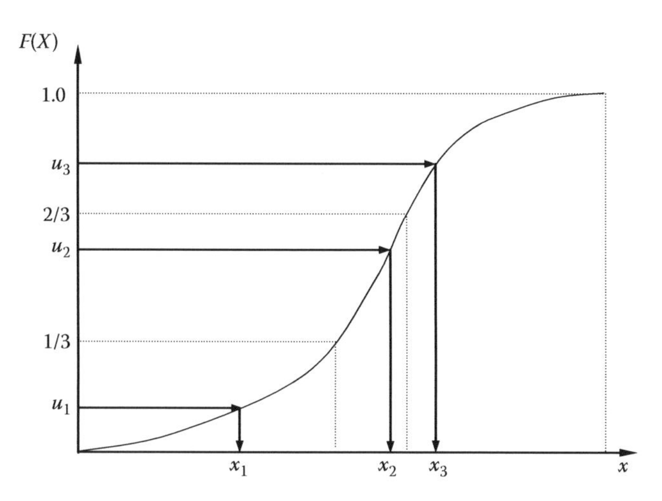

Figure 7.15 Sampling in three equal-probability intervals. (After Mckay, M. D., Sensitivity and uncertainty analysis using a statistical sample of input values, in: Ronen, Y., Ed., *Uncertainty Analysis*, CRC Press, Boca Raton, FL, pp. 145–186, 1988. With permission.)

To draw a sample size of N for n independent random variables, the ith sample element for variable x_j is obtained as (Pebesma and Heuvelink, 1999)

$$x_{ij} = F_j^{-1}\left(\frac{p_{ij} - u_{ij}}{N}\right) \quad (i = 1,\ldots,N \quad \text{and} \quad j = 1,\ldots,n) \tag{7.52}$$

where:
 F_j^{-1} is the inverse CDF of random variable x_j
 p_{ij} is the random permutation of 1, ..., N
 u_{ij} is a uniformly distributed random number in [0, 1]

If the random variables are correlated, the joint multivariate PDF of the random variables are used (Pebesma and Heuvelink, 1999).

Using the LHS, the estimators of the mean and distribution function of the output variable are unbiased (Mckay, 1988). Moreover, when the output function is monotonic with each input random variable, the variances of the estimators are no more than and often much less than the variances when random variables are generated from the conventional Monte Carlo simulation. The variance reduction properties of the LHS technique imply fewer samples or computer runs can achieve a degree of precision comparable to that obtained from a simple random sampling of random variables (Mckay, 1988). Therefore, LHS is more efficient than conventional Monte Carlo simulation when the numerical simulation of the system response is time consuming.

7.3.2.4 Subset simulation

The development of subset simulation (Au and Beck, 2001; Au and Wang, 2014) is one of the most notable progresses in structural reliability in the past decade. During a surprisingly short period, this method has gained popularity in many reliability problems. The subset simulation first decompose the failure domain F into a series of intermediate failure domains F_1, F_2, ..., F_n with $F_n \in F \in F_{n-1} \in \ldots \in F_1$ (Figure 7.16). The failure probability can then be written as

$$P(F) = P(F \mid F_{n-1}) \cdots P(F_3 \mid F_2) P(F_2 \mid F_1) P(F_1) \tag{7.53}$$

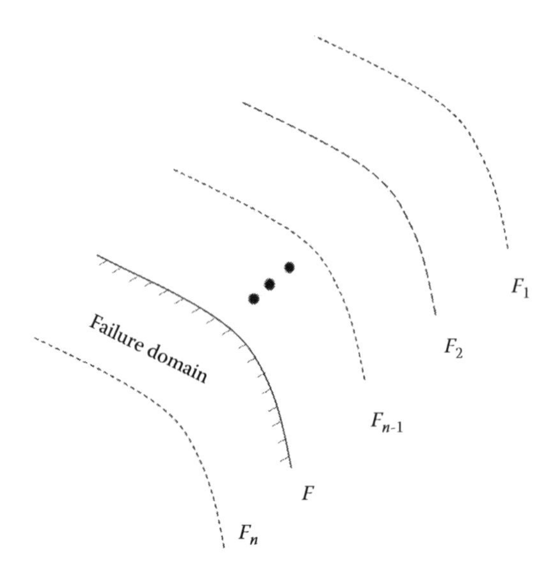

Figure 7.16 Failure domain decomposition scheme used in a subset simulation.

As such, the difficult problem of estimating a small failure probability is transformed to relatively easier problems of estimating a series of larger failure probabilities. In Equation 7.53, the intermediate failure domains are determined by specifying failure probability values for $P(F_1)$ and $P(F_i|F_{i-1})$. Let $P(F_1) = P(F_i|F_{i-1}) = p_{ft}$. The following procedure is typically used to estimate the failure probability based on subset simulation:

1. Draw an initial N samples (for instance, $N = 1000$) according to $f_X(\mathbf{x})$.
2. Among the N samples, collect the $N \cdot p_{ft}$ samples with smallest performance values, which are seeds for exploring the failure region of F_1. With this procedure, $P(F_1) = p_{ft}$.
3. Based on the seeds obtained in Step (2), generate N samples belonging to the failure domain F_1 using the Markov chain Monte Carlo (MCMC) simulation; among the N samples, select $N \cdot p_{ft}$ samples with the smallest performance values, which are seeds for exploring in the failure region of F_2. With this procedure, $P(F_2|F_1) = p_{ft}$. Note MCMC is used here because it is difficult to draw samples from the intermediate failure domain analytically.
4. As the above process continues, the intermediate failure domain will gradually move towards the actual failure domain F. When F_n is a subset of F, as shown in Figure 7.16, the iteration can be terminated. Among the samples generated for F_n, count the ratio of samples belong to F, which is denoted as p_{fn} here. Thus, $P(F|F_n) = p_{fn}$.

Based on the above procedure, the failure probability can be estimated as follows:

$$P(F) = p_{fn} \prod_{i=1}^{n-1} p_{ft} = p_{fn} p_{ft}^{n-1} \tag{7.54}$$

The subset simulation was used in Wang et al. (2011) for reliability analysis of soil slopes. Santoso and Phoon (2011) modified the above subset simulation and applied it to probabilistic prediction of rainfall-induced slope failure.

Example 7.7

Evaluate the failure probability of Example 7.4 with subset simulation. The MATLAB code for subset simulation is shown as follows.

```
%Matlab code for estimating slope failure probability using subset
   simulation%
function Subset
global xm Cx xsd
xm=[100 200 500 10]';
xsd=[30 40 75 1]';
xr=[1 0.6 0 0;
 0.6 1 0 0;
 0 0 1 0.4;
 0 0 0.4 1];
Cx=(xsd*transpose(xsd)).*xr;

thresh=0.1
dx=length(xm);
N=1000;
Ni=N*thresh;
x_next= mvnrnd(xm',Cx,N);

for i=1:N
 y_next(i,1)=gfun(x_next(i,:));
end
[y_next_sort,index]=sort(y_next);
pf_t(1)= max(size(find(y_next<0)))/max(size(y_next));

if pf_t(1)>thresh
 cumpf(1,1)=pf_t(1); boundary(1,1)=0;
else
 boundary(1)=y_next_sort(N*thresh+1);
 Ind_Fail=index(1:N*thresh); x_Fail=x_next(Ind_Fail,:);
 y_Fail=y_next(Ind_Fail); pf_t(1)=thresh;
 i=1;
 BoundaryTouch=0;
 while BoundaryTouch==0
         i=i+1;
      for j=1:length(y_Fail)
         x0=x_Fail(j,:);
         x_next(j*Ni-Ni+1:j*Ni,:)=MCMC(x0,boundary(i-1),Ni);
       end
      for j=1:max(size(x_next))
          y_next(j,1)=gfun(x_next(j,:));
       end
```

(Continued)

```
            [y_next_sort,index]=sort(y_next);
             boundary(i,1)=y_next_sort(N*thresh+1);
       if boundary(i)<0
            pf_t(i,1)= length(find(y_next<0))/length(y_next);
            BoundaryTouch=1; boundary(i)=0;
       else
            Ind_Fail=index(1:N*thresh); x_Fail=x_next(Ind_Fail,:);
            y_Fail=y_next(Ind_Fail); pf_t(i,1)=thresh;
       end
 end
end
pf=prod(pf_t)

function X=MCMC(x0,gi,nsamples)
global xsd
dx=length(x0);
for i=1:nsamples
     x2=x0;
     for k=1:dx
     x2(k)=normrnd(x0(k),xsd(k));
    end
     post0=postpdf(x0,gi); post2=postpdf(x2,gi);
     r=post2/post0; p=rand(1);
     if p<r
         x0=x2;
     else
         x0=x0;
     end
 X(i,:)=x0;
end

function post0=postpdf(x0,gi)
global xm Cx
rej=0;
g0=gfun(x0);
if g0<gi
        like0=1;
else
        like0=0;
end
post0=like0*mvnpdf(x0,xm',Cx);

function y=gfun(x)
y=(10*x(1)+10*x(2))/(3*x(3)+5*x(4))-1;
```

In the above procedure, the key parameters are N and p_{ft}. The accuracy of the subset simulation increases with N and p_{ft}. In this example, $N = 1000$ and $p_{ft} = 0.1$ are adopted. Based on the MATLAB code, the subset simulation terminates at the second stage, and the obtained failure probability is 0.015. The corresponding number of samples used in the subset simulation is thus 2000.

7.3.3 Response surface method

The basic concept of response surface reliability method is to approximate the performance function $g(\mathbf{X})$ by a response surface function $g'(\mathbf{X})$ (e.g., Wong, 1985; Faravelli, 1989; Bucher and Bourgund, 1990). In RSM, a sufficient number of numerical experiments are conducted to obtain the explicit input–output function and to fulfill the requirement that the error of the approximated response surface function is acceptable in the region of interest.

The critical issues related to RSM include the selection of the response function and the determination of sampling points. Usually, the polynomial regression method is adopted. For example, the simplest form of a polynomial response surface is the linear function:

$$g'(\mathbf{X}) = \lambda_0 + \sum_{i=1}^{n} \lambda_i X_i \tag{7.55}$$

where λ_0, λ_i are the coefficients of the response function and n is the number of variables Xi.

The quadratic models can be adopted to include linear terms, cross-product terms, and squared terms:

$$g'(\mathbf{X}) = \lambda_0 + \sum_{i=1}^{n} \lambda_i X_i + \sum_{i=1}^{n} \lambda_{ii} X_i^2 + \sum_{i \neq j}^{n} \lambda_{ij} X_i X_j \tag{7.56}$$

where λ_0, λ_i, λ_{ii}, and λ_{ij} are the coefficients of the response function.

In the above quadratic polynomial equation, the number of unknown coefficients m is $2n + 1 + n(n-1)/2$. A large amount of deterministic analyses may be needed to evaluate the coefficients if the dimension of X is large. To improve the computational efficiency, the quadratic polynomial form excluding the cross-product terms is usually adopted:

$$g'(\mathbf{X}) = \lambda_0 + \sum_{i=1}^{n} \lambda_i X_i + \sum_{i=1}^{n} \lambda_{ii} X_i^2 \tag{7.57}$$

Figure 7.17 shows the Box–Wilson Central Composite Design, which contains a fractional factorial design with center points and is augmented with a group of star points that allow estimation of curvature. The star points represent new extreme values (low and high) for each variable in the design. The distance from the center of the design space to a factorial point is ±1 unit for each variable. The distance from the center of the design space to a star point is $|a|$, which is normally greater than 1. The precise value of a depends on certain properties desired for the design and on the number of factors involved. A common choice of a is

$$a = \left(2^n\right)^{1/4} \tag{7.58}$$

Simulations in the neighborhood of the estimated design points $\boldsymbol{x}^{*\prime}$ are conducted and then regression analysis is used to obtain the coefficients of the response surface function. The response surface function $g'(\mathbf{X})$ can then be used to estimate reliability using the advanced first-order reliability method. As the real design point $\boldsymbol{x}^*$ may differ from

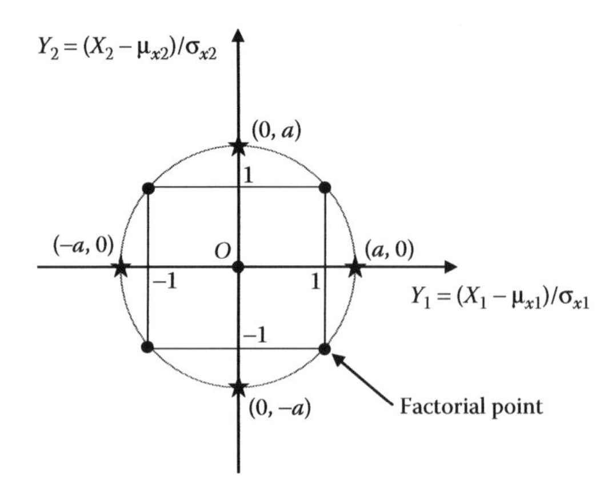

Figure 7.17 Diagram of the central composite design of RSM.

the estimated design point $x^{*\prime}$, iterations may be required until a convergence of reliability index is obtained. The RSM is conceptually simple and is an efficient and practical method for reliability analysis.

The RSM of reliability for reliability analysis usually includes the following steps:

1. Select a response surface function $g'(\mathbf{X})$, which contains m unknown coefficients.
2. Choose an initial design point $x^{*\prime}$, which is often taken as the central point.
3. Sampling the input points in the neighborhood of $x^{*\prime}$. The number of sample points, N should be at least larger than m.
4. Simulations are carried out using these sample points as input and lead to a series of responses as output. This process often requires numerical analysis software, or model experiment.
5. Based on the N sample points and their responses, the regression analysis method, for example, the least-squares method, can be used to solve the m unknown coefficients in $g'(\mathbf{X})$.
6. Using the determined response surface function $g'(\mathbf{X})$ as the performance function, AFORM can be used to calculate the reliability index β and the design point x^*.
7. If $\|x^{*\prime} - x^*\|$ or the difference of reliability indexes in two successive steps is larger than a tolerance ε, x^* will be used as $x^{*\prime}$ and the steps 3–7 will be repeated. Otherwise the iteration can be terminated, and the β obtained in the last iteration will be regarded as the reliability index.

Example 7.8

An example is presented here to illustrate the process of RSM used to solve reliability. Referring to Figure 7.18, the stability of a cut slope composed of silty soil is of concern. The shear strength parameters of the silt soil are obtained in laboratory tests. The cohesion c follows a normal distribution with $\mu_c = 30$ kPa, $\sigma_c = 6$ kPa and the tangent of the friction angle $\tan\phi$ also follows a normal distribution with $\mu_{\tan\phi} = 0.50, \sigma_{\tan\phi} = 0.10$.

RSM is used to solve the reliability of the slope as follows:

1. The performance equation is defined as

$$g = F_s - 1$$

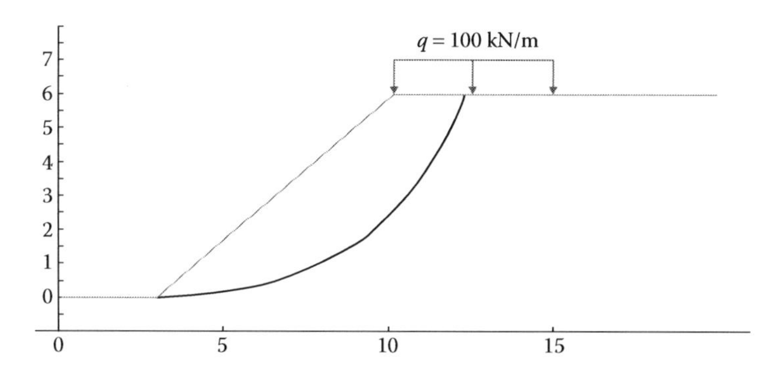

Figure 7.18 A cut slope composed by silty soil.

2. A quadratic polynomial without cross-product terms is used as the response surface function g' as follows:

$$g' = \lambda_0 + \lambda_1 c + \lambda_2 \tan\phi + \lambda_3 c^2 + \lambda_4 (\tan\phi)^2$$

3. The central point of the variables, that is, c and tanf is used as the initial estimated design point $x^{*\prime}$ of reliability analysis.
4. The central composite design with no factorial points is used and the value of a is taken as 1.414. The five sampling points are listed in Table 7.3.
5. EMU 2005 (Chen et al., 2005), a stability analysis software based on limit equilibrium analysis method, is used to calculate the factor of safety of the slope. The values of factor of safety of the slope at the five sampling points are listed in Table 7.3.
6. Least-squares method is used to solve the unknown parameters:

$$\lambda_0 = -0.9643;\ \lambda_1 = 0.0293;\ \lambda_2 = 1.4600;\ \lambda_3 = 0.0000;\ \lambda_4 = 0.0425;$$

7. Using g' as the performance function, AFORM is used to calculate the reliability index β and the design point $\boldsymbol{x}^*$

$$\beta = 2.8784,\ c^* = 16.6199,\ \tan\phi^* = 0.3180$$

8. Using $c^* = 16.6199, \tan\phi^* = 0.3180$ as the NEW central points and repeating Steps 4 to 7. A series of reliability index can be obtained. The iteration is stopped until the difference of reliability index obtained is smaller than 0.01. The reliability indexes and design points obtained in the iterations are listed in Table 7.4.

Table 7.3 Sampling with the central composite design and the stability performance function g (Example 7.8)

Point no.	*c*	*tan* ϕ	F_s	*g*
1	30.0000	0.5000	1.6842	0.6842
2	21.5160	0.5000	1.4215	0.4215
3	30.0000	0.3586	1.4726	0.4726
4	38.4840	0.5000	1.9516	0.9516
5	30.0000	0.6414	1.8975	0.8975

Table 7.4 The iteration process in RSM

	Iteration no.		
	1	*2*	*3*
Coefficients of response surface function g'	$\lambda_0 = -0.9643$ $\lambda_1 = 0.0293$ $\lambda_2 = 1.4600$ $\lambda_3 = 0.0000$ $\lambda_4 = 0.0425$	$\lambda_0 = -0.9829$ $\lambda_1 = 0.0303$ $\lambda_2 = 1.4380$ $\lambda_3 = 0.0000$ $\lambda_4 = 0.1125$	$\lambda_0 = -0.9927$ $\lambda_1 = 0.0308$ $\lambda_2 = 1.4774$ $\lambda_3 = 0.0000$ $\lambda_4 = 0.04$
Design point (c^*, tan ϕ^*)	$c^* = 16.6199$ tan $\phi^* = 0.3180$	$c^* = 16.6768$ tan $\phi^* = 0.3182$	$c^* = 16.6739$ tan $\phi^* = 0.3205$
Reliability index β	$\beta = 2.8784$	$\beta = 2.8546$	$\beta = 2.8560$

The MATLAB code for solving the above reliability problem with RSM is shown below.

```
% Example 7.8 %
% RSM %
clear; clc;
muX = [30;0.2];  sigmaX = [0.5;0.2]; %c, tan(fai)
Amatrix = CentralCompositeDesign(muX, sigmaX, 1.414) %alpha=1.414
Zresponse = [-0.9643;0.0293;1.4600;0.0000;0.0425];
%Z=Fos-1, calculated by EMU2005 in this case
Lambda = Amatrix\Zresponse % inverse(Amatrix)*Zresponse
correlationX = [1,0;0,1];  % the correlation coefficients for c and tan(fai)
[beta, xstar] = ReliabilityFor2ndPolynomial(muX, sigmaX, correlationX,
Lambda)
1-normcdf(beta) % the probability of failure

function [ Amatrix ] = CentralCompositeDesign(muX, sigmaX, f)
% Central Composite Design
x = muX; n = length(muX); diagsigmaX = diag(f*sigmaX);
column1 = ones(2*n+1, 1);
b1 = x'; b2 = (x.*x)'; row1 = [b1, b2];
c1 = repmat(x',n,1); c2 = c1-diagsigmaX; c3 = c2.*c2;
d2 = c1+diagsigmaX; d3 = d2.*d2;
row2n = [c2, c3]; row3n = [d2, d3];
Amatrix = [column1, [row1;row2n;row3n]];  % the 2n+1 design points
end
%% Calculate the reliability index for 2nd order polynomial approximation
function [beta, xstar] = ReliabilityFor2ndPolynomial(muX, sigmaX,
correlationX, lambda)
 covarianceX = diag(sigmaX)*correlationX*diag(sigmaX);
 [orthogonalYtoX, diagsigmaY2] = eig(covarianceX);
 muY = orthogonalYtoX'*muX; sigmaY = sqrt(diag(diagsigmaY2));
 x=muX; y=muY; normX = eps; n = length(muX);
```

(*Continued*)

```
  while abs(norm(x)-normX)/normX > 1e-6
     normX=norm(x);
     g = lambda'*[1;x;x.*x];
     deltagX = lambda(2:n+1)+2*lambda(n+2:2*n+1).*x;
     deltagY = orthogonalYtoX'*deltagX;
     gs = deltagY.*sigmaY;
     alphaY = -gs/norm(gs);
     beta = (g+deltagY'*(muY-y))/norm(gs);
     y = muY + beta*sigmaY.*alphaY;
     x = orthogonalYtoX*y;
  end
  xstar = x;
end
```

7.4 UNCERTAINTIES OF SOIL PROPERTIES

7.4.1 Index, strength, and compressibility parameters

The reported uncertainties of common soil properties for saturated soils from literature are summarized in Tables 7.5 and 7.6. According to the statistical methods adopted in these references, the COVs in the tables are point value. Based on the two tables, some general findings about the uncertainties of soil properties can be listed as follows:

1. The COV of soil unit weight is relatively small, around 5% ~ 10%. The COV of water contents varies from 10% to 25%. The COV of porosity is from 10% to 30%.
2. The uncertainties of the shear strength parameters, the compression indexes and the consolidation indexes are relatively large. The COV values vary from 5% to 50%.
3. The uncertainties of field measured soil properties are greater than those of the lab measured ones.
4. The soil parameters generally follow Normal or lognormal distributions (Baecher and Christian, 2003; Réthâti, 2012).

Table 7.5 Inherent variability of laboratory measured soil properties

Soil property	*Parameter*	*Soil type*	*Distribution*	*COV (%)*	*References*
1. Index	Submerged unit weight	All soils	N	0–10	Lacasse and Nadim (1996)
	Density	All soils	–	5–10	Lumb (1974)
	γ (kN/m^3)	Fine grained	–	3–20	Phoon and Kulhawy (1999)
	γ_d (kN/m^3)	Fine grained	–	2–13	Phoon and Kulhawy (1999)
	D_r (%) (direct method)	Sand		11–36	Phoon and Kulhawy (1999)
	D_r (%) (indirect method)	Sand		49–74	Phoon and Kulhawy (1999)
	Void ratio, porosity	All soils	N	7–30	Lacasse and Nadim (1996)
	Void ratio	All soils	–	15–30	Lumb (1974)

(*Continued*)

Table 7.5 (Continued) Inherent variability of laboratory measured soil properties

Soil property	*Parameter*	*Soil type*	*Distribution*	*COV (%)*	*References*
	w_n (%)	Fine grained	–	7–46	Phoon and Kulhawy (1999)
	w_L (%)	Fine grained	–	7–39	Phoon and Kulhawy (1999)
	w_P (%)			6–34	Phoon and Kulhawy (1999)
	w_L, w_P	Clay	N	3–20	Lacasse and Nadim (1996)
	PI	Fine grained	–	9–57	Phoon and Kulhawy (1999)
	LI	Clay, silt	–	60–88	Phoon and Kulhawy (1999)
2. Strength	Undrained shear strength	Clay (triaxial)	LN	5–20	Lacasse and Nadim (1996)
	Undrained shear strength	Clay	LN	10–35	Lacasse and Nadim (1996)
	Undrained shear strength	Clayey silt	N	10–30	Lacasse and Nadim (1996)
	S_u (UC)	Fine grained	–	6–56	Phoon and Kulhawy (1999)
	S_u (UU)	Clay, silt	–	11–49	Phoon and Kulhawy (1999)
	S_u (CIUC)	Clay	–	18–42	Phoon and Kulhawy (1999)
	S_u/σ'_{V0}	Clay	N/LN	5–15	Lacasse and Nadim (1996)
	Undrained cohesion	Clay	–	20–50	Lumb (1974)
		Sand	N	2–5	Lacasse and Nadim (1996)
	ϕ'	Clay		40	Kotzias et al. (1993)
	ϕ'	Alluvial		16	Wolff (1996)
	ϕ'	Tailings		5–20	Baecher et al. (1983)
	ϕ'	Sand	–	5–11	Phoon and Kulhawy (1999)
	ϕ'	Clay, silt	–	4–50	Phoon and Kulhawy (1999)
	$\tan\phi'$	Sand	–	5–15	Lumb (1974)
	$\tan\phi'$ (TC, DS)	Clay, silt	–	6–46	Phoon and Kulhawy (1999)
3. Compressibility and consolidation	OCR	Clay	N/LN	10–35	Lacasse and Nadim (1996)
	Compressibility	All soils	–	25–30	Lumb (1974)
	C_c, C_r	Bangkok Clay	–	20	Zhu et al. (2001)
		Dredge Spoils	–	35	Thevanayagam et al. (1996)
		Gulf of Mexico Clay	–	25–28	Baecher and Ladd (1997)
		Various soils	–	25–50	Lumb (1974)
	C_v	Ariake Clay	–	10	Tanaka et al. (2001)
		Singapore Clay	–	17	Tanaka et al. (2001)
		Bangkok Clay	–	16	Tanaka et al. (2001)

Notes: γ, unit weight; γ_d, dry unit weight; D_r, relative density; w_n, natural water content; w_L, liquid limit; w_P, plastic limit; PI, plasticity index; LI, liquidity index; S_u, undrained shear strength; ϕ', Effective friction angle; UC, unconfined compression test; UU, unconsolidated-undrained triaxial compression test; CIUC, isotropic consolidated undrained triaxial compression test; TC, triaxial compression test; DS, direct shear test; OCR, Over Consolidation Ratio; C_c, compression index; C_r, recompression index; C_v, coefficient of consolidation.

7.4.2 Soil hydraulic properties

Soil hydraulic properties are important properties in research fields such as soil science and agriculture, hydrology and water resources, and environmental engineering. Many researchers have been working on studying the variability of hydraulic properties for decades, including the spatial variability of field-measured hydraulic properties (Greminger et al., 1985;

Table 7.6 Approximate guidelines for inherent soil variability of field tests

Test type	*Property*	*Soil type*	*Mean*	*Unit*	*COV (%)*
CPT	q_T	Clay	0.5–2.5	MN/m²	<20
	q_c	Clay	0.5–2	MN/m²	20–40
	q_c	Sand	0.5–30	MN/m²	20–60
VST	S_u	Clay	5–400	kN/m²	10–40
SPT	N	Clay and sand	10–70	blows/ft	25–50
PMT	P_L	Clay	400–2800	kN/m²	10–35
	P_L	Sand	1600–3500	kN/m²	20–50
	E_{PMT}	Sand	5–15	MN/m²	15–65

Source: Phoon, K. K., and Kulhawy, F. H., *Canadian Geotechnical Journal*, 36, 612–624, 1999. With permission.

Notes: CPT, cone penetration test; VST, vane shear test; SPT, standard penetration test; PMT, pressuremeter test; q_c, CPT tip resistance; q_T, corrected CPT tip resistance; N, SPT blow count; p_L, PMT limit stress; E_{PMT}, PMT modulus.

Russo and Bouton, 1992; Ragab and Cooper, 1993) and the lab-measured hydraulic properties (Sillers and Fredlund, 2001; Phoon et al., 2010). Others tried to develop statistics of estimated soil hydraulic properties (Carsel and Parrish, 1988; Zapata et al., 2000).

Carsel and Parrish (1988) presented a comprehensive investigation of the PDF of the parameters of the van Genuchten (1980) and Maulem (1976) (VGM) SWCC model. A soil database compiled by Carsel et al. (1988) was used to obtain bulk density, sand and clay contents for the twelve USDA (United States Department of Agriculture) textural classifications (Figure 7.19) including: clay, clay loam, loam, loamy sand, silt, silt loam, silty clay, silty clay loam, sand, sandy clay, sandy clay loam and sandy loam. The saturated water content, the sand contents and clay contents were then used to estimate the parameters for the van Genuchten model and k_s based on a multiple regression model. Table 7.7 lists the descriptive statistics for θ_s, k_s and the parameters α_{vgm} and n_{vgm} in the van Genuchten model for the twelve types of soils. Sillers and Fredlund (2001) used only the data with substantial portion of experimental points in the database. Totally 230 sets of measured soil-water characteristic curves (SWCCs) were collected in the database. The soils were also classified into twelve categories according to USDA soil classification system. A nonlinear least-squares method was used to determine best-fitting parameters for commonly used empirical SWCC models such as the Brooks and Corey (1964) model, the Gardner (1958) model, the van Genuchten (1980) model, and the Fredlund and Xing (1994) model. Tables 7.8 and 7.9 show the statistics of the fitting parameters for the Fredlund and Xing (1994) model and the van Genuchten (1980) model for eight different soils.

Carsel and Parrish (1988) performed goodness-of-fit tests for the van Genuchten (1980) model parameters using four distribution types: NO, normal; LN, lognormal; SB, log ratio; and SU, hyperbolic arcsine. The random variables following the LN, SB, or SU distributions can be transformed to normal distribution as follows:

$$\text{LN}: \quad Y = \ln(X) \tag{7.59}$$

$$\text{SB}: \quad Y = \ln[(X-a)/(\text{b}-X)] \tag{7.60}$$

$$\text{SU}: \quad Y = \sinh^{-1}[U] = \ln[U + (1+U^2)^{1/2}] \tag{7.61}$$

where:

X denotes an untransformed variable with limits of variation from a to b ($a < X < b$)

$U = (X - a)/(b - a)$ a and b are lower and upper limits of X for the log-ratio distribution, respectively

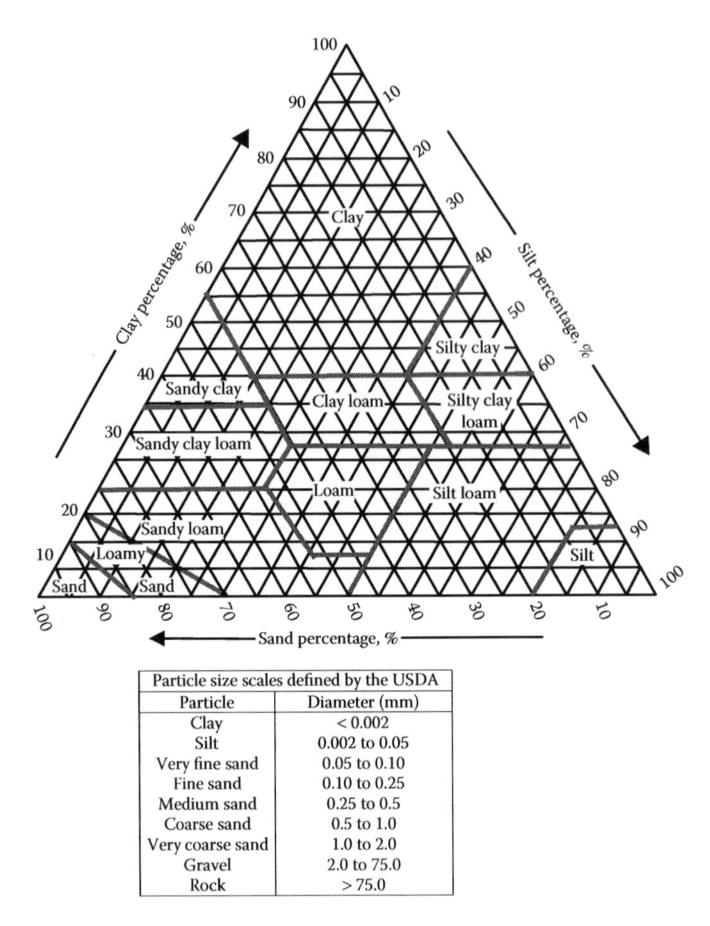

Particle size scales defined by the USDA	
Particle	Diameter (mm)
Clay	< 0.002
Silt	0.002 to 0.05
Very fine sand	0.05 to 0.10
Fine sand	0.10 to 0.25
Medium sand	0.25 to 0.5
Coarse sand	0.5 to 1.0
Very coarse sand	1.0 to 2.0
Gravel	2.0 to 75.0
Rock	> 75.0

Figure 7.19 USDA soil textual triangle showing the 12 major textural classes and particle size scales defined by the USDA.

The LN distribution is defined for all positive values of X, being unbounded above. SB is bounded between limits a and b, while SU generally is unbounded. Here, X corresponds to any of the random variables k_s, θ_r, α_{vgm}, and n_{vgm}. In each case, Y has a normal distribution.

The Kolmogorov–Smirnov (K–S) D statistic (e.g, Ang and Tang, 2007), D_{max}, was used to select the best fitting distribution from among the four candidates (NO, LN, SB, SU). The K–S D statistic D_{max} is the maximum absolute deviation between the empirical and fitted CDF. Therefore, the smallest observed value of D_{max} value implies the most appropriate transformation. Table 7.10 presents the estimates of the distribution mean and standard deviation for the transformed variables, limits of variation for the original variables, and values of the K–S goodness-of-fit statistic. Very few of the data sets could be adequately described by the normal distribution without using one of the transformations. Notably, many data sets were significantly better described by the SB and SU distributions rather than the more commonly used lognormal (Carsel and Parrish, 1988).

Correlations coefficients for the transformed parameters were evaluated based on statistical analyses (Table 7.11). The diagonal and upper triangular entries are the triangular Cholesky decomposition **T** of the covariance matrix **C**, which means that **C** = **T′T**. The lower triangular matrix comprises Pearson product-moment correlation coefficients. As shown in Table 7.11, in most cases, correlations were significant for the van Genuchten model parameters. For example, correlations generally were greater than 0.70 for between k_s and a_{vgm}, and between k_s and n_{vgm}.

Table 7.7 Statistics for saturated water content θ_s, residual water content θ_r, saturated coefficient of permeability k_s, and parameter a_{vgm} and n_{vgm} of van Genuchten (1980) and Mualem (1976) model

	θ_s			θ_r			k_s (cm/h)			a_{vgm} (cm^{-1})			n_{vgm}		
Soil type	*Mean*	*COV (%)*	*Sample* size	*Mean*	*COV (%)*	*Sample* size	*Mean*	*COV (%)*	*Sample* size	*Mean*	*COV (%)*	*Sample* size	*Mean*	*COV(%)*	*Sample* size
Clay	0.38	24.1	400	0.068	49.9	353	0.20	210.3	114	0.008	160.3	400	1.09	7.9	400
Clay loam	0.41	22.4	364	0.095	10.1	363	0.26	267.2	345	0.019	77.9	363	1.31	7.2	364
Loam	0.43	22.1	735	0.078	16.5	735	1.04	174.6	735	0.036	57.1	735	1.56	7.3	735
Loamy sand	0.41	21.6	315	0.057	25.7	315	14.59	77.9	315	0.124	35.2	315	2.28	12.0	315
Silt	0.46	17.4	82	0.034	29.8	82	0.25	129.9	88	0.016	45.0	82	1.37	3.3	82
Silt loam	0.45	18.7	1093	0.067	21.6	1093	0.45	275.1	1093	0.020	64.7	1093	1.41	8.5	1093
Silty clay	0.36	19.6	374	0.07	33.5	371	0.02	453.3	126	0.005	113.6	126	1.09	5.0	374
Silty clay loam	0.43	17.2	641	0.089	10.9	641	0.07	288.7	592	0.010	61.5	641	1.23	5.0	641
Sand	0.43	15.1	246	0.045	22.3	246	29.70	52.4	246	0.145	20.3	246	2.68	20.3	246
Sandy clay	0.38	13.7	46	0.1	12.9	46	0.12	234.1	46	0.027	61.7	46	1.23	7.9	46
Sandy clay loam	0.39	17.5	214	0.1	6.0	214	1.31	208.6	214	0.059	64.6	214	1.48	8.7	214
Sandy loam	0.41	21.0	1183	0.065	26.6	1183	4.42	127.0	1183	0.075	49.4	1183	1.89	9.2	1183

Source: Carsel, R. F., and Parrish, R. S., *Water Resources Research*, 24, 755–769, 1988. With permission.

Table 7.8 Statistics of parameters a_f, n_f, and m_f for the Fredlund and Xing (1994) model with Fredlund and Xing (1994) correction function for eight different soils

		a_f (kPa)				n_f				m_f			
Soil type	*Sample size*	*Mean*	*Standard deviation*	*Median*	*COV (%)*	*Mean*	*Standard deviation*	*Median*	*COV (%)*	*Mean*	*Standard deviation*	*Median*	*COV (%)*
Clay	12	506.3	790.9	30.95	156.2	9.27	27.57	1.075	297.4	0.626	0.648	0.327	103.5
Clay loam	24	172.6	210.3	92.30	121.8	2.42	6.31	0.864	260.9	0.492	0.28	0.535	56.9
Loam	18	20.3	43.1	4.971	212.4	2.37	1.39	1.941	58.9	0.562	0.283	0.413	50.4
Loamy sand	12	6.7	8.3	3.667	123.9	3.40	3.44	2.026	101.1	0.874	0.343	0.836	39.2
Sand	50	17.3	84.5	2.814	489.2	6.54	4.29	5.776	65.6	4.428	10.72	0.885	242.1
Silt	25	188.4	370.7	33.58	196.8	4.29	9.65	1.676	225.2	0.949	1.381	0.714	145.5
Silty loam	23	63.1	153.6	9.656	243.3	2.19	1.99	1.294	90.8	0.665	0.323	0.626	48.6
Sandy loam	36	16.4	21.2	8.953	129.7	5.53	16.44	1.729	297.3	2.608	10.76	0.58	412.6

Source: Sillers, W. S., and Fredlund, D. G., *Canadian Geotechnical Journal*, 38, 1297–1313, 2001. With permission.

Note: A residual suction of 3000 kPa was used when obtaining the best-fit parameters.

Table 7.9 Statistics of parameters a_{vg}, n_{vg}, and m_{vg} for the van Genuchten (1980) model with Fredlund and Xing (1994) correction function for eight different soils

		α_{vg} (cm^{-1})				n_{vg}				m_{vg}			
Soil type	*Sample size*	*Mean*	*Standard deviation*	*Median*	*COV (%)*	*Mean*	*Standard deviation*	*Median*	*COV (%)*	*Mean*	*Standard deviation*	*Median*	*COV(%)*
Clay	12	0.968	1.973	0.07	203.8	9.883	27.43	1.45	277.5	0.902	2.511	0.043	278.4
Clay loam	24	0.7	1.821	0.03	260.1	3.554	7.282	1.4	204.9	0.092	0.07	0.081	76.1
Loam	18	0.427	0.367	0.341	85.9	5.99	5.59	3.252	93.3	0.102	0.106	0.052	104.0
Loamy sand	12	1.188	2.581	0.448	217.3	8.695	10.698	3.242	123.0	0.186	0.174	0.147	93.5
Sand	50	0.418	0.25	0.399	59.8	9.621	7.207	7.441	74.9	2.976	7.843	0.197	263.5
Silt	25	0.095	0.162	0.051	170.5	7.41	8.827	3.35	119.1	2.371	10.3	0.075	434.4
Silty loam	23	0.42	0.468	0.266	111.4	3.323	2.815	2.136	84.7	0.142	0.132	0.08	93.0
Sandy loam	36	0.41	0.609	0.222	148.5	3.899	4.223	2.32	108.3	2.28	11.88	0.156	521.0

Source: Sillers, W. S., and Fredlund, D. G., *Canadian Geotechnical Journal*, 38, 1297–1313, 2001. With permission.

Note: A residual suction ψ_r of 3000 kPa was used when obtaining the best-fitting parameters.

Table 7.10 Distribution and statistics for transformed parameters of VGM model

		Limits of variation			*Statistics of transformed variable*		
Soil texture	*Random variable*	*a*	*b*	*Transformation*	*Mean*	*Standard deviation*	D_{max}
Sand	k_S	0	70	SB	–0.394	1.15	0.045
	θ_r	0	0.1	LN	–3.12	0.224	0.053
	a_{vgm}	0	0.25	SB	0.378	0.439	0.05
	n_{vgm}	1.5	4	LN	0.978	0.1	0.063
Sandy loam	k_S	0	30	SB	2.49	1.53	0.029
	θ_r	0	0.11	SB	0.384	0.7	0.034
	a_{vgm}	0	0.25	SB	–0.937	0.764	0.044
	n_{vgm}	1.35	3	LN	0.634	0.082	0.039
Loamy sand	k_S	0	51	SB	–1.27	1.4	0.036
	θ_r	0	0.11	SB	0.075	0.567	0.043
	a_{vgm}	0	0.25	NO	0.124	0.043	0.027
	n_{vgm}	1.35	5	SB	–1.11	0.307	0.07
Silt loam	k_S	0	15	LN	–2.19	1.49	0.046
	θ_r	0	0.11	SB	0.478	0.582	0.073
	a_{vgm}	0	0.15	LN	–4.1	0.555	0.083
	n_{vgm}	1	2	SB	–0.37	0.526	0.104
Silt	k_S	0	2	LN[a]	–2.2	0.7	0.168
	θ_r	0	0.09	NO[a]	0.042	0.015	0.089
	a_{vgm}	0	0.1	NO	0.017	0.006	0.252
	n_{vgm}	1.2	1.6	NO	1.38	0.037	0.184
Clay	k_S	0	5	SB	–.75	2.33	0.122
	θ_r	0	0.15	SU[a]	0.445	0.282	0.058
	a_{vgm}	0	0.15	SB[a]	–4.145	1.293	0.189
	n_{vgm}	0.9	1.4	LN[a]	2E–04	0.118	0.131
Silty clay	k_S	0	1	LN	–5.69	1.31	0.205
	θ_r	0	0.14	NO	0.07	0.023	0.058
	a_{vgm}	0	0.15	LN	–5.66	0.584	0.164
	n_{vgm}	1	1.4	SB	–1.28	0.821	0.069
Sandy clay	k_S	0	1.5	LN	–4.04	2.02	0.13
	θ_r	0	0.12	SB	1.72	0.7	0.078
	a_{vgm}	0	0.15	LN	–3.77	0.563	0.127
	n_{vgm}	1	1.5	LN	0.202	0.078	0.1
Silty clay loam	k_S	0	3.5	SB	–5.31	1.62	0.049
	θ_r	0	0.115	NO	0.088	0.009	0.056
	n_{vgm}	1	1.5	NO	1.23	0.061	0.082
Clay loam	k_S	0	7.5	SB[a]	–5.87	2.92	0.058
	θ_r	0	0.13	SU	0.679	0.06	0.061
	a_{vgm}	0	0.15	LN	–4.22	0.72	0.052
	n_{vgm}	1	1.6	SB	0.132	0.725	0.035

(Continued)

Table 7.10 (Continued) Distribution and statistics for transformed parameters of VGM model

Soil texture	*Random variable*	*Limits of variation* *a*	*b*	*Transformation*	*Statistics of transformed variable* *Mean*	*Standard deviation*	D_{max}
Sandy clay loam	k_S	0	20	SB	–4.04	1.85	0.047
	θ_r	0	0.12	SB[a]	1.65	0.439	0.077
	a_{vgm}	0	0.25	SB	–1.38	0.823	0.048
	n_{vgm}	1	2	LN	0.388	0.086	0.043
Loam	k_S	0	15	SB	–3.71	1.78	0.019
	θ_r	0	0.12	SB	0.639	0.487	0.064
	a_{vgm}	0	0.15	SB	–1.27	0.786	0.039
	n_{vgm}	1	2	SU	0.532	0.099	0.036

Source: Carsel, R. F., and Parrish, R. S., *Water Resources Research*, 24, 755–769, 1988. With permission.

Note: D_{max} is K–S goodness-of-fit test statistic NO is normal distribution; LN is lognormal distribution; SB is log ratio distribution; SU is hyperbolic arcsine distribution.

[a] Truncated form of the distribution.

Table 7.11 Correlation among transformed hydraulic parameters of VGM model

	k_S	θ_r	a_{vgm}	n_{vgm}
Silt (*N* = 61)				
k_S	0.535	–0.002	0.003	0.01
θ_r	–0.204	0.008	0	–0.015
a_{vgm}	0.984	–0.2	0.001	0.014
n_{vgm}	0.466	–0.61	0.551	0.013
Clay (*N* = 95)				
k_S	1.96	0.07	0.565	0.048
θ_r	0.972	0.017	–0.08	–0.014
a_{vgm}	0.948	0.89	0.172	0.002
n_{vgm}	0.908	0.819	0.91	0.016
Silty clay (*N* = 123)				
k_S	1.25	0.008	0.314	0.367
θ_r	0.949	0.003	0.04	–0.086
a_{vgm}	0.974	0.964	0.06	0.066
n_{vgm}	0.908	0.794	0.889	0.131
Sandy clay (*N* = 46)				
k_S	2.02	0.883	0.539	0.076
θ_r	0.939	0.324	0.063	0.004
a_{vgm}	0.957	0.937	0.15	–0.001
n_{vgm}	0.972	0.928	0.932	0.018
Sand (*N* = 237)				
k_S	1.04	0.883	0.328	0.081
θ_r	–0.515	0.324	0.258	–0.047

(Continued)

Table 7.11 (Continued) Correlation among transformed hydraulic parameters of VGM model

	k_S	θ_r	a_{vgm}	n_{vgm}
a_{vgm}	0.743	0.937	0.143	–0.011
n_{vgm}	0.843	0.928	0.298	0.017
Sandy loam (*N* = 1145)				
k_S	1.6	–0.153	0.037	0.211
θ_r	–0.273	0.538	0.017	–0.194
a_{vgm}	0.856	0.151	0.014	0.019
n_{vgm}	0.686	–0.796	0.354	0.108
Loamy sand (*N* = 313)				
k_S	1.48	–0.201	0.037	0.211
θ_r	–0.359	0.522	0.017	–0.194
a_{vgm}	0.986	–0.301	0.014	0.019
n_{vgm}	0.73	–0.59	0.354	0.108
Silt loam (*N* = 1072)				
k_S	1.478	–0.201	0.525	0.353
θ_r	–0.359	0.522	0.03	–0.17
a_{vgm}	0.986	–0.301	0.082	0.234
n_{vgm}	0.73	–0.59	0.775	0.158
Silty clay loam (*N* = 591)				
k_S	1.612	0.006	0.511	0.049
θ_r	0.724	0.005	0.048	–0.009
a_{vgm}	0.986	0.77	0.073	0.008
n_{vgm}	0.918	0.549	0.911	0.017
Clay loam (*N* = 328)				
k_S	1.92	0.04	0.589	0.542
θ_r	0.79	0.031	–0.062	–0.154
a_{vgm}	0.979	0.836	0.106	0.065
n_{vgm}	0.936	0.577	0.909	0.116
Sandy clay loam (*N* = 212)				
k_S	1.85	0.102	0.784	0.077
θ_r	0.261	0.378	0.122	–0.031
a_{vgm}	0.952	2.392	0.22	–0.008
n_{vgm}	0.909	–0.113	0.787	0.016
Loam (*N* = 328)				
k_S	1.41	–0.1	0.611	0.055
θ_r	0.204	0.478	0.073	–0.055
a_{vgm}	0.982	–0.086	0.093	0.026
n_{vgm}	0.632	–0.748	0.591	0.029

Source: Carsel, R. F., and Parrish, R. S., *Water Resources Research*, 24, 755–769, 1988. With permission.

Note: Entries in the lower triangular portion of the matrix are sample Pearson product-moment correlations. The diagonal and upper triangular entries are triangular Cholesky decomposition **T** of the sample covariance matrix **C**, which means **C** = **T′T**. *N* is sample size.

7.5 EFFECTS OF UNCERTAINTY OF HYDRAULIC PROPERTIES ON INFILTRATION AND SLOPE STABILITY

In this section, the pore-water pressure distributions in the completely decomposed granite (CDG) soil slopes and the completely decomposed volcanic (CDV) soil slopes are investigated by numerical analyses. The effects of uncertainty of soil unsaturated hydraulic properties on the matric suctions in the soil slopes are illustrated.

7.5.1 Uncertainty of hydraulic parameters of CDG and CDV soils

Laboratory studies on the SWCCs of Hong Kong CDG soils and CDV soils have been conducted by Gan and Fredlund (1996), Fung (2001) and Pang (1999). Figure 7.20 shows

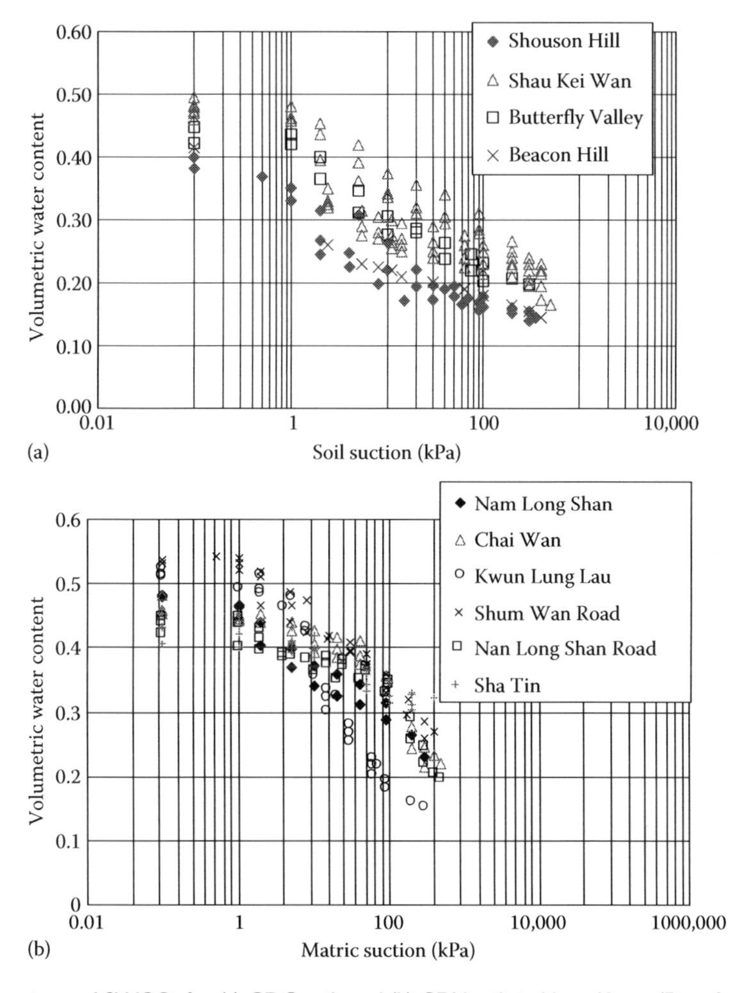

Figure 7.20 Experimental SWCCs for (a) CDG soils and (b) CDV soils in Hong Kong. (Data from Gan, J. K. M., and Fredlund, D. G., *Study of the Application of Soil-Water Characteristic Curves and Permeability Functions to Slope Stability, Part I, Permeability and Soil-Water Characteristic Curve Tests, and the Computation of Permeability Functions*, Department of Civil Engineering University of Saskatchewan, Saskatoon, Saskatchewan, Canada, 1996; Pang, Y. W., *Investigation of Soil-Water Characteristics and Their Implications on Stability of Unsaturated Soil Slopes*, M.Phil Thesis, Hong Kong University of Science and Technology, Hong Kong, 1999; and Fung, W. T., *Experimental Study and Centrifuge Modeling of Loose Fill Slope*, M.Phil Thesis, Hong Kong University of Science and Technology, Hong Kong, 2001.)

experimentally measured SWCCs for CDG soils and CDV soils from different sites in Hong Kong. These measured SWCCs are all drying curves. By the least-squares method, the parameters a_f, n_f, and m_f of the Fredlund and Xing SWCC (1994) model can be estimated. The residual suction ψ_r is assumed to be 3000 kPa when obtaining the best-fitting soil parameters as the SWCC at low suctions is not significantly influenced by ψ_r and the magnitude of ψ_r is generally in the range from 1500 to 3000 kPa (Fredlund and Xing, 1994). The best-fitting curves of the SWCC for CDG soils are presented in Figure 7.21. The K–S test (e.g., Ang and Tang, 2007) is performed to evaluate the possible distribution of a_f and n_f. Normal distributions can be substantiated for both $\ln(a_f)$ and $\ln(n_f)$ at a significance level of 5% as shown in Figure 7.22.

For the experimentally measured SWCCs (Figure 7.20), the corresponding saturated coefficients of permeability are not available. Therefore, the coefficients of saturated permeability are estimated based on the percentages of sand and clay and the porosity using the regression model developed by Rawls and Brakensiek (1985):

$$\begin{aligned} k_s = \exp[&-8.96847 - 0.028212P_{clay} + 19.52348\theta_s + 0.00018107P_{sand}{}^2 \\ &-0.0094125P_{clay}{}^2 - 8.395215\theta_s^2 + 0.077718P_{sand} \cdot \theta_s + 0.0000173P_{sand}{}^2 \cdot P_{clay} \\ &+0.02733P_{clay}{}^2 \cdot \theta_s \\ &+0.001434P_{sand}{}^2 \cdot \theta_s - 0.0000035P_{sand} \cdot P_{clay}{}^2 - 0.00298P_{sand}{}^2 \cdot \theta_s^2 - 0.019492P_{clay}{}^2 \cdot \theta_s^2 \end{aligned} \tag{7.62}$$

where:

k_s is the saturated permeability (cm/h)
P_{sand} is the percentage of sand (%)
P_{clay} is the percentage of clay (%)

The statistics of the hydraulic parameters for CDG and CDV soils are listed in Table 7.12. The mean and standard deviation of the transformed variables and the correlation matrix are presented in Table 7.13 and Table 7.14, respectively.

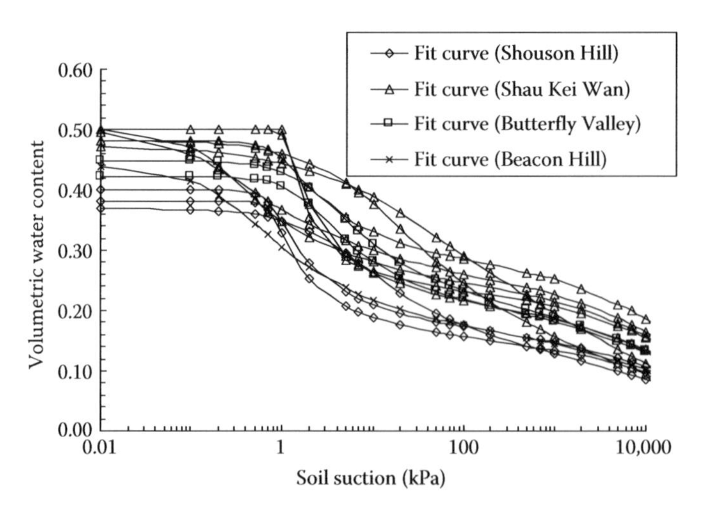

Figure 7.21 Best-fitting curves of the measured SWCCs of CDG soils.

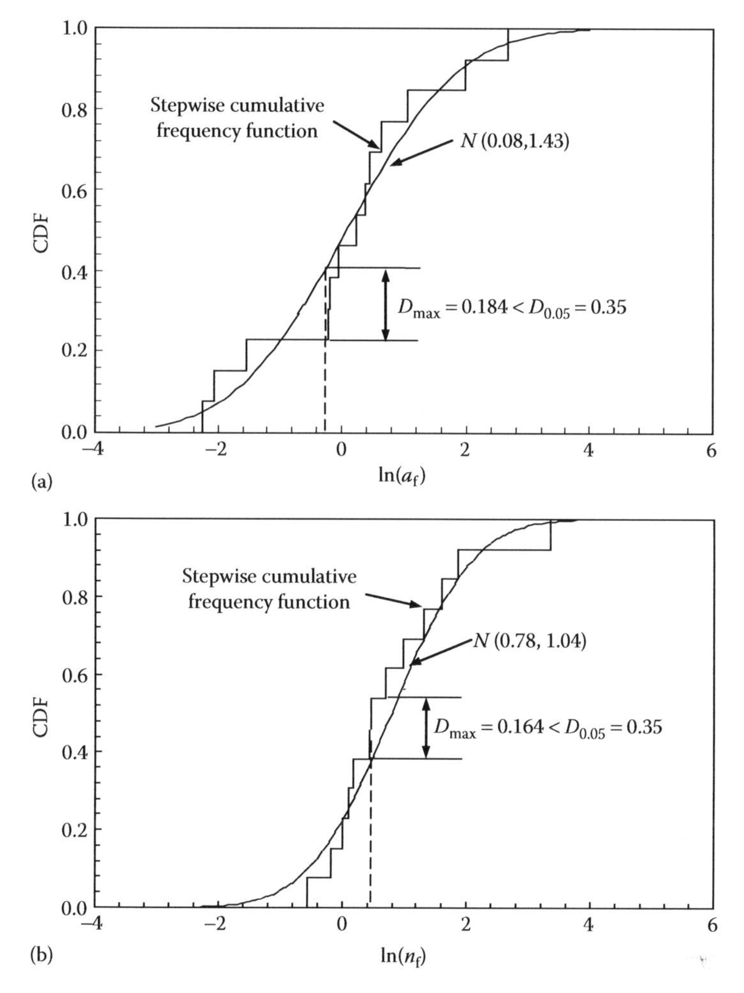

Figure 7.22 K–S test for distribution of (a) $\ln(a_f)$ and (b) $\ln(n_f)$ for Hong Kong CDG soils.

7.5.2 Infiltration in CDG and CDV soil slopes

A 20-m-high slope inclined at 30° shown in Figure 7.23 was used in the numerical modeling study. Along each of the left and right boundaries beneath the ground water table, a constant head was applied. A zero flux boundary was applied along the left and right boundaries above the ground water table. To illustrate the pore-water pressure profiles more clearly, the groundwater table was fixed by applying a constant pressure head equal to zero at the groundwater table. The base of the finite element mesh was assumed to be impermeable. Rainfall was modeled as a moisture flux boundary, q, applied along the slope surface.

Kasim et al. (1997) and Zhang et al. (2004) found that the a_f value of a soil has a greater influence than the soil desaturation parameter, n_f, on the long term matric suction conditions in a soil. Here, only the variation of the parameter a_f is considered for illustrative purpose. The other four parameters, θ_s, n_f, m_f, and the saturated coefficient of permeability k_s, are chosen to be their respective mean values based on the statistical studies. Three SWCCs with the parameter a_f of mean value (μ_a), mean minus one standard deviation ($\mu_a - \sigma$), and mean plus one standard deviation ($\mu_a + \sigma$) for each soil are used in the numerical modeling.

Table 7.12 Basic statistics of hydraulic parameters for CDG and CDV soils

	CDG					*CDV*				
	θ_s	a_f	n_f	m_f	k_s *(m/s)*	θ_s	a_f	n_f	m_f	k_s *(m/s)*
Mean	0.453	2.575	4.247	0.391	2.00×10^{-5}	0.381	38.502	2.147	0.628	1.19×10^{-5}
Standard deviation	0.048	3.963	7.280	0.208	1.55×10^{-5}	0.058	76.609	3.031	0.592	8.32×10^{-6}
COV	10.5%	153.9%	171.4%	53.3%	77.6%	15.1%	199%	1.412	0.944	70.0%
Max	0.500	14.232	27.784	0.860	4.11×10^{-5}	0.446	300.001	13.296	2.116	2.93×10^{-5}
Min	0.369	0.104	0.568	0.160	3.41×10^{-6}	0.275	0.983	0.487	0.097	3.06×10^{-6}
No. of samples	13	13	13	13	13	19	19	19	19	19

Note: ψ_r= 3000 kPa was used when obtaining the best-fitting parameters.

Table 7.13 Mean, standard deviation, and distribution of transformed hydraulic parameters for CDG and CDV soils

	CDG		CDV		
	Mean	*Standard deviation*	*Mean*	*Standard deviation*	*Distribution*
θ_s	0.453	0.048	0.381	0.058	Normal
ln (a_f)	0.078	1.443	2.363	1.593	Normal
ln (n_f)	0.782	1.040	0.282	0.888	Normal
ln (m_f)	−1.052	0.481	−0.842	0.896	Normal
ln(k_s)	−11.161	0.912	−11.606	0.791	Normal

Table 7.14 Correlation matrix for the transformed hydraulic parameters for CDG and CDV soils

CDG	θ_s	*ln*(a_f)	*ln*(n_f)	*ln*(m_f)	*ln*(k_s)
θ_s	1				
ln(a_f)	−0.148	1			
ln(n_f)	0.090	−0.120	1		
ln(m_f)	−0.168	0.485	−0.840	1	
ln(k_s)	0.744	−0.188	0.299	−0.28	1
CDV	θ_s	*ln*(a_f)	*ln*(n_f)	*ln*(m_f)	*ln*(k_s)
θ_s	1				
ln(a_f)	−0.191	1			
ln(n_f)	0.098	−0.537	1		
ln(m_f)	−0.310	0.827	−0.651	1	
ln(k_s)	−0.456	−0.317	0.101	0.126	1

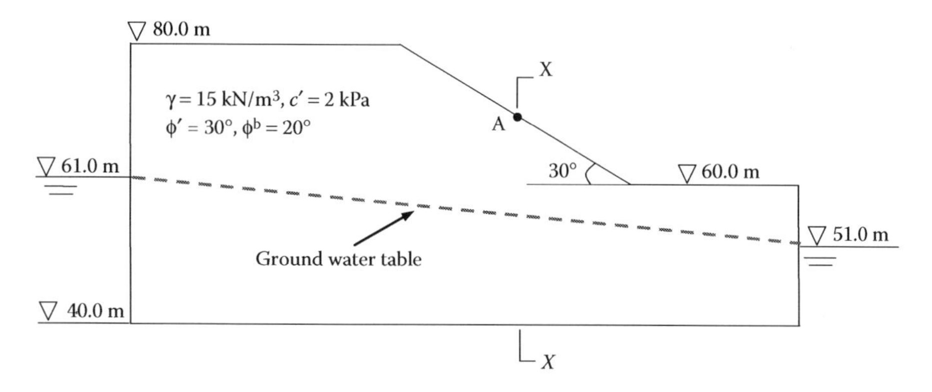

Figure 7.23 Geometry of a simple soil slope.

Figure 7.24(a) shows the mean curves and the one standard deviation bands of the SWCCs for the two soils. The corresponding coefficient of permeability functions for CDG and CDV are shown in Figure 7.24(b). The distribution and statistical parameters of a_f are based on Table 7.13.

Long-term rainfall fluxes q varying from 0.01 k_s to 1.0 k_s were applied on the slope. Figure 7.25 shows the steady state matric suction values at point A (the point at ground surface in Figure 7.23) of the section X–X in the slope versus the ratio of steady state

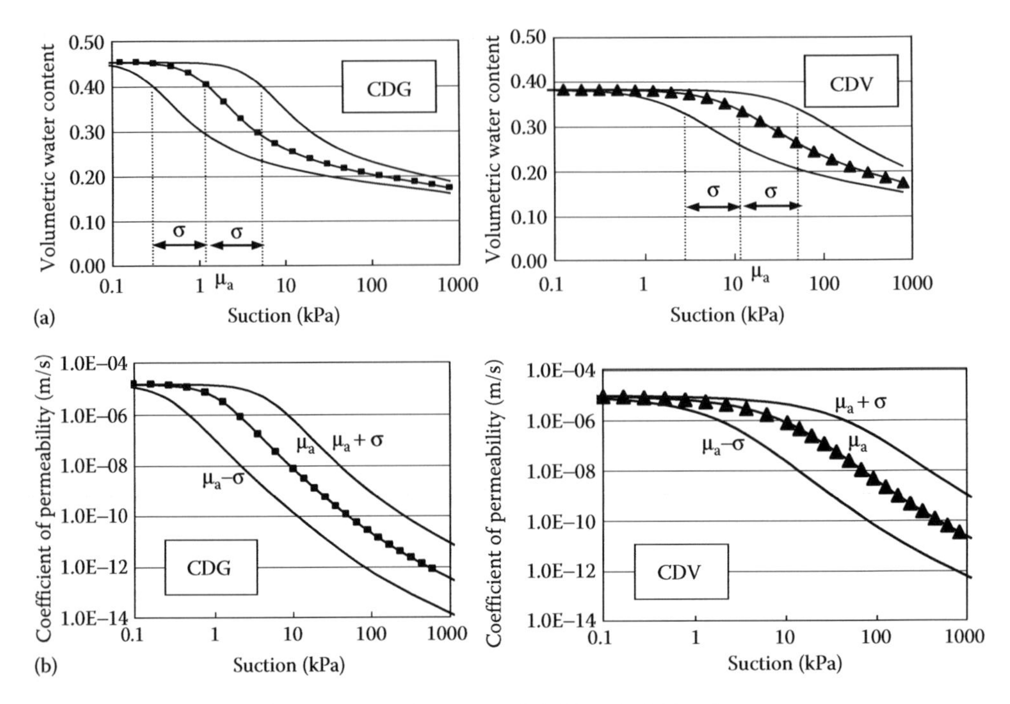

Figure 7.24 (a) SWCC and (b) permeability functions of CDG and CDV soils.

rainfall flux q to k_s for CDG and CDV soil slopes, respectively. The solid lines are the results for soils with the mean values of the parameter a_f. The dotted lines are the one standard deviation bounds of the pore-water pressure. The results illustrate that the decrease of matric suction at the surface of the slope is not linearly related to the ratio of q/k_s. When q is only 1% of k_s, the matric suction at the ground surface of the slope can be as great as 16 kPa for the CDG soil and 110 kPa for the CDV soil. When the ratio of q/k_s is greater than 0.3, the matric suction values at the surface of the slope decrease to only a few kPa. As the mean value of a_f for the CDG soil is about 10% of that of the CDV soil, the matric suction values at point A in the CDG soil slope are much less than those in the CDV soil slope. The variation of a_f for the CDG soil is also smaller than that of the CDV soil, therefore the range of matric suctions in the CDG soil slope is smaller than that in the CDV soil slope.

In the transient seepage analyses, the initial pore-water pressure profile in the slope is assumed to be hydrostatic and the rainfall fluxes applied in the slope are assumed to be equal to the coefficients of saturated permeability of the soils. Figure 7.26 shows the pore-water pressure profiles at the cross-section X–X in a slope after hours of rainfall. The graphs show the wetting fronts for CDV soils are not as distinct as those in CDG soil slopes. Here, the average depth of the curved wetting front is taken as the depth of wetting front. Figure 7.27 shows the advance of wetting fronts with the duration of rainfall. The advance of the wetting front in the CDV soil is on average much faster. For the CDV soil with the mean a_f value, the depth of wetting front is greater than 2 m after 12 h rainfall. As the mean value of a_f of the CDG soil is small, the water storage capacity of the soil is very large. Therefore, the advance of the wetting front is slow. After 50 hours rainfall, the depth of wetting front is only about 0.4 m for the CDG soil with the mean a_f value. After the same duration of rainfall infiltration,

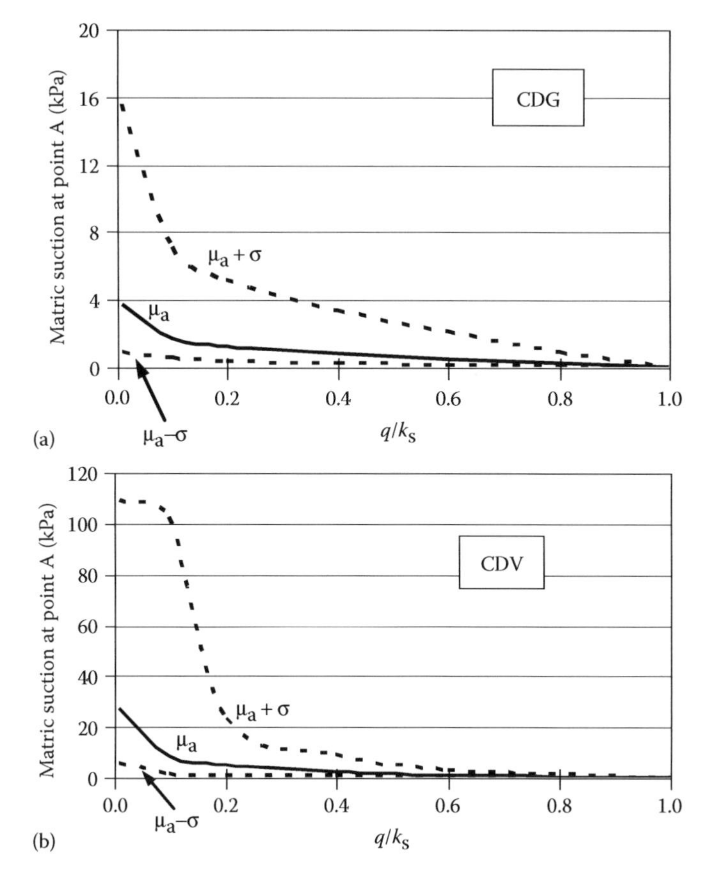

Figure 7.25 The steady state matric suctions at the point A versus the ratio of q/k_s for (a) CDG and (b) CDV soil slope.

the uncertainty bounds of the wetting front for CDG soil slopes are much narrower than those of the CDV soils as shown in Figure 7.27. This is consistent with the observations steady state infiltration condition (Figure 7.25).

7.5.3 Effects of correlation of hydraulic properties on slope reliability

Correlations among the parameters of hydraulic properties exist because of the physical relationships among soil properties. The correlation matrix of the transformed hydraulic parameters is obtained by a correlation analysis as shown in Table 7.14. For CDG soils, the coefficient of correlation between $\ln(k_s)$ and θ_s is 0.744, which is reasonable because the larger the void space in a soil, the larger the saturated permeability of the soil. $\ln(k_s)$ and θ_s for CDV soils are negatively correlated, which may be due to the estimation error of k_s. Generally, the correlations of θ_s with the three SWCC parameters, that is, $\ln(a_f)$, $\ln(n_f)$, and $\ln(m_f)$ are not significant for both CDG and CDV soils. However, the correlations among the three SWCC parameters themselves are considerable. The correlations among soil parameters for CDV soils are generally consistent with those of CDG soils.

To illustrate the effect of correlations on the reliability of slope, three sets of random SWCCs, and permeability functions of CDG soils with different assumptions of correlation

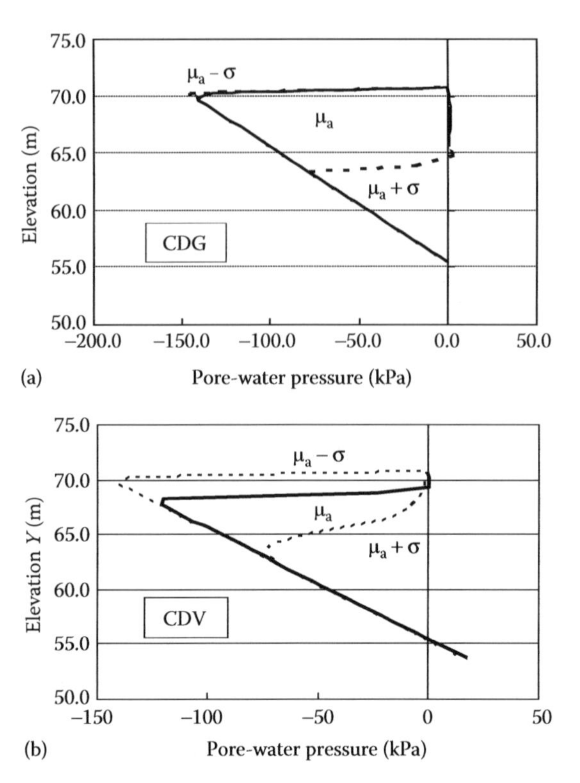

Figure 7.26 Pore-water pressure profiles at section X–X in the slope under transient condition with (a) CDG soil after 32 hours of rainfall (b) CDV soil after 12 hours of rainfall.

coefficients were generated using the LHS technique. For the first set of samples, the five random variables (i.e., $\ln[k_s]$, θ_s, $\ln[a_f]$, $\ln[n_f]$, $\ln[m_f]$) are assumed to be statistically independent. For the second set of samples, the correlation matrix in Table 7.14 is used. For the third set of samples, the coefficients of correlation for the four random variables are assumed to be either –1 or +1 based on the signs in Table 7.14. In other words, the four random variables are perfectly correlated.

Figure 7.28 shows the generated samples of SWCC and permeability functions with different assumptions of correlation. A significant difference was observed in the generated SWCC curves and permeability functions. The patterns of the SWCCs based on the correlation matrix (Figure 7.28b) are quite similar to the fitted SWCC curves shown in Figure 7.21 The assumption of statistical independence among parameters increased scatter in the generated SWCCs and permeability functions while the assumption of perfectly correlated parameters reduced the scatter of the SWCCs and permeability functions.

A long-term rainstorm with a density of 10^{-8} m/s is applied on the slope surface of the simple slope (Figure 7.23). Figure 7.29 illustrates the effect of correlation on the long-term maintained suction at point A. Figure 7.30 illustrates the effect of correlation on the safety factor of the slope under the long-term rainfall. Table 7.15 presents the statistics of factor of safety, the reliability index and the probability of failure for the slope. The reliability index β and probability of failure of the slope are calculated based on Equations 7.12 and 7.4, respectively.

As shown in the above results, the assumption of statistically independent parameters exaggerates uncertainties in suction in the slope and hence the safety factor of the slope. The assumption of perfectly correlation reduces the uncertainty of safety factor of the slope and

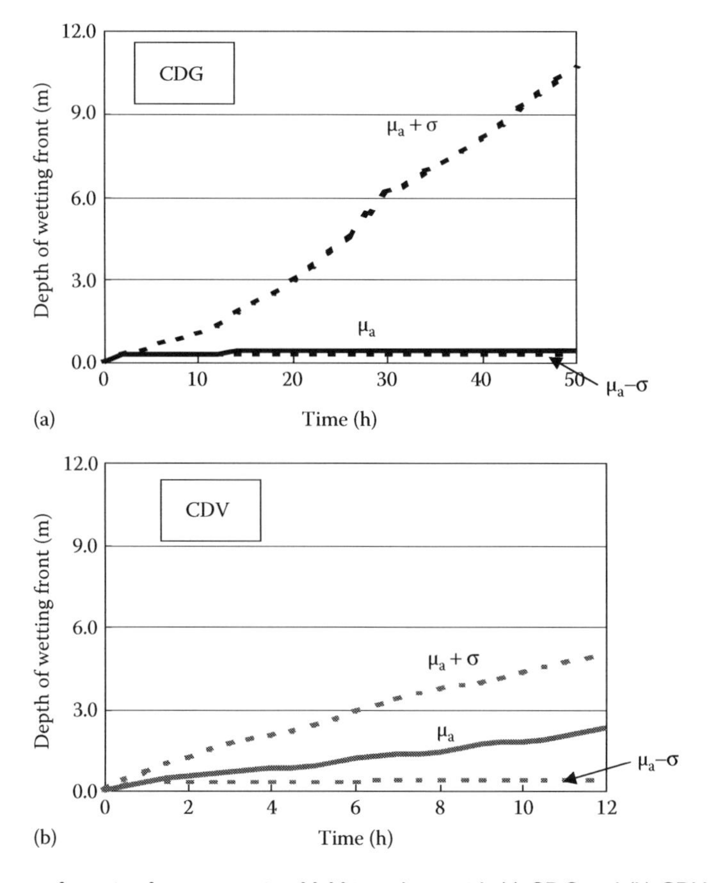

Figure 7.27 Advance of wetting front at section X–X in a slope with (a) CDG and (b) CDV soils.

in turn may lead to an unconservative estimation of reliability. The results imply that a good estimation of correlation coefficients for soil hydraulic parameters should be obtained and used in a reliability analysis of slope stability.

7.6 RELIABILITY ANALYSIS OF RAINFALL-INDUCED SLOPE FAILURE: THE SAU MAU PING LANDSLIDE

7.6.1 Statistics of random variables

In the deterministic coupled hydro-mechanical modeling and slope stability analyses for the Sau Mau Ping slope failure in Chapter 4, soil models were proposed to describe the soil volume change, hydraulic property, and shear strength. In this section, considering both the uncertainties of soil shear strength parameters and the parameters of hydraulic properties, eight parameters, that is, the soil porosity θ_s, the saturated permeability k_s, the parameters a_f, and n_f in the Fredlund and Xing SWCC model and the shear strength parameters, M, M_{col}, Γ, and λ, are selected as the random variables in reliability analysis. These parameters are considered as important parameters that may influence the pore pressure distributions in a slope and the stability of the slope during a rainstorm. The uncertainty of void ratio state surface is omitted due to lack of experimental data.

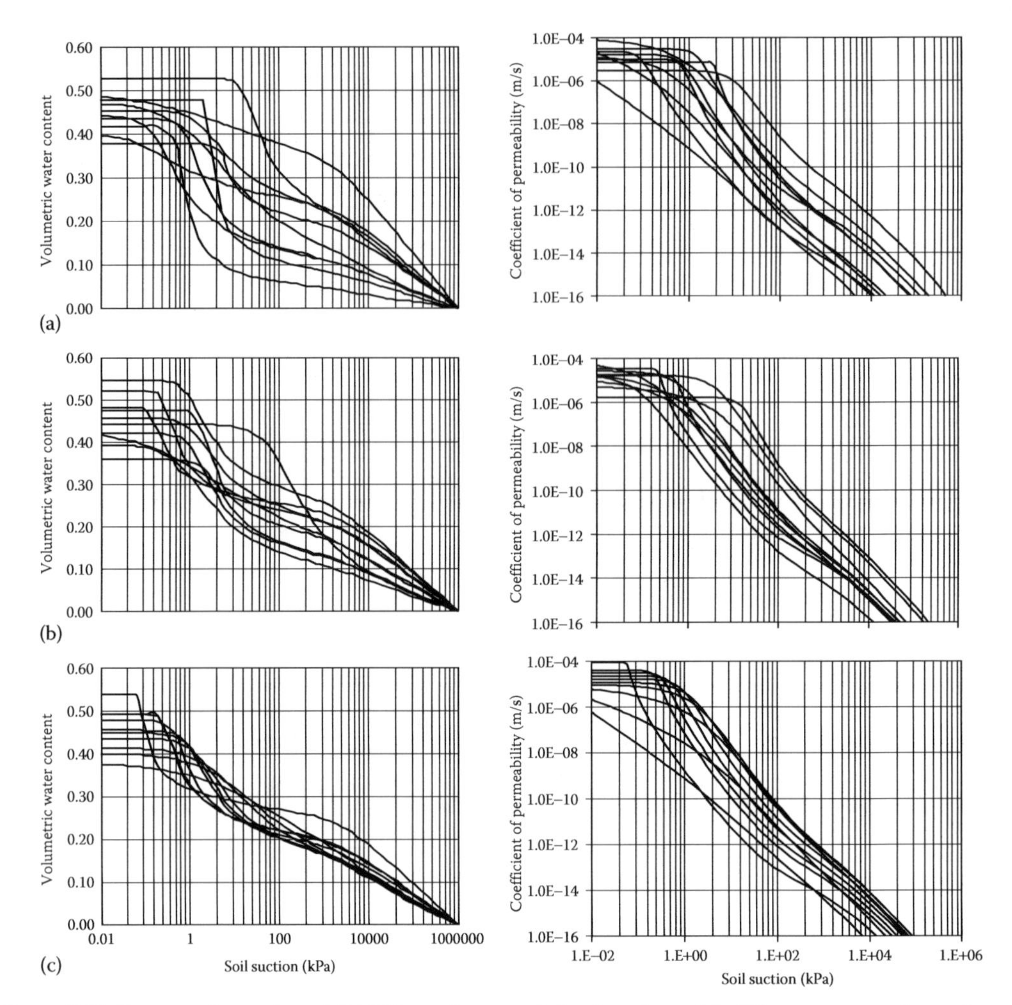

Figure 7.28 Generated samples of SWCCs and permeability function with different assumptions of correlation coefficients: (a) statistically independent, (b) correlated, and (c) perfectly correlated.

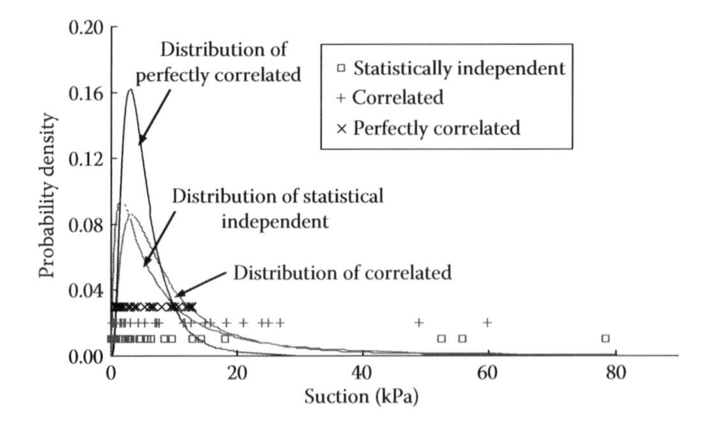

Figure 7.29 Effect of correlation among SWCC parameters on soil suction maintained in a soil slope (Point A) under a long-term rainfall with density of 10^{-8} m/s.

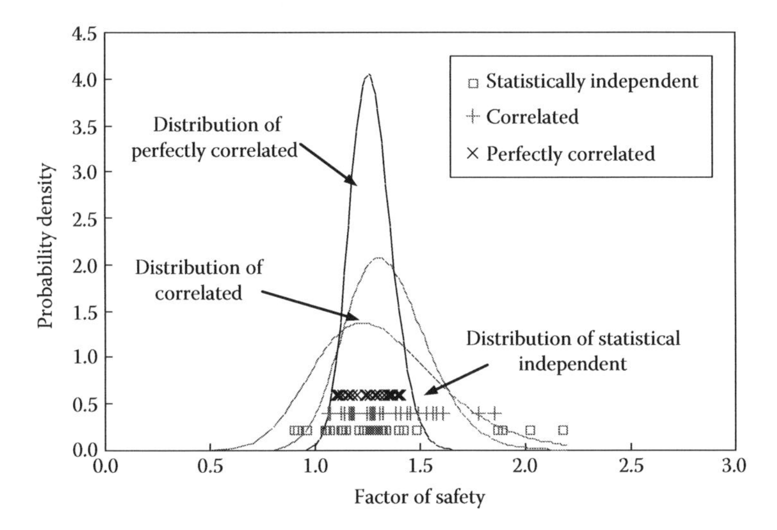

Figure 7.30 Effect of correlation among SWCC parameters on safety factor of a soil slope under a long-term rainfall with density of 10^{-8} m/s.

Table 7.15 Effect of correlation among hydraulic parameters on reliability of a simple soil slope under a long-term rainfall with density of 10^{-8} m/s

	Statistically independent	*Correlated*	*Perfectly correlated*
Mean safety factor	1.33	1.35	1.27
Standard deviation of safety factor	0.31	0.20	0.10
COV of safety factor	23%	15%	8%
Reliability index β	1.11	1.97	2.99
Probability of failure	0.1326	0.0244	0.0014

Note: $\gamma = 15$ kN/m^3, $c' = 2$ kPa, $\phi' = 30°$, $\phi^b = 20°$.

Table 7.16 shows the critical state parameters M, M_{col}, Γ, and λ of CDG soils from nine sites in Hong Kong based on laboratory triaxial tests results. In the reliability analysis, the distributions of M, M_{col}, Γ, and $\ln(\lambda)$ are assumed to be normal. The mean values of the four random variables are the same as the parameters in the deterministic study of Chapter 4. The COV of M, M_{col}, is assumed to be 15%. The COV of Γ is assumed to be 10% and the standard deviation of $\ln(\lambda)$ is assumed to be 0.2 based on the reported critical state parameters of sandy soils in the literature, which was summarized by Zhang (2005).

The mean values of the four random variables $\ln(k_s)$, θ_s, $\ln(a_f)$, and $\ln(n_f)$ are the same as the parameters in the deterministic study in Chapter 4. The probability distribution of in situ measured θ_s is shown in Figure 7.31. The mean and the standard deviation of the in situ porosity are 0.47 and 0.05, respectively. The calculated Anderson–Darling value by the Anderson–Darling goodness-of-fit test (Anderson and Darling, 1954) of normal distribution is 0.6, which is smaller than the critical value, 0.714, at 5% significance level. Therefore, the PDF of θ_s is substantiated to be normal. Figure 7.32 illustrates the frequency diagram and pdf of $\ln(k_s)$ obtained from 26 soil samples taken from bore holes at Sau Mau Ping (k_s in m/s). The PDF for $\ln(k_s)$ is fitted using normal distribution. At the Sau Mau Ping site, the variability of the $\ln(k_s)$ is significant. The standard deviation of the in situ $\ln(k_s)$ is 1.13. $\ln(a_f)$ and $\ln(n_f)$ are both assumed normal. The standard deviations of $\ln(a_f)$ and $\ln(n_f)$ are

Table 7.16 Critical state parameters of CDG soils in Hong Kong

Data sources	*Soil type*	*Location*	λ	Γ	*M*	M_{col}
Sun (1999)	Granitic saprolite	Sau Mau Ping	0.123	2.23	1.50	
	Granitic saprolite	Shouson Hill	0.078	1.99	1.58	
	Granitic saprolite	King's Park	0.079	1.956	1.50	
Fung (2001)	CDG	Shau Kee Wan	0.093	2.112	1.58	1.12
		Shau Kee Wan	0.107	2.166	1.61	1.23
		Shau Kee Wan	0.099	2.136	1.67	1.16
		Shau Kee Wan	0.094	2.108	1.63	1.13
	CDG	Beacon Hill	0.052	1.173	1.60	1.05
HKIE (1998)	CDG fill	King's Park	0.084		1.46	
		Chi Lin Ching Yuen	0.098		1.55	
		Valley Road Estate	0.052		1.64	
		Diamond Hill	0.107		1.64	
		Diamond Hill	0.100		1.64	
		Diamond Hill	0.094		1.64	
HKIE (2003)	CDG fill	Ho Man Tin	0.048			
		Stubbs Road	0.048			
		Lai Ping Road	0.130			
Chiu (2001); Ng and Chiu (2003)	CDG fill	Cha Kwo Ling	0.110	2.334	1.55	

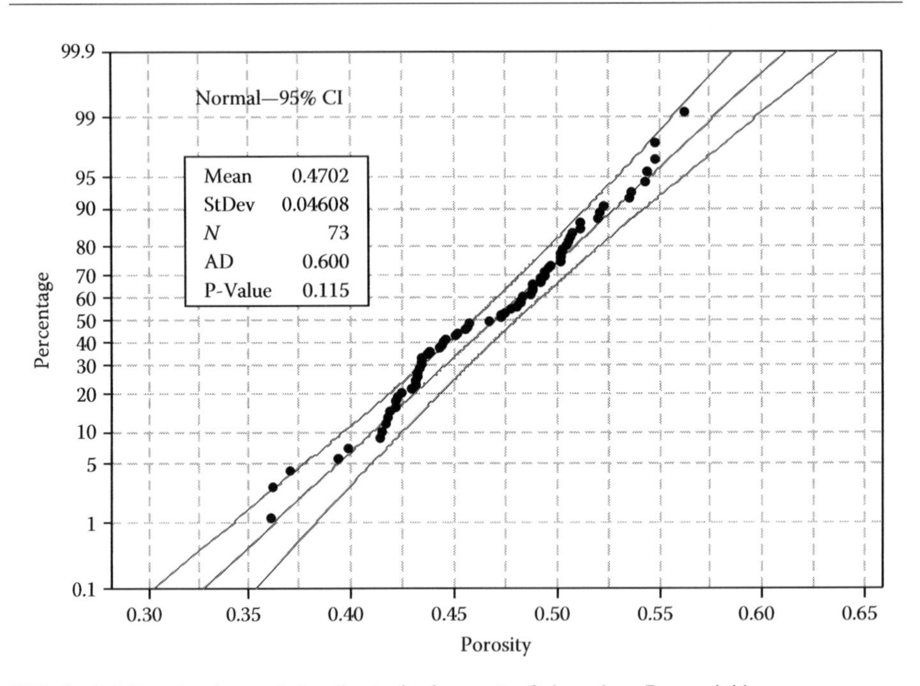

Figure 7.31 Probability plot (normal distribution) of porosity θ_s based on Figure 4.46.

assumed to be 1.43 and 1.04, respectively, as in Table 7.13. The statistics and distribution of the random variables are listed in Table 7.17. In this study, only the correlations among the hydraulic properties (Table 7.14) are considered. The correlations between the hydraulic parameters and the shear strength parameters and those among the shear strength parameters themselves are not considered due to lack of experimental data.

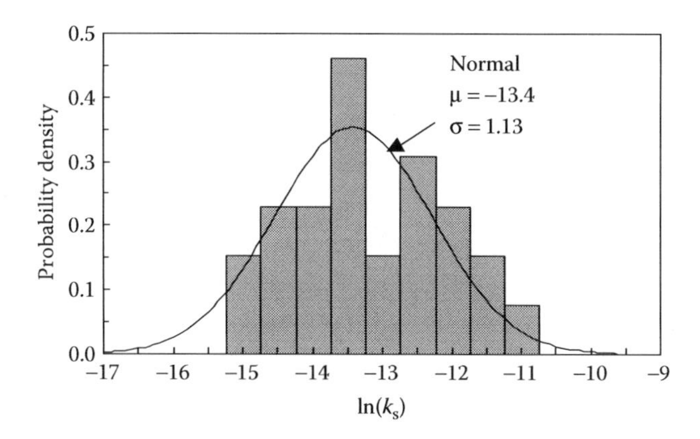

Figure 7.32 Frequency diagram and PDF of $\ln(k_s)$ based on Figure 4.47 (k_s in m/s).

Table 7.17 Mean, standard deviation, and distribution of random variables

			Distribution	*Sources of data*
$\ln(k_s)$	−13.4	1.13	Normal	Hong Kong Government (1977)
θ_s	0.47	0.05	Normal	
$\ln(a_f)$	1.22	1.43	Normal	Gan and Fredlund (1996), Fung (2001)
$\ln(n_f)$	0.26	1.04	Normal	
M	1.50	0.23	Normal	Sun (1999)
M_{col}	0.98	0.15	Normal	Zhang (2005)
Γ	2.22	0.22	Normal	Sun (1999)
$\ln(\lambda)$	−2.09	0.20	Normal	Sun (1999)

7.6.2 Methodology of reliability analysis

The reliability of the Sau Mau Ping slope under the August 24–25, 1976, rainstorm is investigated using the LHS technique. Sixteen samples of soil parameters are generated by the LHS technique based on the distributions and correlation matrix of the random variables (Table 7.18). For each set of soil parameters, the SWCC of the soil is first determined based on the Fredlund and Xing (1994) SWCC model. Then, the corresponding permeability function is estimated from the SWCC and k_s using the Fredlund et al. (1994) prediction method.

The coupled hydro-mechanical analyses and slope stability analyses, which have been used to analyze the Sau Mau Ping slope and the Kadoorie field test in Chapter 4, are conducted for each of the 16 sets of soil parameters. The initial condition and boundary conditions are the same as those in the deterministic study of the Sau Mau Ping slope. The rainstorm at Sau Mau Ping during August 24–25, 1976 (Figure 4.45) is applied to the slope in the coupled seepage and deformation modeling. The uncertainties of initial conditions and boundary conditions are not considered.

7.6.3 Results and discussion

7.6.3.1 Uncertainty of pore-water pressure profiles and wetting fronts

Figure 7.33 illustrates the pore-water pressure profiles at cross-section 2–2 in the slope at different moments of the rainstorm. The initial pore-water pressure distribution is hydrostatic with a maximum suction value of 10 kPa. After 8 hours of rainfall, the wetting fronts

Table 7.18 Generated samples of the random variables for reliability analysis

Sample no.	*M*	M_{col}	Γ	λ	a_f	n_f	k_s *(m/s)*	q_s
1	1.84	0.80	2.46	0.17	0.76	1.79	1.20E–06	0.46
2	1.71	1.19	2.12	0.13	5.58	2.14	6.21E–07	0.49
3	1.63	1.10	2.16	0.12	3.02	0.49	9.34E–07	0.46
4	1.59	0.91	2.50	0.11	9.05	4.75	3.77E–06	0.44
5	1.52	1.00	2.32	0.07	1.57	0.44	5.82E–07	0.39
6	1.86	0.96	2.41	0.15	1.85	18.98	4.41E–06	0.53
7	1.36	1.23	1.91	0.11	8.39	0.62	2.74E–06	0.52
8	1.34	1.05	2.27	0.14	1.24	4.56	6.36E–08	0.30
9	1.49	1.14	2.10	0.10	7.50	2.41	5.21E–06	0.52
10	1.45	1.08	2.35	0.09	0.18	1.81	3.52E–06	0.54
11	1.69	0.87	2.19	0.14	48.50	0.18	1.17E–06	0.52
12	1.10	0.70	2.05	0.12	0.33	0.72	3.97E–06	0.49
13	1.19	0.92	1.74	0.12	8.30	0.84	6.22E–07	0.45
14	1.42	0.83	2.25	0.10	7.70	0.39	6.78E–07	0.44
15	1.54	0.97	2.58	0.13	5.20	0.62	3.52E–06	0.49
16	1.29	1.01	1.98	0.16	10.14	2.25	1.93E–06	0.45

are generally shallow. The mean depth of wetting front is 0.54 m and the standard deviation is 0.31 m. The mean depth of wetting front increases with time and the standard deviation of the depth also increases due to the uncertainties of soil hydraulic properties. The standard deviation of the wetting front depth increases to 1.41 m after 32 hours of rainfall. As the depths of wetting front become deeper, the shapes of the wetting front evolve into two different forms: sharp horizontal ones and smooth ones with large transition zones. As shown in Figure 7.33(c), for soils with small water storage capacity and large k_s (e.g., Samples 4, 9, and 16), the dissipation of suction at shallow depth of the slope can be very fast and the ground water table rises. However, for the soils with relative large water storage capacity or small k_s (e.g., Samples 8, 12, etc.), the movement of the wetting front can be very slow.

7.6.3.2 Uncertainty of safety factor

Figure 7.34 illustrates the safety factor at different moments of rainfall. At a specific time during the storm, 16 values of safety factor can be obtained from the 16 random samples with soil parameters generated by the LHS technique. The mean of the safety factor and the one standard deviation bands of the safety factor are shown in Figure 7.34. The mean and standard deviation of the safety factor are shown in Figure 7.35. At the beginning of the rainstorm, the mean of the safety factor is 1.37 and the standard deviation is 0.27. Only the uncertainties of the shear strength parameters are propagated to the uncertainty of safety factor at this moment. After the rainstorm starts, with the movement of the wetting fronts, the uncertainties of the hydraulic properties start to influence the pore-water pressure distributions. Therefore, the standard deviation of safety factor is increased. From t = 12 h to 16 h, the rainfall flux is decreasing gradually. The mean of safety factor is increased to 1.19 and the standard deviation of safety factor is reduced to 0.37 at t = 16 h. From t = 16 h to 20 h, although the rainfall flux increases again, the mean of safety factor continues to increase to a value of 1.24. The standard deviation of safety factor is reduced to 0.29 at t = 20 h. After t = 20 h, the lag of response in the safety factor is eliminated and the safety factor keeps to decrease because of the extremely high rainfall flux. Finally, the mean of safety factor is 1.10 and the standard

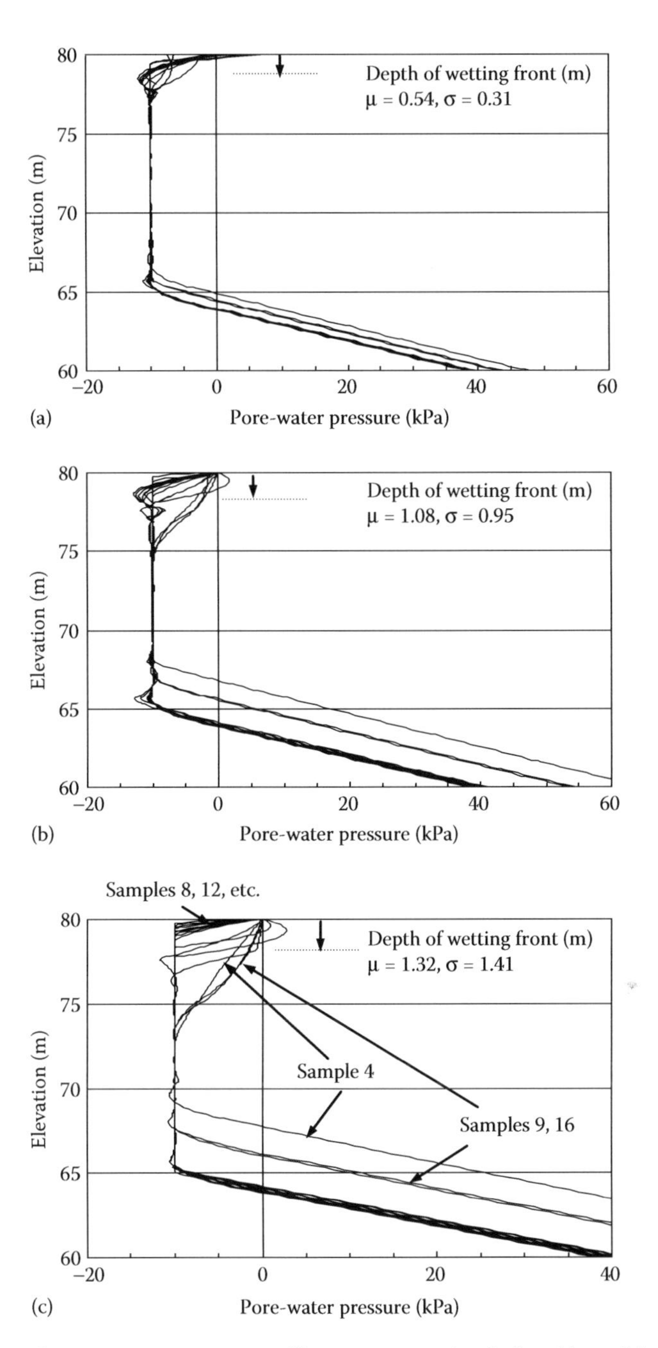

Figure 7.33 Variation of pore-water pressure profiles at cross-section 2–2 at (a) $t = 8$ hours, (b) 24 hours, and (c) $t = 32$ hours.

deviation of safety factor is 0.38 after 32 hours of rainfall. Figure 7.36 presents the COV of safety factor during the rainstorm. The COV of the safety factor is 20% before the rainstorm started. At $t = 8$ h, the COV of the safety factor increases to 34.6%. Then, the COV is reduced but it is still larger than the initial one. At the end of the rainstorm, the COV is 34.5%.

Figure 7.37 illustrate that the distribution of the safety factor at $t = 32$ h can be substantiated by normal distribution at a significance level of 5%. Therefore, the probability of failure can be

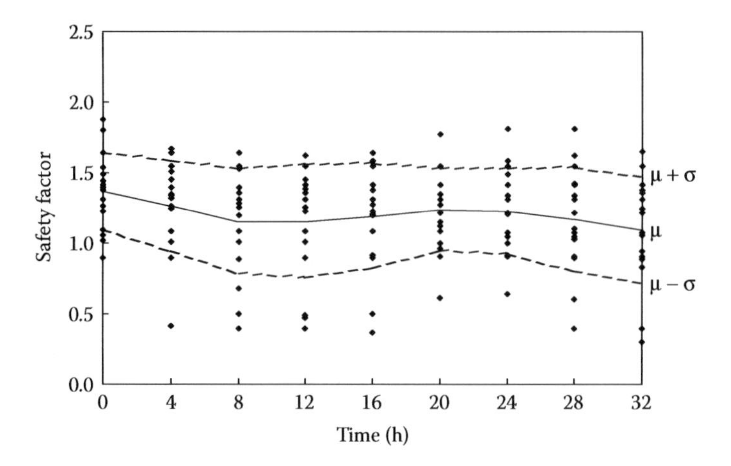

Figure 7.34 Variation of safety factors with time.

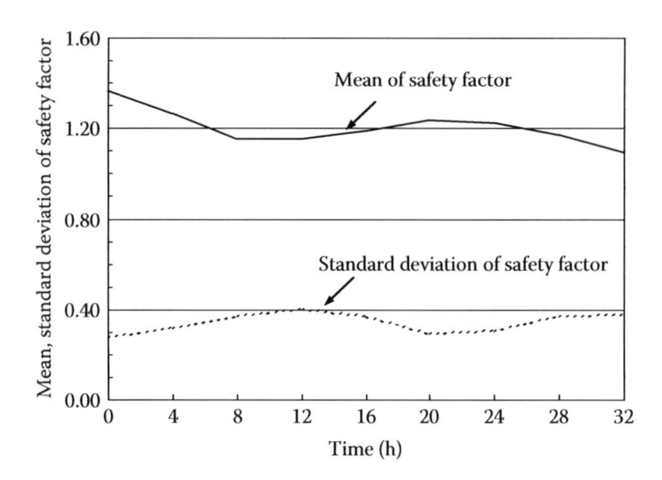

Figure 7.35 Mean and standard deviation of safety factor versus time.

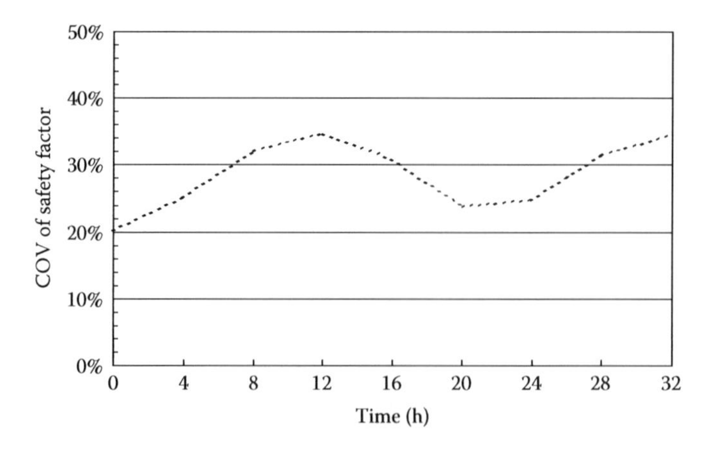

Figure 7.36 COV of safety factor versus time.

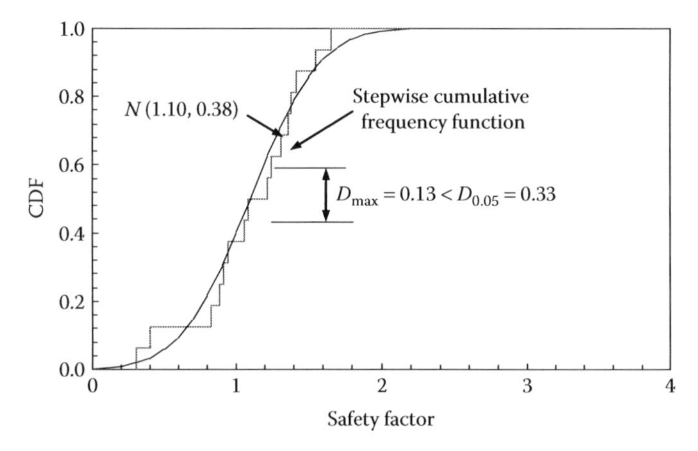

Figure 7.37 K–S test for safety factor at $t = 32$ hours.

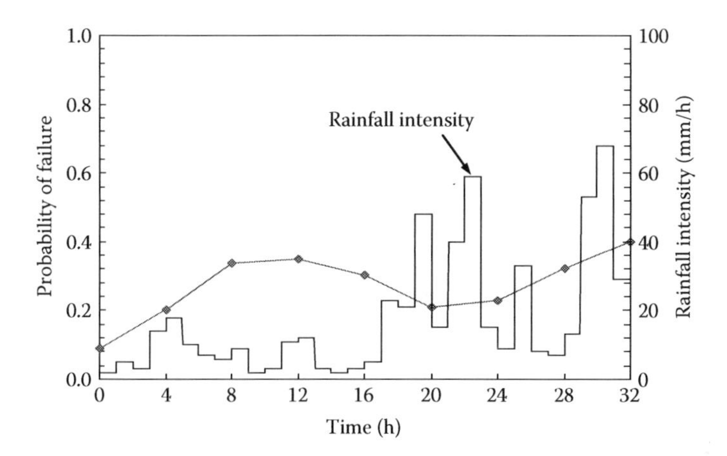

Figure 7.38 Variation of probability of failure with time.

estimated using Equation 7.4. Figure 7.38 illustrates the change of probability of failure of the slope during the process of the rainstorm. At the beginning of the rainstorm, the probability of failure of the slope is 0.09. With increasing of rainfall flux rate, the probability of failure also increases. With decreasing of rainfall flux, the probability of failure decreases. However, similar to the mean of safety factor, the response of the probability of failure shows a time lag with respect to the change of rainfall flux. This is probably because when the rainfall flux rate starts to decrease, the matric suction in the shallow depth of the slope increases but at deeper depth the wetting front continues to move down. As a result, the global safety factor keeps decreasing for some time. On the other hand, when the rainfall flux starts to increase, the matric suction in the shallow depth of the slope can dissipate quickly but at deeper depth the wetting front is not yet influenced immediately. Therefore, the global safety factor may stay stable or even continues to increase if there is a sufficiently long period of no rainfall before the rainstorm.

If only considering the major intense rainstorm before the landslide, for example, 16 hours before the landslide and assuming the same initial condition as in Figure 7.38, the probability of failure continues to increase with time as shown in Figure 7.39. Comparing Figures 7.38 and 7.39, the antecedent rainfall can further increase the probability of failure of the slope.

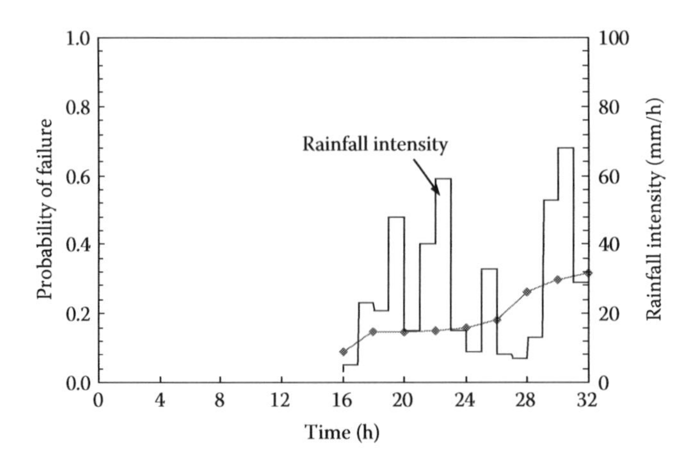

Figure 7.39 Variation of probability of failure with time (no antecedent precipitation before the major rainstorm).

7.6.3.3 Uncertainty analysis of horizontal displacement

Figure 7.40 shows the change of the maximum horizontal displacement. The horizontal displacement keeps on increasing despite the variation of rainfall flux with time. The mean of the displacement increases from 0 to 169 mm. The standard deviation of the maximum horizontal displacement increases to 60 mm. At $t = 32$ h, the three largest values of displacement are of Samples 4, 9, and 12 as shown in Figure 7.40. For the soils with parameters of Samples 4 and 9 (Table 7.18), they have a relative large k_s and a small water storage capacity. Hence, the movements of wetting fronts are relatively fast as shown in Figure 7.33. When the soils at shallow depth of the slope become saturated, the shear strengths of the soils are reduced. Consequently, large movements occur in the slope. Figure 7.41 illustrates the displacements in the slope with parameters of Sample 4 at $t = 32$ h. The maximum displacement occurs near the toe of the slope.

For the soil slope with parameters of Sample 12, the shear strength of the soil is extremely small ($M = 1.10$ or $\phi'_{cs} = 27.7°$ and $M_{col} = 0.70$ or $\phi'_{col} = 18.3°$ as shown in Table 7.18).

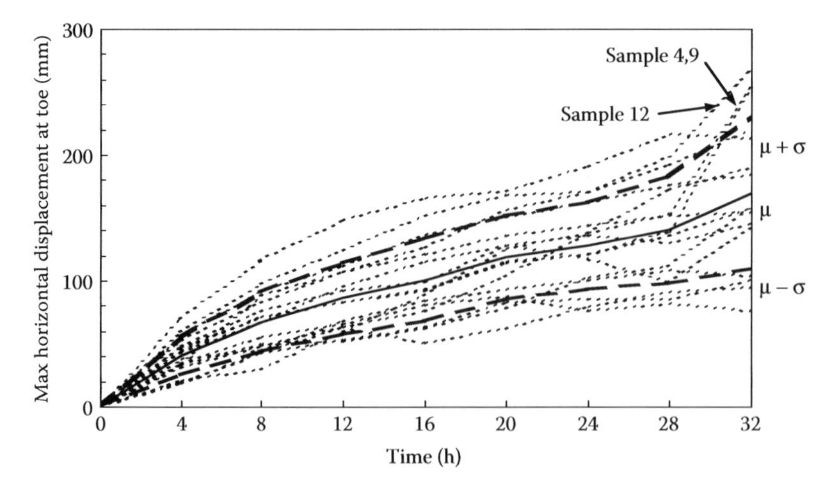

Figure 7.40 Probability band of maximum horizontal displacement.

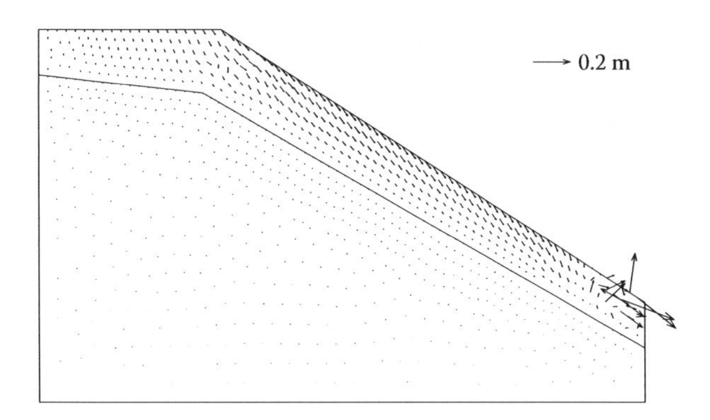

Figure 7.41 Displacement vector plot in the slope at $t = 32$ h (Sample 4).

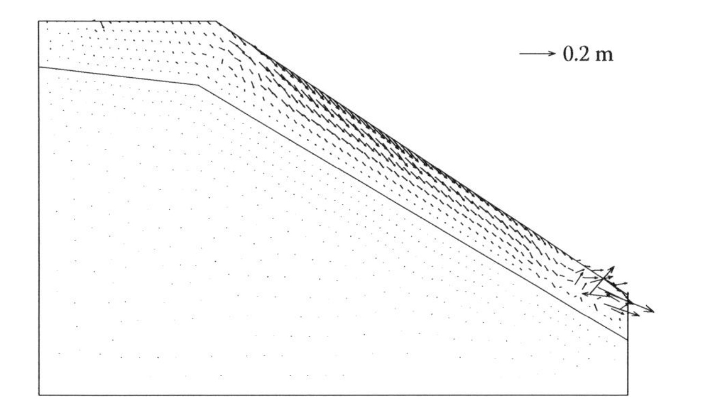

Figure 7.42 Displacement vector plot in the slope at $t = 32$ h (Sample 12).

The safety factor before the rain starts is 1.02. In other words, the slope is marginally stable. When the rainstorm starts, the penetration of wetting front is not deep because the soil has a large water storage capacity (Figure 7.33). However, the increase of soil weight due to wetting can induce large deformation in the slope (Figure 7.42).

7.7 QUANTITATIVE RISK ASSESSMENT OF LANDSLIDE AND RISK ACCEPTANCE CRITERIA

7.7.1 Concept of quantitative risk assessment

The assessment of landslide hazard is dominated with uncertainties. There are many definitions for risk. For the landslide hazard, IUGS Working Group on Landslides, Committee on Risk Assessment (1997) defined that risk is a measure of the probability and severity of an adverse effect to health, property, or the environment. As risk not only measures the likelihood of the hazard, but also considers the possible consequence caused by the hazard, it has been recognized as a more plausible measure of hazard. The risk of a landslide can be expressed both qualitatively and quantitatively (Dai et al., 2002). It has been suggested that quantitative risk analysis (QRA) for natural hazards is to be preferred over qualitative analysis whenever possible, as it allows for a more objective output and an improved basis for communication between the various categories involved in technical and political decision-making (Uzielli et al., 2008). In recent years, QRA has become increasingly attractive as it provides a convenient tool for measuring the threat posed by the landslide hazard that is easy to understand for decision makers (e.g., ERM Hong Kong Ltd, 1998; Aleotti and Chowdhury, 1999; Cheung and Shiu, 2000; Guzzetti, 2000; Lo and Cheung, 2004; Wong et al., 2006; Remondo et al., 2008).

The likelihood of the landslide, the elements that are threatened by the landslide, and vulnerability of the elements at risk are three components for estimating the risk. There are various definitions of vulnerability, as reviewed in Fuchs et al. (2007). Based on UNDRO (1979), vulnerability is defined as the degree of loss of a given element at risk as a result from the occurrence of a natural phenomenon of a given intensity, ranging between 0 (no damage) and 1 (total loss) (UNDRO, 1979; Fell et al., 2008). As the definition of risk varies, the method for evaluating risk also varies. The following equation is often used to calculate the risk of landslide (e.g., Bell and Glade, 2004)

$$\text{Risk} = P(\text{Hazard}) \times \text{Vulnerability} \times \text{Elements at risk} \tag{7.63}$$

The probability of a landslide can be calculated using various reliability methods as described previously in this chapter. The consequence of the landslide, however, often involves substantial judgement and is less addressed in the literature. In the following, we will briefly introduce how QRA can be evaluated.

7.7.2 Method for quantitative risk assessment

7.7.2.1 Likelihood of landslide

Currently, two types of methods are often employed for evaluating the likelihood of landslide: (1) empirical method developed based on historical slope failure data; and (2) reliability method developed based on mechanics. In the literature, various methods, such as multivariate regression, logistic regression, and artificial neural network, have been used for developing empirical models for evaluating the likelihood of landslides. In these models, the input parameters are often the rainfall information and geometry parameters of slopes. The soil properties, such as the cohesion and the friction angle of the soil, are generally not considered. The empirical methods are applicable when very limited information about the slope is available, and when a reliable past failure database can be compiled. Currently, the empirical method is often used for regional landslide hazard management.

When sufficient site-specific data are available, the factor of safety of the slope can be readily calculated, and the failure probability of the slope can be evaluated by considering the variability of the input parameters and modeling errors. The mechanics-based method requires very detailed site-specific data, and is more suitable for site-specific landslide analysis when detailed soil data are available. As a slope may have many potential slip surfaces, the reliability of a slope is a typical system reliability problem (Zhang et al., 2013a). In recent years, substantial progress has been made on evaluating the system reliability of soil slopes. To consider the spatial variability of soil properties, random field models can be employed to assess the reliability of slopes. Currently, there are several studies focusing on slope reliability analysis considering the spatial variability of soil properties, but such researches are still rather limited. Simulation of spatial variability of soil properties and the stability of spatially heterogeneous soil skope will be discussed in Chapter 8.

7.7.2.2 Vulnerability analysis

Mathematically, the vulnerability measures the degree of loss related to the intensity of the hazard. For the landslide problem, the intensity can be related to a set of spatially distributed parameters describing the destructiveness of a landslide, such as the maximum velocity, total displacement, differential displacement (relative to points adjacent to the point under consideration), and depth of the moving mass (Hungr, 1997). For debris-flow-type landslide, the thickness of the debris is often used to measure the intensity of the hazard (e.g., Fuchs et al., 2007). The relationship between intensity and degree of loss is often developed based on historical data. For instance, Equation 7.64 shows the vulnerability function suggested by Fuchs et al. (2007) based on data at a catchment in the Eastern Alps of Austria, and Equation 7.65 shows the vulnerability function suggested by Akbas et al. (2009) based on data at an alpine valley in the Lombardy region of Italy.

$$V = 0.11{t_{\text{debris}}}^2 - 0.02t_{\text{debris}} \tag{7.64}$$

$$V = 0.17{t_{\text{debris}}}^2 - 0.03t_{\text{debris}} \tag{7.65}$$

where:

V is the vulnerability
t_{debris} is the thickness of debris

Figure 7.43 compares the two vulnerability curves based on Equations 7.64 and 7.65. We can see that the two curves are similar in shape but are not the same, indicating the vulnerability curves developed in one region may not be applicable to another region. To make sure that the vulnerability is within 0 and 1, vulnerability models have also been suggested based on the CDF of a random variable (Totschnig et al., 2011). In addition to the univariate intensity model, multivariate models for landslide intensity and vulnerability analyses have also been developed (e.g., Kaynia et al., 2008; Uzielli et al., 2008).

7.7.2.3 Elements at risk

The determination of elements at risk is often related to runout analysis, that is, to what extent the landslide process will affect the neighboring objects. The method for runout analysis often includes the empirical methods developed based on historical analysis of past

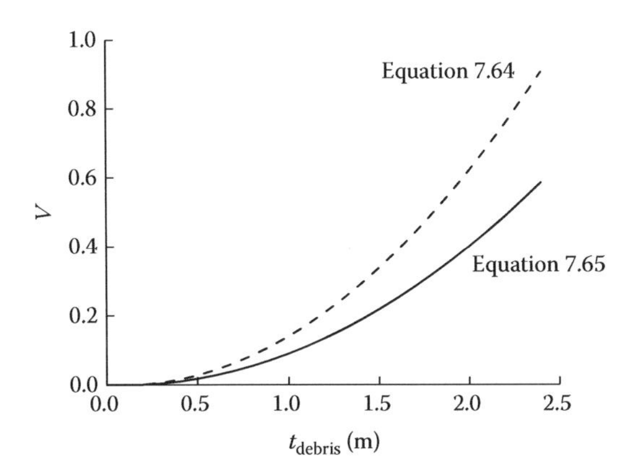

Figure 7.43 Comparison of vulnerability functions.

data, or approaches based on detailed mechanics analysis. The empirical methods are generally easy to apply, but are only applicable to a future case, which is similar to those in the database for model development. For instance, noticing that large landslides tend to have larger mobility, Corominas (1996) suggested the following relationship to predict the travel length of the landslide mass based on regression analysis:

$$L_m = 1.03V_m^{0.105}H_m, 10^2 \text{ m}^3 \le V_m \le 10^{10} \text{ m}^3 \quad (7.66)$$

where:

H_m is the elevation difference between the starting point and lowest point of deposition of the mass movement (m)
L_m is the corresponding horizontal distance (m)
V_m is the volume of the sliding mass (m^3)

While being simple to apply, most empirical models can only predict the travel distance of a landslide. The mechanics-based methods (e.g., Hungr, 1995; Li et al., 2010) can provide more comprehensive information about the landslide process, such as the depth of the sliding mass and the velocity of the sliding process. The major challenge in applying the mechanics-based method resides in the difficulties in choosing the model input parameters appropriately. In practice, the most reliable way to determine such parameters seems to be back analysis of case histories (e.g., Ho and Ko, 2009). The readers can refer to Rickenmann (2005) for a comprehensive review of various methods for runout analysis.

7.7.3 Risk acceptance criteria

As noticed by Fell (1994), deciding the acceptance criteria is often affected by the legal, political, social, and financial issues. The decision will usually be made by owners, politicians, or others, not by the person doing the risk analysis. However, it is the engineers' responsibility to present the risk in an accessible way to help the decision-making process. When analyzing the landslide risk, the risk can often analyzed from two perspectives, that is, the maximum risk posed to an individual, and risk posed to a group of people exposed to the hazard, which are often called individual risk and societal risk, respectively (e.g., Porter and Morgenstern, 2013).

For the individual risk, two definitions are typical. In Demark, the individual risk is defined as the risk associated with a person who is continually present and unprotected at a given location (Duijm, 2009). For comparison, the individual risk is defined as the risk associated with a person at a given location accounting for temporal factors and protection in Hong Kong (PD, 2014). While it is usually difficult to decide how much risk one can accept, it is logical to argue that the risk posed by a landslide hazard should not be significantly higher than that posed by other hazards that people are living with. As such, a useful clue to find the acceptance criteria for landslide risk is to compare the risk posed by a landslide with those associated with other hazards. Table 7.19 shows the risk statistics for persons exposed to various hazards. In this table, voluntary risk refers to risk associated with activities in which a person participates by choice, and involuntary risks are associated with activities, conditions, or events to which a person might be exposed without his consent (Starr, 1969). In general, people are more tolerable to voluntary risk than involuntary risk. Table 7.20 shows the Qualitative Descriptors for Risk of Loss of Life (AGS, 2007), where the definition of individual risk employed is similar to that in PD (2014).

Table 7.19 Risk statistics for persons exposed voluntarily or involuntarily to various hazards

Cause	*Risk* ($\times 10^{-6}$)
Building hazards (UK) (Construction Industry and Information Association, 1977)	0.14
Building fire (Austrilian Government statistics)	4
Nature hazards (USA)	
Hurricane (1901–1972)	0.4
Tornado (1953–1971)	0.4
Lightning (1969)	0.5
Earthquake (California) (Kleetz, 1976)	2
General accidents (USA, 1979)	
Railway travel	4
Electrocution	6
Air travel	9
Water transport	9
Poisoning	20
Drowning	30
Fires and burns	40
Falls	90
Road accidents	300
Occupation (UK) (Royal Society Study Group, 1983)	
Clothing manufacturing	5
Vehicle manufacturing	15
Chemical and allied industries	85
Shipbuilding and marine engineering	105
Agriculture	110
Construction industries	150
Railways	180
Coal mining	210
Quarrying	295
Mining	750
Offshore oil and gas (1967–1976)	1650
Deep-sea fishing (1959–1968)	2800
Sports (Royal Society Study Group, 1983)	
Cave exploration (USA, 1970–1978)	45
Glider flying (USA, 1970–1978)	400
Scuba diving (USA, 1970–1978)	420
Power boat racing (USA, 1970–1978)	800
Hang gliding (USA, 1977–1979)	1500
Parachuting (USA, 1978)	1900

Source: Reid, S. G., *Risk Assessment Research Report R591*. School of Civil and Mining Engineering, University of Sydney, Australia, 1989. With permission.

Table 7.20 Qualitative descriptors for risk of loss of life

Annual probability of loss of life for the individual most at risk	*Qualitative descriptor*
>1:1000	Very high
1:1000 to 1:10,000	High
1:10,000 to 1:100,000	Moderate
1:100,000 to 1:1,000,000	Low
<1:1,000,000	Very low

Source: Australian Geomechanics Society (AGS) Sub-Committee on Landslide Risk Management, A national landslide risk management framework for Australia. *Australian Geomechanics*, 42, 1–36, 2007. With permission.

Societal risk expresses the risk that a group of people are simultaneously exposed to the consequences of an accident. When analyzing the societal risk, the *F–N* curve as introduced in the nuclear engineering (Kendall et al., 1977) is often used, in which *F* denotes the expected frequency of the accident and *N* denotes the number of people who will die (or be injured) as a result of the accident. As an example, Figure 7.44 shows the societal risk criteria employed in Hong Kong (Ho and Ko, 2009). In this figure, the horizontal axis denotes the number of fatalities in an accident, and the vertical axis denotes the frequency of the accident per year. The lines passing through this figure is called *F–N* curve. Above the *F–N* curve indicates unacceptable risk. Below the line is a region where the risk should be controlled as low as reasonably practicable (ALARP). There is a small zone between the unacceptable region and the ALARP region when the number of fatalities is in between 1000 and 5000. In this region, the allowable frequency of the accident must be decided with intense scrutiny. We can see that as the number of fatalities increase, the acceptable frequency of the accident is decreased. In Figure 7.44, the slope of the *F–N* curve is –1. However, a community may have more difficulty in accepting one major accident than several smaller accidents, even if the total loss of life is the same. To reflect the above risk aversion attitude, a slope of –2 is adopted for the *F–N* curves in the Netherland and Denmark (Duijm, 2009).

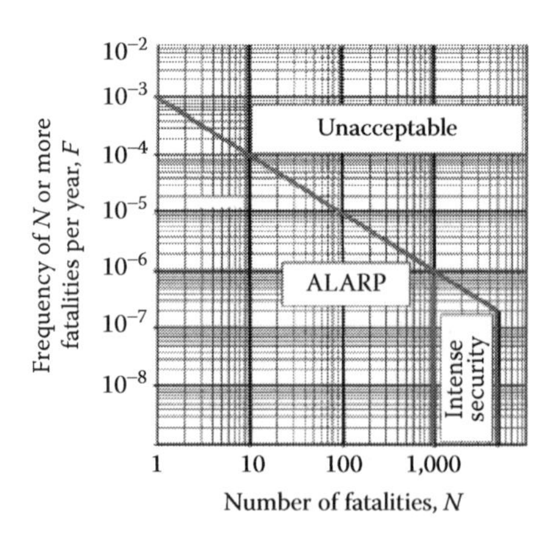

Figure 7.44 *F–N* curve adopted in Hong Kong for societal risk assessment. (From Ho, K. K. S., and Ko, F. W. Y., *Georisk*, 3, 134–146, 2009. With Permission.)

REFERENCES

Akbas, S. O., Blahut, J., and Sterlacchini, S. (2009). Critical assessment of existing physical vulnerability estimation approaches for debris flows. In: Malet, J., Remaître, A., and Bogaard, T. (eds.), *Landslide Processes: From Geomorphological Mapping to Dynamic Modelling*. CERG Editions, Strasbourg, pp. 229–233.

Aleotti, P., and Chowdhury, R. (1999). Landslide hazard assessment: summary review and new perspectives. *Bulletin of Engineering Geology and the Environment*, 58(1), 21–44.

Anderson, T. W., and Darling, D. A. (1954). A Test of Goodness-of-Fit. *Journal of America Statistical Association*, 49, 765–769.

Ang, A. H. S., and Tang, W. H. (1975). *Probability Concepts in Engineering Planning and Design*. Wiley, New York.

Ang, A. H. S., and Tang, W. H. (1984). *Probability Concepts in Engineering Planning and Design: Decision, Risk and Reliability, Vol. 2*. Wiley, New York.

Ang, A. H. S., and Tang, W. H. (2007). *Probability Concepts in Engineering: Emphasis on Applications to Civil and Environmental Engineering, Vol. 1, 2nd ed.* Wiley, New York.

Au, S. K., and Beck, J. L. (2001). Estimation of small failure probabilities in high dimensions by subset simulation. *Probabilistic Engineering Mechanics*, 16(4), 263–277.

Au, S. K., and Wang, Y. (2014). *Engineering Risk Assessment with Subset Simulation*. John Wiley & Sons, Singapore, 330 pp.

Australian Geomechanics Society (AGS) Sub-Committee on Landslide Risk Management. (2007). A national landslide risk management framework for Australia. *Australian Geomechanics*, 42(1), 1–36.

Baecher, G., and Ladd, C. (1997). Formal observational approach to staged loading. *Transportation Research Record. Journal of the Transportation Research Board*, 1582, 49–52.

Baecher, G. B., and Christian, J. T. (2003). *Reliability and Statistics in Geotechnical Engineering*. Wiley, New York.

Baecher, G. B., Marr, W. A., Lin, J. S., and Consla, J. (1983). *Critical Parameters for Mine Tailings Embankments*. US Bureau of Mines, Denver, CO.

Bell, R., and Glade, T. (2004). Quantitative risk analysis for landslides? Examples from Bildudalur, NW-Iceland. *Natural Hazards and Earth System Sciences*, 4 (1), 117–131.

Bhattacharya, G., Jana, D., Ojha, S., and Chakraborty, S. (2003). Direct search for minimum reliability index of earth slopes. *Computers and Geotechnics*, 30(6), 455–462.

Brooks, R. H., and Corey, A. T. (1964). Hydraulic properties of porous medium. *Hydrology Paper No. 3*. Civil Engineering Department, Colorado State University.

Bucher, C. G., and Bourgund, U. (1990). A fast and efficient response surface approach for structural reliability problems. *Structural Safety*, 7 (1), 57–66.

Carsel, R. F., and Parrish, R. S. (1988). Developing joint probability distributions of soil water retention characteristics. *Water Resources Research*, 24(5), 755–769.

Carsel, R. F., Parrish, R. S., Jones, R. L., Hansen, J. L., and Lamb, R. L. (1988). Characterizing the uncertainty of pesticide movement in agricultural soils. *Journal of Contaminant Hydrology*, 2(2), 111–124.

Chen, Z. Y., Sun, P., Wang, Y. J., and Yang, J. (2005). *The User Manual of EMU 2005*. China Institute of Hydropower and Water Resource. Beijing, China.

Cheung, W. M., and Shiu, Y. K. (2000). *Assessment of Global Landslide Risk Posed by Pre-1978 Man-Made Slope Features: Risk Reduction from 1977 to 2000 Achieved by the LPM Programme* (GEO Report No. 125). Geotechnical Engineering Office, Hong Kong.

Ching, J., Phoon, K. K., and Hu, Y. G. (2009). Efficient evaluation of reliability for slopes with circular slip surfaces using importance sampling. *Journal of Geotechnical and Geoenvironmental Engineering*, 135(6), 768–777.

Chiu, C. F. (2001). *Behaviour of Unsaturated Loosely Compacted Weathered Materials*, Ph.D Thesis, Hong Kong University of Science and Technology, Hong Kong.

Chowdhury, R. N., Tang, W. H., and Sidi, I. (1987). Reliability model of progressive slope failure. *Géotechnique*, 37(4), 467–481.

Christian, J. T., Ladd, C. C., and Baecher, G. B. (1994). Reliability applied to slope stability analysis. *Journal of Geotechnical Engineering*, 120(12), 2180–2207.

Corominas, J. (1996). The angle of reach as a mobility index for small and large landslides. *Canadian Geotechnical Journal*, 33 (2), 260–271.

Dai, F. C., Lee, C. F., and Ngai, Y. Y. (2002). Landslide risk assessment and management: An overview. *Engineering Geology*, 64 (1), 65–87.

Ditlevsen, O. (1981). *Uncertainty Modeling: With Applications to Multidimensional Civil Engineering Systems*. McGraw-Hill, New York.

Duijm, N. J. (2009). *Acceptance Criteria in Denmark and the EU. Environmental Project No. 1269*. Environmental Protection Agency, Danish Ministry of Environment.

El-Ramly, H., Morgenstern, N. R., and Cruden, D. M. (2002) Probabilistic slope stability analysis for practice. *Canadian Geotechnical Journal*, 39 (3), 665–683.

ERM Hong Kong Ltd. (1998). *Landslides and Boulder Falls from Natural Terrain: Interim Risk Guidelines* (GEO Report No. 75). Report prepared for the Geotechnical Engineering Office, Hong Kong.

Faravelli, L. (1989). Response-surface approach for reliability Analysis. *Journal of Engineering Mechanics, ASCE*, 115(12), 2763–2781.

Fell, R. (1994). Landslide risk assessment and acceptable risk. *Canadian Geotechnical Journal*, 31(2), 261–272.

Fell, R., Corominas, J., Bonnard, C., Cascini, L., Leroi, E., and Savage, W. Z. (2008). Guidelines for landslide susceptibility, hazard and risk zoning for land use planning. *Engineering Geology*, 102(3–4), 85–98.

Fredlund, D. G., and Xing, A. Q. (1994). Equations for the soil-water characteristic curve. *Canadian Geotechnical Journal*, 31(4), 521–532.

Fredlund, D. G., Xing, A. Q., and Huang, S. Y. (1994). Predicting the permeability function for unsaturated soils using the soil-water characteristic curve. *Canadian Geotechnical Journal*, 31(4), 533–546.

Fuchs, S., Heiss, K., and Hübl, J. (2007). Towards an empirical vulnerability function for use in 795 debris flow risk assessment. *Natural Hazard and Earth System Science*, 7 (5), 495–506.

Fung, W. T. (2001). *Experimental Study and Centrifuge Modeling of Loose Fill Slope*, M.Phil Thesis, Hong Kong University of Science and Technology, Hong Kong.

Gan, J. K. M., and Fredlund, D. G. (1996). *Study of the Application of Soil-Water Characteristic Curves and Permeability Functions to Slope Stability, Part 1, Permeability and Soil-Water Characteristic Curve Tests, and the Computation of Permeability Functions*, Department of Civil Engineering University of Saskatchewan, Saskatoon, Saskatchewan, Canada.

Gardner, W. R. (1958). Some steady state solutions of the unsaturated moisture flow equation with application to evaporation from a water table. *Soil Science*, 85(4), 228–232.

Gilbert, R. B., and Tang, W. H. (1989). Progressive failure probability of soil slopes containing geologic anomalies. *Proceedings of ICOSSAR '89, the 5th International Conference on Structural Safety and Reliability*, Part I, San Francisco, CA, pp. 255–262.

Greminger, P. J., Sud, Y. K., and Nielsen, D. R. (1985). Spatial variability of field-measured soil-water characteristic. *Soil Science Society of America Journal*, 49(5), 1075–1082.

Griffiths, D. V., Huang, J., and Fenton, G. A. (2009) Influence of spatial variability on slope reliability using 2-D random fields. *Journal of Geotechnical and Geoenvironmental Engineering*, 135(10), 1367–1378.

Guzzetti, F. (2000). Landslide fatalities and the evaluation of landslide risk in Italy. *Engineering Geology*, 58(2), 89–107.

Harbitz, A. (1986). An efficient sampling method for probability of failure calculation. *Structural Safety*, 3(2), 109–115.

Hasofer, A. M., and Lind, N. C. (1974). Exact and invariant second-moment code format. *Journal of the Engineering Mechanics Division*, 100(1), 111–121.

Hassan, A. M., and Wolff, T. F. (1999). Search algorithm for minimum reliability index of earth slopes. *Journal of Geotechnical and Geoenvironmental Engineering*, 125(4), 301–308.

Ho, K. K. S., and Ko, F. W. Y. (2009). Application of quantified risk analysis in landslide risk management practice: Hong Kong experience. *Georisk*, 3(3), 134–146.

Hong Kong Government. (1977). *Report on the Slope Failures at Sau Mau Ping, August 1976*. Hong Kong Government Printer, Hong Kong.

Hong Kong Institution of Engineers (HKIE). (1998). *Soil Nails in Loose Fill Slopes, a Preliminary Study, Draft for Comment*, Hong Kong Institution of Engineers, Hong Kong.
Hong Kong Institution of Engineers (HKIE). (2003). *Soil Nails in Loose Fill Slopes, a Preliminary Study, Final Report*, Hong Kong Institution of Engineers, Hong Kong, 98 pp.
Hu, S. (2000). *Reliability of Slope Stability Considering Infiltration through Surface Cracks*, M.Phil Thesis, Hong Kong University of Science and Technology, Hong Kong.
Hungr, O. (1995). A model for the runout analysis of rapid flow slides, debris flows, and avalanches. *Canadian Geotechnical Journal*, 32, 610–623.
Hungr, O. (1997). Some methods of landslide intensity mapping. In: Cruden, D. M., and Fell, R. (eds.), *Landslide Risk Assessment—Proceedings of the International Workshop on Landslide Risk Assessment, Honolulu, 19–21 February 1997*. Balkema, Rotterdam, the Netherlands, pp. 215–226.
IUGS Working Group on Landslides, Committee on Risk Assessment. (1997). Quantitative risk assessment for slopes and landslides—the state of the art. In: Cruden, D., and Fell, R. (eds.), *Landslide Risk Assessment—Proceedings of the International Workshop on Landslide Risk Assessment, Honolulu, 19–21 February 1997*. Balkema, Rotterdam, the Netherlands, pp. 3–12.
Kasim, F. B. (1997). *Effects of Steady State Rainfall on Long Term Matric Suction Conditions in Slopes*. Unsaturated Soils Group, University of Saskatchewan, Saskatoon, Saskatchewan, Canada.
Kaynia, A. M., Papathoma-Köhle, M., Neuhäuser, B., Ratzinger, K., Wenzel, H., and Medina-Cetina, Z. (2008). Probabilistic assessment of vulnerability to landslide: Application to the village of Lichtenstein, Baden-Württemberg, Germany. *Engineering Geology*, 101(1), 33–48.
Kendall, H. W., Hubbard, R. B., Minor, G. C., and Bryan, W. M. (1977). *Union of Concerned Scientists, The Risks of Nuclear Power Reactors: a Review of the NRC Reactor Safety Study*. WASH-1400, Cambridge.
Kotzias, P. C., Stamatopoulos, A. C., and Kountouris, P. J. (1993). Field quality control on earth-dam: Statistical graphics for gauging. *Journal Geotechnical Engineering, ASCE*, 119(5), 957–964.
Lacasse, S., and Nadim, F. (1996). Uncertainties in characterizing soil properties. In: Shackelford, C. D., Nelson, P. P., Roth, M. J. S. (Eds.). *Uncertainty in the Geologic Environment, Proceedings of Uncertainty '96, Geotechnical Special Publications (GSP) GSP 58*. Madison, Wisconsin, July 31–August 3, 1996. ASCE, New York, pp. 49–75.
Li, D. Q., Chen, Y. F., Lu, W. B., and Zhou, C. B. (2011). Stochastic response surface method for reliability analysis of rock slopes involving correlated non-normal variables. *Computers and Geotechnics*, 38(1), 58–68.
Li, D. Q., Jiang, S. H., Cao, Z. J., Zhou, W., Zhou, C. B., and Zhang, L. M. (2015). A multiple response-surface method for slope reliability analysis considering spatial variability of soil properties. *Engineering Geology*, 187, 60–72.
Li, K. S., and Lumb, P. (1987). Probabilistic design of slopes. *Canadian Geotechnical Journal*, 24(4), 520–531.
Li, Z., Huang, H., Nadim, F., and Xue, Y. (2010). Quantitative risk assessment of cut-slope projects under construction. *Journal of Geotechnical and Geoenvironmental Engineering*, 136(12), 1644–1654.
Liang, R. Y., Nusier, O. K., and Malkawi, A. H. (1999). A reliability based approach for evaluating the slope stability of embankment dams. *Engineering Geology*, 54 (3), 271–285.
Liu, N., Tang, W. H., and Ng, C. W. W. (2001). Probabilistic FEM for reliability of strain softening media. *Finite Elements in Analysis and Design*, 37(8), 603–619.
Lo, D. O. K., and Cheung, W. M. (2004). *Assessment of Landslide Risk of Man-made Slopes in Hong Kong* (GEO Report No. 177). Geotechnical Engineering Office, Hong Kong.
Low, B. K., Gilbert, R. B., and Wright, S. G. (1998). Slope reliability analysis using generalized method of slices. *Journal of Geotechnical and Geoenvironmental Engineering*, 124(4), 350–362.
Low, B. K., and Tang, W. H. (1997a). Efficient reliability evaluation using spreadsheet. *Journal of Engineering Mechanics, ASCE*, 123(7), 749–752.
Low, B. K., and Tang, W. H. (1997b). Probabilistic slope analysis using Janbu's generalized procedure of slices. *Computers and Geotechnics*, 21(2), 121–142.
Low, B. K., and Tang, W. H. (1997c). Reliability analysis of reinforced embankments on soft ground. *Canadian Geotechnical Journal*, 34(5), 672–685.
Low, B. K., and Tang, W. H. (2004). Reliability analysis using object-oriented constrained optimization. *Structural Safety*, 26(1), 69–89.

Lumb, P. (1974). Application of statistics in soil mechanics. In: Lee, I. K. (Ed.). *Soil Mechanics: New Horizons*. Butterworth and Company Publishers Limited, London, pp. 44–111.

Mckay, M. D. (1988). Sensitivity and uncertainty analysis using a statistical sample of input values. In: Ronen, Y. (Ed.). *Uncertainty Analysis*. CRC Press, Boca Raton, FL, pp. 145–186.

Mualem, Y. (1976). A new model for predicting the hydraulic conductivity of unsaturated porous media. *Water Resources Research,* 12(3), 593–622.

Ng, C. W., and Chiu, A. C. (2003). Laboratory study of loose saturated and unsaturated decomposed granitic soil. *Journal of Geotechnical and Geoenvironmental Engineering,* 129(6), 550–559.

Pang, Y. W. (1999). *Investigation of Soil-Water Characteristics and Their Implications on Stability of Unsaturated Soil Slopes*, MPhil Thesis, Hong Kong University of Science and Technology, Hong Kong.

Park, H. J., Lee, J. H., and Woo, I. (2013). Assessment of rainfall-induced shallow landslide susceptibility using a GIS-based probabilistic approach. *Engineering Geology,* 161, 1–15.

Pebesma, E. J., and Heuvelink, G. B. M. (1999). Latin hypercube sampling of Gaussian random fields. *Technometrics,* 41(4), 303–312.

Penalba, R. F., Luo, Z., and Juang, C. H. (2009). Framework for probabilistic assessment of landslide: a case study of El Berrinche. *Environmental Earth Sciences,* 59(3), 489–499.

Phoon, K. K. (2008). *Reliability-Based Design in Geotechnical Engineering: Computations and Applications*. Taylor & Francis Group, Singapore.

Phoon, K. K., and Kulhawy, F. H. (1999). Characterization of geotechnical variability. *Canadian Geotechnical Journal,* 36(4), 612–624.

Phoon, K. K., Santoso, A., and Quek, S. T. (2010). Probabilistic analysis of soil-water characteristic curves. *Journal of Geotechnical and Geoenvironmental Engineering,* 136(3), 445–455.

Planning Department (PD). (2014). *Hong Kong Planning Standards and Guidelines*. Government of Hong Kong SAR, Hong Kong.

Porter, M., and Morgenstern, N. (2013). *Landslide Risk Evaluation—Canadian Technical Guidelines and Best Practices Related to Landslides: A National Initiative for Loss Reduction*. Geological Survey of Canada, Open File 7312, 21p. doi:10.4095/292234.

Ragab, R., and Cooper, J. D. (1993). Variability of unsaturated zone water transport parameters: implications for hydrological modelling. 2. Predicted vs. in situ measurements and evaluation of methods. *Journal of Hydrology,* 148, 133–147.

Rawls, W. J., and Brakensiek, D. L. (1985). Prediction of soil water properties for hydrologic modelling. *Watershed Management in the Eighties, Proceedings of Symposium on Watershed Management, Denver, Colorado*, April 30-May 1. pp. 293–299.

Reid, S. G. (1989). *Risk Assessment Research Report R591*. School of Civil and Mining Engineering, University of Sydney, Australia.

Remondo, J., Bonachea, J., and Cendrero, A. (2008). Quantitative landslide risk assessment and mapping on the basis of recent occurrences. *Geomorphology,* 94(3), 496–507.

Réthàti, L. (2012). *Probabilistic Solutions in Geotechnics*. Elsevier, Amsterdam, the Netherlands.

Rickenmann, D. (2005) Runout prediction methods. In: Jakob, M., and Hungr, O. (Eds.). *Debris-flow hazards and related phenomena*. Springer, Berlin, Germany.

Russo, D., and Bouton, M. (1992). Statistical analysis of spatial variability in unsaturated flow parameters. *Water Resources Research,* 28(7), 1911–1925.

Santoso, A. M., Phoon, K. K., and Quek, S. T. (2011). Effects of soil spatial variability on rainfall-induced landslides. *Computers and Structures,* 89(11–12), 893–900.

Sillers, W. S., and Fredlund, D. G. (2001). Statistical assessment of soil-water characteristic curve models for Geotechnical Engineering. *Canadian Geotechnical Journal,* 38(6), 1297–1313.

Sivakumar Babu, G. L., and Murthy, D. S. N. (2005). Reliability analysis of unsaturated soil slopes. *Journal of Geotechnical and Geoenvironmental Engineering,* 131(11), 1423–1428.

Starr, C. (1969), Social benefit versus technological risk. *Science,* 165(September), 1232–1238.

Stein, M. (1987). Large sample properties of simulations using Latin hypercube sampling. *Technometrics,* 29(2), 149–151.

Sun, H. W. (1999). *Review of Fill Slope Failures in Hong Kong*. Geotechnical Engineering Office, Civil Engineering Deptartment, Hong Kong Government, Hong Kong.

Tan, X. H., Hu, N., Li, D., Shen, M. F., and Hou, X. L. (2013). Time-variant reliability analysis of unsaturated soil slopes under rainfall. *Geotechnical and Geological Engineering,* 31(1), 319–327.

Tanaka, H., Loat, J., Shibuya, S., Soon, T. T., and Shiwakoti, D. (2001). Characterization of Singapore, Bangkok, and Ariake clays. *Canadian Geotechnical Journal,* 38, 378–400.

Tang, W. H., Yucemen, M. S., and Ang, A. S. (1976). Probability-based short term design of soil slopes. *Canadian Geotechnical Journal,* 13(3), 201–215.

Thevanayagam, S., Kavazanjian, E., Jacob, A., and Nesarajah, S. (1996). Deterministic and probabilistic analysis of preload design at a hydraulic fill site. *Uncertainty in the Geologic Environment, from Theory to Practice, Proceedings of Uncertainty 96,* Madison, WI. Geotechnical Special Publication (GSP) 58, ASCE, Reston, VA, pp. 1417–1431.

Totschnig, R., Sedlacek, W., and Fuchs, S. (2011). A quantitative vulnerability function for fluvial sediment transport. *Natural Hazards,* 58, 681–703.

Tung, Y. K., and Chan, G. C. C. (2003). Stochastic analysis of slope stability considering uncertainties of soil-water retention characteristics. *Proceedings of ICASP9,* San Francisco, pp. 1409–1414.

Tung, Y. K., Yen, B. C., and Melching, C. S. (2006). *Hydrosystems Engineering Reliability Assessments and Risk Analysis,* McGraw-Hill, 495 pp.

UNDRO. (1979). *Natural Disasters and Vulnerability Analysis.* Department of Humanitarian Affairs/ United Nations Disaster Relief Office, Geneva, p. 53.

Uzielli, M., Nadim, F., Lacasse, S., and Kaynia, A. M. (2008). A conceptual framework for quantitative estimation of physical vulnerability to landslides. *Engineering Geology,* 102(3), 251–256.

van Genuchten, M. T. (1980). A close form equation for predicting the hydraulic conductivity of unsaturated soils. *Soil Science Society of America Journal,* 44(5), 892–898.

Vanmarcke, E. H. (1977). Reliability of earth slopes. *Journal of the Geotechnical Engineering Division,* ASCE, 103(GT11), 1247–1265.

Wang, Y., Cao, Z., and Au, S. K. (2011). Practical reliability analysis of slope stability by advanced Monte Carlo simulations in a spreadsheet. *Canadian Geotechnical Journal,* 48(1), 162–172.

Wolff, T. F. (1996). Probabilistic slope stability in theory and practice. *Uncertainty in the Geologic Environment, from Theory to Practice, Proceedings of Uncertainty 96,* Madison, WI. Geotechnical Special Publication (GSP) 58, ASCE, Reston, VA, pp. 419–433.

Wong, F. S. (1985). Slope reliability and response surface method. *Journal of Geotechnical Engineering, ASCE,* 111(1), 32–53.

Wong, H. N., Ko, F. W. Y., and Hui, T. H. H. (2006). *Assessment of Landslide Risk of Natural Hillsides in Hong Kong* (GEO Report No. 191). Geotechnical Engineering Office, Hong Kong, 117 pp.

Wu, T. H., and Kraft, L. M. (1970). Safety analysis of slopes. *Journal of Soil Mechanics and Foundation Division, ASCE,* SM2, 609–630.

Xue, J. F., and Gavin, K. (2007). Simultaneous determination of critical slip surface and reliability index for slopes. *Journal of Geotechnical and Geoenvironmental Engineering,* 137(7), 878–886.

Zapata, C. E., Houston, W. N., Houston, S. L., and Walsh, K. D. (2000). Soil-water characteristic curve variability. *Advances in Unsaturated Geotechnics: Proceedings of Sessions of Geo-Denver 2000,* Denver, CO, pp. 84–124.

Zhang, J., Huang, H. W., Juang, C. H., and Li, D. Q. (2013a). Extension of Hassan and Wolff method for system reliability analysis of soil slopes. *Engineering Geology,* 160, 81–88.

Zhang, J., Huang, H. W., and Phoon, K. K. (2013b). Application of the Kriging-based response surface method to the system reliability of soil slopes. *Journal of Geotechnical and Geoenvironmental Engineering,* 139(4), 651–655.

Zhang, J., Huang, H. W., Zhang, L. M., Zhu, H. H., and Shi, B. (2014). Probabilistic prediction of rainfall-induced slope failure using a mechanics-based model. *Engineering Geology,* 168, 129–140.

Zhang, J., Zhang, L. M., and Tang, W. H. (2011). Slope reliability analysis considering site-specific performance information. *Journal of Geotechnical and Geoenvironmental Engineering,* 137(3), 227–238.

Zhang, L. L. (2005). *Probabilistic Study of Slope Stability under Rainfall Condition.* PhD Thesis. Hong Kong University of Science and Technology.

Zhang, L. L., Fredlund, D. G., Zhang, L. M., and Tang, W. H. (2004). Numerical study of soil conditions under which matric suction can be maintained. *Canadian Geotechnical Journal,* 41(4), 569–582.
Zhang, L. L., Zhang, L. M., and Tang, W. H. (2005). Rainfall-induced slope failure considering variability of soil properties. *Géotechnique,* 55(2), 183–188.
Zhu, G., Yin, J. H., and Graham, J. (2001). Consolidation modelling of soils under the test embankment at Clek Lap Kok International Airport in Hong Kong using a simplified finite element model. *Canadian Geotechnical Journal,* 38(2), 349–363.

Chapter 8

Probabilistic assessment of randomly heterogeneous soil slopes

8.1 SPATIAL VARIABILITY OF SOILS

Soils are geological materials formed by weathering processes, transported by physical means to their present locations. They have been subject to various stresses, pore fluids, and physical and chemical changes. Thus, it is not surprising that the physical properties vary from place to place, even in one soil stratum. Researchers have identified and classified major components of uncertainties associated with the estimation of soil properties (Vanmarcke, 1977; Tang, 1984; Phoon and Kulhawy, 1999; Baecher and Christian, 2003; Zhang et al., 2003). Vanmarcke (1977) presented basic concepts and methods for describing the spatial variability of a soil stratum within random field framework. Three major sources of uncertainties were identified for stochastic modeling of soil profiles. The first is natural heterogeneity or in situ variability of the soil, which is caused by variation in mineral composition, varying depths of strata during soil formation and stress history. The second is attributed to limited availability of information about subsurface conditions. The third is measurement errors due to sample disturbance, test imperfections, human factors, and estimating soil properties through correlation with index properties. Kulhawy (1992) classified three primary sources of geotechnical uncertainties: inherent variability, measurement uncertainties, and transformation uncertainties (Figure 8.1), which are basically the same as those defined by Vanmarcke (1977). The uncertainties caused by equipment, operator, random test effects, and statistical uncertainties were all included in measurement errors. The transformation uncertainties meant the uncertainty introduced when field or laboratory measurements are transformed into design soil properties using empirical or other correlation models.

The spatial variability in a soil profile may be either continuous or discrete (JCSS 2002). The continuous spatial variability may be characterized by an average trend and continuous fluctuations around the average trend (Figure 8.2). Usually, this type of variability is modeled as a stationary random field. Soils may be mixed with dislocations such as faults and lenses depending on the geological history. This type of spatial variability is defined as discrete spatial variability. Though local of nature, these features may have a significant influence on the performance of a geotechnical engineering structure. In this book, we will focus on the continuous spatial variability.

The effect of inherent spatial variability of soil properties on the performance of geotechnical works has received considerable attention in recent years. Some researchers have considered the spatial variability for the problem of flow in unsaturated soil and slope stability. Zhang (1999), Tartakovsky et al. (2003), Yang et al. (2004), Lu and Zhang (2007), Li et al. (2009), Mousavi Nezhad et al. (2011), Cho (2012), Le et al. (2012), and Meftah et al. (2012) among others, analyzed unsaturated flow in heterogeneous soils with spatially distributed uncertain hydraulic parameters. Cho (2007), Jiang et al. (2014a,b) Le (2014), Li et al. (2014; 2015),

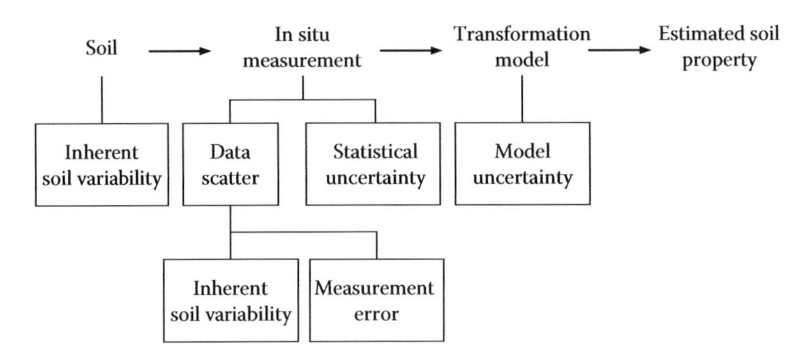

Figure 8.1 Uncertainties in estimation of soil properties. (From Phoon, K. K., and Kulhawy, F. H., *Geotechnical Journal*, 36, 612–624, 1999. With permission.)

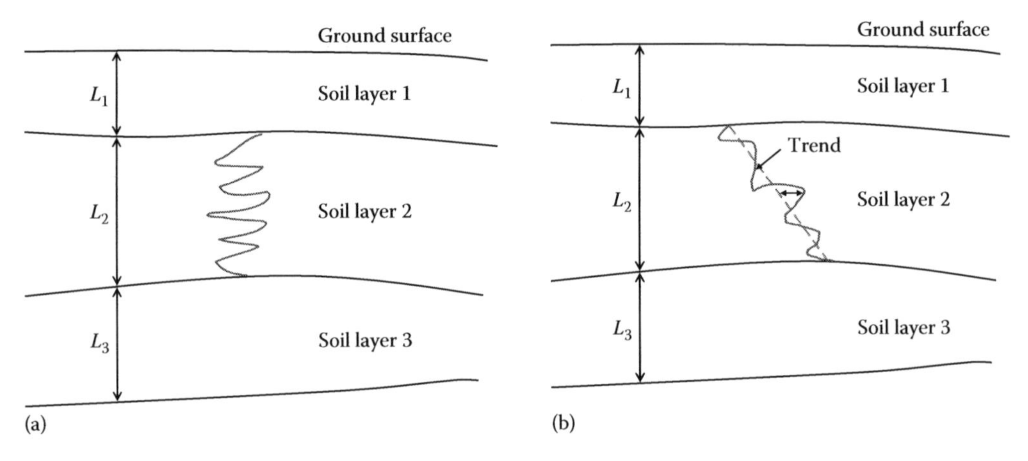

Figure 8.2 Inherent spatial variability of a soil property (a) without a trend and (b) with a trend.

and Li and Jiang (2015) assessed the effect of inherent variability on slope stability. Srivastava et al. (2010), Santoso et al. (2011), and Zhu et al. (2013) simulated permeability as a spatially correlated lognormally distributed random variable and studied its influence on water flow and slope stability. Le et al. (2013) investigated rainfall-induced differential settlement of a strip foundation on an unsaturated soil with spatially varying values of either preconsolidation stress or porosity.

In this chapter, the basic theory and concepts of random field are presented in Section 8.2. Some common methods of random field generation, such as matrix decomposition method, fast Fourier transformation method, turning bands method (TBM), and so forth are reviewed in Section 8.3. An example of stochastic modeling for seepage and stability of a two-dimensional random heterogonous soil slope is presented in Section 8.4.

8.2 RANDOM FIELD THEORY

8.2.1 Concept of random field

A random field is a conceivable model to characterize continuous spatial fluctuations of a soil property within a soil unit. The actual value of a soil property at each location within the unit is assumed to be a realization of a random variable. Field can be modeled as a location-dependent random variable function $Z(\mathbf{x})$, where $\mathbf{x}$ is the location coordinates vector ($\mathbf{x}$ can be along any direction and is not necessarily along x-axis). The covariance/autocovariance between two locations $\mathbf{x}$ and $\mathbf{x}'$ is

$$C(\mathbf{x},\mathbf{x}') = E\{Z(\mathbf{x})Z(\mathbf{x}')\} - E[Z(\mathbf{x})]E[Z(\mathbf{x}')] \quad (8.1)$$

where:
$E[.]$ is the expectation operator

If the random field is stationary, the covariance of the soil property values at two locations is only a function of the separation distance

$$C(\tau) = C(\mathbf{x},\mathbf{x}+\tau) = E\{Z(\mathbf{x})Z(\mathbf{x}+\tau)\} - E[Z(\mathbf{x})]^2 \quad (8.2)$$

where:
$\boldsymbol{\tau}$ is the lag vector representing separation between two spatial locations

The normalized form of the covariance function is known as the autocorrelation function (ACF):

$$\rho(\tau) = \frac{C(\tau)}{C(0)} \quad (8.3)$$

where:
$C(0)$ is the autocovariance function at zero separation distance, which is the stationary variance of data

The above autocovariance function and ACF are usually estimated from soil samples from a site. Analytical models are fitted to the sample ACFs using regression analysis. One of the commonly used autocorrelation models is the exponential model, in which the autocorrelation decays exponentially with the separation distance. The one-dimensional form of the exponential model is

$$\rho(\tau) - \exp\left(-\frac{2\tau}{\delta}\right) \quad (8.4)$$

where:
τ is the separation distance between two points and δ is the scale of fluctuation

The frequently used autocorrelation models are presented in Table 8.1. The scale of fluctuation δ is an important indicator of soil spatial variability describing the distance within which the spatially random values will tend to be significantly correlated (i.e., by more than about 10% [Fenton and Griffiths, 2008]). That is to say, within a range of δ, soil property values are highly correlated; but in a range over δ, soil property values are irrelevant. Usually, the scale of fluctuation is estimated using the moving window method or by curve fitting of empirical autocorrelation models.

8.2.2 Spatial-averaged soil properties

Parameters in a geotechnical analysis usually refer to averages of soil properties over some area or some volume; for example, average shear strength along a sliding surface or average stiffness of a volume affected by loading. Hence, geotechnical analysis concerns mostly the uncertainties in the averaged properties over specific surfaces or volumes.

Table 8.1 Common autocorrelation functions and variance reduction functions

Type	*Autocorrelation function*	*Variance reduction function*
Exponential	$\rho(\tau)=e^{-\tau/a}, a=\delta/2$	$\Gamma^2(L)=\frac{2a^2}{L^2}\left(\frac{L}{a}-1+e^{-L/a}\right)$
Gaussian	$\rho(\tau)=e^{-(\tau/b)^2}, b=\delta/\sqrt{\pi}$	$\Gamma^2(L)=\frac{b^2}{L^2}\left[\frac{L}{b}\sqrt{\pi}\mathrm{erf}(L/b)-1+e^{-L^2/b^2}\right]$
Spherical	$\rho(\tau)=\begin{cases}1-1.5(\tau/d)+0.5(\tau/d)^3 & (\tau\le d)\\ 0 & (\tau>d)\end{cases}$	$\Gamma^2(L)=2-1.5L/d+0.25(L/d)^3$
Uniform type I	$\rho(\tau)=\begin{cases}1 & \tau\le\delta\\ 0 & \tau>\delta\end{cases}$	$\Gamma^2(L)=\begin{cases}1 & L\le\delta\\ \delta/L & L>\delta\end{cases}$
Uniform type II	$\rho(\tau)=\begin{cases}1 & \tau\le\delta/2\\ 0 & \tau>\delta/2\end{cases}$	$\Gamma^2(L)=\begin{cases}1 & L\le\delta/2\\ \frac{\delta}{L}(1-\delta/4L) & L>\delta/2\end{cases}$
Triangular	$\rho(\tau)=\begin{cases}1-\tau/\delta & \tau\le\delta\\ 0 & \tau>\delta\end{cases}$	$\Gamma^2(L)=\begin{cases}1-L/(3\delta) & L\le\delta\\ \frac{\delta}{L}(1-\delta/3L) & L>\delta\end{cases}$

Notes: δ is scale of fluctuation; erf() is the error function; a, b, and d are coefficients of correlation function; L is the domain size for the spatial-averaged soil property.

The "point-to-point" variation forms the basis for quantitative assessment of uncertainties of averaged soil parameters. Assume the variance of a soil property is var_P. The larger the domain size over which the property is averaged, the larger the reduction in the variance of the spatial averaged soil property will be. The variance of the spatial averaged soil property, var_A, can be expressed as the product of the variance reduction factor function and the point variance var_P (Vanmarcke, 1977):

$$\mathrm{var}_A = \Gamma^2(\tau)\mathrm{var}_P \tag{8.5}$$

where:

$\Gamma^2(\tau)$ is the function of variance reduction factor, which can be determined based on the ACF

The spatial averaged coefficient of variation (COV), cov_A, over a domain can be expressed as

$$\mathrm{cov}_A = \Gamma(\tau)\mathrm{cov}_P \tag{8.6}$$

where:

cov_P is the point COV of a soil property

The variance reduction functions for the common ACFs are shown in Table 8.1.

8.2.3 Definitions in geostatistics

In geostatistics, the spatial variability of soil properties are described by variogram or semi-variogram function (Deutsch and Journel, 1997). The variogram is defined as the variance of the difference $\{Z(\mathbf{x}+\tau)-Z(\mathbf{x})\}$.

$$2\gamma(\tau) = \mathrm{var}\{Z(\mathbf{x}+\tau)-Z(\mathbf{x})\} \tag{8.7}$$

where:

var[] is the variance operator

The semivariogram function is defined as half of the variogram and can be expressed as the difference of the stationary variance and covariance:

$$\gamma(\tau) = C(0) - C(\tau) \tag{8.8}$$

By definition, $\gamma(\tau)$ is semivariogram and the variogram is $2\gamma(\tau)$. However, the terms semivariogram and variogram are often used interchangeably.

The relationship between the variogram $\gamma(\tau)$ and covariance $C(\tau)$ is shown in Figure 8.3. In general, $\gamma(\tau)$ increases from its initial value at $\gamma(0)$ as τ increases. For continuous random variables, the variogram function levels off and becomes stable about a limiting value called the sill, which is generally at, or near, the variance of the stochastic process. The separation distance where the variogram function approaches the sill is called the range of influence. In some cases, the behavior of the variogram near the origin ($\tau = 0$) can have a discontinuity. The discontinuity near the origin is called the nugget effect, and produces an apparent intercept at zero separation distance termed the nugget. The nugget effect can result from small-scale effects or measurement errors.

Based on the variogram function or the ACF, the autocovariance distance, sometimes named as correlation length, autocorrelation length, or correlation distance, is defined as the distance at which the spatial variance/autocorrelation has decayed by 1/e (37%). The scale of fluctuation is usually between 1.4 and 2.0 times the correlation length for exponential, Gaussian, and spherical ACFs (Vanmarcke, 1983). In addition, the ACF is named correlogram in geostatistics.

8.2.4 Reported statistics of spatial variability

Table 8.2 summarizes reported values of scale of fluctuation for various soil properties from literature. As shown in the table, for shear-strength-related parameters such as cone tip resistance by Cone Penetration Test (CPT), Standard Penetration Test (SPT) N-value, undrained shear strength S_u, the vertical scale of fluctuation is usually less than 5 m; however,

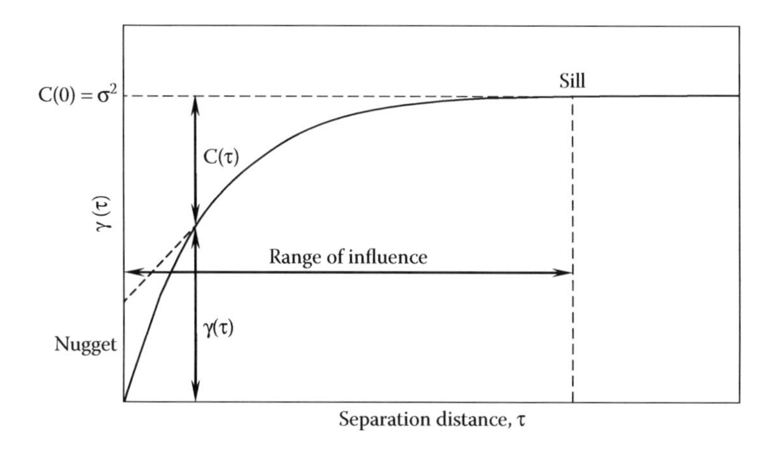

Figure 8.3 Relationship between autocovariance function $C(\tau)$ and variogram $\gamma(\tau)$.

Table 8.2 Scale of fluctuation for various soil properties

Soil property	*Scale of fluctuation* δ	*Soil*	*Source*	*Test*
Average cone resistance	$\delta_h = 35$–60 m (different levels)	Marine clay	Tang (1979)	CPT
Average cone resistance	$\delta_h = 55$ m (0–3 m below sea bottom)			
CPT tip resistance, q_c	$\delta_v = 2$ m	Sensitive clay	Chiasson et al. (1995)	CPT
CPT tip resistance, q_c	$\delta_h = 20.0$–35.0 m	Glacial sands	Vrouwenvelder and Calle (2003)	CPT
CPT tip resistance, q_c	$\delta_v = 0.8$–1.8 m		Popescu et al. (1995)	
CPT tip resistance, q_c	$\delta_v = 1$ m	Clay	Vanmarcke (1977)	
CPT tip resistance, q_c	$\delta_h = 14$–38 m	Offshore soils	Keaveny et al. (1989)	CPT
CPT tip resistance, q_c	$\delta_h = 5$–12 m $\delta_v = 1$ m	Silty clay	Lacasse and de Lamballerie (1995)	CPT
CPT tip resistance, q_c	$\delta_v = 3$ m $\delta_v = 1$ m	Clean sand Mexico clay	Alonzo and Krizek (1975)	CPT
CPT tip resistance, q_c	$\delta_v = 0.5$ m	Copper tailings	Baecher and Christian (2003)	CPT
CPT tip resistance, q_c	$\delta_v = 1.6$ m	Clean sand	Kulatilake and Ghosh (1988)	CPT
CPT tip resistance, q_c	$\delta_v = 0.1$–2.2 m $\delta_h = 3.0$–80.0 m	Sand, clay	Phoon and Kulhawy (1996)	CPT
Corrected CPT tip resistance, q_T	$\delta_v = 0.2$–0.5 m $\delta_h = 23.0$–66.0 m	Clay	Phoon and Kulhawy (1996)	CPT
Penetrometer resistance	$\delta_h = 40.0$–70.0 m	Sandy clay	Mulla (1988)	
DMT P_0	$\delta_h = 0.5$–2.5 m	Sand	Jaksa et al. (2004)	DMT
DMT P_0	$\delta_v = 1$ m	Varved clay	DeGroot (1996)	DMT
Undrained shear strength S_u	$\delta_v = 2$ m		Chiasson et al. (1995)	FVT
Undrained shear strength	$\delta_v = 2.5$–6 m	Clay	Asaoka and A-Grivas (1982)	FVT
Undrained shear strength S_u	$\delta_h = 23$ m	Sensitive clay	DeGroot and Baecher (1993)	FVT
Undrained shear strength S_u	$\delta_v = 1$ m	Sensitive clay	Baecher (1982)	FVT
Undrained shear strength S_u	$\delta_v = 0.5$ m	Chicago clay	Wu (1974)	Unconfined compression test
Undrained shear strength S_u	$\delta_v = 0.3$–0.6 m	Offshore soils	Keaveny et al. (1989)	Triaxial tests and DSS tests
Undrained shear strength, S_u	$\delta_v = 0.8$–6.1 m	Clay	Phoon and Kulhawy (1996)	Laboratory test
Undrained shear strength, S_u	$\delta_v = 2$–6.2 m $\delta_h = 46$–60 m	Clay	Phoon and Kulhawy (1996)	VST

(*Continued*)

Table 8.2 (Continued) Scale of fluctuation for various soil properties

Soil property	*Scale of fluctuation* δ	*Soil*	*Source*	*Test*
Unconfined compressive strength	$\delta_v = 4$ m, $\delta_h = 80.0$ m		Honjo and Kuroda (1991)	
Shear strength	$\delta_v = 2.0$ m	Clay	Ronold (1990)	
Shear strength	$\delta_v = 2.0$ m, $\delta_h = 20.0$ m	Clay	Soulié et al. (1990)	
N-value	$\delta_v = 2.4$ m	Sand	Phoon and Kulhawy (1996)	SPT
N-value	$\delta_h = 20$ m	Dune sand	Hilldale-Cunningham (1971)	SPT
N-value	$\delta_h = 17$ m	Alluvial sand	DeGroot (1996)	SPT
Permeability $\ln(k_{unsaturated})$	$\delta_h = 12–16$ m		Unlu et al. (1990)	
$\ln(k)$	$\delta_v = 3.2$ m, $\delta_h = 25.0$ m		Rehfeldt et al. (1992)	Flowmeter
$\ln(k)$	$\delta_v = 1.5–3.0$ m, $\delta_h = 25.0–50.0$ m		Rehfeldt et al. (1992)	
$\ln(k)$	$\delta_v = 0.2–1$ m, $\delta_h = 2–10$ m		Hess et al. (1992)	Large-scale tracer test
k	$\delta_h = 1500$ m	Salt dome	Ditmars et al. (1988)	
k	$\delta_h = 0.5–2$ m	Compacted clay	Benson (1991)	
k	$\delta_h = 1–2.5$ m	Sand	Bjerg et al. (1992)	
Natural water content, w_n	$\delta_v = 1.6–12.7$ m	Clay, loam	Phoon and Kulhawy (1996)	
Natural water content, w_n	$\delta_h = 170.0$ m	Clay	Phoon and Kulhawy (1996)	
Liquid limit, w_L	$\delta_v = 1.6–8.7$ m	Clay, loam	Phoon and Kulhawy (1996)	
Total unit weight, γ	$\delta_v = 2.4–7.9$ m	Clay, loam	Phoon and Kulhawy (1996)	
Effective unit weight, γ'	$\delta_v = 1.6$ m	Clay	Phoon and Kulhawy (1996)	
Sand content	$\delta_h = 60.0–80.0$ m	Sandy clay	Mulla (1988)	
Clay content	$\delta_h = 40.0–60.0$ m	Sandy clay	Mulla (1988)	
Natural water content	$\delta_h = 40.0–60.0$ m	Sandy clay	Mulla (1988)	
Thickness of natural deposit	$\delta_h = 750$ m		Rosenbaum (1987)	

Notes: CPT, Cone Penetration Test; SPT, Standard Penetration Test; DMT, Dilatometer Test; FVT, Field Vane Test; DSS, Direct Simple Shear test.

the horizontal one is generally greater than 10 m. For basic soil properties such as water content and unit weight, the vertical scale of fluctuation is slightly larger but basically is less than 10 m. The horizontal scale of fluctuation can be greater than 100 m. For permeability of a soil, the scale of fluctuation for $\ln(k)$ in vertical direction can be less than 1 m and the horizontal scale of fluctuation is usually greater than 10 m.

8.3 MODELING OF RANDOM FIELD

Different algorithms are available for simulation of random field. The most popular methods include the spectral methods based on Fourier transformation (Cooley and Tukey, 1965; Mejia and Rodriguez-Iturbe, 1974; Dietrich and Newsam, 1996), the moving average method (Matérn, 1986; Yaglom, 1987; Oliver, 1995; Chiles and Delfiner, 1999; Cressie and Pavlicová, 2002), the local average subdivision (LAS) method (Fenton and Vanmarcke, 1990; Fenton and Griffths, 2008), the covariance matrix decomposition method (Davis, 1987), the turning bands method (TBM) (Matheron, 1973; Mantoglou and Wilson, 1982), and the sequential simulation methods (Bellin and Rubin, 1996; Deutsch and Journel, 1997). If property values are known or measured at specific points, the conditional random field simulation methods (Davis, 1987; Gutjahr et al., 1997; Strebelle, 2002; Emery, 2008) can be applied. Comprehensive detailed review on the various methods can be found in Stefanou (2009), Deutsch and Journel (1997) and Fenton and Griffiths (2008), etc.

8.3.1 Covariance matrix decomposition method

The covariance matrix decomposition method is a direct method to produce a random field assuming that the parameters at different locations in the field are correlated random variables. This method is applicable to any configuration of the simulated locations and any covariance model.

For simplicity, consider a one-dimensional problem. The one-dimensional space has first been discretized into n points with the soil property at each point is defined as a random variable. Assume that the prescribed covariance matrix for these correlated random variables is C. If C is positive definite, then a correlated standard normal random field **Z** can be produced using independent standard normal random variables according to the following equation:

$$\mathbf{Z} = \mathbf{L}\mathbf{U} \tag{8.9}$$

where:

L is a lower triangular matrix satisfying $\mathbf{L}\mathbf{L}^T = \mathbf{C}$
U is a vector of n-independent standard normal random variables
Z is a vector of n-correlated standard normal random variables

The lower triangular matrix **L** is typically obtained using the Cholesky decomposition method. If a soil property **Z′** of the random field is normally distributed with a known mean and variance, then a realization for the soil property can be obtained using the realization of the correlated standard normal variate **Z** as follows:

$$\mathbf{Z}' = \mu_{Z'}\mathbf{I} + \sigma_{Z'}\mathbf{Z} \tag{8.10}$$

where:

I is the unit matrix
$\mu_{Z'}$ and $\sigma_{Z'}$ are the mean value and standard variation of the normally distributed **Z′**

Non-normal random fields can be obtained through a suitable transformation of a normally distributed random field. For example, a lognormally distributed random field can be obtained from

$$\mathbf{Z}' = \exp\left(\mu_{\ln Z'}\mathbf{I} + \sigma_{\ln Z'}\mathbf{Z}\right) \tag{8.11}$$

where $\mu_{\ln Z'}$ and $\sigma_{\ln Z'}$ are the mean and standard variation of $\ln(\mathbf{Z}')$, respectively.

Compared with the TBM, spectral fast Fourier method and the sequential Gaussian simulation (SGS) method, this method shows the least artificial bias (Harter 1994). It is usually used for small fields with small size of discrete points as it requires the decomposition of a covariance matrix. For random fields with many realizations, this is indeed a very effective way of random field generation, because the covariance matrix must only be decomposed once for an entire simulation (Harter, 1994). Each realization then simply requires the generation of random numbers and the multiplication with the **L** matrix (Equation 8.9). It should yet be noted that considerable round-off error may be induced in the Cholesky decomposition when covariance matrices are poorly conditioned and become numerically singular.

8.3.2 Fourier transformation methods

The Fourier transform method are based on the spectral representation of continuous random field, $Z(\mathbf{x})$, which can be expressed as follows (Yaglom, 1962):

$$Z(\mathbf{x}) = \int_{-\infty}^{\infty} e^{ix\omega} dW(\omega) \tag{8.12}$$

where:

Z is the soil property
$\mathbf{x}$ is the vector of spatial variable
ω is the angular frequency
$dW(\omega)$ is an interval white noise process with mean zero and variance of $S(\omega)d\omega$

$S(\omega)$ is the spectral density function of the random field. $Z(\mathbf{x})$ and is simply the Fourier transform of the covariance of $Z(\mathbf{x})$ (Harter, 1994). The term $S(\omega)d\omega$ is a measure of the contribution of the amplitude of a frequency ω to the random field.

The above integral is usually expressed as a summation. For illustration purpose, only one-dimensional case will be presented. Based on the discrete Fourier transform (DFT) method, the random field in one dimension that is discretized into $2K$+1 frequencies can be expressed as

$$Z(x) = \sum_{k=-K}^{K} A_k \cos(\omega_k x) + B_k \sin(\omega_k x) \tag{8.13}$$

where:

x is the location coordinate
ω_k is the kth discrete angular frequency
A_k and B_k are mean zero, mutually independent, normal distributed random numbers

The variance of A_k and B_k can be obtained from the spectral density function $S(\omega)$. Hence, the spectral representation of a single random field realization can be intuitively understood as a field of amplitudes, where the coordinates are the frequencies of sine–cosine waves.

In the fast Fourier transform (FFT) method (Cooley and Tukey, 1965), both space and frequency are discretized into a series of points. Assume the numbers of spatial and frequency discretization points are the same and the random field is periodic (i.e., $Z(x_j) = Z(x_{K+j})$. For discrete $Z(x_j)$, $j = 1, 2, \ldots, K$, the Fourier transform can be evaluated as

$$Z(x_j) = \sum_{k=0}^{K} \chi_k e^{i(2\pi jk/K)} \tag{8.14}$$

where χ_k are the Fourier coefficients with $\chi_k = A_k - iB_k$. A_k and B_k are mean zero, mutually independent, normal distributed random numbers, whose variance can be obtained from the spectral density function $S(\omega)$. A_k and B_k are symmetric due to the fact that Z is real.

The simulation process of DFT and FFT methods is as follows.

1. Decide how to discretize the space and frequency of spectral density function.
2. Generate independent normally distributed realizations of A_k and B_k.
3. Use the symmetry relationships to produce the Fourier coefficients (only in FFT).
4. Produce the field realization using Equation 8.13 or 8.14.

The major shortcoming of DFT is its computational demand. The DFT approach is basically only computational practical for one-dimensional problem (Fenton and Griffiths, 2008). The FFT is generally much faster than DFT and can be adopted for two-dimensional or three-dimensional problems. However, the covariance function obtained from the FFT random field simulation is always symmetric about the midpoint of the field. This deficiency can be overcome by generating a field twice as long as required in each coordinate direction and keeping only the first quadrant of the field (Fenton and Griffiths, 2008).

8.3.3 Turning bands method

The TBM is a multidimensional random field generation method that performs simulations along one-dimensional lines (bands) instead of synthesizing the multidimensional field directly. The TBM generally includes the following steps (Matheron, 1973; Mantolou and Wilson, 1982; Fenton and Griffiths, 2008):

1. Define the multidimensional covariance model $C(\boldsymbol{\tau})$ of the random field in which $\boldsymbol{\tau}$ is the distance vector between the two points in space.
2. Determine the one-dimensional covariance $C_1(\tau)$ of the random field. Mantoglou and Wilson (1982) and Bras and Rodriyguez-Iturbe (1985) review and summarize several multidimensional covariance models and the corresponding unidimensional covariance functions.
3. Choose an arbitrary origin within or near the domain of the field to be generated. Select L lines crossing the domain with the ith line having a direction given by the unit vector $\mathbf{u}_i$.
4. Generate a realization of one-dimensional random field, $Z_i(\xi)$, along the ith line base on the one-dimensional covariance function $C_1(\tau)$, where ξ is the coordinate along the line.
5. Project one field point $\mathbf{x}_k$ orthogonally onto each line to define the coordinate ξ_{ki} ($\xi_{ki} = \mathbf{x}_k \cdot \mathbf{u}_i$) and obtain the component of the one-dimensional random field realization value $Z_i(\xi_{ki}) = Z_i(\mathbf{x}_k \cdot \mathbf{u}_i)$.
6. Summing the contributions of all lines and normalize to the field value $Z(\mathbf{x}_k)$ by the factor $\sqrt{L}$, that is, $Z(\mathbf{x}_k) = 1/\sqrt{L}\sum_{i=1}^{L} Z_i(\mathbf{x}_k \cdot \mathbf{u}_i)$ for each $\mathbf{x}_k$.

The TBM method is not a fundamental random field generator and requires an existing one-dimensional generator such as FFT. Therefore, the TBM method can be classified based on different approaches of generation one-dimensional line processes: (1) the approach that uses a moving average method combined with weighting function approximation (Bras et al., 1985), (2) the approach that evaluates the integration of unidimensional spectral density function by a standard integration approximation method (Mantolou and Wilson,

1982; Vanmarcke, 1983), and (3) the approach that applies an FTT algorithm to generate the unidimensional line processes (Tompson et al., 1989; Kottegoda and Kassim, 1991).

8.3.4 Karhunen–Loeve expansion method

The Karhunen–Loeve (KL) expansion of a random field (Ghanem and Spanos, 1991; Huang et al., 2001; Phoon et al., 2002a,b; Yang et al., 2004) is based on the spectral decomposition of the covariance function:

$$Z(\mathbf{x},\theta) = \mu(\mathbf{x}) + \sum_{i=1}^{\infty} \sqrt{\lambda_i}\,\varphi_i(\mathbf{x})\varepsilon_i(\theta) \tag{8.15}$$

where:

$\mathbf{x}$ is the coordinate of a point in the continuous domain Ω
θ is the coordinate in the sample space Θ
$\mu(\mathbf{x})$ is the mean function of the random field
$\varepsilon_i(\theta)$ is uncorrelated standard normal random variables
φ_i and λ_i denote the eigenfunctions and eigenvalues of the covariance matrix obtained from solving the homogeneous Fredholm integral equation of the second kind:

$$\int_{\Omega} C(\mathbf{x},\mathbf{x}')\varphi_i(\mathbf{x}')d\mathbf{x}' = \lambda_i\varphi_i(\mathbf{x}) \tag{8.16}$$

The approximate random field is defined by truncating the ordered series:

$$Z(\mathbf{x},\theta) = \mu(\mathbf{x}) + \sum_{i=1}^{M} \sqrt{\lambda_i}\,\varphi_i(\mathbf{x})\varepsilon_i(\theta) \tag{8.17}$$

where M is the truncating level.

The truncated KL expansion underestimates the true variability of the original random field. The truncating level M to be chosen strongly depends on the desired accuracy and the covariance function of the random field. In addition, the truncated KL expansion of homogeneous random fields is only approximately homogeneous, because the standard deviation function of the truncated field varies in space (Stefanou and Papadrakakis, 2007; Betz et al., 2014).

8.3.5 LAS method

The LAS method proposed by Fenton and Vanmarcke (1990) is a fast and accurate method of producing realizations of a random process. The motivation of the method arose out of the fact that most engineering measurements are only defined over some finite domain and thus represent a local average of the property. The method was based on the stochastic subdivision methods (Carpenter, 1980; Fournier et al., 1982) and incorporates the concept of local averaging. The procedure of LAS for a one-dimensional stationary random field is presented here to illustrate the basic concept of this method.

The construction of a local average process is essentially a top–down recursive fashion (Figure 8.4). In Stage 0, a global average is generated for the process. In Stage 1, the domain is subdivided into two regions whose local averages must in turn average to the global (parent) value. Subsequent stages are obtained by subdividing each parent cell and

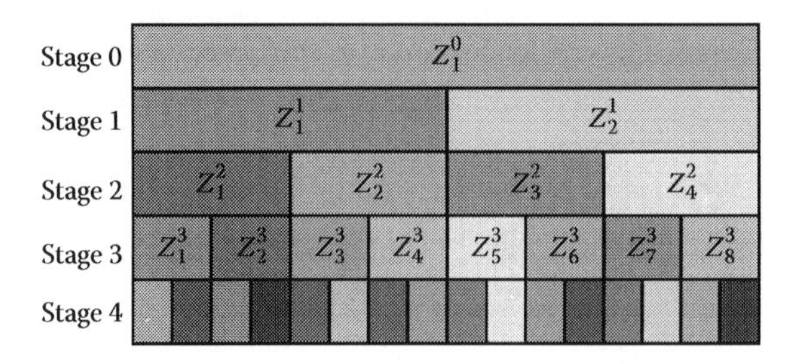

Figure 8.4 Top-down approach of the local average subdivision (LAS) method. (Adopted from Fenton, G. A., and Vanmarcke, E. H., *Journal of Engineering Mechanics*, 116, 1733–1949, 1990. With permission.)

generating values for the resulting two regions while preserving upwards averaging. The global average remains constant throughout the subdivision. The detailed procedures are as follows.

1. Generate a normally distributed global average Z_1^0 with mean zero and variance obtained from local averaging theory.
2. Subdivide the field into two equal parts and generate two normally distributed values of Stage 1, Z_1^1 and Z_2^1, whose means and variances satisfy the following three criteria:
   a. Z_1^1 and Z_2^1 show the correct variance according to local averaging theory.
   b. Z_1^1 and Z_2^1 average to the parent value, that is, $1/2\left(Z_1^1 + Z_2^1\right) = Z_1^0$. In other words, the distributions of Z_1^1 and Z_2^1 are conditioned on the value of Z_1^0.
   c. Z_1^1 and Z_2^1 are properly correlated.
3. Subdivide each cell in Stage 1 into equal parts. Generate normally distributed values, Z_1^2, Z_2^2, Z_3^2, and Z_4^2 of Stage 2. The means and variances of Z_1^2, Z_2^2, Z_3^2, and Z_4^2 satisfy the above three criteria just like in Stage 1. In addition, Z_1^2 and Z_2^2 are properly correlated with Z_3^2 and Z_4^2, which can be satisfied approximately by conditioning the distributions of Z_1^2 and Z_2^2 also on Z_2^1.

8.3.6 Sequential simulation method

Different sequential simulation methods use basically the same algorithm and can be classified based on different data types. Sequential Gaussian simulation (SGS) is a simulation method for continuous variables. Sequential indicator simulation (SIS) simulates discrete variables, using SGS methodology to create a grid of zeros and ones. The general sequential simulation procedure is as follows (Deutsch and Journel, 1997).

1. Perform a normal transformation of the raw data into data with standard normal cumulative density function (CDF).
2. Define a random path that visits each node of the grid (not necessary regular) once.
3. Select one node that is not yet simulated in the grid. Construct the conditional probability distribution function by kriging.
4. Draw a simulated value from the conditional probability distribution function.
5. Include the newly simulated value in the data set and check whether the results honor the data and the desired spatial variability.
6. Repeat until all grid nodes have simulated values.

8.4 SEEPAGE AND STABILITY OF A RANDOMLY HETEROGENEOUS SLOPE UNDER RAINFALL INFILTRATION

The permeability of unsaturated soil may change spatially due to uncertainties in soil fabric. It is of great significance to perform probabilistic infiltration and stability analysis of slopes considering the permeability function as a random field. Few attempts have been made to study the variations in groundwater table in a spatial varied soil slope under rainfall infiltration. The statistical response of pore-water pressures in two-dimensional unsaturated heterogeneous slopes has not been sufficiently investigated. Also the critical hydraulic conditions that may lead to failure of heterogeneous slopes are not well known. The objective of this section is to illustrate how the variability of permeability function propagates to the variability of hydraulic conditions (i.e., pore-water pressure and groundwater table) induced by steady rainfall infiltration and its effect on slope stability.

8.4.1 Slope geometry and boundary conditions

Numerical modeling of seepage and slope stability of a hypothetical slope is conducted. Figure 8.5 shows the geometry of the slope. The hydraulic heads on both side boundaries which are below the groundwater table (BC and DE) are constant. The bottom boundaries as well as the side boundaries above the water table (AB and EF) are assumed to be impermeable. The uncertainties of boundary conditions are not considered.

8.4.2 Soil properties and generation of random field

The Fredlund and Xing (1994) SWCC model is used in this example. The exponential equation proposed by Leong and Rahardjo (1997) is employed for the unsaturated permeability function,

$$k = k_s \left(\frac{\theta_w - \theta_r}{\theta_s - \theta_r} \right)^{\delta_{lr}} \tag{8.18}$$

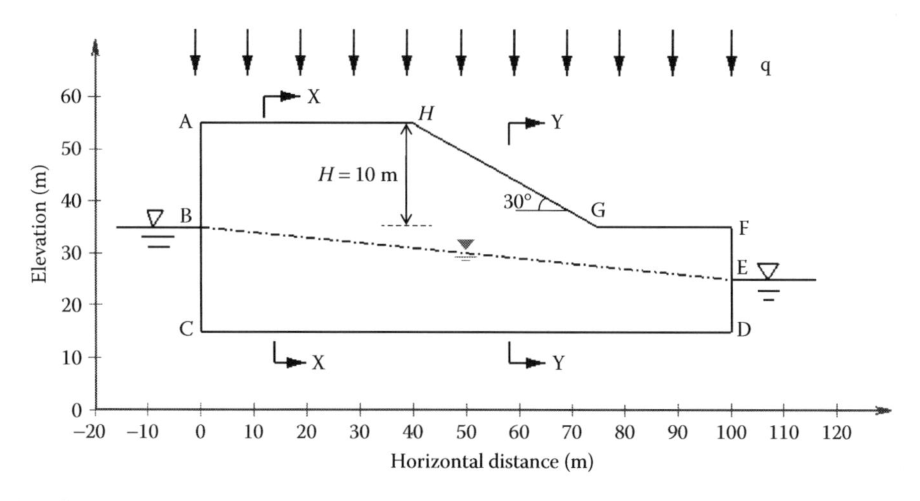

Figure 8.5 Geometry of a hypothetic slope in the numerical study.

where:

k_s is the saturated permeability
θ_w is the volumetric water content
θ_s is the saturated volumetric water content
θ_r is the residual water content
$(\theta_w-\theta_r)/(\theta_s-\theta_r)$ is the normalized water content
δ_{lr} is a constant depending on the soil type

The SWCC and unsaturated permeability function are presented in Figure 8.6 with the soil parameters listed in Table 8.3. The soil within the slope is assumed to be a fine sand with the mean value of k_s is equal to 2×10^{-5} m/s. The δ_{lr} value in the permeability function is assumed to be 3 (Tami et al., 2004). The vertical flux applied to the slope surface is given as 2×10^{-7} m/s (moderate rainfall intensity), that is, the ratio between the rainfall intensity and the saturated permeability is 0.01. This small value is shown to be able to maintain a constant matric suction within the unsaturated zone (Zhang et al., 2004).

As the most important parameter in rain infiltration in unsaturated soil slope is the saturated permeability, the spatial variability of k_s is considered and the other soil hydraulic parameters remain constant for different locations within the slope. The shear strength parameters are also considered to be deterministic.

The generation of a random field normally requires a mean value (μ), a standard deviation or COV and a correlation structure. Based on field measurements, previous studies have

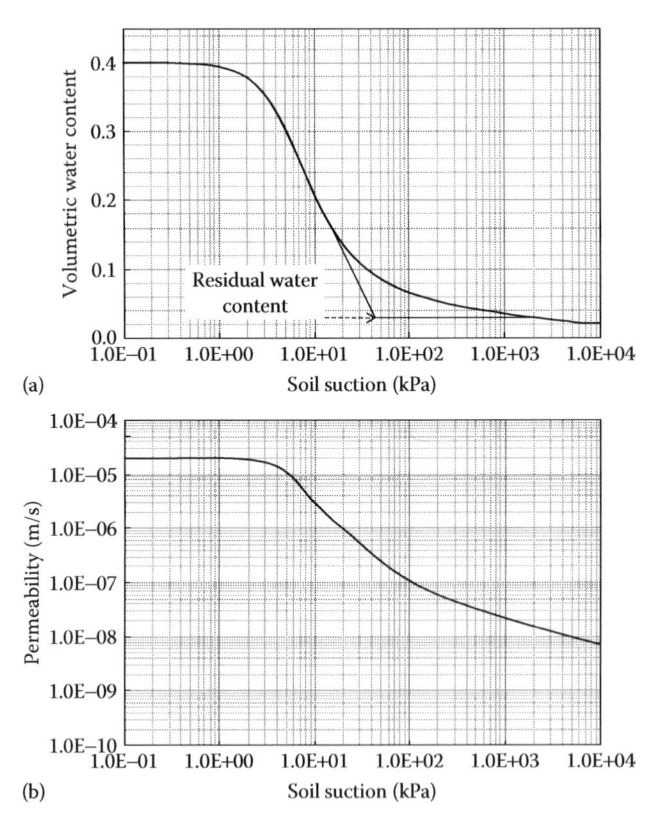

Figure 8.6 Soil property functions used in the numerical model: (a) soil–water characteristic curve and (b) permeability function. (From Zhu, H. et al., *Computers and Geotechnics*, 48, 249–259, 2013. With permission.)

Table 8.3 Parameters for the hypothetic study cases

Parameters	*Definition*	*Value*
k_s	Saturated coefficient of permeability	$\mu_{k_s} = 2 \times 10^{-5}$ m/s COV = 60%–100% Correlation length (d): 0.5, 8, 100, 1000, 2000 m Normalized correlation length = d/H: 0.025, 0.4, 5, 10, 100 m
θ_s	Saturated volumetric water content	0.4
θ_r	Residual volumetric water content	0.02
a_f	SWCC parameter	5
n_f	SWCC parameter	2
m_f	SWCC parameter	1
δ_{lr}	Permeability function parameter	3
q	Vertical infiltration flux	2×10^{-7} m/s
c'	Effective cohesion	1 kPa
ϕ'	Effective angle of internal friction	30°
ϕ^b	Angle indicating the rate of increase of shear strength related to soil suction	15°

shown that k_s can be modeled as a lognormal random field (Whitman, 2000), with a mean of μ_{k_s} and a standard deviation of σ_{k_s}. Thus, ln k_s is assumed to follow a normal distribution with a mean of $\mu_{\ln k_s}$ and a variance of, $\sigma^2_{\ln k_s}$ where

$$\sigma^2_{\ln k_s} = \ln[1 + (\sigma_{k_s})^2 / (\mu_{k_s})^2] \tag{8.19}$$

$$\mu_{\ln k_s} = \ln(\mu_{k_s}) - \frac{1}{2}\sigma^2_{\ln k_s} \tag{8.20}$$

In Equations 8.19 and 8.20, $\sigma^2_{\ln k_s}$ is dimensionless. In this study, an isotropic spherical autocorrelation model is assumed.

A generic range of COV of k_s has been suggested as 60%–100% in the literature (Duncan, 2000), and the correlation length of ln k_s is often taken as 0.1–5 times the slope height (Srivastava et al., 2010). In order to comprehensively investigate the perfect correlation scenario where correlation length is sufficiently high to yield a nearly homogeneous domain, the correlation length is extended to a range of 0.5–2000 m or 0.025–100 times the slope height, as shown in Table 8.3. To avoid the effect of slope dimension, a normalized correlation length (m) is used, which is defined as $m = d/H$ (H is the slope height; d is the coefficient of the spherical correlation model that represents the correlation length as shown in Table 8.1).

In order to create a random field of a soil parameter, a rectangular random field is first generated. The FFT code modified from Kozintsev (1999) is adopted. A sequence of data, representing a realization of the Gaussian random field, is first generated with parameters represented by zero mean, unit variance, and a spherical correlation structure. A lognormally distributed random field of k_s is then transformed from the Gaussian random field. Figure 8.7 presents a series of realizations for the Gaussian field, the random fields of ln k_s and k_s. The random fields of the Gaussian field and ln k_s are linearly related. The random field of k_s is truncated to the geometry of the slope for numerical modeling as shown in Figure 8.8.

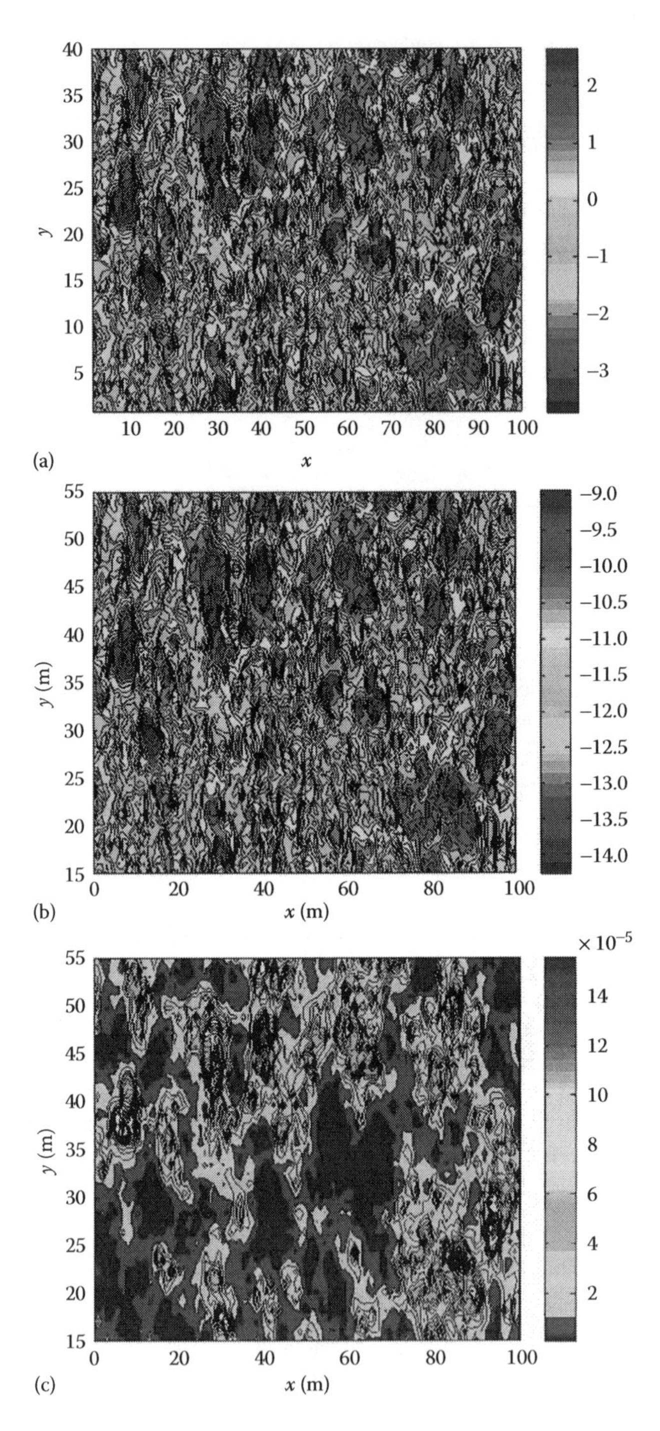

Figure 8.7 Realizations of random field ($\mu_{k_s} = 2 \times 10^{-5}$ m/s, $d = 8$ m, COV = 100%): (a) Gaussian field, (b) log-permeability random field, and (c) permeability random field. (From Zhu, H. et al., *Computers and Geotechnics*, 48, 249–259, 2013. With permission.)

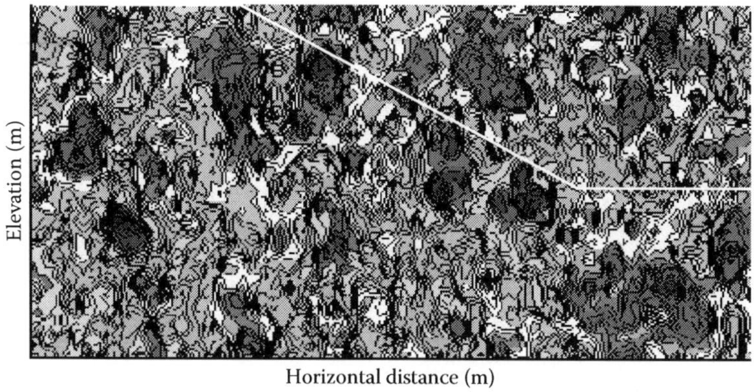

Figure 8.8 Schematic of mapping the k_s random field to a slope profile. (From Zhu, H. et al., *Computers and Geotechnics*, 48, 249–259, 2013. With permission.)

8.4.3 Methodology of stochastic modeling

8.4.3.1 Seepage and slope stability analysis

The PDE based finite element modeling software FlexPDE (PDE solutions Inc., 2015) is utilized to solve the governing equation of infiltration. The software allows users to input variables in tabulated files by which stochastic finite element analyses are performed repeatedly. The slope stability analysis is conducted using the program SVSlope (SoilVision System Ltd., 2010) in which the pore-water pressure distribution are imported from seepage analysis. The Bishop's simplified method with the extended Mohr–Coulomb criterion for unsaturated soil by Fredlund et al. (1978) is used in the slope stability analysis. The shape of the slip surface is assumed to be circular. The critical slip surface corresponding to the minimum factor of safety is searched in each slope stability analysis.

8.4.3.2 Mapping of saturated permeability on finite element seepage analysis

The k_s value at each grid point of the rectangular random field grid is transferred into the finite element seepage analysis as an input soil parameter. Bilinear interpolation is performed to map from the rectangular random field grid to the finite element mesh.

Figure 8.9 shows that the range of k_s within a slope varies with the correlation length. Taking an arbitrary realization for example, the values of k_s range from 0.02×10^{-4} m/s to 2.42×10^{-4} m/s at a correlation length of 0.5 m, and from 1.01×10^{-6} m/s to 3.53×10^{-6} m/s at a correlation length of 1000 m. A larger correlation length leads to a smaller spatial difference in the values of k_s.

8.4.3.3 Stochastic analysis by Monte Carlo simulation technique

In this example, the Monte Carlo simulation (MCS) technique is adopted for stochastic analysis. Realizations of random field are first generated by the FFT random field generation method. Then the deterministic problem is solved for each realization of the random field

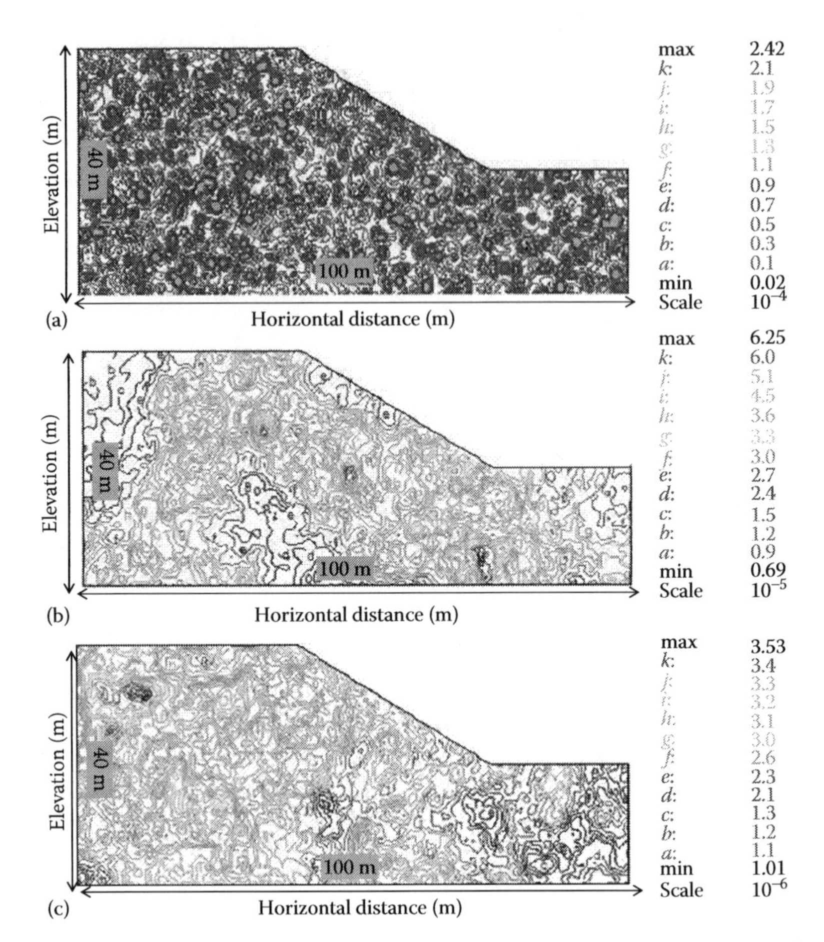

Figure 8.9 Typical realizations of random fields with various correlation lengths ($\mu_{k_s} = 2\times10^{-5}$m/s, COV = 100%): (a) d = 0.5 m, (b) d = 100 m, and (c) d = 1000 m. (From Zhu, H. et al., *Computers and Geotechnics*, 48, 249–259, 2013. With permission.)

and a population of the random response quantities is obtained. This population can then be used to obtain statistics of the prediction. In this example, the number of realizations by the MC simulation is 1000.

A parametric analysis is carried out to study the sensitivity of pore-water pressure and factor of safety to the correlation length of ln k_s. A base case (i.e., a homogeneous profile with the mean value of k_s) is analyzed for comparison purposes. In the graphs of results in the following section, the results of the base case is denoted as deterministic.

8.4.4 Results and discussion

8.4.4.1 Influence of soil spatial variability on pressure head profiles

Figures 8.10 and 8.11 show the estimated quantiles of the pore-water pressure head profiles along cross section X–X at the crest and section Y–Y at the middle of the slope (Figure 8.5), respectively. It can be seen that a higher correlation length produces a smaller 25% quantile of the pressure head profiles, but larger 50% and 75% quantiles. The quantiles of pressure head at normalized correlation lengths of 50 and 100 m are almost the same.

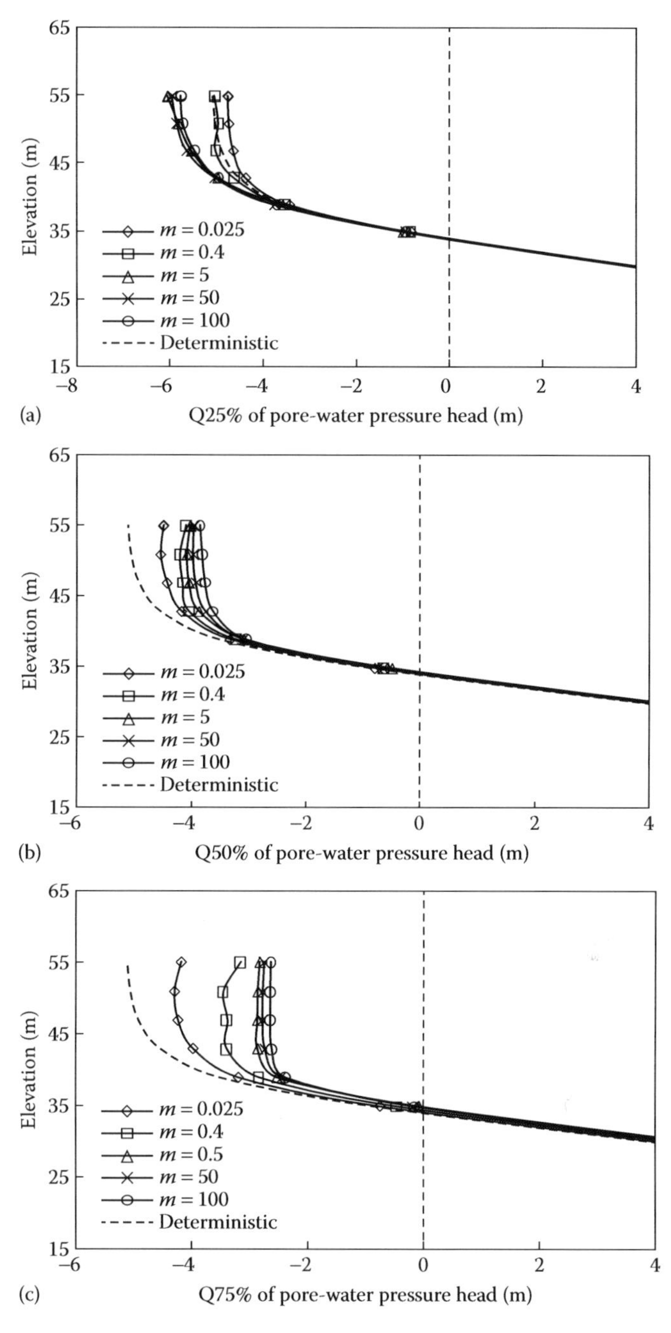

Figure 8.10 Pressure head profiles along section X–X at steady state for various correlation lengths of ln k_s: (a) 25% quantile, (b) 50% quantile, and (c) 75% quantile. (From Zhu, H. et al., *Computers and Geotechnics*, 48, 249–259, 2013. With permission.)

Hence, a normalized correlation length of 100 represents a reasonable value that represents perfect correlation of spatial variability. The depth within which a constant suction can be maintained under steady-state infiltration increases with increasing of the correlation length. Santoso et al. (2011) conducted a one-dimensional stochastic infiltration analysis for an infinite slope assuming a constant groundwater table. They found that the pore-water

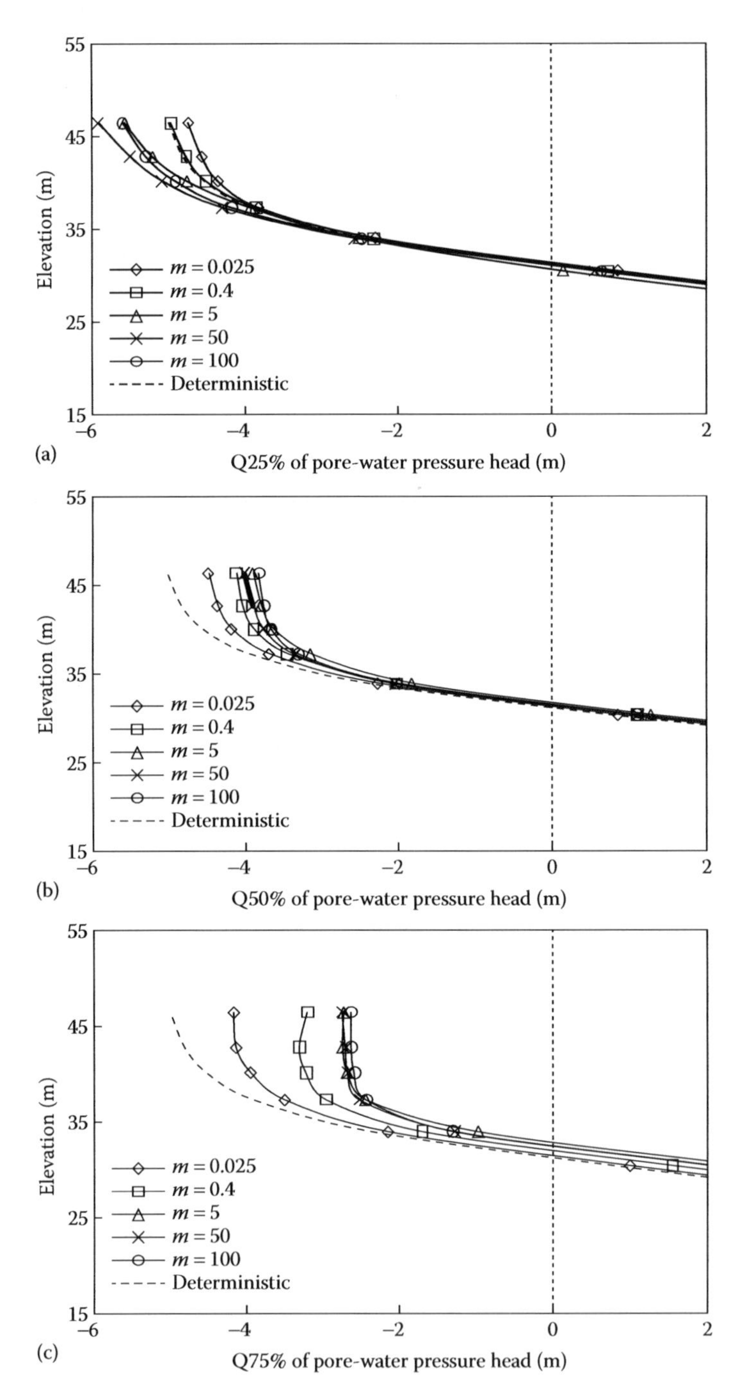

Figure 8.11 Pressure head profiles along section Y–Y at steady state for various correlation lengths of ln k_s: (a) 25% quantile, (b) 50% quantile, and (c) 75% quantile. (From Zhu, H. et al., *Computers and Geotechnics*, 48, 249–259, 2013. With permission.)

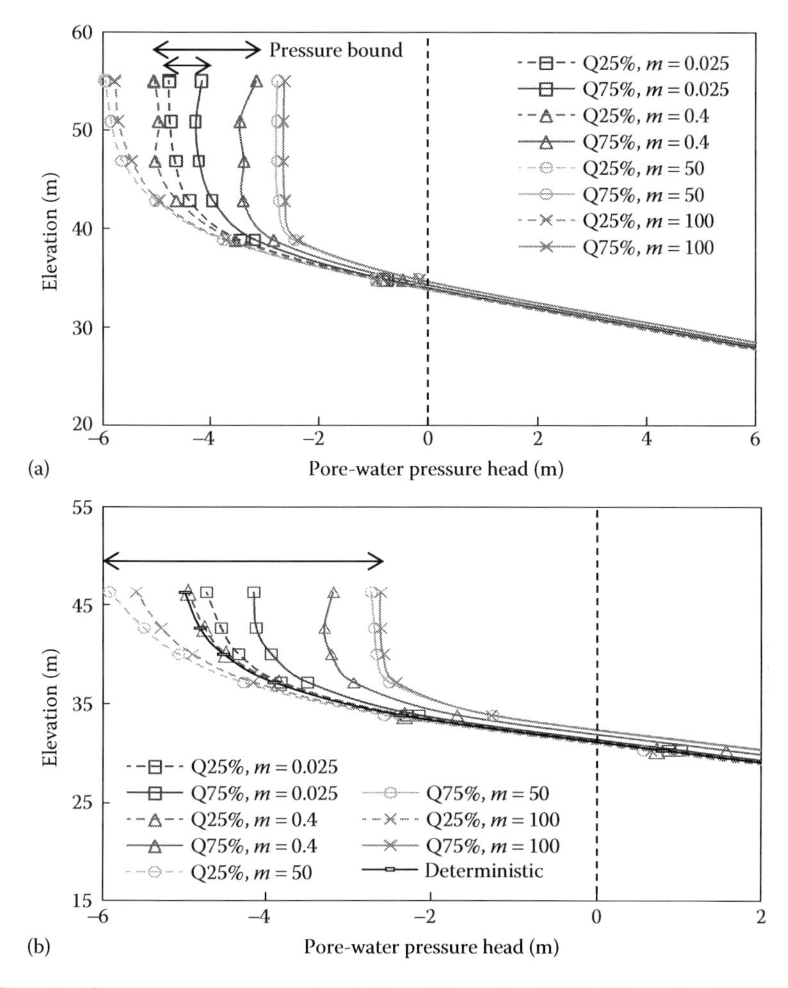

Figure 8.12 Bounds of pore-water pressure head along: (a) section X–X, (b) section Y–Y. (From Zhu, H. et al., *Computers and Geotechnics*, 48, 249–259, 2013. With permission.)

pressure profile changes with the statistical parameters of k_s and finally reaches a perfect correlation state when the correlation length is sufficiently long.

Another important indicator of variability in hydraulic conditions is the bound of pore-water pressure head (Figure 8.12), which represents the range of pore-water pressure head in the stochastic simulation. The Q25% and Q75% bounds of the pressure head for cross sections X–X and Y–Y are shown in Figure 8.12. The spatial variability of k_s can induce a variation in pore pressure head of more than 3 m. A higher correlation length produces more widely distributed negative pressures, which implies larger uncertainty. As shown in Figure 8.12(b), the matric suctions in the random heterogeneous slopes are within the range of 50% to 125% of those in the deterministic slope.

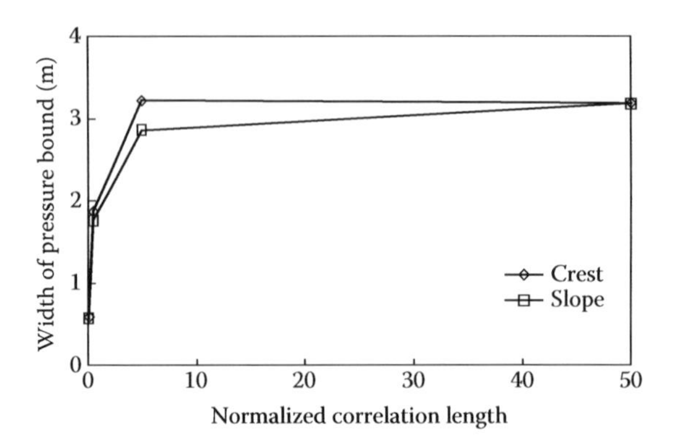

Figure 8.13 Width of pressure bound versus normalized correlation length of ln k_s.

The relationships between the width of the pressure head bound and the correlation length of ln k_s are plotted in Figure 8.13. The bound width increases sharply with increasing normalized correlation length when m is less than 5. Relatively little change is observed when the normalized correlation length is greater than 5. It can be noted that the pressure bounds at the crest (cross section X–X) are slightly wider than those at the middle of the slope (cross section Y–Y).

8.4.4.2 *Influence of soil spatial variability on variations of groundwater table*

In addition to irregularity of pressure distribution in random heterogeneous soils, the variation of groundwater table is another indicator of the performance of the slope. The shape of the groundwater table in a heterogeneous soil differs from that in a homogeneous soil. Figure 8.14 shows two boundary envelopes that cover the possible locations of the groundwater tables obtained from 1000 realizations for case with m = 50 and COV = 100%. The upper bound represents the highest position that the free water can reach, and the lower bound is the lowest one. The variability of groundwater table is characterized

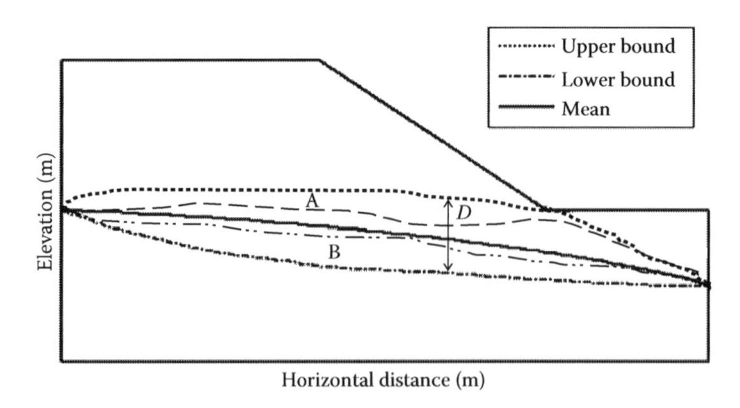

Figure 8.14 Envelopes of the groundwater table for a case with m = 50, COV = 100%. (From Zhu, H. et al., *Computers and Geotechnics*, 48, 249–259, 2013. With permission.)

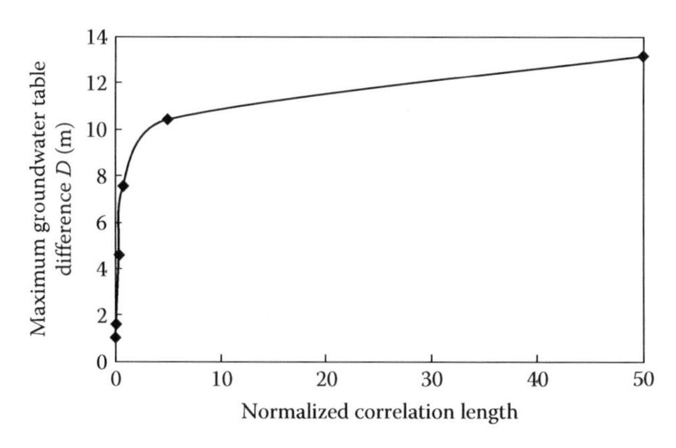

Figure 8.15 Variation of the maximum difference of groundwater table with the correlation length of ln k_s.

by the maximum difference (D) between these two curves in Figure 8.14. A larger D value implies a higher degree of variation.

Figure 8.15 shows the relationship between the variability of the groundwater table and the correlation length of ln k_s. It is noted that D increases sharply with small values of correlation lengths and changes slowly at high correlation length values. The D value is very small at small correlation lengths, indicating that the groundwater table in this case is similar to the deterministic case and the fluctuations can be ignored. This is because the local averaging effect is significant at very small correlation lengths. Any point at which the permeability is large will be surrounded by points where the permeability values are small. As the correlation length increases, the groundwater table becomes wavy due to the presence of a larger area of zones with similar soil property.

8.4.4.3 Effects of correlation length of ln k$_s$ on factor of safety

Figure 8.16(a) presents the variation of mean F_s with correlation length by analyzing 50, 100, 150, and 200 realizations (N = number of realizations). The trend is found to be insensitive to the number of realizations and the mean F_s approaches the deterministic value at a normalized correlation length of 0.025. Hence, 200 simulations are considered sufficient.

In Figure 8.16(a), the factor of safety calculated in the deterministic case is higher than the mean factors of safety considering the spatial variability of permeability function. This is in good agreement with Griffiths and Fenton (2004) that probabilistic slope stability analysis that ignores the spatial variability of soil properties underestimates the probability of the slope failure. The greatest reduction in μ_{F_S} occurs when d is of the same order of the slope height, H. It is hypothesized that $d \approx 5H$ leads to the greatest reduction in μ_{F_S} because this d value allows enough spatial averaging along a failure surface.

In Figure 8.16(b), the COV of factor of safety is sensitive to changes in the correlation length of ln k_s when the correlation length is small, and increases gently when correlation lengths five times longer than the slope height. At a high level of correlation length, the area within which the permeability functions are highly correlated covers the most part of the slope profile. Therefore, the large variation of pore-water pressure distributions leads to a large value of COV for the factor of safety.

The distribution of the factors of safety from 200 simulations is presented in Figure 8.17(a). The frequency histogram and the theoretically fitted lognormal probability density function

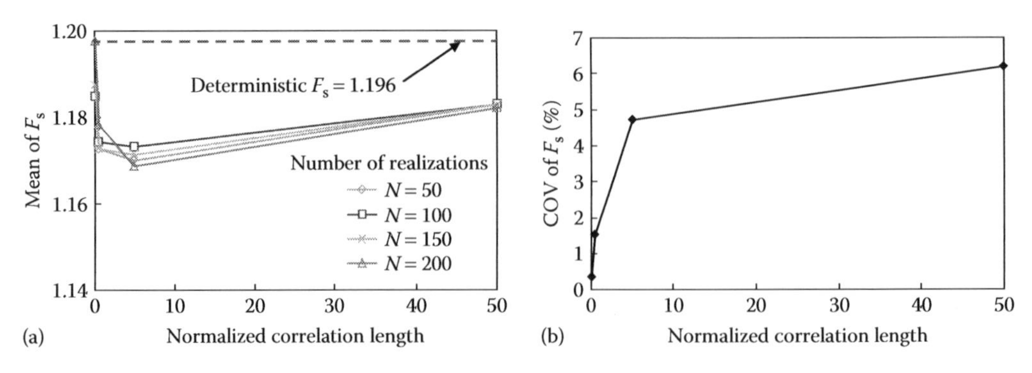

Figure 8.16 Statistical results of factor of safety with spatially variable saturated permeability parameter (COV = 100%): (a) variation of mean F_s with the correlation length of ln k_s and (b) change in COV of F_s with the correlation length of ln k_s.

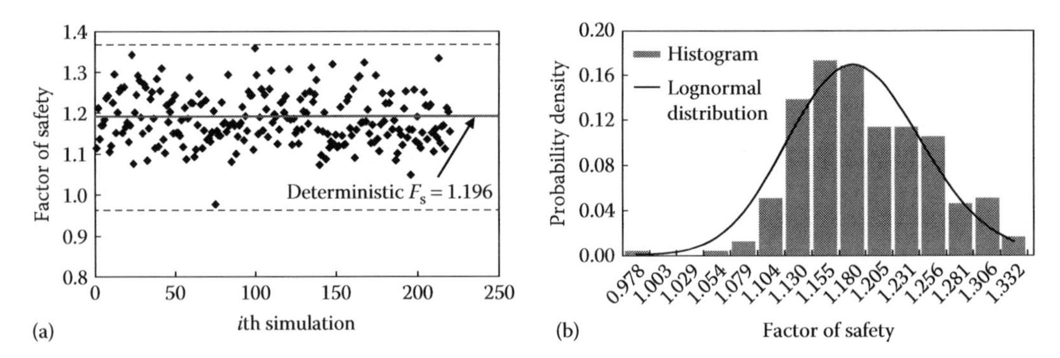

Figure 8.17 Distributions of factor of safety at a correlation length of m = 50: (a) upper and lower bound of F_s and (b) frequency histogram of F_s. (Modified from Zhu, H. et al., *Computers and Geotechnics*, 48, 249–259, 2013. With permission.)

of F_s are shown in Figure 8.17(b). The data of the factor of safety are tested by the Kolmogorov–Smirnov (K–S) goodness-of-fit test, in which the empirical cumulative frequency and the corresponding theoretical CDF for the normal and lognormal distributions are calculated. The maximum discrepancies of the two cumulative frequencies are D_{max} = 0.079 for the normal distribution and D_{max} = 0.070 for the lognormal distribution. At the 5% significance level, the critical value of $D_{0.05}$ is 0.096. As the maximum discrepancies for both normal and lognormal distributions are less than 0.096, the two distribution types are both acceptable. The lognormal distribution is better because the D_{max} value is smaller.

Comparing the results from all the realizations, the values of F_s of the soil slope varies from 0.909 to 1.357, which implies a range of –24% to +13.5% error in estimation of F_s if the spatial variability is not considered for this specific slope (30° in slope angle, 20 m in height and 2×10^{-5} m/s in mean saturated permeability). Previous studies (Tartakovsky et al., 2003; Srivastava et al., 2010; Santoso et al., 2011) focus on the mean factor of safety to illustrate the impact of soil spatial variability. The extreme values of F_s can demonstrate the variability of slope safety more clearly. The results imply that design approaches using mean values tend to be risky by ignoring the influence of spatially variable permeability functions.

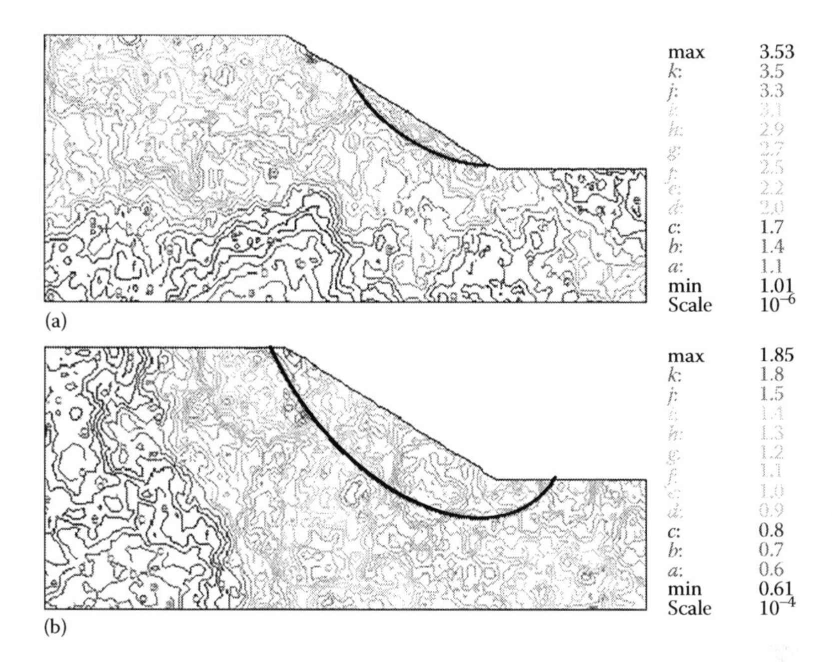

Figure 8.18 Realizations of k_s and critical slip surfaces corresponding to: (a) minimum factor of safety and (b) maximum factor of safety (m = 50, COV = 100%). (From Zhu, H. et al., *Computers and Geotechnics*, 48, 249–259, 2013. With permission.)

8.4.4.4 Groundwater conditions corresponding to the maximum and minimum factors of safety

Certain hydraulic scenarios under steady-state flow conditions induce the maximum and minimum factors of safety. Referring to Figure 8.14, the groundwater table which corresponds to the maximum factor of safety is denoted as curve *B*, while curve *A* corresponds to the minimum factor of safety (the most unfavorable groundwater table). The spatial distributions of saturated permeability of these two extreme scenarios are displayed in Figure 8.18. The range of k_s is from 1.01×10^{-6} m/s to 3.53×10^{-6} m/s in Figure 8.18(a) and from 0.61×10^{-4} m/s to 1.85×10^{-4} m/s in Figure 8.18(b). There is a difference of two orders of magnitude in k_s between these two scenarios although both follow the same prescribed distribution and correlation structure of k_s. Indeed spatially variable permeability function can induce very different pore-water pressures and groundwater tables, which in turn leads to differences of critical slip surface.

REFERENCES

Alonzo, E. E., and Krizek, R. J. (1975). Stochastic formulation of soil properties. *Proceedings of the Second Conference on Application of Probability and Statistics to Soil and Structural Engineering, Vol. 2*. Aachen, Germany, pp. 99–32.

Asaoka, A., and A-Grivas, D. (1982). Spatial variability of the undrained strength of clays. *Journal of the Geotechnical Engineering Division, ASCE*, 108, 743–756.

Baecher, G. B. (1982). Simplified geotechnical data analysis. In: Thoft-Christensen, P. (Ed.). *Reliability Theory and Its Applications in Structural and Soil Engineering*, Reidal Publishing, Dordrecht, the Netherlands.

Baecher, G. B., and Christian, J. T. (2003). *Reliability and Statistics in Geotechnical Engineering*. Wiley, New York.

Bellin, A., and Rubin, Y. (1996). HYDRO GEN: A spatially distributed random field generator for correlated properties. *Stochastic Hydrology and Hydraulics*, 10(4), 253–278.

Benson, C. (1991). Predicting excursions beyond regulatory thresholds of hydraulic conductivity using quality control measurements. *Proceedings of the First Canadian Conference on Environmental Geotechnics*, Montreal, pp. 14–17.

Betz, W., Papaioannou, I., and Straub, D. (2014). Numerical methods for the discretization of random fields by means of the Karhunen–Loeve expansion. *Computer Methods in Applied Mechanics and Engineering*, 271, 109–129.

Bjerg, P.L., Hinsby, K., Christensen, T.H. and Gravesen, P. (1992). Spatial variability of hydraulic conductivity of an unconfined sandy aquifer determined by a mini slug test. *Journal of Hydrology*, 136(1–4), 107–122.

Bras, R. L., and Rodriyguez-Iturbe, I. (1985). *Random Functions and Hydrology*. Addison-Wesley Publishing Company, Reading, MA.

Carpenter, L. C. (1980). *Computer Rendering of Fractal Curves and Surfaces*. ACM SIGGRAPH Computer Graphics, 14(3), 9–15.

Chiasson, P., Lafleur, J., Soulié, M., and Haw, K. T. (1995). Characterizing spatial variability of clay by geostatistics. *Canadian Geotechnical Journal*, 32(1), 1–10.

Chiles, J. P., and Delfiner, P. (1999). *Geostatistics: Modeling Spatial Uncertainty*. Wiley, New York.

Cho, S. E. (2007). Effects of spatial variability of soil properties on slope stability. *Engineering Geology*, 92(3–4), 97–109.

Cho, S. E. (2012). Probabilistic analysis of seepage that considers the spatial variability of permeability for an embankment on soil foundation. *Engineering Geology*, 133–134, 30–39.

Cooley, J. W., and Tukey, J. W. (1965). An algorithm for the machine calculation of complex Fourier series. *Mathematics of Computation*, 19(90), 297–301.

Cressie, N., and Pavlicová, M. (2002). Calibrated spatial moving average simulations. *Statistical Modelling*, 2(4), 267–279.

Davis, M. W. (1987). Production of conditional simulations via the LU triangular decomposition of the covariance matrix. *Mathematical Geology*, 19(2), 91–98.

DeGroot, D.J. (1996). Analyzing spatial variability of insitu soil properties. *Uncertainty in the Geologic Environment, from Theory to Practice, Proceedings of Uncertainty 96*, Madison, WI. Geotechnical Special Publication (GSP) 58, ASCE, Reston, VA, pp. 210–238.

DeGroot, D. J., and Baecher, G. B. (1993). Estimating autocovariance of in-situ soil properties. *Journal of Geotechnical Engineering*, ASCE, 119(GT1), 147–166.

Deutsch, C. V., and Journel, A. G. (1997). *GSLIB: Geostatistical Software Library and User's Guide (Applied Geostatistics)*. Oxford University Press, Oxford, 384pp.

Dietrich, C. R., and Newsam, G. N. (1996). A fast and exact method for multidimensional Gaussian stochastic simulations: extension to realizations conditioned on direct and indirect measurements. *Water Resources Research*, 32(6), 1643–1652.

Ditmars, J.D., Baecher, G.B., Edgar, D.E. and Dowding, C.H. (1988). *Radioactive Waste Isolation in Salt: A Method for Evaluating the Effectiveness of Site Characterization Measurements* (Final report ANL/EES-TM-342), Argonne National Laboratory, Lemont, IL.

Duncan, J. M. (2000). Factors of safety and reliability in geotechnical engineering. *Journal of Geotechnical and Geoenvironmental Engineering*, 126(4), 307–316.

Emery, X. (2008). A turning bands program for conditional co-simulation of cross-correlated Gaussian random fields. *Computers and Geotechnics*, 34(12), 1850–1862.

Fenton, G. A., and Griffiths, D. V. (2008). *Risk Assessment in Geotechnical Engineering*. Wiley, New York.

Fenton, G. A., and Vanmarcke, E. H. (1990). Simulation of random fields via local average subdivision. *Journal of Engineering Mechanics*, 116(8), 1733–1949.

Fournier, A., Fussell, D., and Carpenter, L. (1982). Computer rendering of stochastic models. *Communications of the ACM*, 25(6), 371–384.

Fredlund, D. G., Morgenstern, R. A., and Widger, R. A. (1978). The shear strength of unsaturated soils. *Canadian Geotechnical Journal*, 15(3), 313–321.

Fredlund, D. G., and Xing, A. Q. (1994). Equations for the soil-water characteristic curve. *Canadian Geotechnical Journal*, 31(4), 521–532.

Ghanem, R. G., and Spanos, P. D. (1991). *Stochastic Finite Element—A Spectral Approach*. Springer, New York.

Griffiths, D. V., and Fenton, G. A. (2004). Probabilistic slope stability analysis by finite element. *Journal of Geotechnical and Geoenvironmental Engineering*, 130(5), 507–518.

Gutjahr, A., Bullard, B., and Hatch, S. (1997). General joint conditional simulations using a fast Fourier transform method. *Mathematical Geology*, 29(3), 361–389.

Harter, T. (1994). *Unconditional and Conditional Simulation of Flow and Transport in Heterogeneous, Variably Saturated Porous Media*. Ph.D. Thesis, University of Arizona, Tucson, Arizona.

Hess, K. M., Wolf, S. H., and Celia, M. A. (1992). Large-scale natural gradient tracer test in sand and gravel, Cape Cod, Massachusetts: 3. hydraulic conductivity variability and calculated macrodispersivities. *Water Resources Research*, 28(8), 2011–2027.

Hilldale-Cunningham, C. (1971). *A Probabilistic Approach to Estimating Differential Settlement*, M.S. Thesis, Massachusetts Institute of Technology, Cambridge, MA.

Honjo, Y., and Kuroda, K. (1991). A new look at fluctuating geotechnical data for reliability design. *Soils and Foundations*, 31(1), 110–120.

Huang, S. P., Quek, S. T., and Phoon, K. K. (2001). Convergence study of the truncated Karhunen–Loeve expansion for simulation of stochastic processes. *International Journal for Numerical Methods in Engineering*, 52(9), 1029–1043.

Jaksa, M. B., Yeong, K. S., Wong, K. T., and Lee, S. L. (2004). Horizontal Spatial Variability of Elastic Modulus in Sand from the Dilatometer. *Proceedings of the 9th Australia New Zealand Conference on Geomechanics, Vol. 1*. Auckland, pp. 289–294.

JCSS (2002). *Probabilistic Model Code*. Joint Committee on Structural Safety. Technical University of Denmark, Lyngby, Denmark.

Jiang, S. H., Li, D. Q., Cao, Z. J., Zhou, C. B., and Phoon, K. K. (2014). Efficient system reliability analysis of slope stability in spatially variable soils using Monte Carlo simulation. *Journal of Geotechnical and Geoenvironmental Engineering*, 141(2), 04014096.

Jiang, S. H., Li, D. Q., Zhang, L. M., and Zhou, C. B. (2014). Slope reliability analysis considering spatially variable shear strength parameters using a non-intrusive stochastic finite element method. *Engineering Geology*, 168, 120–128.

Keaveny, J. M., Nadim, F., and Lacasse, S. (1989). Autocorrelation functions for offshore geotechnical data. *Proceedings of the Fifth International Conference on Structural Safety and Reliability, Vol. 1*. Vancouver, BC, pp. 263–270.

Kottegoda, N. T., and Kassim, A. H. M. (1991). The turning bands method with the fast-Fourier transform as an aid to the determination of storm movement. *Journal of Hydrology*, 127(1–4), 55–69.

Kozintsev, B. (1999). *Computations with Gaussian Random Fields*. Ph.D Thesis, University of Maryland.

Kulatilake, P. H. S. W. and Ghosh, A. (1988). An investigation into accuracy of spatial variation estimation using static cone penetrometer data. *Proceedings of the First International Conference on Penetration Testing*, Orlando, FL, pp. 815–821.

Kulhawy, F. H. (1992). On the evaluation of static soil properties. *Stability and Performance of Slopes and Embankment II*, GSP 31, 95–115. ASCE, Reston, VA.

Lacasse, S., and de Lamballerie, J. Y. N. (1995). Statistical treatment of CPT data. *Proceedings of International Symposium on Cone Penetration Testing*, Linkoping, Sweden, pp. 369–380.

Le, T. M. H. (2014). Reliability of heterogeneous slopes with cross-correlated shear strength parameters. *Georisk: Assessment and Management of Risk for Engineered Systems and Geohazards*, 8(4), 250–257.

Le, T. M. H., Gallipoli, D., Sanchez, M., and Wheeler, S. J. (2012). Stochastic analysis of unsaturated seepage through randomly heterogeneous earth embankments. *International Journal for Numerical and Analytical Methods in Geomechanics*, 36(8), 1056–1076.

Le, T. M. H., Gallipoli, D., Sanchez, M., and Wheeler, S. (2013). Rainfall-induced differential settlements of foundations on heterogeneous unsaturated soils. *Géotechnique*, 63(15), 1346–1355.

Leong, E. C., and Rahardjo, H. (1997). Permeability functions for unsaturated soils. *Journal of Geotechnical and Geoenvironmental Engineering, ASCE*, 123(12), 1118–1126.

Li, D. Q., Jiang, S. H., Cao, Z. J., Zhou, W., Zhou, C. B., and Zhang, L. M. (2015). A multiple response-surface method for slope reliability analysis considering spatial variability of soil properties. *Engineering Geology*, 187, 60–72.

Li, D. Q., Qi, X. H., Phoon, K. K., Zhang, L. M., and Zhou, C. B. (2014). Effect of spatially variable shear strength parameters with linearly increasing mean trend on reliability of infinite slopes. *Structural Safety*, 49, 45–55.

Li, W., Lu, Z., and Zhang, D. (2009). Stochastic analysis of unsaturated flow with probabilistic collocation method. *Water Resources Research*, 45, W08425, DOI:10.1029/2008WR007530.

Lu, Z., and Zhang, D. (2007). Stochastic simulations for flow in nonstationary randomly heterogeneous porous media using a KL-based moment-equation approach. *Multiscale Modeling & Simulation*, 6(1), 228–245.

Mantolou, A., and Wilson, J. L. (1982). The turning bands method for simulation of random fields using line generation by a spectral method. *Water Resources Research*, 18(5), 1379–1394.

Matérn, B. (1986). *Spatial Variation*, 2nd ed. Springer-Verlag, Berlin.

Matheron, G. (1973). The intrinsic random functions and their applications. *Advances in Applied Probability*, 5(3), 439–468.

Meftah, F., Dal Pont, S., and Schrefler, B.A. (2012). A three-dimensional staggered finite element approach for random parametric modeling of thermo-hygral coupled phenomena in porous media. *International Journal for Numerical and Analytical Methods in Geomechanics*, 36(5), 574–596.

Mejia, J., and Rodriguez-Iturbe, I. (1974). On the synthesis of random field sampling from the spectrum: An application to the generation of hydrologic spatial processes. *Water Resources Research*, 10(4), 705–711.

Mousavi Nezhad, M., Javadi, A. A., and Abbasi, F. (2011). Stochastic finite element modelling of water flow in variably saturated heterogeneous soils. *International Journal for Numerical and Analytical Methods in Geomechanics*, 35(12), 1389–1408.

Mulla, D. J. (1988). Estimating spatial patterns in water content, matric suction and hydraulic Conductivity. *Soil Science Society*, 52(6), 1547–1553.

Oliver, D. S. (1995). Moving averages for Gaussian simulation in two and three dimensions. *Mathematical Geology*, 27(8), 939–960.

PDE Solutions Inc. (2015). *FlexPDE 6 User Manual Version 6.37*. PDE Solutions Inc. Washington, USA.

Phoon, K. K., Huang, S. P., and Quek, S. T. (2002a). Implementation of Karhunen–Loeve expansion for simulation using a wavelet-Galerkin scheme. *Probabilistic Engineering Mechanics*, 17(3), 293–303.

Phoon, K. K., Huang, S. P., and Quek, S. T. (2002b). Simulation of second-order processes using Karhunen–Loeve expansion. *Computers & Structures*, 80(12), 1049–1060.

Phoon, K. K., and Kulhawy, F. H. (1996). On quantifying inherent soil variability. *Uncertainty in the Geologic Environment, from Theory to Practice, Proceedings of Uncertainty 96*, Madison, WI. Geotechnical Special Publication (GSP) 58, ASCE, Reston, VA, pp. 326–340.

Phoon, K. K., and Kulhawy, F. H. (1999). Characterization of geotechnical variability. *Canadian Geotechnical Journal*, 36(4), 612–624.

Popescu, R., Prevost, J. H., and Vanmarcke, E. H. (1995). Numerical simulations of soil liquefaction using stochastic input parameters. *Proceedings of the 3rd International Conference on Recent Advances in Geotechnical Earthquake Engineering and Soil Dynamics*, St. Louis, MO, pp. 275–280.

Rehfeldt, K. R., Boggs. J. M., and Gelhar, L. W. (1992). Field study of dispersion in a heterogeneous aquifer, 3-d geostatistical analysis of hydraulic conductivity. *Water Resources Research*, 28(12), 3309–3324.

Ronold, M. (1990). Random field modeling of foundation failure modes. *Journal of Geotechnical Engineering*, 166(4), 554–570.

Rosenbaum, M. S. (1987). The use of stochastic models in the assessment of a geological database. *Quarterly Journal of Engineering Geology & Hydrogeology*, 20(1), 31–40.

Santoso, A. M., Phoon, K. K., and Quek, S. T. (2011). Effects of soil spatial variability on rainfall-induced landslides. *Computers and Structures*, 89(11–12), 893–900.

SoilVision System Ltd (2010). *SVSlope User's Manual*. SoilVision System Ltd, Saskatoon, Canada.
Soulié, M., Montes, P., and Silvestri, V. (1990). Modeling spatial variability of soil parameters. *Canadian Geotechnical Journal*, 27(5), 617–630.
Srivastava, A., Sivakumar Babu, G. L., and Haldar, S. (2010). Influence of spatial variability of permeability property on steady state seepage flow and slope stability analysis. *Engineering Geology*, 110(3–4), 93–101.
Stefanou, G. (2009). The stochastic finite element method: past, present and future. *Computer Methods in Applied Mechanics and Engineering*, 198(9), 1031–1051.
Stefanou G., and Papadrakakis M. (2007). Assessment of spectral representation and Karhunen–Loeve expansion methods for the simulation of Gaussian stochastic fields. *Computer Methods in Applied Mechanics and Engineering*, 196(21), 2465–2477.
Strebelle, S. (2002). Conditional simulation of complex geological structures using multiple-point statistics. *Mathematical Geology*, 34(1), 1–21.
Tami, D., Rahardjo, H., and Leong, E. C. (2004). Effects of hysteresis on steady-state infiltration in unsaturated slopes. *Journal of Geotechnical and Geoenvironmental Engineering*, 130(9), 956–967.
Tang, W. H. (1979). Probabilistic evaluation of penetration resistances. *Journal of the Geotechnical Engineering Division, ASCE*, 105(GT10), 1173–1191.
Tang, W. H. (1984). Principles of probabilistic characterization of soil properties. *Proceedings of Symposium on Probabilistic Characterization of Soil Properties*, ASCE National Convention, Atlanta, pp. 74–89.
Tartakovsky, D. M., Lu, Z., Guadagnini, A., and Tartakovsky, A. M. (2003). Unsaturated flow heterogeneous soils with spatially distributed uncertain hydraulic parameters. *Journal of Hydrology*, 275(3–4), 182–193.
Tompson, A. F. B., Ababou, R., and Gelhar, L. W. (1989). Implementation of the three-dimensional turning bands random field generator. *Water Resources Research*, 25(10), 2227–2243.
Ünlü, K., Nielsen, D. R., Biggar, J. W., and Morkoc, F. (1990). Statistical parameters characterizing the spatial variability of selected soil hydraulic properties. *Soil Science Society of America Journal*, 54(6), 1537–1547.
Vanmarcke, E. (1983). *Random Fields: Analysis and Synthesis*. MIT Press, Cambridge, MA.
Vanmarcke, E. H. (1977). Probabilistic modeling of soil profiles. *Journal of the Geotechnical Engineering Division, ASCE*, 103(GT11), 1227–1246.
Vrouwenvelder, T., and Calle, E. (2003). Measuring spatial correlation of soil properties. *Heron*, 48(4), 297–311.
Whitman, R. V. (2000). Organizing and evaluating uncertainty in geotechnical engineering. *Journal of Geotechnical and Geoenvironmental Engineering*, 126(7), 583–593.
Wu, T. H. (1974). Uncertainty, safety, and decision in soil engineering. *Journal of Geotechnical Engineering*, 100(3), 329–348.
Yaglom, A. M. (1962). *An Introduction to the Theory of Stationary Random Functions*. Dover Publications Inc., Mineola, NY.
Yaglom, A. M. (1987). *Correlation Theory of Stationary and Related Random Functions. Vol. I: Basic Results*. Springer, Berlin.
Yang, J., Zhang, D., and Lu, Z. (2004). Stochastic analysis of saturated–unsaturated flow in heterogeneous media by combining Karhunen-Loeve expansion and perturbation method. *Journal of Hydrology*, 294(1), 18–38.
Zhang, D. (1999). Nonstationary stochastic analysis of transient unsaturated flow in randomly heterogeneous media. *Water Resources Research*, 35(4), 1127–1141.
Zhang, L. L., Fredlund, D. G., Zhang, L. M., and Tang, W. H. (2004). Numerical study of soil conditions under which matric suction can be maintained. *Canadian Geotechnical Journal*, 41(4), 569–582.
Zhang, L. M., Zheng, Y. R., and Wang, J. L. (2003). Errors in calculating hydrodynamic pressures for stability analysis of soil slopes subject to rainfall. *Proceedings of ICASP9*, San Francisco, CA, pp. 1431–1438.
Zhu, H., Zhang, L. M., Zhang, L. L., and Zhou, C. B. (2013). Two-dimensional probabilistic infiltration analysis with a spatially varying permeability function. *Computers and Geotechnics*, 48, 249–259.

Chapter 9

Probabilistic model calibration

9.1 INTRODUCTION

A slope failure implies that the factor of safety of the slope at the moment of failure is unity. Based on this information, the traditional back analysis of a slope failure is often carried out to improve knowledge on slope stability parameters, such as soil shear strength parameters and pore-water pressure parameters at the moment of slope failure. Using back analysis, important factors that may not be well represented in laboratory testing, such as soil heterogeneity and the influence of fissures and structural fabric on soil shear strength, can be incorporated.

In this chapter, we use a more general term, "model calibration," to define the process in which input parameters and the model error of a prediction model are estimated or determined based on the observed responses of a system. Model calibration can be done either manually or automatically through certain mathematical algorithms. The calibrated results of input parameters and model error can be either deterministic (Wesley and Leelaratnam, 2001; Tiwari et al., 2005) or probabilistic (Luckman et al., 1987; Gilbert et al., 1998; Chowdhury et al., 2004; Zhang et al., 2010a,b; 2013; 2015). In a deterministic approach, the slope stability model is usually believed or assumed accurate, and a set of parameters that would result in the slope failure is estimated. The traditional back analysis in geotechnical engineering can be classified as a deterministic model calibration approach. In a probabilistic approach, however, it is recognized that the slope stability model may not be perfectly accurate and numerous combinations of slope stability parameters may result in slope failure.

In this chapter, a probabilistic framework of model calibration based on the Bayesian theory is presented to incorporate various sources of information in a consistent way. Multiple soil parameters and model error together with their uncertainties can be characterized individually based on the observed information such as a slope failure state. For the problem of rainfall-induced slope failure, time varying soil responses are especially important and valuable as the variation of pore-water pressure or displacement during a rainstorm actually reflects the field soil properties that can be quite different from the laboratory measured ones. The probabilistic model calibration framework is extended to incorporate the time-varied measurement into model calibration assuming that the time-varied model errors are mutually independent and Gaussian-distributed with a constant variance.

As the prediction model of slope stability is highly nonlinear and usually involves multiple input parameters and the prior information of input parameters could be non-normal, the probabilistic model calibration is conducted using the Markov chain Monte Carlo (MCMC) simulation method (Gelman et al., 2004). Three example problems are presented to illustrate the methodology of probabilistic model calibration. The first example is a cut slope failure. In this example, the effects of the jumping distribution and the number of samples on the efficiency of Markov chains are studied. The effect of prior distribution on the back calculated statistics of uncertain parameters and the remediation design of the slope are presented. The second example is a case study of a rainfall-induced slope failure occurred at Lai Ping Road, Sha Tin, Hong Kong on July 2, 1997. The objective of this example is to investigate the allocation of information for shear strength parameters, hydraulic parameters, and model error by the probabilistic model calibration. The third example is an instrumented natural terrain site in Hong Kong with field measurements of rainfall and pore-water pressures. In this example, the effects of uncertainty reduction of soil hydraulic properties on the predicted pore-water pressures and the safety factor of the slope are illustrated and discussed.

9.2 PROBABILISTIC MODEL CALIBRATION WITHIN BAYESIAN FRAMEWORK

9.2.1 Parameter estimation with known model error

First, a residual ε is used to characterize the model error, which is defined as the difference between the actual performance and the model simulation:

$$y = g(\theta) + \varepsilon \tag{9.1}$$

where:

y is the performance of a geotechnical structure
g is the function of a prediction model
$\theta = \{x_1, \ldots, x_n\}$ is the vector of n uncertain input model parameters
$g(\boldsymbol{\theta})$ is the simulated performance using the prediction model

Assume that ε follows the normal distribution with a mean of μ_ε and a standard deviation of σ_ε. If μ_ε and σ_ε are known and the actual observed performance is equal to $\hat{y}$, the likelihood function, that is, the conditional probability density function of $\boldsymbol{\theta}$ given the observed performance $\hat{y}$, can be written as follows:

$$L(\theta \mid y = \hat{y}) = \phi\left(\frac{\hat{y} - g(\theta) - \mu_\varepsilon}{\sigma_\varepsilon}\right) \tag{9.2}$$

where ϕ is the probability density function (PDF) of a standard normal variable $N(0,1)$.

Based on the Bayes' theorem, the posterior PDF of $\boldsymbol{\theta}$ is

$$f(\theta \mid y = \hat{y}) = k\phi\left(\frac{\hat{y} - g(\theta) - \mu_\varepsilon}{\sigma_\varepsilon}\right) f(\theta) \tag{9.3}$$

where:

k is a normalization constant
$f(\boldsymbol{\theta})$ denotes a prior PDF of $\boldsymbol{\theta}$

For the problem of slope failure, the prediction model would be the slope stability analysis model such as the limit equilibrium method or numerical models. Theoretically, a slope failure implies that the factor of safety of the slope at the moment of failure is equal to unity. In practice, there might be uncertainties in defining and identifying slope failures. In this chapter, it is assumed that the slope failure is well defined and identified such that the uncertainties associated with slope failure definition and identification are minimized. Hence, y is the factor of safety of a slope, F_s, and the observation of a slope failure means that F_s equals 1.0. Hence, the likelihood function of the slope failure event is therefore

$$L(\theta \mid F_s = 1) = \phi\left(\frac{1 - g(\theta) - \mu_\varepsilon}{\sigma_\varepsilon}\right) \tag{9.4}$$

9.2.2 Simultaneous estimation of model error and input parameters

Assume that ε follows a normal distribution with unknown mean μ_ε and unknown standard deviation σ_ε. Therefore, the likelihood function of y given $\boldsymbol{\theta}$, μ_ε, and σ_ε is

$$L\left(\mu_\varepsilon, \sigma_\varepsilon, \theta \mid y = \hat{y}\right) = \frac{1}{\sqrt{2\pi}\sigma_\varepsilon} \exp\left[-\frac{\left(\hat{y} - g(\theta) - \mu_\varepsilon\right)^2}{2\sigma_\varepsilon^2}\right] \tag{9.5}$$

Let $f(\mu_\varepsilon,\sigma_\varepsilon)$ and $f(\boldsymbol{\theta})$ denote the prior probability density functions of $\{\mu_\varepsilon, \sigma_\varepsilon\}$ and $\boldsymbol{\theta}$, respectively. Assume that $\{\mu_\varepsilon, \sigma_\varepsilon\}$ and the input parameters $\boldsymbol{\theta}$ are independent. According to the Bayes' theorem, the posterior joint probability density function of $\{\mu_\varepsilon, \sigma_\varepsilon, \theta\}$ is

$$f\left(\mu_\varepsilon, \sigma_\varepsilon, \theta \mid y = \hat{y}\right)$$

$$= kf\left(\mu_\varepsilon, \sigma_\varepsilon\right) \frac{1}{\sqrt{2\pi}\sigma_\varepsilon} \exp\left[-\frac{\left(\hat{y} - g(\theta) - \mu_\varepsilon\right)^2}{2\sigma_\varepsilon^2}\right] f(\theta) \tag{9.6}$$

To assess the statistics of model uncertainty, one may make use of case histories at m sites where information on the input parameters $\boldsymbol{\theta}$ and the performance y are available. Let $\boldsymbol{\theta}_j$ and $\hat{y}_j$ represent $\boldsymbol{\theta}$ and $\hat{y}$ at the jth site ($j = 1, \ldots, m$), respectively. Assuming that the observations from all m sites are statistically independent, the likelihood function can be written as the product of the probabilities of the observations at all the sites:

$$L\left(\mu_\varepsilon, \sigma_\varepsilon, \Theta \mid \hat{Y}\right) = \prod_{j=1}^{m} \frac{1}{\sqrt{2\pi}\sigma_\varepsilon} \exp\left[-\frac{\left(\hat{y}_j - g(\theta_j) - \mu_\varepsilon\right)^2}{2\sigma_\varepsilon^2}\right] \tag{9.7}$$

where:

$\boldsymbol{\Theta} = \{\boldsymbol{\theta}_1, \boldsymbol{\theta}_2, \ldots, \boldsymbol{\theta}_m\}$
$\hat{\mathbf{Y}} = \{\hat{y}_1, \hat{y}_2, \ldots, \hat{y}_m\}$

Let $f(\mu_\varepsilon,\sigma_\varepsilon)$ and $f(\boldsymbol{\theta}_j)$ represent the prior probability density functions of $\{\mu_\varepsilon, \sigma_\varepsilon\}$ and $\boldsymbol{\theta}_j$, respectively. $\{\mu_\varepsilon, \sigma_\varepsilon\}$ and $\boldsymbol{\theta}_j$ are independent. According to the Bayes' theorem, the joint posterior probability density function of $\{\mu_\varepsilon, \sigma_\varepsilon, \boldsymbol{\Theta}\}$ is

$$f\left(\mu_\varepsilon, \sigma_\varepsilon, \Theta \mid \hat{Y}\right)$$

$$= kf\left(\mu_\varepsilon, \sigma_\varepsilon\right) \prod_{j=1}^{m} \frac{1}{\sqrt{2\pi}\sigma_\varepsilon} \exp\left[-\frac{\left(\hat{y}_j - g(\theta_j) - \mu_\varepsilon\right)^2}{2\sigma_\varepsilon^2}\right] f(\theta_j) \tag{9.8}$$

9.2.3 Probabilistic parameter estimation based on time-varied measurement

Assume that the prediction model g can simulate the time-dependent responses of a geotechnical system. The time-varied response of a slope during rainfall can be pore-water pressure, water content, displacement, and so on. Consider the vector of simulated outputs $\mathbf{g} = \{g(\theta)_1, \ldots, g(\theta)_d\}$ of the prediction model g from time t_1 to time t_d, where d is the number of data points which is a time series. The difference between the vector of model simulated outputs g with the vector of d observed response $\widehat{\mathbf{Y}} = \{\widehat{y}_1, \ldots, \widehat{y}_d\}$ is the vector of residual errors $\varepsilon = \{\varepsilon_1, \ldots, \varepsilon_d\}$ with

$$\varepsilon_i = \widehat{y}_i - g_i(\theta) \tag{9.9}$$

The closer to zero are the residuals, the better the model simulates the observed data. However, due to errors in the initial and boundary conditions, structural inadequacies in the model, uncertainties of input model parameters, and measurement errors, the residual values of the prediction model are not expected to be equal to zero.

Assuming that the residuals at different time in Equation 9.9 are mutually independent and Gaussian-distributed with a mean of zero and a constant variance, σ_ε^2, the likelihood function is

$$L(\theta \mid \widehat{\mathbf{Y}}) = \prod_{i=1}^{d} \frac{1}{\sqrt{2\pi\sigma_\varepsilon^2}} \exp\left(-\frac{(\widehat{y}_i - g_i(\theta))^2}{2\sigma_\varepsilon^2} \right) \tag{9.10}$$

For simplicity and numerical stability, it is convenient to estimate the logarithm of the likelihood function rather than the likelihood function itself. The log-likelihood of Equation 9.10 is

$$\log\left[L(\theta \mid \widehat{\mathbf{Y}}) \right] = -\frac{d}{2}\ln(2\pi) - \frac{d}{2}\ln(\sigma_\varepsilon^2) - \frac{1}{2}\sigma_\varepsilon^{-2} \sum_{i=1}^{d} (\widehat{y}_i - g_i(\theta))^2 \tag{9.11}$$

The posterior probability density function of $\boldsymbol{\theta}$ can then be written as follows:

$$f(\theta \mid \widehat{\mathbf{Y}}) = k \cdot f(\theta) \cdot L(\theta \mid \widehat{\mathbf{Y}}) \tag{9.12}$$

where:

k is a normalizing constant

$f(\theta)$ is the prior joint probability density function of $\boldsymbol{\theta}$

With the specification of a prior distribution of parameters, Equation 9.12 can be used to calculate the posterior distribution.

For prediction purpose, without the information of the observed data, the prior distribution of the predicted model response, y_i, at time t_i, can be evaluated as

$$f(y_i) = \int \cdots \iint \frac{1}{\sqrt{2\pi\sigma_\varepsilon^2}} \exp\left(-\frac{[\widehat{y}_i - g_i(\theta)]^2}{2\sigma_\varepsilon^2} \right) f(\widehat{y}_i \mid \theta) \, d\theta \tag{9.13}$$

Given the observed data $\widehat{\mathbf{Y}}$, the posterior prediction of y_i at time t_i is

$$f(y_i \mid \widehat{\mathbf{Y}}) = \int \cdots \iint \frac{1}{\sqrt{2\pi\sigma_\varepsilon^2}} \exp\left(-\frac{[\widehat{y}_i - g_i(\theta)]^2}{2\sigma_\varepsilon^2} \right) f(\widehat{y}_i \mid \theta) \, d\theta \tag{9.14}$$

In Section 9.2.1, the residual or model error ε is assumed to follow a normal distribution with given mean and variance. The model calibration process is basically a probabilistic back analysis. In Section 9.2.2, the mean and variance of ε are unknown and are estimated together with the input parameters of a prediction model. Here, the variance of the residual errors σ_ε^2 is not explicitly estimated. It is mainly because σ_ε^2 can be determined after the posterior samples of $\boldsymbol{\theta}$ are obtained using the random sampling methods. Given that the other parameters are known, the posterior distribution of σ_ε^2 is an inverse chi-square with d degrees of freedom (number of data points) and scale of s with (Vrugt et al. 2009)

$$s^2 = \frac{1}{d}\left(\sum_{i=1}^{d} \varepsilon_i^2\right) \tag{9.15}$$

9.3 MARKOV CHAIN MONTE CARLO SIMULATION METHOD

When conjugate priors can be adopted, considerable mathematical simplification of Bayesian updating can be achieved. A list of conjugate distributions is summarized in Ang and Tang (2007). However, the method of conjugate priors is only applicable to specific prior distributions and likelihood functions. For many problems with nonlinear prediction models such as slope stability under rainfall, the posterior distribution function of input parameters cannot be derived through analytical means or by analytical approximation. Random sampling methods such as Monte Carlo sampling are therefore needed to generate samples from the posterior distribution function. In this chapter, the random sample generation from the posterior distribution is efficiently done using the Markov Chain Monte Carlo (MCMC) simulation.

The basic idea of the MCMC simulation is that drawing samples from an arbitrary distribution and then correcting those samples to better approximate and finally converge to the target posterior distribution. The samples are drawn sequentially with the distribution of the current value depending on the last value drawn, which forms a Markov chain. The Gibbs sampling algorithm and the Metropolis algorithm are two basic ways to build Markov chains (Thomas and John, 1999; Gill, 2002). Compared with analytical methods and analytical approximation methods such as the maximum posterior density (MPD) method (Gelman et al., 2004) and system identification method (Wu et al., 2007), the advantages of the MCMC simulation include: (1) it does not require that the number of observed data is far larger than the number of variables to be updated; (2) it can consider any type of prior information; and (3) the samples from MCMC simulation can be directly used for model prediction, as will be shown in the illustrative examples.

9.3.1 Metropolis algorithm

1. Let $\boldsymbol{\theta}_0$ denote the starting point of a Markov chain. $\boldsymbol{\theta}_0$ can be randomly drawn from a starting distribution or simply chosen deterministically.
2. For $t = 1, 2, \ldots, N_{\text{chain}}$ (where t is the tth generation and N_{chain} is the length or number of samples of a Markov chain).
   (a) Sample a candidate $\boldsymbol{\theta}^*$ from a distribution $J(\boldsymbol{\theta}^*|\boldsymbol{\theta}_{(t-1)})$, where $\boldsymbol{\theta}^*$ is dependent on $\boldsymbol{\theta}_{(t-1)}$. This distribution $J(\cdot|\cdot)$ is called the jumping distribution or transition kernel of the Markov chain. The jumping distribution is symmetric in the Metropolis algorithm.
   (b) Calculate the ratio of the densities.

$$r = \frac{q(\boldsymbol{\theta}^*)}{q(\boldsymbol{\theta}_{(t-1)})} \tag{9.16}$$

where $q(\boldsymbol{\theta})$ is the target distribution function. In this chapter, the target distribution function is the unnormalized posterior density function.

(c) Set $\boldsymbol{\theta}_{(t)} = \boldsymbol{\theta}^*$ with a probability of $\min(r, 1)$; otherwise set $\boldsymbol{\theta}_{(t)} = \boldsymbol{\theta}_{(t-1)}$, which means that the jump is not accepted and this counts as an iteration in the algorithm.

(d) Stop the iteration if $t = N_{\text{chain}}$; otherwise $t = t + 1$ and go to Step (a).

It can be proven that the Markov chain sample $\boldsymbol{\theta}_{(t)}$ at the tth step approaches the target distribution $q(\boldsymbol{\theta})$ as $t \rightarrow \infty$ (Gelman et al., 2004).

The aforementioned Step (c) is basically the acceptance rule for the Metropolis algorithm. If the jump increases the posterior density, the density ratio is larger than 1. Therefore, $\boldsymbol{\theta}^*$ is accepted as $\boldsymbol{\theta}_{(t)}$ with a probability of 1. If the jump decreases the posterior density, the density ratio is smaller than 1. A random number in the interval of [0, 1] is generated. If the random number is less than the density ratio r, then $\boldsymbol{\theta}^*$ is accepted as $\boldsymbol{\theta}_{(t)}$. Otherwise, $\boldsymbol{\theta}^*$ is rejected and $\boldsymbol{\theta}_{(t)}$ is set to be $\boldsymbol{\theta}_{(t-1)}$. In other words, the algorithm always accepts steps that increase the posterior density but only sometimes accepts downward steps. A good jumping distribution should be easily sampled and the jumps should not be rejected too frequently, otherwise the random walk wastes too much time standing still. How to select the appropriate jumping distribution function for efficient probabilistic back analysis of slope failure will be demonstrated in the illustrated example in Section 9.5.

9.3.2 Differential evolution adaptive metropolis algorithm

For a highly nonlinear problem with a large number of input parameters that are potentially correlated, the direct random walk algorithm may be inefficient. Vrugt et al. (2008) proposed a novel adaptive MCMC algorithm to efficiently estimate the posterior distribution of parameters in complex high-dimensional sampling problems. The algorithm, titled Differential Evolution Adaptive Metropolis algorithm (DREAM), uses differential evolution (Storn and Price, 1997) algorithm for population evolution, with a Metropolis selection rule to decide whether candidate points should replace their respective parents or not. It runs multiple chains simultaneously for global exploration and automatically tunes the scale and orientation of the jumping distribution during evolution to the posterior distribution. The algorithm shows excellent efficiency on complex, highly nonlinear, and multimodal target distributions. The algorithm can be described as follows (Vrugt et al., 2008).

1. In the algorithm, M different Markov chains are run simultaneously in parallel. Draw an initial population $\boldsymbol{\theta}_0$ of size M, where M is the number of Markov chains and typically $M = n$ or $2n$. The symbol n represents the number of input parameters to be estimated or the dimension of $\boldsymbol{\theta}$.
2. At the current time (the tth generation), M samples (each represents a sample for one Markov chain) form a population, conveniently stored as an $M \times n$ matrix $\boldsymbol{\theta}_{(t)}$. The current state of the ith chain is $\boldsymbol{\theta}^i_{(t)}$ $(i = 1,\ldots, M)$ and the jth element of the vector $\boldsymbol{\theta}^i_{(t)}$ is $(\boldsymbol{\theta}^i_{(t)})_j$ $(j = 1, \ldots, n)$. In the following paragraphs, we drop the subscript of generation, t, for simplicity and convenience.
3. FOR $i = 1, 2, \ldots, M$ (CHAIN EVOLUTION)
   Jumps in each chain are generated by the differential evolution strategy and the Metropolis selection rule with replacement.

(a) Using the differential evolution strategy to generate a candidate point $\mathbf{z}^i$ in chain i,

$$\mathbf{z}^i = \theta^i + \gamma(\delta)\cdot\left(\sum_{j_1=1}^{\delta}\theta^{r_1(j_1)} - \sum_{j_2=1}^{\delta}\theta^{r_2(j_2)}\right) \tag{9.17}$$

where δ is the number of pairs used to generate the candidate point, and $r_1(j_1), r_2(j_2)$ are random permutation of $\{1,\ldots,M\}$ with $r_1(j_1) \neq r_2(j_2) \neq i$. γ is the scaling factor for mutation. The value of γ depends on the number of pairs to create the candidate point. Compared with the random walk metropolis algorithm, the value of γ is set to be equal to $2.38/\sqrt{2\delta n_{\text{eff}}}$, with $n_{\text{eff}} = n$ initially, but potentially decreased in the next step. This choice is expected to yield an acceptance probability of 0.44 for $n = 1$, 0.28 for $n = 5$, and 0.23 for large n.

(b) Replace each element of the proposal $\mathbf{z}_j^i$, $j = 1, \ldots, n$, with θ_j^i using a binomial scheme with crossover probability CR,

$$\mathbf{z}_j^i = \begin{cases} \theta_j^i & \text{if } U \leq 1 - CR, \quad n_{\text{eff}} = n_{\text{eff}} - 1 \\ \mathbf{z}_j^i & \text{otherwise} \end{cases} \quad (j = 1,\ldots,n) \tag{9.18}$$

where $U \in [0,1]$ is a random draw from a uniform distribution.

(c) Compute the density functions of $f(\theta^i \mid \widehat{\mathbf{Y}})$ and $f(\mathbf{z}^i \mid \widehat{\mathbf{Y}})$.

(d) Accept the candidate point with Metropolis acceptance probability, $\alpha(\theta^i, \mathbf{z}^i)$,

$$\alpha(\theta^i, \mathbf{z}^i) = \begin{cases} \min\left(\dfrac{f(\mathbf{z}^i \mid \widehat{\mathbf{Y}})}{f(\theta^i \mid \widehat{\mathbf{Y}})}, 1\right) & \text{if } f(\theta^i \mid \widehat{\mathbf{Y}}) > 0 \\ 1 & \text{if } f(\theta^i \mid \widehat{\mathbf{Y}}) = 0 \end{cases} \tag{9.19}$$

(e) If the candidate point is accepted, move the chain, that is, set $\theta^i = \mathbf{z}^i$; otherwise remain at the old location, θ^i. END FOR (CHAIN EVOLUTION).

4. Remove outlier chains using the inter-quartile-range (IQR) statistic. The IQR is computed as Q_3–Q_1, in which Q_1 and Q_3 denote the lower and upper quartile of the N different chains. This step does not maintain detailed balance and can therefore only be used during burn-in. If an outlier chain is detected, then apply another burn-in period.
5. Compute the Gelman–Rubin convergence diagnostic (Gelman and Rubin, 1992) R_{stat} for each dimension $j = 1, \ldots, n$ using the last 50% of the samples in each chain. If $R_{\text{stat}} < 1.2$ for all j, stop, otherwise go to Step 3.
6. Stop iteration if the maximum number of samples for the simulation is reached.

The convergence diagnostic R_{stat} is calculated based on the within and between chain variance of each parameter as follows (Gelman and Rubin, 1992):

$$R_{\text{stat}} = \sqrt{\left[\frac{N_{\text{chain}} - 1}{N_{\text{chain}}} + \frac{M+1}{M \cdot N_{\text{chain}}}\frac{B}{W}\right]\frac{df}{df-2}} \tag{9.20}$$

where:

N_{chain} is the number of samples in each chain
M is the number of chains

B is the variance between the mean values of M chains
W is the average of the M within-chain variances
df is the degrees of freedom of the approximate Student's t distribution

For notational simplicity, the input parameter θ is assumed to be single dimensional. B and W can calculated as follows:

$$W = \frac{1}{M}\sum_{i=1}^{M}\sigma_i^2 \tag{9.21}$$

$$B = \frac{N_{\text{chain}}}{M-1}\sum_{i=1}^{M}(\bar{\theta}_i - \bar{\bar{\theta}})^2 \tag{9.22}$$

where:
σ_i^2 is the within-chain variance of the ith chain
$\bar{\theta}_i$ is the mean value of samples in the ith chain
$\bar{\bar{\theta}} = \frac{1}{M}\sum_{i=1}^{M}\bar{\theta}_i$ is the average of $\bar{\theta}_i$

9.4 PROCEDURES FOR PROBABILISTIC MODEL CALIBRATION AND PREDICTION

9.4.1 Approximation of the implicit prediction model using response surface

A possible limitation of MCMC simulation is that it requires iterative evaluation of the prediction model, which could be computationally intensive. For implicit prediction models such the prediction models by limit equilibrium methods or the coupled seepage and deformation models, the computation cost will be extraordinarily high.

To overcome this problem, the response surface method (Wong, 1985; Babu and Srivastava, 2007; Mollon et al., 2009) can be used to approximate the implicit prediction model, and then MCMC simulation is carried out based on the fitted response surface. By coupling the MCMC simulation with the response surface methodology, the model calibration becomes computationally efficient.

For example, the prediction model $g(\boldsymbol{\theta})$ can be approximated with a second-order polynomial response surface function:

$$g'(\boldsymbol{\theta}) = b_0 + \sum_{i=1}^{n} b_i\theta_i + \sum_{i=1}^{n} b_{n+i}\theta_i^2 \tag{9.23}$$

where:
n is the dimension of $\boldsymbol{\theta}$
$b_0, b_1, \ldots, b_{2n}$ are the regression coefficients

The regression coefficients can be obtained by curve fitting. For different geotechnical problems, the forms of the response surface should be verified for accuracy of approximating the original prediction model.

After the response surface function is obtained, it can be used in MCMC simulation instead of the original prediction model for evaluating the likelihood function. The integration of response surface method and the MCMC simulation for model calibration will be illustrated in the examples.

9.4.2 General procedures of probabilistic model calibration and prediction

The general procedures of probabilistic model calibration and prediction using the MCMC simulation method are summarized as follows.

1. Assume or determine the prior distribution of input parameters $f(\theta)$.
2. Based on the characteristics of observations and model error of prediction, determine the analytical formulation of the likelihood function $L(\theta \mid \text{observation})$ and the posterior density function $f(\theta \mid \text{observation})$.
3. Obtain random samples of posterior distribution $f(\theta \mid \text{observation})$ or log of $f(\theta \mid \text{observation})$ using the MCMC method.
4. Determine the burn-in length of Markov chains based on the Gelman–Rubin convergence diagnostic value. Ignore the samples before convergence and use the posterior samples after convergence as samples from the stationary posterior distribution.
5. Estimate the statistics of the posterior distribution such as mean, variance, maximum posterior density (MPD) values and calculate the confidence intervals based on the posterior samples.
6. Predict the response of the slope using the posterior samples directly or the posterior statistics.

The procedure of determine the confidence intervals are as follows. After convergence has been achieved to a stationary distribution, the samples from the stationary distribution of $\boldsymbol{\theta}$ are drawn and used to evaluate the model output $\mathbf{Y}$ for each $\boldsymbol{\theta}$. The obtained predictive values of $\mathbf{Y}$ are summarized to calculate the 2.5% and 97.5% percentiles. This predictive uncertainty bounds only includes the effect of parameter uncertainty of $\boldsymbol{\theta}$. The remaining error is assumed to be additive. The total predictive uncertainty bounds are obtained as follows. Draw a random value z from a chi-square distribution with d degrees of freedom and calculate $\sigma_\varepsilon^2 = \frac{d}{z} s^2$. For each model prediction $\mathbf{Y}$, the residual error $\varepsilon \sim N(0, \sigma_\varepsilon^2)$ is then added to the prediction. The 2.5% and 97.5% percentiles of the total predictive uncertainty are then estimated in a similar way as described before.

9.5 EXAMPLE 1: A CUT SLOPE FAILURE

A cut slope failure reported by Duncan (1999) is adopted here as an illustrative example. The slope was excavated by trimming back the original hill slope to a steeper configuration. The ground investigation showed the presence of layers of sandy clay, highly plastic clay, and dense to very dense sand. The soil profile used in design is shown in Figure 9.1. Properties of sandy clay and highly plastic clay based on laboratory tests (Duncan, 1999) are summarized in Table 9.1. The design ground water table shown in Figure 9.1 was based on the water level observed in exploratory boring. Investigation after the slope failure indicated that the slip surface passed through the bottom of the highly plastic clay layer. Measurements from the piezometers installed after the slope failure showed that the ground water level was essentially as assumed in the design. For the conditions shown in Figure 9.1 and using the shear strength parameters shown in Table 9.1, the computed factor of safety is 1.41, but the slope failed (Duncan, 1999).

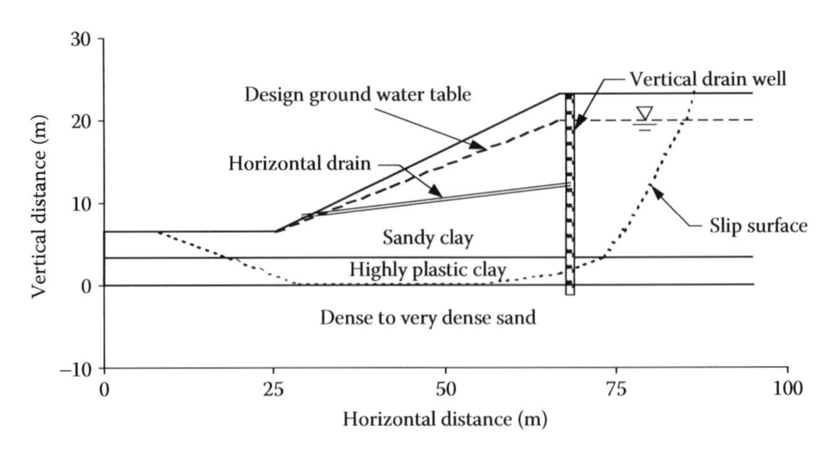

Figure 9.1 Cross section of a cut slope. (From Zhang, L. L. et al., *Computers and Geotechnics*, 37, 905–912, 2010b. With permission.)

Table 9.1 Material properties of the cut slope

Sandy clay			*Highly plastic clay*		
c_1 (kPa)	ϕ_1 (°)	γ_1 (kN/m³)	c_2 (kPa)	ϕ_2 (°)	γ_2 (kN/m³)
14.4	35.00	20.8	0.00	25.00	20.8

Source: Duncan, J. M., *Civil Engineering Practice*, Spring/Summer, 75–91.

9.5.1 Model uncertainty and prior knowledge about uncertain parameters

In this example, a probabilistic back analysis is conducted for this slope. First, the shear strength parameters of the two soil layers, c_1, ϕ_1, and ϕ_2, are considered as uncertain variables, that is, $\boldsymbol{\theta} = \{c_1, \phi_1, \phi_2\}$. The mean values of the three uncertain variables are assumed to be the same as those design values based on laboratory tests. The variability of these variables is not reported in Duncan (1999). According to Phoon and Kulhawy (1999), the coefficient of variation (COV) of the friction angle for sandy soils and clayey soils is between 5% and 20%. The COV of soil cohesion is usually greater than 20% (Réthāti, 2012). In this example, the prior COV values of c_1, ϕ_1, and ϕ_2 are assumed to be 0.20, 0.15, and 0.15, respectively, as summarized in Table 9.2. It is assumed that c_1, ϕ_1, and ϕ_2 are statistically independent and follow the multivariate lognormal distribution. A sensitivity analysis will be carried out to study the effect of prior distribution on back analysis.

The method of Morgenstern and Price (1965) is used to calculate the factor of safety of this cut slope. As the emphasis of this paper is to improve the knowledge on slope stability parameters through back analysis, the probability density function of the model uncertainty variable ε is not updated in the back analysis. According to Christian et al. (1994), the model uncertainty of the simplified method of Bishop (1955) has a mean of 0.05 and a standard

Table 9.2 Prior statistics of random variables

	c_1 (*kPa*)	ϕ_1 (°)	ϕ_2 (°)
Mean	14.4	35.0	25.0
COV	0.20	0.15	0.15
Standard deviation	2.88	5.25	3.75

deviation of 0.07. Because the results from Bishop's simplified method are usually close to those from the Morgenstern and Price method, it is assumed that the model uncertainty of the Morgenstern and Price method can also be modeled as a normally distributed random variable with a mean of 0.05 and a standard deviation of 0.07, that is, $\mu_\varepsilon = 0.05$ and $\sigma_\varepsilon = 0.07$.

It should be noted the back analysis results depend on the accuracy of the identified soil profile including the accurate location of the slip surface as well as the understanding of the failure mechanism. The back analysis results may be meaningless if the analysis is carried out based on an unrealistic profile or based on a wrong failure mechanism. It is assumed in this example that both the soil profile relevant to the back analysis and the location of the slip surface are well defined through post-failure investigation. Moreover, the slope stability model can correctly reflect the failure mechanism.

9.5.2 Construction of response surface

To construct a second-order polynomial function to approximate the slope stability model, the factors of safety of the slope along the failure surface are first estimated at the following seven design points: $\{\mu_{c1}, \mu_{\phi1}, \mu_{\phi2}\}$, $\{\mu_{c1} \pm 3\sigma_{c1}, \mu_{\phi1}, \mu_{\phi2}\}$, $\{\mu_{c1}, \mu_{\phi1} \pm 3\sigma_{\phi1}, \mu_{\phi2}\}$, and $\{\mu_{c1}, \mu_{\phi1}, \mu_{\phi2} \pm 3\sigma_{\phi2}\}$. The slope stability model can then be approximated by the following equation:

$$\begin{aligned} g'(\theta) = {} & 5.767\times10^{-2} + 8.816\times10^{-3}c_1 + 5.132\times10^{-3}\phi_1 + 3.561\times10^{-2}\phi_2 \\ & -6.698\times10^{-6}c_1^2 + 1.148\times10^{-4}\phi_1^2 + 3.793\times10^{-4}\phi_2^2 \end{aligned} \tag{9.24}$$

To check the adequacy of the slope stability model, 16 more points are randomly drawn from the space of $\boldsymbol{\theta}$, and the factors of safety corresponding to these parameters are evaluated with both the Morgenstern–Price method and with the second-order polynomial function, as shown in Figure 9.2. The results from the two approaches are very close. Hence, the following probabilistic back analysis is based on the second-order polynomial response surface function in Equation 9.24.

9.5.3 Back analysis using the MCMC simulation

In this example, the Metropolis algorithm is employed for the MCMC simulation of the posterior distribution (9.3). As presented in Section 9.3.1, the jumping function $J(\boldsymbol{\theta}^*|\boldsymbol{\theta}^{(t-1)})$ should first be determined. In this example, a multivariate normal distribution is used as the

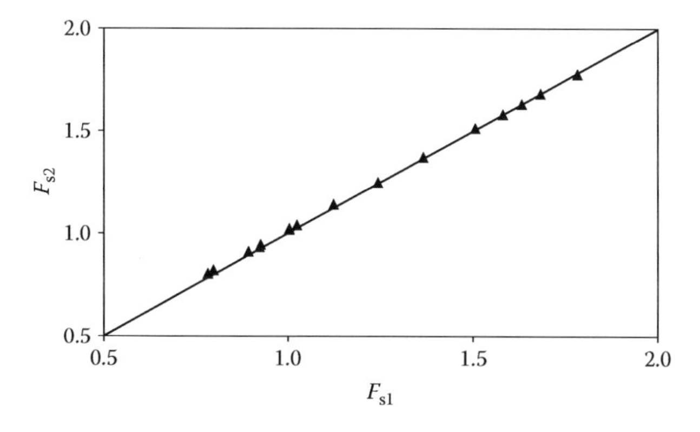

Figure 9.2 Comparison of factors of safety calculated using the response surface function (F_{s1}) with those calculated using the Morgenstern–Price method (F_{s2}).

jumping distribution. Its mean point is the current point in the Markov chain, and the covariance matrix $\mathbf{C}_{\boldsymbol{\theta}t}$ is equal to $\xi\mathbf{C}_{\boldsymbol{\theta}}$, where $\mathbf{C}_{\boldsymbol{\theta}}$ is the covariance matrix of the prior distribution of $\boldsymbol{\theta}$ and ξ is a scaling factor. To study the effect of scaling factor ξ, we established three Markov chains with $\xi = 0.01$, $\xi = 0.5$, and $\xi = 10$, respectively. As an example, the values of 2000 samples of c_1 generated from the three chains are shown in Figure 9.3a–c. It can be seen that ξ has an important effect on the behavior of the Markov chain. When $\xi = 0.01$ as shown in Figure 9.3a, the Markov chain moves slowly in the posterior space, taking a long time to travel from one side of the posterior space to the other side. When $\xi = 10$, there are many horizontal traces in Figure 9.3c, indicating that the current point in a Markov chain is the same as its previous point, that is, the chain does not move. This is because when the value of ξ is too large, the random samples drawn from the jumping function will frequently fall into regions with a small posterior density, and hence the suggested samples are frequently rejected. For comparison, when $\xi = 0.5$, the Markov chain can move actively in the posterior space as shown in Figure 9.3b.

To investigate the effect of ξ on posterior statistics, 10 Markov chains are generated with $\xi\iota = 0.01$ from different initial points. Based on the samples from each chain, the posterior statistics of c_1, ϕ_1, and ϕ_2, that is, the mean values, standard deviations and correlation coefficients, can be calculated. It is expected that, when the number of samples drawn from each chain is sufficiently large, the inferred posterior statistics from different chains will be similar. As at the early stage the Markov chain may not reach the equilibrium or stationary state, it is often suggested that the first few samples should be discarded and not used for posterior statistics inference (burn-in). It is also suggested that one does not

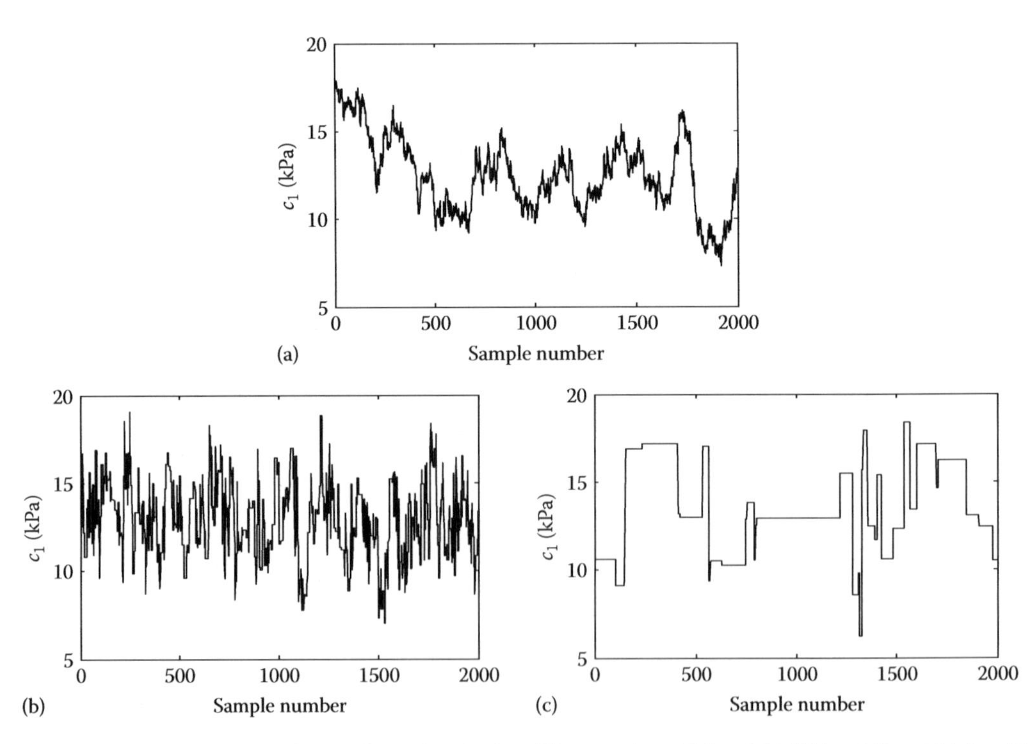

Figure 9.3 Samples for c_1 in a Markov chain when ξ takes different values: (a) $\xi = 0.01$, (b) $\xi = 0.5$, and (c) $\xi = 10$. (From Zhang, L. L. et al., *Computers and Geotechnics*, 37, 905–912, 2010b. With permission.)

indeed need to discard the initial samples in a Markov chain, because when the number of samples drawn from a Markov chain increases, the effects of the first few samples on the posterior inference results gradually diminish (Driscoll and Maki, 2007). In this example, the first 500 samples are discarded in each chain. Table 9.3 compares the mean values and standard deviations of the posterior statistics of c_1, ϕ_1, and ϕ_2 inferred from the 10 Markov chains when the number of samples drawn from each chain are 5000, 10,000, 20,000, 40,000, and 80,000, respectively. As the number of samples increases, the standard deviations of the posterior statistics decrease as expected. It means the estimates of the posterior statistics become more accurate when the number of the samples in a chain increases. For comparison, 10 Markov chains with $\xi = 0.5$ and 10 Markov chains with $\xi = 10$ are also established. The mean values and standard deviations of the posterior statistics inferred from the Markov chains when $\xi = 0.5$ and $\xi = 10$ are summarized in Tables 9.4 and 9.5, respectively. It can be seen that the estimated posterior statistics when ξ takes different values are generally consistent, indicating that the Metropolis algorithm is quite robust. It can also be observed that, when the numbers of samples are the same, the posterior statistics inferred from the Markov chains with $\xi = 0.5$ have smaller variability. Thus, to achieve the same accuracy with a Markov chain with $\xi = 0.5$, more samples should be drawn from Markov chains when $\xi = 0.01$ or $\xi = 10$ are adopted. In other words, the Markov chain with $\xi = 0.5$ is more efficient in collecting representative samples from the posterior distribution.

An appropriate value of ξ for the jumping distribution is in fact closely related to the acceptance rate, which is defined as the proportion of samples accepted in the Markov chain. Gelman et al. (2004) recommended that when the acceptance rate is around 20%–40%, the Markov chain is most efficient in collecting samples of the posterior density function. For this example, the relationship between acceptance rate and ξ is plotted in Figure 9.4. It can be seen that the acceptance rate decreases with ξ. When the ξ values adopted are 0.01, 0.5, and 10, the corresponding acceptance rates are about 0.86, 0.30, and 0.02, respectively. The efficiency of the Markov chain with $\xi = 0.5$ is higher than those based on $\xi = 0.01$ and $\xi = 10$ as shown in Tables 9.3 through 9.5. Therefore, it is consistent with the suggestions made by Gelman et al. (2004). In practical application of an MCMC simulation, one could first determine the relationship between ξ and the acceptance rate first, and select a ξ that can result in an acceptance rate in the range of 0.2–0.4 for the MCMC simulation.

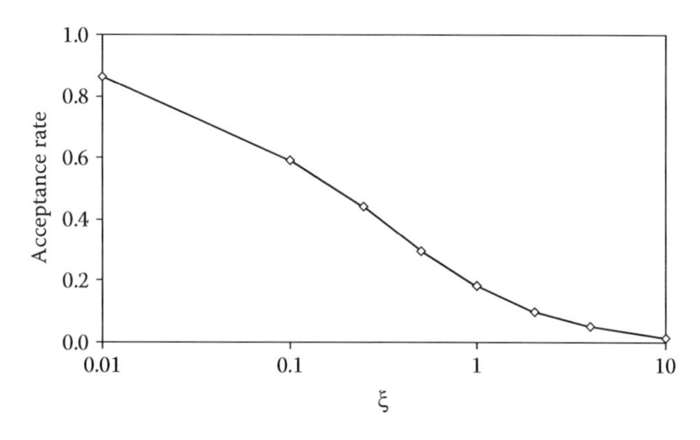

Figure 9.4 Effect of the scaling factor of the jumping distribution on the acceptance rate of MCMC. (From Zhang, L. L. et al., *Computers and Geotechnics*, 37, 905–912, 2010b. With permission.)

Table 9.3 Mean values and standard deviations of the posterior statistics estimated from 10 Markov chains with $\xi = 0.01$

		μ_{c1}''	$\mu_{\phi1}''$	$\mu_{\phi2}''$	σ_{c1}''	$\sigma_{\phi1}''$	$\sigma_{\phi2}''$	$\rho_{c1,\phi1}''$	$\rho_{c1,\phi2}''$	$\rho_{\phi1,\phi2}''$
5,000	Mean	13.05	28.31	15.91	2.34	3.18	1.25	−0.06	−0.19	−0.37
	Std	0.86	0.75	0.23	0.58	0.33	0.08	0.25	0.10	0.11
10,000	Mean	12.90	28.65	15.89	2.38	3.47	1.29	−0.06	−0.18	−0.40
	Std	0.50	0.03	0.03	0.04	0.02	0.02	0.02	0.01	0.01
20,000	Mean	12.82	28.67	15.87	2.32	3.49	1.31	−0.03	−0.18	−0.42
	Std	0.28	0.34	0.09	0.08	0.10	0.05	0.09	0.07	0.04
40,000	Mean	12.99	28.61	15.89	2.42	3.52	1.32	−0.08	−0.18	−0.41
	Std	0.15	0.30	0.06	0.09	0.18	0.04	0.06	0.04	0.03
80,000	Mean	12.81	28.70	15.88	2.42	3.51	1.31	−0.06	−0.18	−0.40
	Std	0.12	0.20	0.05	0.11	0.20	0.02	0.03	0.03	0.03

Note: Std denotes the standard deviation.

Table 9.4 Mean values and standard deviations of the posterior statistics estimated from 10 Markov chains with $\xi = 0.5$

		μ_{c1}''	$\mu_{\phi1}''$	$\mu_{\phi2}''$	σ_{c1}''	$\sigma_{\phi1}''$	$\sigma_{\phi2}''$	$\rho_{c1,\phi1}''$	$\rho_{c1,\phi2}''$	$\rho_{\phi1,\phi2}''$
5,000	Mean	12.76	28.75	15.89	2.29	3.46	1.28	−0.05	−0.17	−0.41
	Std	0.10	0.21	0.04	0.09	0.10	0.03	0.05	0.05	0.03
10,000	Mean	12.82	28.65	15.91	2.38	3.48	1.31	−0.07	−0.18	−0.41
	Std	0.11	0.13	0.05	0.08	0.07	0.02	0.03	0.03	0.02
20,000	Mean	12.84	28.64	15.90	2.41	3.47	1.30	−0.06	−0.18	−0.39
	Std	0.09	0.12	0.03	0.06	0.07	0.02	0.02	0.02	0.01
40,000	Mean	12.84	28.65	15.89	2.38	3.47	1.29	−0.06	−0.18	−0.40
	Std	0.07	0.03	0.03	0.04	0.02	0.02	0.02	0.01	0.01
80,000	Mean	12.85	28.64	15.89	2.40	3.49	1.30	−0.06	−0.18	−0.40
	Std	0.05	0.03	0.01	0.04	0.04	0.01	0.01	0.01	0.01

Table 9.5 Mean values and standard deviations of the posterior statistics estimated from 10 Markov chains with $\xi = 10$

		μ_{c1}''	$\mu_{\phi1}''$	$\mu_{\phi2}''$	σ_{c1}''	$\sigma_{\phi1}''$	$\sigma_{\phi2}''$	$\rho_{c1,\phi1}''$	$\rho_{c1,\phi2}''$	$\rho_{\phi1,\phi2}''$
5,000	Mean	12.87	28.54	16.01	2.38	3.46	1.35	−0.11	−0.19	−0.38
	Std	0.33	0.76	0.11	0.29	0.26	0.17	0.11	0.13	0.15
10,000	Mean	12.81	28.45	15.94	2.32	3.39	1.30	−0.08	−0.19	−0.37
	Std	0.29	0.37	0.19	0.17	0.24	0.09	0.09	0.07	0.08
20,000	Mean	12.81	28.45	15.94	2.32	3.39	1.30	−0.08	−0.19	−0.37
	Std	0.29	0.37	0.19	0.17	0.24	0.09	0.09	0.07	0.08
40,000	Mean	12.90	28.67	15.85	2.41	3.49	1.31	−0.10	−0.17	−0.40
	Std	0.15	0.18	0.06	0.11	0.11	0.04	0.08	0.05	0.04
80,000	Mean	12.84	28.73	15.90	2.40	3.52	1.29	−0.07	−0.19	−0.40
	Std	0.11	0.15	0.05	0.09	0.08	0.05	0.03	0.03	0.02

Table 9.4 also shows that when the number of samples is 80,000, the variability of the posterior statistics inferred from different Markov chains is very small. Hence, 80,000 samples are sufficient to obtain a robust estimation of the posterior statistics. The Markov chains with 80,000 samples when $\xi = 0.5$ are shown in Figure 9.5. The histograms of these samples are presented in Figure 9.6. The mean values of the three soil parameters all decrease, indicating that the prior distributions overestimate the soil strength parameters. To confirm this observation, the failure probability of the slope with the ground water table at the moment of failure is calculated based on the prior distribution of $\boldsymbol{\theta}$. With a Monte Carlo simulation with 1×10^7 samples, the prior probability of failure is found to be 7.7×10^{-5}, indicating that the slope failure is very unlikely. In reality, however, the slope failed. Thus, the prior distribution of $\boldsymbol{\theta}$ used for reliability analysis overestimates the soil strength parameters. Based on the posterior samples of MCMC simulation, the failure probability of the slope is 9%, which is far larger than the prior probability of failure. Therefore, with the information of slope failure, the estimated probability of failure of the slope is increased, which is more consistent with the actual situation.

9.5.4 Application in remediation design

As suggested in Duncan (1999), one possible way to stabilize the slope is to lower down the ground water table using vertical drain wells and horizontal drains to drain the water from the slope by gravity. Based on the MCMC samples of the posterior distributions, the reliability indices of the slope with different levels beneath the slope crest are calculated and shown in Figure 9.7 (the case of Prior 1). The assumptions and results of cases Prior 2 and Prior 3 will be presented in the next section. Fell (1994) reviewed

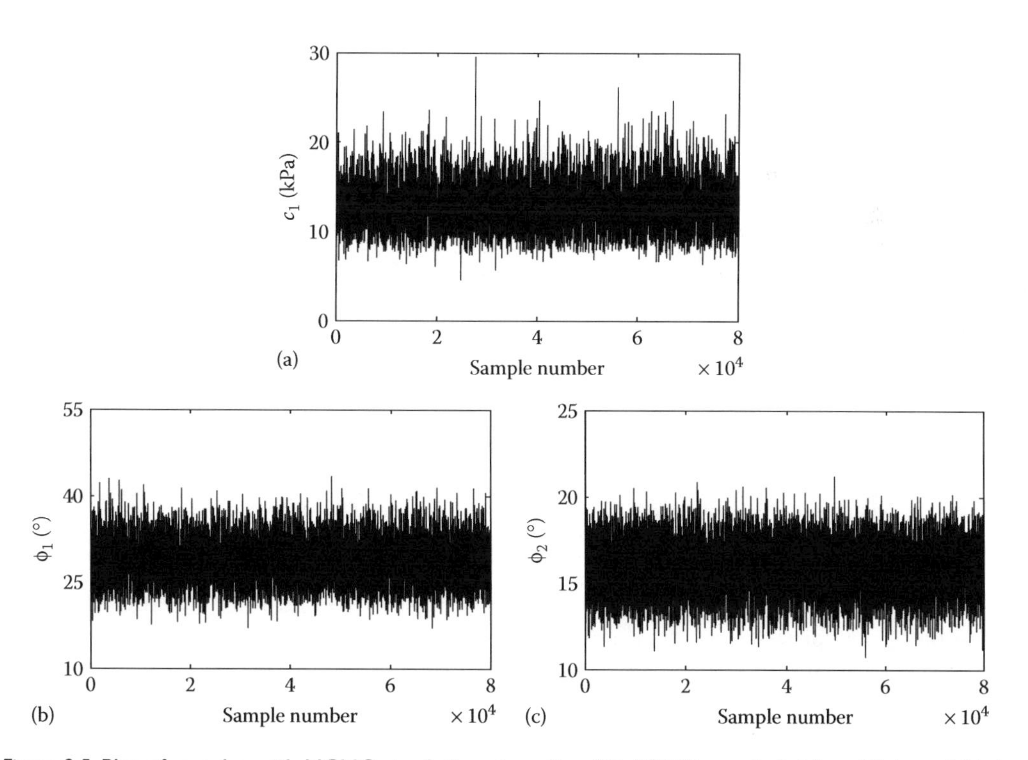

Figure 9.5 Plot of samples with MCMC simulation steps ($\xi = 0.5$, 80000 samples): (a) c_1, (b) ϕ_1, and (c) ϕ_2. (From Zhang, L. L. et al., *Computers and Geotechnics*, 37, 905–912, 2010b. With permission.)

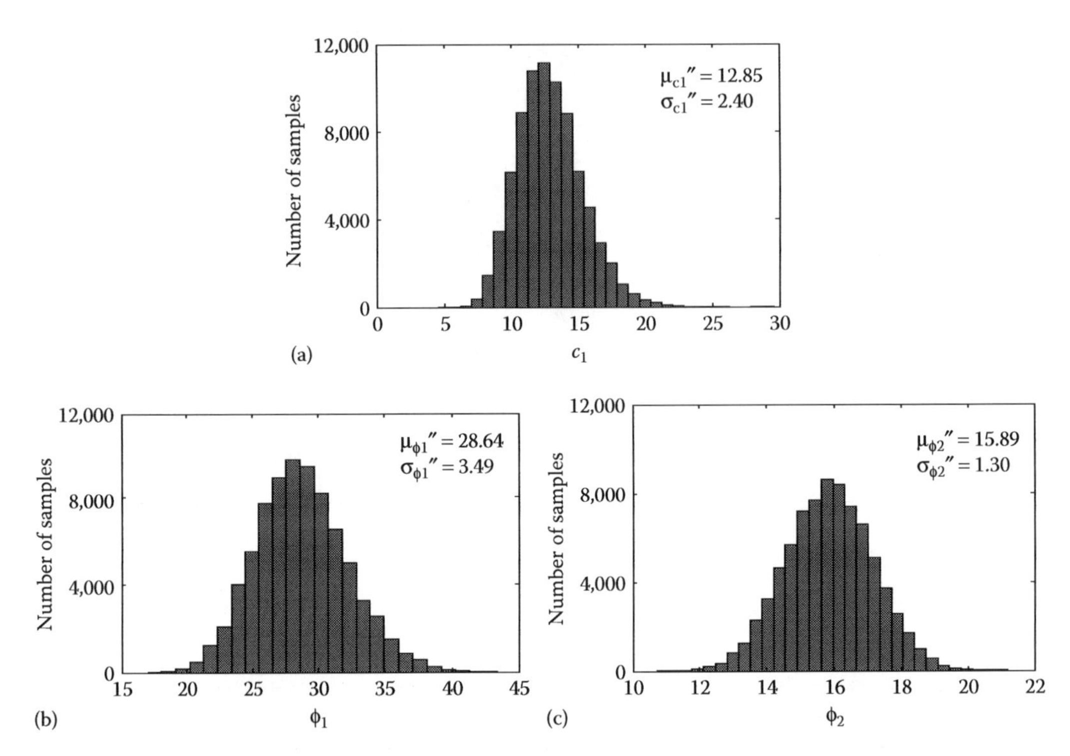

Figure 9.6 Histograms of posterior distributions: (a) c_1, (b) ϕ_1, and (c) ϕ_2 ($\xi = 0.5$, 80,000 samples). (From Zhang, L. L. et al., *Computers and Geotechnics*, 37, 905–912, 2010b. With permission.)

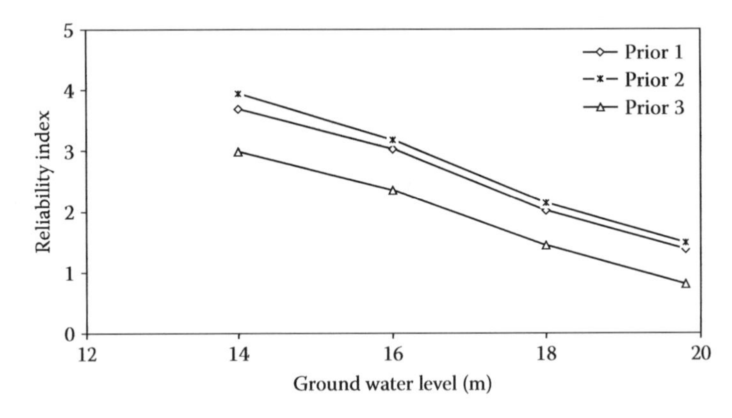

Figure 9.7 Reliability indices of the slope at different ground water levels beneath the crest of the slope. (From Zhang, L. L. et al., *Computers and Geotechnics*, 37, 905–912, 2010b. With permission.)

landslide risk assessment and acceptable risk and concluded that for natural landslides a community may accept an annual specific voluntary risk of loss of life in the order 10^{-3} to 10^{-4}. For involuntary risks, particularly man-made risks, a risk of 10^{-5} to 10^{-6} may be a limiting value. If a target reliability index of 3.0 (probability of failure 0.00135) is desired, the ground water table needs to be lowered to about 16.5 m. As a comparison, Duncan (1999) suggested that the ground water table be lowered to 15.6 m based on the results of a deterministic back analysis.

9.5.5 Effect of prior distribution

In the previous sections, c_1, ϕ_1, and ϕ_2 are assumed to be statistically independent and follow the multivariate lognormal distribution. In the literature (Rethati, 1988; Cherubini, 2000), the cohesion and friction angles of soil are sometimes reported to be negatively correlated. In some cases, the distributions of soil cohesion and friction angles are suggested to follow the normal distribution. To examine the effect of prior distribution, the slope is back analyzed with different assumptions about prior distribution. Denote the previous case where prior c_1, ϕ_1, and ϕ_2 are assumed to be statistically independent and follow the multivariate lognormal distribution as Prior 1. For the second case (Prior 2), it is assumed that the three variables follow the multivariate lognormal distribution with the correlation coefficient between c_1 and ϕ_1 equal to –0.5. For the third case (Prior 3), the three variables follow the multivariate normal distribution and are statistically independent. The back calculated values of the mean, standard deviation, and correlation coefficients of the three random variables with different assumptions about their prior distributions are summarized in Table 9.6. It can be seen that the back calculated result for Prior 2 are very close to that for Prior 1. However, notable difference can be observed comparing the results for Prior 3 and Prior 1. It shows that the correlation of cohesion and friction angles of soil does not affect the posterior statistics significantly, while the assumption of the type of the prior distribution seems to have much influence on the posterior statistics of the random variables. The back calculated results change when the prior distribution changes, indicating that in the probabilistic back analysis the prior knowledge about the parameters also has an important effect on the back analysis results. Thus, it is important to obtain high-quality prior information.

To examine the effect of prior distribution on the remediation design, the reliability indices of the slope at different levels of ground water table are also calculated when different prior distributions are used for back analysis. The results are also plotted in Figure 9.7. It can be seen that the reliability indices of the slope calculated with and without considering the correlation in the prior distribution are similar, and hence the correlation coefficient between c_1 and ϕ_1 does not affect the remediation design significantly. It can also be observed that, when the multivariate normal distribution is assumed as the prior distribution (Prior 3), the calculated reliability indices at different levels of ground water table are much smaller than those with a prior distribution of multivariate lognormal distribution (Prior 1). The multivariate normal distribution seems to be a conservative assumption. The reason for this is that with the assumption of a multivariate normal distribution, the input parameters have more chances to take smaller values. Hence, the mean factor of safety is smaller and the probability of failure is greater.

9.5.6 Comparison with the method based on sensitivity analysis

Zhang et al. (2010a) suggested a simplified method based on sensitivity analysis for probabilistic back analysis of slope failure. The simplified method is basically a system identification method and can yield analytical solutions for posterior mean and covariance matrix of random variables if the prior distribution is multivariate normal and the prediction model

Table 9.6 Comparison of posterior statistics with different prior distributions

	μ_{c1}''	$\mu_{\phi 1}''$	$\mu_{\phi 2}''$	σ_{c1}''	$\sigma_{\phi 1}''$	$\sigma_{\phi 2}''$	$\rho_{c1,\phi 1}''$	$\rho_{c1,\phi 2}''$	$\rho_{\phi 1,\phi 2}''$
Prior 1	12.85	28.64	15.89	2.40	3.49	1.30	–0.06	–0.18	–0.40
Prior 2	14.53	29.50	15.57	2.60	3.56	1.26	–0.46	–0.03	–0.33
Prior 3	13.31	29.74	14.50	2.84	4.70	1.76	–0.04	–0.23	–0.57

Table 9.7 Comparison of prior and posterior statistics calculated using MCMC and a simplified method based on sensitivity analysis

	μ_{c1}''	$\mu_{\phi 1}''$	$\mu_{\phi 2}''$	σ_{c1}''	$\sigma_{\phi 1}''$	$\sigma_{\phi 2}''$	$\rho_{c1,\phi 1}''$	$\rho_{c1,\phi 2}''$	$\rho_{\phi 1,\phi 2}''$
MCMC (Prior 3)	13.31	29.74	14.50	2.84	4.70	1.76	−0.04	−0.23	−0.57
Simplified method	13.47	30.26	14.99	2.86	5.00	1.67	−0.03	−0.37	−0.64

is linear. In this example, Prior 3 assumes a multivariate normal prior distribution. As a comparison, the results obtained using the simplified method is shown in Table 9.7. It shows that the results from the MCMC simulation and the simplified method are consistent. The differences between the results from the MCMC simulation and the simplified back analysis method based on sensitivity analysis are mainly due to nonlinearity of the slope stability model.

9.6 EXAMPLE 2: 1997 LAI PING ROAD LANDSLIDE

9.6.1 Introduction of the 1997 Lai Ping Road landslide

On July 2, 1997, a landslide occurred at a roadside cut slope at Lai Ping Road, Sha Tin, Hong Kong. The incident comprised several discrete failures along a 135-m-long section of the cut slope and completely blocked the Lai Ping Road. The total volume of this large-scale landslide is estimated to be about 100,000 m^3. The principal trigger of the landslide is likely the build-up of pore-water pressure due to elevation of the main groundwater table in a complex hydro geological regime following heavy rain fall (Sun and Campbell, 1999).

Based on the site investigation, the geological profile of the site is characterized by a layer of saprolite overlying a rock mass comprising mainly slightly to moderately decomposed tuff as shown in Figure 9.8. The highest recorded ground water level and the slip surface of the main scar are also presented in Figure 9.8.

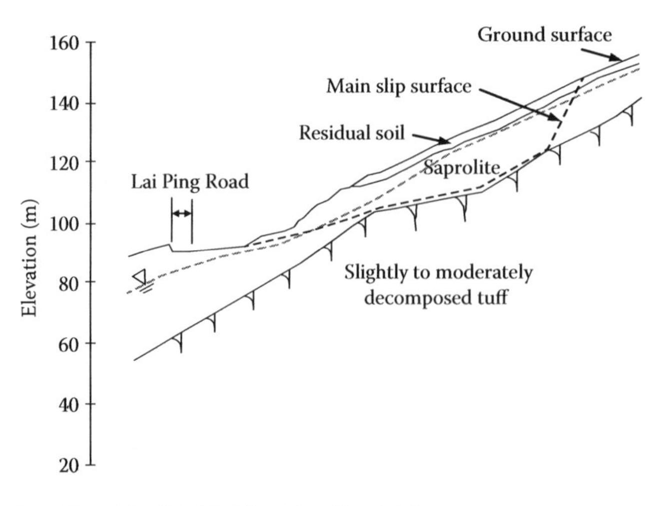

Figure 9.8 Geological profile of the 1997 Lai Ping Road landslide.

9.6.2 Back analysis using the MCMC simulation

Chan (2001) performed a stochastic analysis for the 1997 Lai Ping Road landslide using Latin Hypercube sampling technique. The slope model is the simplified Bishop's method of slope stability analysis using SLOPE/W with pore-water pressures in the slope estimated by a finite element seepage analysis for unsaturated soils using SEEP/W. Response surface functions for the minimum safety factor of the slope after 3, 6, 9, and 12 hours of rainfall are obtained. For example, the response surface function for the safety factor after 6 hours of rainfall is as follows.

$$F_s = 1.1727 + 0.00879Y_1 + 0.22037Y_2 + 0.05862Y_3 - 0.03609Y_4Y_5 - 0.00917Y_6Y_7 \quad (9.25)$$

where Y_i is the standardized random variable for X_i in which $X_1 = k_{s1}$; $X_2 = \phi_1'$; $X_3 = \phi_2'$; $X_4 = \theta_{s1}$; $X_5 = \alpha_{vgm1}$; $X_6 = \phi_3'$; and $X_7 = c_3'$. k_{s1} is the saturated permeability of Soil Layer 1. ϕ_1', ϕ_2', and ϕ_3' are the effective friction angles of Soil Layer 1, 2, and 3, respectively. Here Soil Layer 1 represents the soil layer of saprolite as shown in Figure 9.8. Soil Layer 2 and 3 are layers of decomposed tuff. θ_{s1} is the porosity of Soil Layer 1. a_{vgm1} is the VGM (van Genuchten, 1980; Mualem, 1976) SWCC model parameter related to air-entry value for Soil Layer 1.

Considering that k_{s1}, ϕ_1', and ϕ_2' are most important random variables on the uncertainty of safety factor, these three soil parameters are back analyzed using the probabilistic model calibration method with the MCMC simulation. The prior distributions of the three random variables are all assumed to be normal distributions. The mean values and standard deviations of the three parameters are listed in Table 9.8. 30,000 samples are used to determine the posterior distribution after a convergence of the Markov chain has been achieved. The other soil parameters in Equation 9.25 are assumed to be deterministic ($\theta_{s1} = 0.499$; $a_{vgm1} = 0.496$; $c_3' = 5$ kPa; $\phi_3' = 35°$).

To study the effect of model error on back analysis results, two cases with different assumption about the model error are analyzed. In Case 1, the model error ε of the response surface model is assumed to be known with a mean of 0 and a standard deviation of 0.05. The distribution of model error ε will not be updated. The soil parameter will be updated based on the likelihood function (Equation 9.4) and the posterior distribution in Equation 9.3. In Case 2, the prior distribution for μ_ε is assumed to be a normal distribution with a mean of 0 and a standard deviation of 0.5, while the prior distribution for σ_ε is assumed to be a log-normal distribution with a mean of 0.5 and a standard deviation of 1.0. The prior distributions adopted here are thought to be wide enough to cover possible values of μ_ε and σ_ε. Posterior distributions of μ_ε, σ_ε, k_{s1}, ϕ_1', and ϕ_2' are evaluated based on Equation 9.6.

9.6.3 Effect of uncertainty of model error

Figure 9.9 illustrates the prior distributions and posterior histograms of the three soil parameters for Case 1. As shown in Figure 9.9a, there is no significant difference between the prior and posterior distributions for k_{s1}. However, the variance of ϕ_1' and ϕ_2' are greatly

Table 9.8 Prior distribution and statistics of soil parameters

Parameter	*Mean*	*Std*	*Distribution*
k_{s1} (cm/h)	14.0	1.50	Normal
ϕ_1' (°)	30.0	6.24	Normal
ϕ_2' (°)	34.0	7.07	Normal

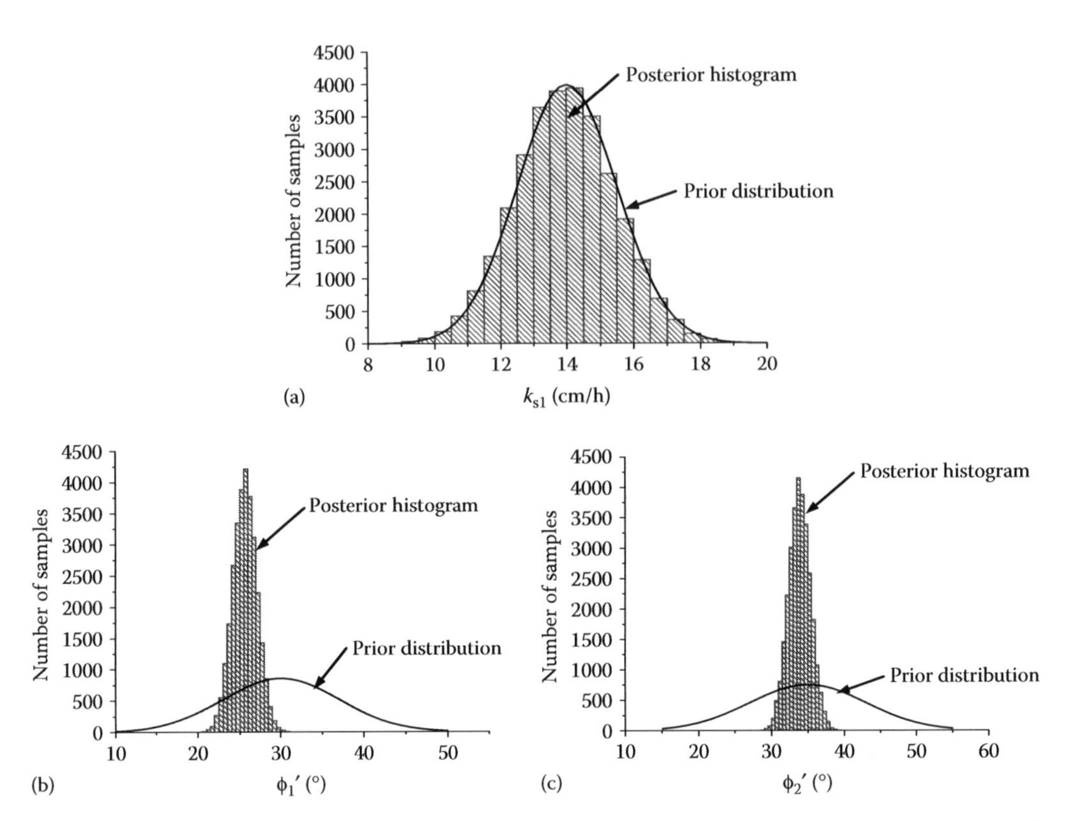

Figure 9.9 Prior distribution and posterior histogram of soil parameters (Case 1): (a) k_{s1}, (b) ϕ_1', and (c) ϕ_2'.

reduced. The COV of ϕ_1' is reduced from 20.8% to 5.6%. The COV of ϕ_2' is reduced from 20.8% to 4.4%. The result indicates that when only soil parameters are back analyzed, the shear strength parameters are updated more than the soil permeability.

Table 9.9 presents the posterior statistics for the three soil parameters in Case 1. The posterior mean of ϕ_1' is 25.64, which is much less than the prior mean value. The posterior mean value of ϕ_2' is 33.86, which is slightly smaller the prior mean. The safety factor equation is not conservative as the calculated safety factor for the slope is 1.173 while the slope actually failed after the rainfall. Therefore, the back analyzed mean values of ϕ_1' and ϕ_2' are less than the prior ones. In terms of the change of mean value, the friction angle of the soil layer near the ground surface ϕ_1' is updated more than ϕ_2'. However, the reduction of uncertainty for ϕ_1' and ϕ_2' are comparable.

Table 9.9 Posterior statistics of input parameters

	Case 1		*Case 2*	
Parameter	*Mean*	*Std*	*Mean*	*Std*
k_{s1} (cm/h)	13.94	1.484	14.01	1.484
ϕ_1' (°)	25.64	1.438	33.70	5.925
ϕ_2' (°)	33.86	1.483	34.99	1.483
μ_ε	–	–	–0.230	0.301
σ_ε	–	–	0.308	0.446

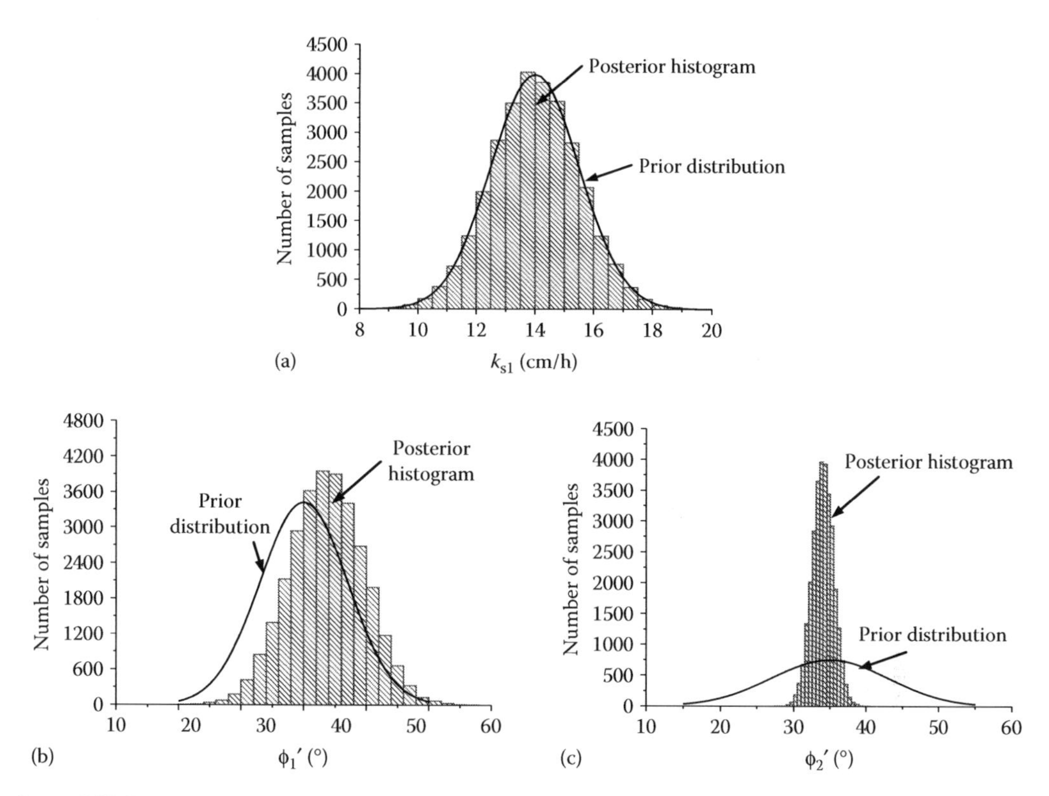

Figure 9.10 Prior distribution and posterior histogram of soil parameters (Case 2): (a) k_{s1}, (b) ϕ_1', and (c) ϕ_2'.

Figure 9.10 presents the prior distributions and posterior histograms of soil parameters for Case 2. The posterior statistics of the updated soil parameters for Case 2 are presented in Table 9.9. Comparing Figure 9.10 with Figure 9.9, we can observe that considering the model uncertainty in back analysis, the updating of the soil input parameters in Case 2 are less significant than that in Case 1. The posterior mean values of ϕ_1' and ϕ_2' are even greater than the prior mean values. The COV of ϕ_1' is reduced from 20.8% to 17.4%. The COV of ϕ_2' is reduced from 20.8% to 4.2%.

Figure 9.11 shows the prior distribution and posterior histogram for μ_ε and σ_ε. The posterior mean value of μ_ε is −0.23 indicating that the estimated safety factor using Equation 9.25

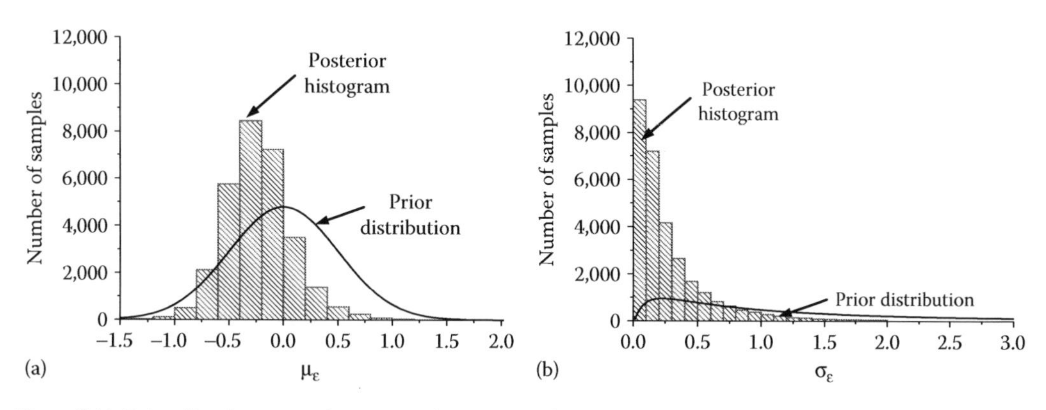

Figure 9.11 Prior distribution and posterior histogram of model error parameters (Case 2): (a) μ and (b) σ.

is not conservative. The standard deviation of μ_ε is reduced from 0.5 to 0.3 and the standard deviation of σ_ε is reduced from 1.0 to 0.4. This indicates that the information of slope failure helps to reduce the uncertainty of model error more than those of soil parameters.

9.6.4 Posterior analysis

Without the information of slope failure, the mean and standard deviation of the safety factor of the Lai Ping Road slope are 1.173 and 0.228, respectively. The reliability and probability of failure of the slope are 0.757 and 0.225, respectively. Using the 30,000 posterior samples, the reliability and probability of failure of the Lai Ping Road can be re-evaluated. The results are presented in Table 9.10. Without back analysis of model error (Case 1), the mean safety factor is reduced to 1.009. The reliability and probability of failure of the slope are 0.180 and 0.428, respectively. For Case 2, the mean safety factor is 1.035. The reliability and probability of failure of the slope are 0.142 and 0.444, respectively. These results show that after model calibration, the estimation of the slope stability model is more reasonable.

9.7 EXAMPLE 3: AN INSTRUMENTED SITE OF NATURAL TERRAIN IN HONG KONG

9.7.1 Introduction of the instrumented site

Evans and Lam (2003) reported a field study which was conducted in a well-instrumented site of natural terrain in Hong Kong with long time monitoring pore-water pressures, groundwater and ground movements. The objective of the field monitoring work was to improve understanding of the factors affecting stability of natural terrain, and to assist with the development of landslip warning schemes and natural terrain hazard assessment methodologies. The site is located above the North Lantau Expressway in the east of Tung Chung on Lantau Island, Hong Kong (see Figure 9.12). In this example, the site is referred to as Tung Chung East (TCE). The TCE site comprises approximately 2.5 ha of terrain, with slopes of 30° to 40° in weathered volcanic rocks. The regolith comprises weathered rock (Grade IV and V), residual soil, colluvium and debris flow deposits at the base of the slope (see Figure 9.13). Based on the laboratory test results, the Grade V material or completely decomposed volcanic (CDV) soil is a clayey silt of low to intermediate compressibility (LL = 30–45) with a clay content of 5%–40%. The residual soil is a clayey silt of intermediate compressibility with a clay content of generally 25%–40%. The colluvium material is a silty clay or clayey silt of intermediate to high compressibility (LL = 40–60). The clay content is generally 20%–50% with an average value of 36%.

Two automatic tipping-bucket rain gauges with internal data loggers were installed during the winter of 1999/2000. The rain gauges (which tip every 0.5 mm of rainfall) are located at the upper and lower margins of the site, at elevations of about 95 mPD and 10 mPD,

Table 9.10 Prior and posterior reliability and probability of failure

	μ_{FS}	σ_{FS}	β	p_f
Prior	1.173	0.228	0.757	0.225
Case 1	1.009	0.048	0.180	0.428
Case 2	1.035	0.247	0.142	0.444

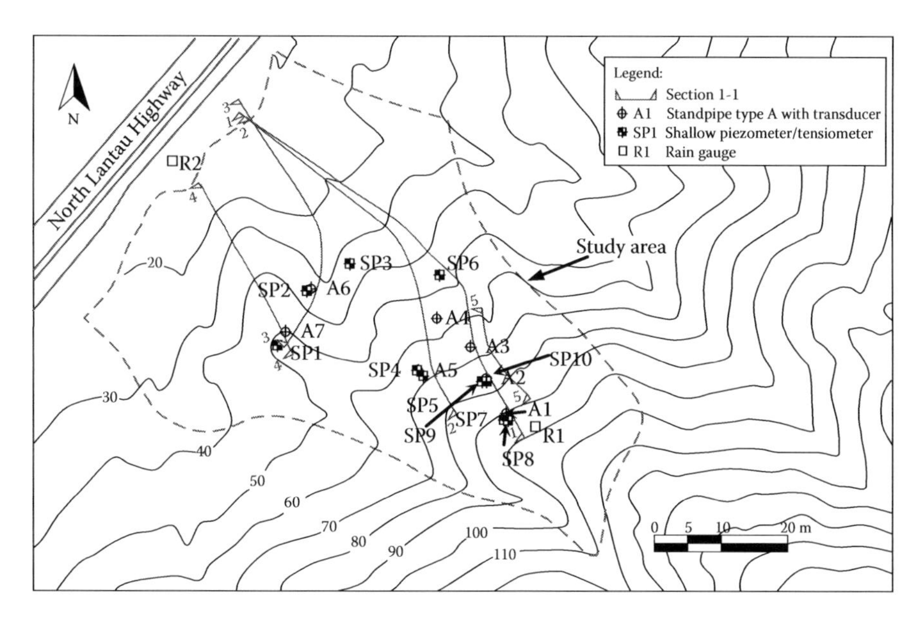

Figure 9.12 Locations of instruments and cross sections of Tung Chung East site.

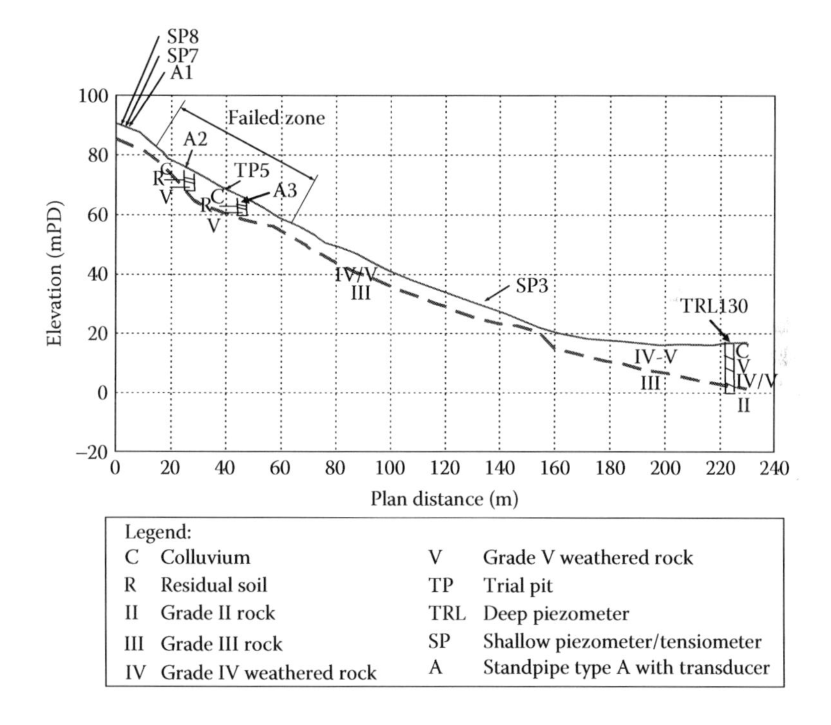

Figure 9.13 Geological profile and locations of instruments at cross section 1–1.

respectively. The rain gauges record rainfall every five minutes. Ten automatic recording piezometers (SP1–SP10), which can measure pore pressures continuously over the range +65 kPa to –100 kPa, were installed at shallow depths (less than 3 m below ground surface) above the natural water table. The depth of the tip of piezometers and the soil type at the tip are listed in Table 9.11. The data recording frequency of these piezometers was 1 hour.

Table 9.11 Depth and ground condition of the 10 piezometers at Tung Chung East site

	Tip depth (mbgl)	*Surface elevation (mPD)*	*Ground condition*	*Soil type*	*Dry season suction (kPa)*
SP1	2.5	50	U	CDV	70
SP2	3.0	32	D	Col	30
SP3	2.0	33	U	CDV	80
SP4	2.73	66	D	CDV	30
SP5	1.53	66	D	CDV	80
SP6	3.0	48	U	CDV	25
SP7	2.62	91	U	CDV	~10
SP8	1.0	91	U	Col	>100
SP9	3.0	74	D	CDV	25
SP10	1.15	74	D	Residual	>100

Notes: D, disturbed ground; U, undisturbed ground; CDV, completely decomposed volcanic, Col, colluvium; Residual, residual soil; mbgl, meters below ground level; mPD, meters above the Principle Datum.

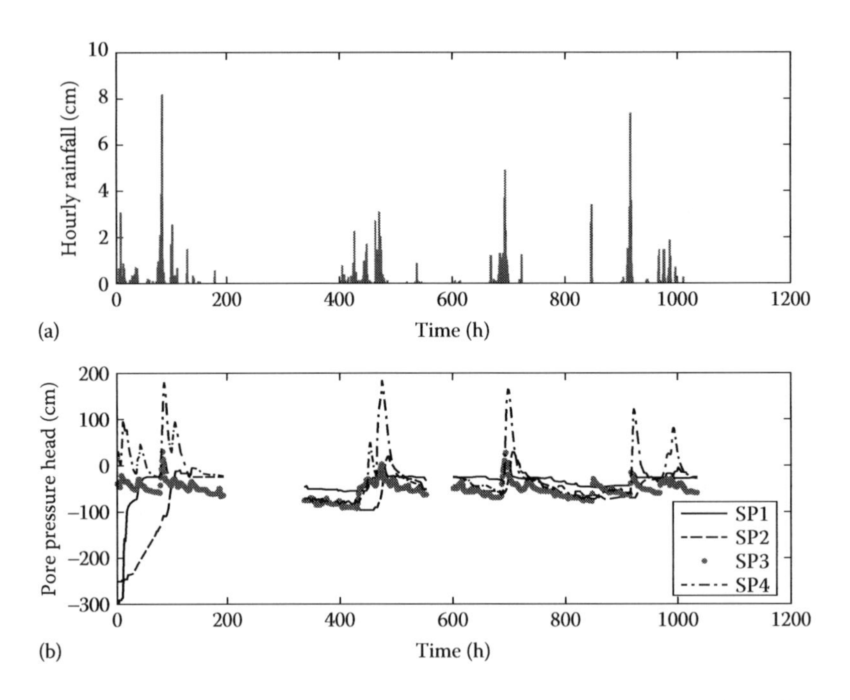

Figure 9.14 (a) Recorded rainfall and (b) measurements of pore-water pressure head from June 8 to July 20, 2001. The shaded rectangle covers a fraction of the period when the measurement data are unavailable. (From Zhang, L. L. et al., *Computers and Geotechnics*, 48, 72–81, 2013. With permission.)

Figure 9.14 shows the pore-water pressure measurements at SP1–SP4 during the period from June 8, 2001 to July 20, 2001. The measured hourly rainfall is also presented. The shaded rectangle covers a fraction of the period when the measurement data are unavailable. According to Evans and Lam (2003), soil suction during the wet season was often eliminated during rainstorms. In general, at depths of 12 m below ground level, suctions are low during the wet season and recover rapidly to high values during dry periods. In contrast, at depths of 23 m below ground level, suctions tend to remain higher and respond less to rainfall during the wet season, and do not recover as rapidly or to such high values during the dry season.

9.7.2 Probabilistic parameter estimation

In this example, the analytical solution of 1D infiltration with time-varied rainfall by Yuan and Lu (2005) (presented in Section 2.4.2) is the prediction model. The field pore-water pressure measurements of the piezometer SP3, which was installed 2.0 m below ground surface and the soil type at tip is CDV soil, is used to calibrate the analytical solution and estimate the soil parameters of hydraulic properties. Data from Period 1 (hour 1 to hour 193, i.e., June 8, 2001, to June 16, 2001, as shown in Figure 9.14), were used as the calibration dataset whereas data from Periods 2 to 4 (hour 337 to 1034, i.e., June 22 to July 20, 2001, as shown in Figure 9.14) were used as the validation data set.

The three soil parameters ($\theta_s - \theta_r$), $\log_{10}(k_s)$, and $\log_{10}(\alpha_{sy})$, which are considered important for infiltration in unsaturated soils, are considered as random input parameters. Here a single parameter ($\theta_s - \theta_r$) instead of θ_s and θ_r separately, is adopted to avoid unreasonable back calculated combinations of θ_s and θ_r, where θ_r is greater than θ_s. The log of k_s and α_{sy} are adopted in the parameter estimation as k_s and α_{sy} are generally small values and follow log-normal distributions (Zhang et al., 2005; Scharnagl et al., 2011).

The prior distributions and basic statistics of the soil parameters for calibration are listed in Table 9.12. According to Scharnagl et al. (2011), a uniform prior distribution (noninformative), which is often assumed in practice, is insufficient to reliably estimate all soil hydraulic parameters. An informative prior distribution, for example based on the soil texture information, can significantly reduce the uncertainty of soil hydraulic parameters in back analysis and even if the prior is biased, the result is not distorted. In this example, as there are no data of measured soil hydraulic properties for this site, the mean values and standard deviations of the three parameters are assumed based on the summarized statistics of estimated soil hydraulic parameters for the 12 soil texture types in United States Department of Agriculture textural classification by Carsel and Parish (1988). According to the laboratory test results of grain size distribution in the TCE site, the CDV soil can be classified to be silt loam or loam. The mean values of k_s for loam and silt loam are 1.04 cm/h and 0.45 cm/h, respectively. The mean values of the parameter a_{vgm} in the van Genuchten model for loam and silt loam are 0.036 cm^{-1} and 0.02 cm^{-1}, respectively. Therefore, the statistics for the parameters in the van Genuchten model is used as reference values. Here, the mean value of $\log_{10}(k_s)$ is taken to be 0 (i.e., k_s is 1 cm/h) and the mean value for $\log_{10}(\alpha_{sy})$ is assumed to be -2.0 (i.e., α_{sy} is 0.01 cm^{-1}). The COV for soil hydraulic parameters are usually much greater than those of the shear strength parameters. Based on Section 7.4, the COV for k_s is usually greater than 100% and the COV of the parameter that represents the inverse of the air-entry value is normally greater than 50%. For the saturated volumetric water content θ_s, the COV value is normally larger than 10%. It should be noted that the COV values reported in literature is mostly estimated from database with soils from various sites while the within-site variability should be smaller than the cross-site variability (Zhang et al., 2004b). Here, the

Table 9.12 Mean values, standard deviations, distributions, and upper and lower bounds of parameters in calibration

Parameters	*Unit*	*Mean*	*Standard deviation*	*Distribution*	*Lower bound*	*Upper bound*
$\theta_s - \theta_r$	–	0.30	0.05	Normal	0.2	0.4
$\log_{10}(\alpha_{sy})$	cm^{-1}	−2.0	0.2	Normal	−3.0	−1.0
$\log_{10}(k_s)$	cm/h	0	0.5	Normal	−4.0	2.0
L	cm	500	50	Normal	300	700
h_l	cm	0	144.34	Uniform	−250	250
q_0	cm/h	0	28.87	Uniform	−50	50

standard deviations for $(\theta_s - \theta_r)$, $\log_{10}(k_s)$, and $\log_{10}(\alpha_{sy})$ are assumed to be 0.05, 0.5, and 0.2, respectively. The upper and lower bounds of the soil parameters are assumed based on the reported statistics of soil hydraulic properties in the literature (Carsel et al., 1988; Zapata et al., 2000; Sillers et al., 2001).

Numerous field investigations and numerical studies of rainfall infiltration in unsaturated soil slopes have illustrated that initial conditions and boundary conditions can significantly influence the pore-water pressure profiles and variation (Ng et al., 1998; Zhang et al., 2004a, 2011; Rahimi et al., 2010). Considering the uncertainties of the domain size of the analytical model, the variation of ground water table, and the antecedent surface flux, in this example the total depth of soil layer L, the pressure head at the low boundary h_1, and the initial flux q_0, are also considered as random variables, where L is assumed to be normal distribution while h_1 and q_0 are assumed to be uniform distribution as no prior information is available. The upper and lower bounds and prior distributions of these three input parameters for calibration are also listed in Table 9.12. The mean value of L and h_1 is assumed based on the soil profiles and ground water level from ground investigation. The mean of antecedent surface flux is assumed to be zero, which means the long-term evaporation and rainfall flux applied on the slope surface is in equilibrium.

Six chains with total number of evaluations equal to 200,000 are used in the MCMC simulation with the DREAM algorithm. Based on the Gelman and Rubin R_{stat} values for the model parameters, 150,000 samples are enough to ensure convergence within the chains to a stable posterior distribution (Figure 9.15). Therefore, the last 50,000 samples of the Markov chains were used as stationary samples from posterior distributions and for prediction inference.

9.7.3 Results and discussions

9.7.3.1 *Posterior distribution of input parameters*

Figure 9.16 illustrates the prior distribution and posterior histograms of the model parameters. Compared with the prior distribution of the parameters, the posterior distributions of all the parameters except $(\theta_s - \theta_r)$ are well identified within a small region. The posterior

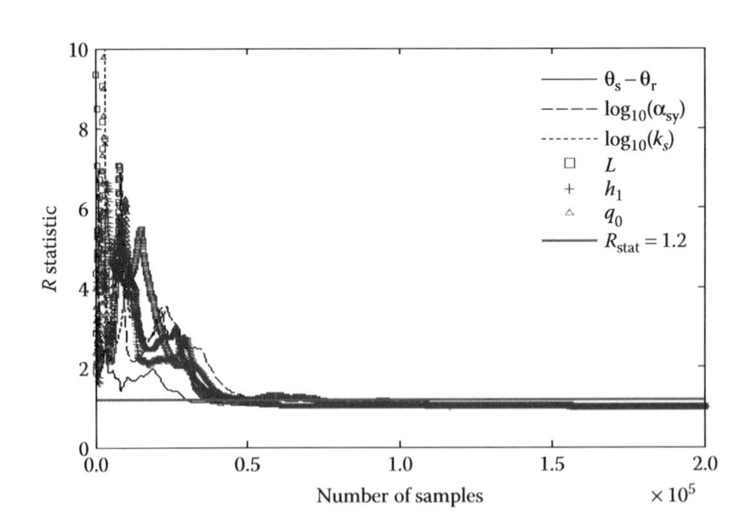

Figure 9.15 Gelman and Rubin convergence diagnostic R_{stat} value for parameter estimation. (From Zhang, L. L. et al., *Computers and Geotechnics*, 48, 72–81, 2013. With permission.)

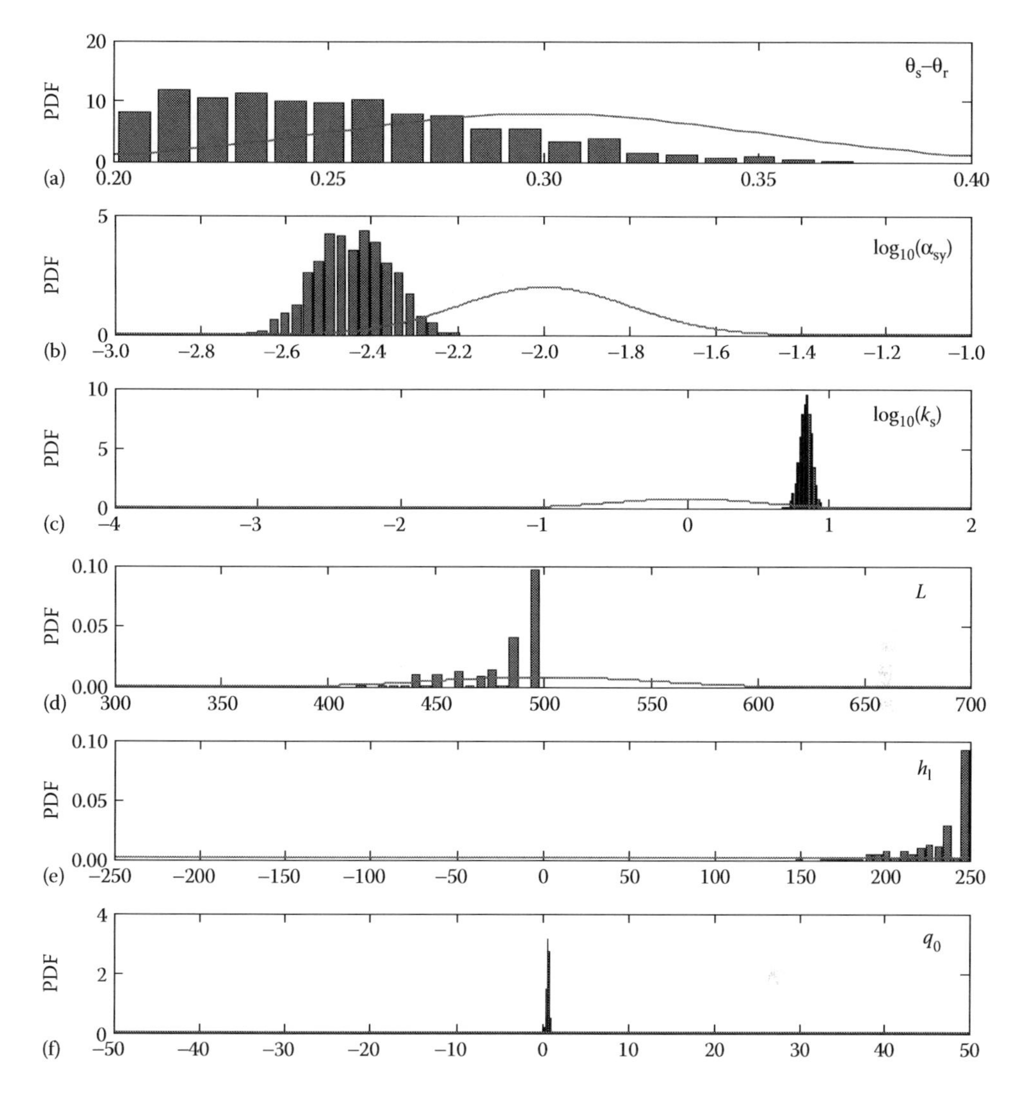

Figure 9.16 Prior distributions and posterior histograms of the six input parameters: (a) $\theta_s-\theta_r$, (b) $\log_{10}(\alpha_{sy})$, (c) $\log_{10}(k_s)$, (d) L, (e) h_l, and (f) q_0. (From Zhang, L. L. et al., *Computers and Geotechnics*, 48, 72–81, 2013. With permission.)

distributions of the hydraulic parameters $\log_{10}(k_s)$ and $\log_{10}(\alpha_{sy})$ show a normal shape. The antecedent surface flux before the test period q_0 is almost equal to zero. Table 9.13 summarizes the mean, standard deviation, MPD value of the six parameters. According to Table 9.13, the standard deviation of $(\theta_s - \theta_r)$ is reduced slightly from 0.05 to 0.035. The standard deviation of $\log_{10}(\alpha_{sy})$ is reduced from 0.2 to 0.087. The standard deviation of $\log_{10}(k_s)$ is reduced significantly from 0.5 to 0.04. The variations of the boundary parameters are also greatly reduced. The posterior standard deviations of L, h_l, and q_0, are 18.2 cm, 18.1 cm, and 0.14 cm/h, respectively. Small differences can be found when comparing the posterior mean and the MPD values of the six parameters. In Table 9.13, the correlation coefficients that represent linear relationship between a pair of parameters are presented. The correlation coefficient between $\log_{10}(k_s)$ and $\theta_s - \theta_r$ is 0.44. This means when the water storage capacity $\theta_s - \theta_r$ is larger, the saturated permeability of soil is also larger. This is

Table 9.13 Summary of statistics of posterior distributions based on the 50,000 stationary posterior samples

Parameters	*Unit*	*MPD*	*Mean*	*Standard deviation*	*Correlation coefficients* $\theta_s - \theta_r$	$\log_{10}(\alpha_{sy})$	$\log_{10}(k_s)$	L	h_l	q_0
$\theta_s - \theta_r$	–	0.275	0.253	0.035	1	–0.87	0.44	0.04	0.03	0.07
$\log_{10}(\alpha_{sy})$	cm^{-1}	–2.51	–2.44	0.087		1	–0.51	–0.02	–0.02	–0.05
$\log_{10}(k_s)$	cm/h	0.85	0.83	0.042			1	0.29	0.32	0.24
L	cm	496.85	483.12	18.22				1	0.99	0.02
h_l	cm	246.98	233.72	18.13					1	0.07
q_0	cm/h	0.54	0.60	0.14						1

Note: MPD means maximum posterior density.

consistent with the findings by statistical analysis of laboratory measured data (Zapata et al., 2000; Sillers et al., 2001; Zhang et al., 2005). The correlation coefficient between $\log_{10}(\alpha_{sy})$ and $\theta_s - \theta_r$ and the correlation coefficient between $\log_{10}(\alpha_{sy})$ and $\log_{10}(k_s)$ are –0.87 and –0.51, respectively. The correlations between the soil parameters and boundary condition parameters are generally less than 0.5.

The 95% uncertainty bounds of the permeability functions and soil–water characteristic curves (SWCCs) corresponding to the prior and posterior distributions of soil parameters are presented in Figure 9.17. The permeability function and SWCC corresponding to the MPD values are also presented in Figure 9.18. In the graph, the residual water content θ_r is assumed to be 0.1. It shows that the posterior uncertainty bounds of the permeability functions are much narrower than the prior uncertainty bounds. The difference between the posterior and prior uncertainty bounds for the SWCCs is less significant. As noted in Zhang et al. (2010b), the variables that contribute more uncertainty to the observed data will receive more information during model calibration and hence are more updated. The fact that the SWCC is less updated here may be due to the fact that the SWCC parameters such as water storage capacity $\theta_s - \theta_r$ and $\log_{10}(\alpha_{sy})$ contribute less uncertainty into the pore-water estimation than the saturated permeability, as shown in the previous reliability studies of slope stability under rainfall condition (Zhang et al., 2005).

9.7.3.2 Uncertainty in prediction for the calibration periods

Figure 9.18a compares the simulated pore-water pressure head using the MPD values of the input parameters with the field measurement for the calibration period (June 8–16, 2001). It shows that the MPD simulation agrees well with the field measured data. The root-mean-square error (RMSE) between the simulation and the measurement is 6.2, which means that the average difference between the simulated and measured pore-water pressure is only about 6.2 cm. The coefficient of determination, R^2, is estimated to be 0.718.

The prediction uncertainty bounds of the calibration period are presented in Figure 9.18b. In this graph, the solid circles denote the pore pressure head observations. The solid lines denote the 95% confidence intervals of simulation only due to parameter uncertainty, whereas the dashed lines represent the 95% confidence intervals of simulation due to total uncertainty. As shown in Figure 9.18b, the 95% total uncertainty bounds are relatively narrow, indicating a good model performance. The coverage, which measures the percentage of field observations contained in the 95% total uncertainty bounds, is 96.4% for the calibration period. This confirms the overall good performance for calibration.

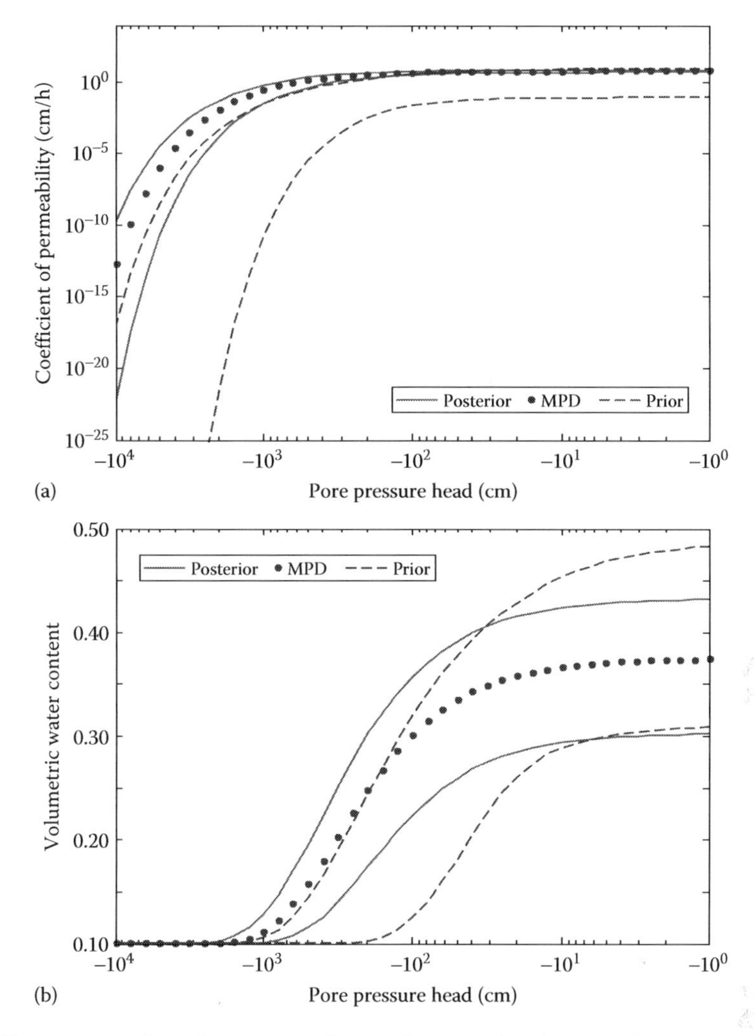

Figure 9.17 95% uncertainty bounds corresponding to the prior distributions, the posterior distributions, and the MPD value of (a) permeability function and (b) the SWCC. (From Zhang, L. L. et al., *Computers and Geotechnics*, 48, 72–81, 2013. With permission.)

9.7.3.3 Uncertainty in prediction for the validation periods

As shown in Figure 9.19b, for the validation periods (Period 2.4), the MPD predictions of pressure head generally follow the variation of rainfall intensity. However, the discrepancy between the predicted and measured pore pressure head is much larger than that of the calibration period. The RMSE value for the MPD predictions is 8.34 for the whole validation period. The R^2 is estimated to be 0.375. To compare the performance of the model during different periods in the whole validation period, the coefficients of determination and RMSE values for Period 2–4 are calculated separately and summarized in Table 9.14. It can be seen that the performance of the model prediction for Period 3 are generally better than Periods 2 and 4. Figure 9.19c illustrates the uncertainty bounds for the validation period. For the validation period, the 95% total predictive uncertainty bounds only brackets 72% of the observations. For Periods 2–4, the values of coverage are 52%, 79%, and 79%, respectively.

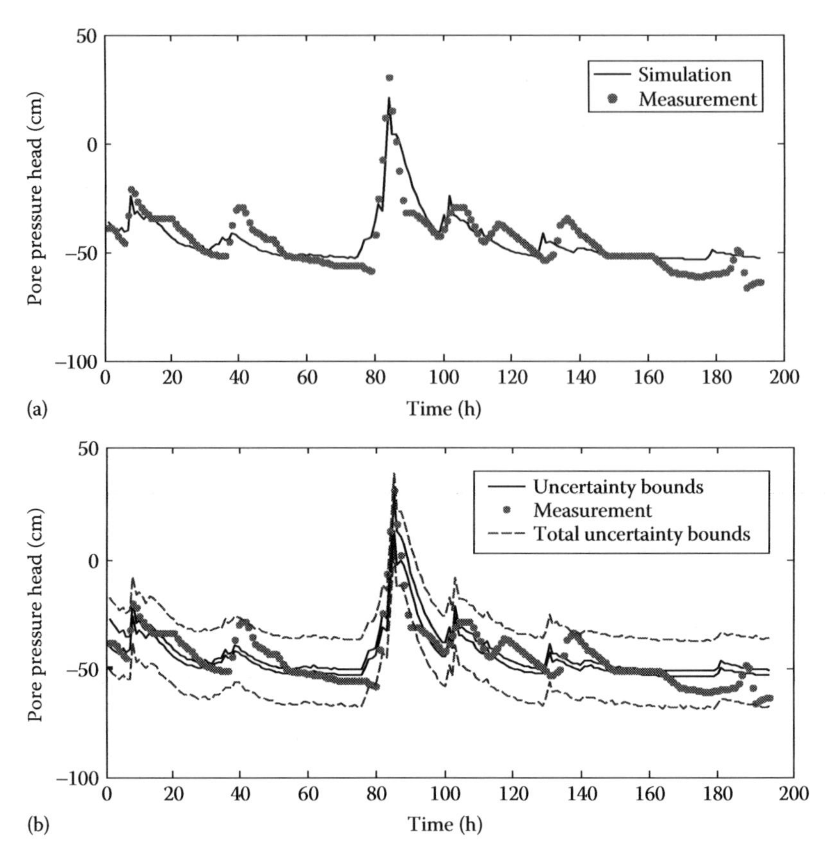

Figure 9.18 (a) MPD simulation and (b) prediction uncertainty bounds for pore-water pressure head during the calibration period (June 8–16, 2001). The solid circles denote the pore pressure head observations. The solid lines in the second graph denote the 95% confidence intervals of simulation only due to parameter uncertainty, whereas the dash lines represent the 95% confidence intervals of simulation due to total uncertainty. (From Zhang, L. L. et al., *Computers and Geotechnics*, 48, 72–81, 2013. With permission.)

Table 9.14 Summary of statistics of the pore-water pressure head predictions

		Total uncertainty bounds	*MPD prediction*	
		Coverage	R^2	*RMSE*
Calibration	Period 1	0.964	0.718	6.22
Validation	Periods 2–4	0.719	0.375	8.34
	Period 2	0.516	0.367	6.96
	Period 3	0.793	0.530	7.76
	Period 4	0.788	0.458	8.88

The large discrepancy between the model prediction and the measured data is mainly attributed to nonconsideration of evaporation flux boundary for the validation period. As shown in Figure 9.14, the rainy hours of the calibration period (Period 1) are continuous and the no-rain hours are short. However, for the validation periods, there is a long no-rain period before each rainfall event. In the prediction model, the surface flux for no-rain period is set to zero, which is a typical assumption in many slope stability analyses. In reality, there

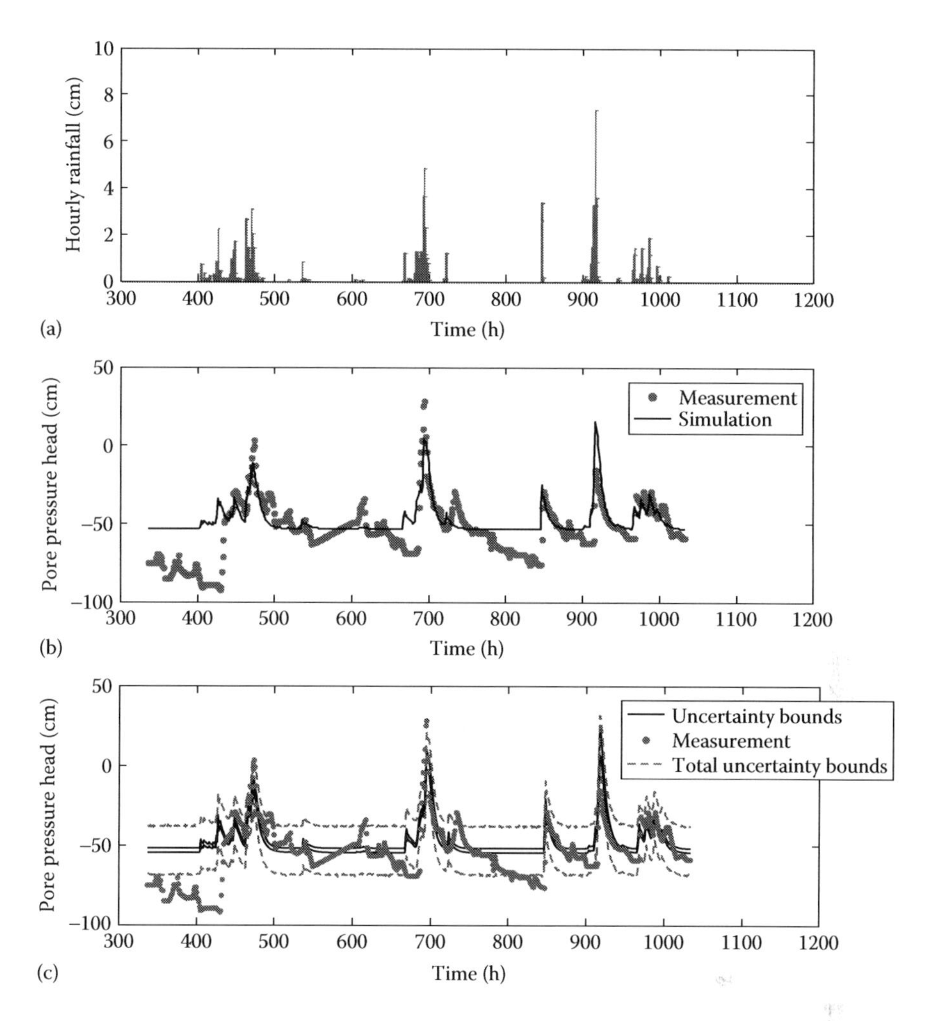

Figure 9.19 (a) Rainfall intensity, (b) MPD simulation, and (c) prediction uncertainty bounds of pore-water pressure head for the validation period (June 22–July 20, 2001). The shaded rectangle covers a fraction of the period when the measurement data are unavailable. The solid circles denote the pore pressure head observations. The solid lines denote the 95% confidence intervals of simulation only due to parameter uncertainty, whereas the dashed lines represent the 95% confidence intervals of simulation due to total uncertainty. (From Zhang, L. L. et al., *Computers and Geotechnics,* 48, 72–81, 2013. With permission.)

is evaporation on the ground surface. Hence, the measured pressure head decreases during the no-rain hours while the predicted pressure shows no variation during the no-rain hours. Therefore, significant difference can be found during the no-rain hours and the initial several rainy hours of a rainfall event. In addition, a calibration period that is not long enough may also lead to large discrepancy between model prediction and field observation (Scharnagl et al., 2011).

9.7.3.4 Uncertainty in predicted safety factor of the slope

Using the prior and posterior evaluation of pore-water pressure at the location of SP3, the safety factor for the infinite slope can be estimated. As shown in Chapter 3, the slope angle, the depth of slip surface, the shear strength parameters, and pore-water pressure can all influence the safety factor of the slope. However, for the problem of slope stability under rainfall, we usually focus on a slope that is stable before rainfall and may be unstable due to infiltration. Hence, the reduction of safety factor instead of the absolute value of the safety factor is more important for assessment of slope stability. For a slope with certain slope geometry and shear strength parameters, the reduction of safety factor corresponding to a slip surface at certain depth is affected mostly by the variation of pore-water pressure. Therefore, in the example, it is assumed that the slope angle, the depth of slip surface, and the shear strength parameters are constant values. The slope angle is 40° and the depth of the slip surface is equal to the depth of pore-water pressure measurement point. The cohesion, ϕ', and ϕ^b are assumed to be 5 kPa, 35° and 20°, respectively, based on typical laboratory test results for the Hong Kong CDV soil. The factor of safety for infinite slope stability analysis in Equation 3.4 is adopted.

Figure 9.20a shows the 95% predictive intervals for the safety factor of the slope. As shown in the graph, the safety factor of the slope decreases with the infiltration of rainfall and increases slightly when the rain stops. The posterior uncertainty bounds are much narrower than the prior ones, with the standard deviation of safety factor reduced significantly from about 0.35 to only 0.01 as shown in Figure 9.20b. This implies that with the reduction of uncertainty in soil hydraulic parameters, the uncertainty in prediction of slope stability can be significantly reduced.

9.7.3.5 *Predicted pore-water pressure at different locations*

To illustrate the robustness of the estimated parameters, the MPD parameters from the measurement of SP3 are used to estimate the pore-water pressure responses at other locations. As the initial conditions and boundary conditions can influence the pore pressure responses significantly, the parameter h_1 which represents the pressure head at the low boundary is

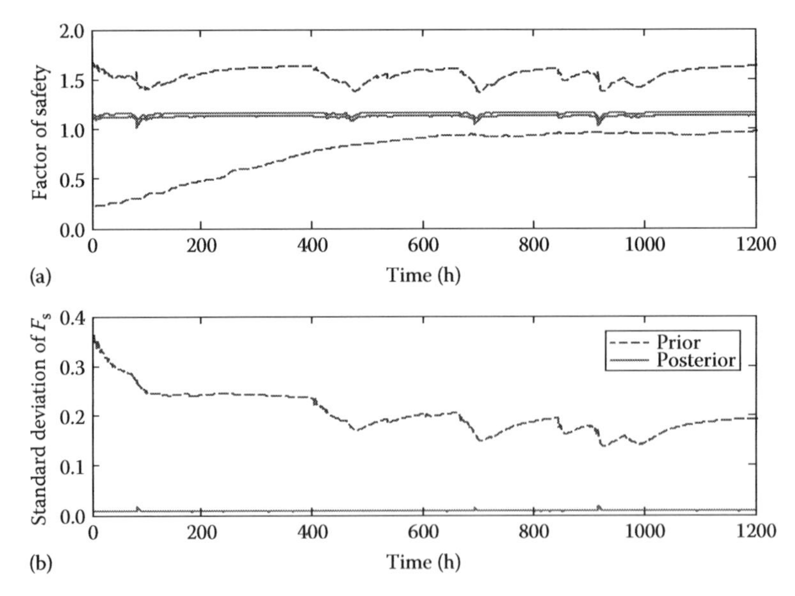

Figure 9.20 (a) Uncertainty bounds and (b) standard deviation of safety factor of the slope ($a_s = 40°$, $c' = 5$ kPa, $\phi' = 35°$, $\phi^b = 20°$, $\gamma_t = 20$ kN/m³, $D = 2$ m). (From Zhang, L. L. et al., *Computers and Geotechnics*, 48, 72–81, 2013. With permission.)

adjusted so that the initial pore pressures in the simulation are the same as the measured values. Figure 9.21 shows the predicted and measured pore pressure responses for SP5 and SP8. The simulation agrees well with the measurement especially the time to reach a peak value of pore-water pressure. The RMSE values for SP5 and SP8 are 23.2 cm and 11.6 cm, respectively. There are significant differences between the simulated and measured data during the no-rain hours. As discussed in the previous section, this may due to assumption of zero-flux boundary condition in the prediction model for the no-rain hours.

It should be noted that the pore-water pressure responses for the 10 piezometers at different locations are quite different, even the transducers are all installed within 3 meters depth of the soil slope. As shown in Figure 9.14, the responses of SP1 and SP2 are quite different from the responses of SP3 and SP4, especially during 0 to 150 hours. The peak values of SP3 and SP4 follows the rainfall intensity very well and when the rain stops, the pore pressure values decrease accordingly. However, during the first 150 hours, the pore-water pressures at SP1 and SP2 continue to increase. The significantly different responses of piezometers at

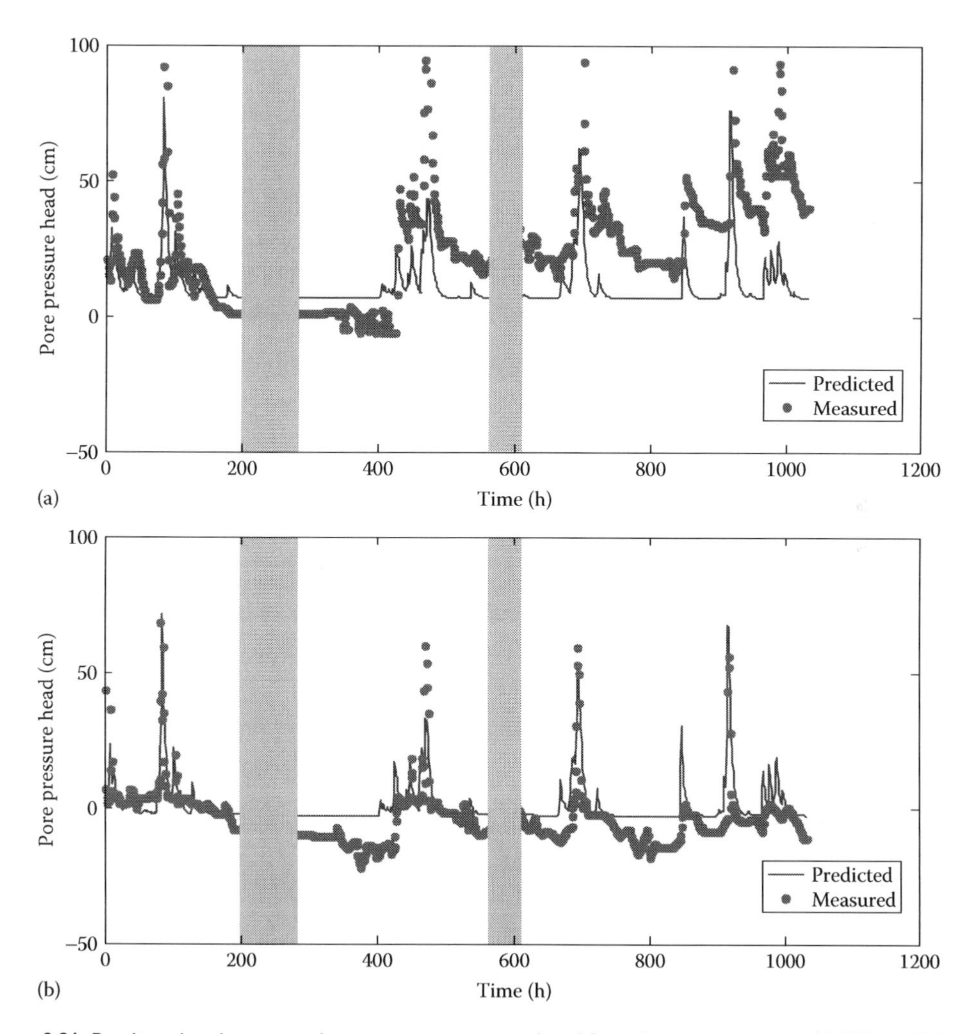

Figure 9.21 Predicted and measured pore-water pressure head for other piezometers: (a) SP5 and (b) SP8. The shaded rectangle covers a fraction of the period when the measurement data are unavailable. (From Zhang, L. L. et al., *Computers and Geotechnics,* 48, 72–81, 2013. With permission.)

different locations may due to the spatial heterogeneity of soils. The estimated parameters based on the measurement at SP3 cannot match the responses of those piezometers that have different soil properties.

9.7.4 Limitations and discussion

It should be noted that there are several limitations in this illustrative example. First, the difference between the measurements and calculated model outputs includes different types of uncertainties, such as model structural error, measurement error, uncertainties of model input, boundary conditions and initial conditions. In this example we use a total residual error instead of separating out various error sources. This approach provides less insight into the sources of error but can yield practical estimates of the parameter and total prediction uncertainties.

Second, we assume that the slope is composed of a homogenous soil layer in the prediction model and explicitly ignore spatial variability of the hydraulic parameters. This simplifies the prediction model and reduces the computation load of back analysis. To investigate the spatially varied soil properties using information from observations at multiple locations, a more complicated prediction model, for example, a numerical model with multiple soil layers, may be needed and the computation load for probabilistic back analysis will increase dramatically.

The factor of safety of a slope under rainfall infiltration is dependent on both soil hydraulic properties and soil shear strength parameters. In the example, only six random variables including three soil hydraulic parameters and three boundary parameters of the infiltration analysis are selected for model calibration. The uncertainties of shear strength parameters are not considered. This is because the observation used in the calibration is only pore-water pressure responses. To estimate shear strength parameters, measured data such as slope displacement or the performance of slope stability should also be provided.

REFERENCES

Ang, A. H. S., and Tang, W. H. (2007). *Probability Concepts in Engineering: Emphasis on Applications to Civil and Environmental Engineering*. Wiley, New York.

Babu, G. L. S., and Srivastava, A. (2007). Reliability analysis of allowable pressure on shallow foundation using response surface method. *Computers and Geotechnics*, 34(3), 187–194.

Bishop, A. W. (1955). The use of the slip circle in the stability analysis of slopes. *Géotechnique*, 5(1), 7–17.

Carsel, R. F., and Parrish, R. S. (1988). Developing joint probability distributions of soil water retention characteristics. *Water Resources Research*, 24(5), 755–769.

Chan, C. C. (2001). *Stochastic Analysis of Infiltration and Slope Stability Considering Uncertainties in Soil-water Retention and Shear Strength Parameters*. M.Phil Thesis. Hong Kong University of Science and Technology, Hong Kong.

Cherubini, C. (2000). Reliability evaluation of shallow foundation bearing capacity on c', ϕ' soils. *Canadian Geotechnical Journal*, 37(1), 264–269.

Chowdhury, R., Zhang, S., and Flentje, P. (2004). Reliability updating and geotechnical back-analysis. In: Jardine, R. J., Potts, D. M., and Higgins, K. G. (Eds.). *Advances in Geotechnical Engineering: The Skempton Conference*, Thomas Telford, London, pp. 815–821.

Christian, J., Ladd, C., and Baecher, G. (1994). Reliability applied to slope stability analysis. *Journal of Geotechnical Engineering*, 120(12), 2180–2207.

Driscoll, T. A., and Maki, K. L. (2007). Searching for rare growth factors using multicanonical Monte Carlo methods. *Society for Industrial and Applied Mathematics Review*, 49(4), 673–92.

Duncan, J. M. (1999). The use of back analysis to reduce slope failure risk. *Civil Engineering Practice*, Spring/Summer, 75–91.

Evans, N. C., and Lam, J. S. (2003). *Tung Chung East Natural Terrain Study Area Ground Movement and Groundwater Monitoring Equipment and Preliminary results* (GEO Report No. 142). Geotechnical Engineering Office, Hong Kong SAR.

Fell, R. (1994). Landslide risk assessment and acceptable risk. *Canadian Geotechnical Journal*, 31(2), 261–272.

Gelman, A., and Rubin, D. B. (1992). Inference from iterative simulation using multiple sequences. *Statistical Science*, 7, 457–472.

Gelman, B. A., Carlin, B. P., Stem, H. S., and Rubin, D. B. (2004). *Bayesian Data Analysis*. Chapman & Hall/CRC, London.

Gilbert, R. B., Wright, S. G., and Liedtke, E. (1998). Uncertainty in back analysis of slopes: Kettleman Hills case history. *Journal of Geotechnical and Geoenvironmental Engineering*, 124(12), 1167–1176.

Gill, J. (2002). *Bayesian Methods: A Social and Behavioral Sciences Approach*. Chapman & Hall/CRC, London.

Luckman, P. G., Der Kiureghian, A., and Sitar, N. (1987). Use of stochastic stability analysis for Bayesian back calculation of pore pressures acting in a cut at failure. In: Lind, N. (Ed.). *Proc 5th International Conference on Application of Statistics and Probability in Soil and Structural Engineering*. University of Waterloo Press, Ontario, pp. 922–929.

Mollon, G., Dias, D., and Soubra, A. H. (2009). Probabilistic analysis of circular tunnels in homogeneous soil using response surface methodology. *Journal of Geotechnical and Geoenvironmental Engineering*, 135(9), 1314–1325.

Morgenstern, N. R., and Price, V. E. (1965). The analysis of the stability of general slip surfaces. *Géotechnique*, 15(1), 79–93.

Mualem, Y. (1976). A new model for predicting the hydraulic conductivity of unsaturated porous media. *Water Resources Research*, 12(3), 593–622.

Ng, C. W. W., and Shi, Q. (1998). Numerical investigation of the stability of unsaturated soil slopes subjected to transient seepage. *Computers and Geotechnics*, 22(1), 1–28.

Phoon, K. K. and Kulhawy, F. H. (1999). Characterization of geotechnical variability. *Canadian Geotechnical Journal*, 36(4), 612–624.

Rahimi, A., Rahardjo, H., and Leong, E. C. (2010). Effect of hydraulic properties of soil on rainfall-induced slope failure. *Engineering Geology*, 114(3–4), 135–143.

Réthάti, L. (2012). *Probabilistic Solutions in Geotechnics*. Elsevier, Amsterdam, The Netherlands.

Scharnagl, B., Vrugt, J. A., Vereecken, H., and Herbst, M. (2011). Bayesian inverse Modelling of in-situ soil water dynamics: Using prior information about the soil hydraulic properties. *Hydrology and Earth System Sciences Discussions*, 8, 2019–2063.

Sillers, W. S., and Fredlund, D. G. (2001). Statistical assessment of soil-water characteristic curve models for geotechnical engineering. *Canadian Geotechnical Journal*, 38(6), 1297–1313.

Storn, R., and Price, K. (1997). Differential evolution—A simple yet efficient heuristic for global optimization over continuous spaces. *Journal of Global Optimization*, 11, 341–359.

Sun, H. W., and Campbell, S. D. G. (1999). *The Lai Ping Road Landslide of 2 July 1997* (Geotechnical Report No. 95). Geotechnical Engineering Office, The Government of Hong Kong SAR.

Thomas, L., and John, S. J. H. (1999). *Bayesian Methods: An Analysis for Statisticians and Interdisciplinary Researchers*. Cambridge University Press, New York.

Tiwari, B., Brandon, T. L., Marui, H., and Tuladhar, G. R. (2005). Comparison of residual shear strengths from back analysis and ring shear tests on undisturbed and remolded specimens. *Journal of Geotechnical and Geoenvironmental Engineering*, 131(9), 1071–1079.

van Genuchten, M. T. (1980). A close form equation for predicting the hydraulic conductivity of unsaturated soils. *Soil Science Society of America Journal*, 44(5), 892–898.

Vrugt, J. A., ter Braak, C. J. F., Clark, M. P., Hyman, J. M., and Robinson, B. A. (2008). Treatment of input uncertainty in hydrologic modeling: Doing hydrology backward with Markov Chain Monte Carlo simulation. *Water Resources Research*, 44, W00B09, doi:10.1029/2007WR006720.

Vrugt, J. A., ter Braak, C. J., Gupta, H. V., and Robinson, B. A. (2009). Equifinality of formal (DREAM) and informal (GLUE) Bayesian approaches in hydrologic modeling?. *Stochastic Environmental Research and Risk Assessment*, 23(7), 1011–1026.

Wesley, L. D., and Leelaratnam, V. (2001). Shear strength parameters from back-analysis of single slips. *Géotechnique*, 51(4), 373–374.

Wong, F. S. (1985). Slope reliability and response surface method. *Journal of Geotechnical Engineering*, 111(1), 32–53.

Wu, T. H. Zhou, S. Z., and Gale, S. M. (2007). Embankment on sludge: Predicted and observed performances. *Canadian Geotechnical Journal*, 44(5), 545–563.

Yuan F., and Lu Z. (2005). Analytical solutions for vertical flow in unsaturated, rooted soils with variable surface fluxes. *Vadose Zone Journal*, 4(4), 1210–1218.

Zapata, C. E., Houston, W. N, Houston, S. L., and Walsh, K. D. (2000). Soil-water characteristic curve variability. *Advances in Unsaturated Geotechnics, Proceedings of Geo-Denver 2000*. ASCE, Denver, pp. 84–124.

Zhang, J., Tang, W. H., and Zhang, L. M. (2010a). Efficient probabilistic back-analysis of slope stability model parameters. *Journal of Geotechnical and Geoenvironmental Engineering*, 136(1), 99–109.

Zhang, L. L., Fredlund, D. G., Zhang, L. M., and Tang, W. H. (2004a). Numerical study of soil conditions under which matric suction can be maintained. *Canadian Geotechnical Journal*, 41(4), 569–582.

Zhang, L. L., Zhang, J., Zhang, L. M., and Tang, W. H. (2010b). Back analysis of slope failure with Markov chain Monte Carlo simulation. *Computers and Geotechnics*, 37(7–8), 905–912.

Zhang, L. L., Zhang, J., Zhang, L. M., and Tang, W. H. (2011). Stability analysis of rainfall-induced slope failure: A review. *Proceedings of the Institution of Civil Engineers-Geotechnical Engineering*, 164(5), 299–316.

Zhang, L. L., Zhang, L. M., and Tang, W. H. (2005). Rainfall-induced slope failure considering variability of soil properties. *Géotechnique*, 55(2), 183–188.

Zhang, L. L., Zuo, Z. B., Ye, G. L., Jeng, D. S., and Wang, J. H. (2013). Probabilistic parameter estimation and predictive uncertainty based on field measurements for unsaturated soil slope. *Computers and Geotechnics*, 48, 72–81.

Zhang, L. M., Tang, W. H., Zhang, L. L., and Zheng, J. (2004b). Reducing uncertainty of prediction from empirical correlations. *Journal of Geotechnical and Geoenvironmental Engineering*, 130(5), 526–534.

Index

Note: Page numbers followed by f and t refer to figures and tables, respectively.